AF410826

Sustainable Organic Agriculture for Developing Agribusiness Sector

Sustainable Organic Agriculture for Developing Agribusiness Sector

Editors

Nikola Puvača
Vincenzo Tufarelli

MDPI • Basel • Beijing • Wuhan • Barcelona • Belgrade • Manchester • Tokyo • Cluj • Tianjin

Editors
Nikola Puvača
University Business Academy in Novi Sad
Serbia

Vincenzo Tufarelli
University of Bari Aldo Moro
Italy

Editorial Office
MDPI
St. Alban-Anlage 66
4052 Basel, Switzerland

This is a reprint of articles from the Special Issue published online in the open access journal *Sustainability* (ISSN 2071-1050) (available at: www.mdpi.com/journal/sustainability/special_issues/organic_agriculture_agribusiness_sector).

For citation purposes, cite each article independently as indicated on the article page online and as indicated below:

LastName, A.A.; LastName, B.B.; LastName, C.C. Article Title. *Journal Name* **Year**, *Volume Number*, Page Range.

ISBN 978-3-0365-1219-8 (Hbk)
ISBN 978-3-0365-1218-1 (PDF)

Contents

About the Editors

Nikola Puvača is a Researcher and Associate Professor at the Department of Engineering Management in Biotechnology of the University Business Academy in Novi Sad, Serbia. He is a multidisciplinary scientist with Ph.D. diplomas in Biotechnology, Veterinary Medicine, and Human Medicine, and a postdoctoral diploma in Toxicology and Molecular Genetics. He has significant experience in animal and poultry science, with a major interest in nutrition, feed quality and safety, natural alternatives for antibiotics, and antimicrobial resistance. He is involved in many research collaborations, including with international institutions, in various science fields. He is serving as an Editorial Board Member and peer reviewer for many indexed journals, and he is the author of more than 200 scientific papers published in international journals and the proceedings of national and international conferences.

Vincenzo Tufarelli is Associate Professor of Animal Nutrition at the Department of Emergency and Organ Transplants (DETO), Section of Veterinary Science and Animal Production of the University of Bari "Aldo Moro", Italy. He has considerable experience in animal and poultry science, with a particular interest in nutrition and feed technology. He is involved in many research collaborations, even with international institutions, in the field of animal science and feed quality. He serves as editorial board member and peer reviewer for many indexed journals and he is the author of more than 190 scientific papers published in international journals and proceedings of national and international conferences.

Preface to "Sustainable Organic Agriculture for Developing Agribusiness Sector"

This book is focused on the latest knowledge and innovations on sustainable organic agriculture, rural development, agricultural economy, policy and management, sustainable food technology, and food safety principles.

Nikola Puvača, Vincenzo Tufarelli
Editors

 *sustainability*

Article

A Profile of Organic Food Consumers—Serbia Case-Study

Vuk Radojević, Mirela Tomaš Simin *, Danica Glavaš Trbić and Dragan Milić

Faculty of Agriculture, University of Novi Sad, TrgDositejaObradovića ε, 21000 Novi Sad, Serbia; rvuk@polj.uns.ac.rs (V.R.); danicagt@polj.uns.ac.rs (D.G.T.); dragan.milic@polj.edu.rs (D.M.)
* Correspondence: mirela.tomas@polj.edu.rs; Tel.: +381-62-240-123

Abstract: In this paper, the authors analyze products from the organic farming system from consumption and consumers. The research aimed to determine the characteristics of the Serbian organic market, discover attitudes, practices, and features of different organic food consumers and identify factors that influence organic products' purchase. This was done following the theoretical framework of green marketing, which refers to the holistic management process responsible for identifying, anticipating, and satisfying customers and society's needs for profitably and sustainability. The research for this study was conducted in Novi Sad and Belgrade in late 2015 and early 2016. The sample included 496 respondents over 18 years of age, varying levels of education, marital status, and other sociodemographic characteristics. The sample was divided into three internally homogeneous yet mutually heterogeneous clusters according to three criteria: factors that generally influence their food purchasing decisions, their opinions regarding characteristics of food products and their eating habits, and their sociodemographic characteristics. The analysis shows that organic consumers can be divided into three clusters with corresponding components related to aspects of products that are sold in the Serbian market. In Cluster 1, respondents who assign the least importance to whether they eat "healthy" products and to the number of calories in those products are presented. Cluster 2 respondents care most about what they consume, and in Cluster 3, respondents assign importance to eat "healthy" food. The findings of this study show that the decision whether to buy organic products or not is predominantly determined by the price and quality of products (which is also related to the socio-economic characteristics of consumers) so that eco-marketing should therefore be more directed towards those consumers who are already "more environmentally and health-conscious", because in this way, the desired results of improving the domestic market of organic products will be achieved.

Keywords: organic production; consumption; sustainability; green marketing; organic consumer

check for **updates**

Citation: Radojević, V.; Tomaš Simin, M.; Glavaš Trbić, D.; Milić, D. A Profile of Organic Food Consumers—Serbia Case-Study. *Sustainability* **2020**, *13*, 131. https://dx.doi.org/10.3390/su13010131

Received: 13 October 2020
Accepted: 16 December 2020
Published: 25 December 2020

Publisher's Note: MDPI stays neutral with regard to jurisdictional claims in published maps and institutional affiliations.

1. Introduction

Agriculture is considered the most significant strategic economic branch that aims to produce food that is of good quality and safe to consume. However, the question of how healthy and safe the produced food is for human consumption is not easy to answer. The growing needs of the population and rapid modernization of agriculture have transformed the ecosystem and its utilization in agricultural production, intensifying the exploitation of the soil, water, forests, and other natural resources. In turn, this has caused excessive use—frequently unchecked—of agrichemicals to protect plants and enhance their growth and fertility, thereby posing a threat to the production of safe and healthy food of high quality, human and animal health, and natural cycles critical for the survival of all life on Earth [1,2].

Organic agriculture has emerged as a response to a high degree of chemicalization in the primary agricultural production, a loss in biodiversity, and an increasing number of consumers who suffer adverse health effects, predominantly in highly industrialized countries [3]. As organic agriculture is gaining prominence, there is a need to compare organic and conventional farming methods to their impact on the environment [4,5]. Since organic

1

farming aims at sustainability, it is considered less damaging for the environment than conventional farming, which relies more heavily on external inputs [6].

Unlike conventional agricultural production, organic food production is based on the biological balance of the soil–plant–animal–human system. As a sustainable agriculture system, organic food production observes ecological principles. It aims to produce food with high nutritional value, the development of sustainable agriculture coupled with ecosystem preservation and soil fertility improvement [7]. A system of organic production includes the uttermost application of renewable energy sources, safeguarding genetic diversity of the agroecosystem, conservation of the environment, and the reduction of all types of pollution caused by agricultural production to create conditions in which agricultural producers can satisfy their basic needs and make an adequate profit [1,2]. For all these reasons, the study of the organic food market is highly significant because of the great demand for these products, which correlates with consumers' increased awareness and their attentiveness to health and environmental concerns [8]. Furthermore, there are also differences in the purchasing power and lifestyle between groups of consumers who can and want to afford these products.

As general concern about the environment and sustainability has grown, so has the popularity of "green marketing". Sarkar [9] argues that sustainability is not merely "an internal reform movement" within companies. Still, it serves as a "bridge between business and green [10] ensuring a better quality of life for everyone, now and for generations to come." Green marketing has twin objectives of minimizing environmental harm while securing economic benefits. "Green marketing" is defined by Peattie [11] as "the holistic management process responsible for identifying, anticipating and satisfying the needs of customers and society, profitably and sustainably." Sarkar [9] suggests that the objective of green marketing is to educate and encourage people to "go green", leading them to change their lifestyle and behavior. In their analysis of the term "green", Simula and Lehtimäki [12] claim that it is widely used to describe new technologies and new products which incorporate the principle of sustainability and have a positive impact on the environment. "Green", "pro-environmental", "sustainability", "environmentally friendly", and "ecology" are commonly used to indicate that a company's products and production processes are energy-efficient, recycling-oriented, produce less waste and pollution, and conserve the environment.

The consumption of organic food products is predicated on public awareness about growing environmental problems. It is thought that this fact, coupled with the introduction into people's lifestyle of strategies to encourage them to purchase organic products, will contribute to increased profits of economic subjects. Eco-marketing does not have the same effect on all buyers. Therefore, it is necessary to target particular markets and direct promotional activities at specific groups of people who are "concerned" about their health and the environment. Green marketing offers a possibility for people to be active and promote "green" lifestyles. On the other hand, companies can explore innovative business solutions to secure both economic success and consumers' trust [2,9].

For green marketing to achieve its above goals, it is necessary to research and define organic products' consumers. In their study, Feil et al. [13] state that the motives for consuming organic products are complex, especially concerning the sociodemographic characteristics of consumers. In their research, Falguera et al. [14] state that people with higher incomes (and married couples) show more positive attitudes and prefer to consume organic food. Similarly, in a study conducted in Switzerland [15], one of the starting assumptions was that increased daily consumption of organic products is influenced by sociodemographic variables such as higher incomes and higher education levels. Confirming these findings, Chekima et al. [16,17] draw attention to the fact that it is necessary to pay additional attention and investigate the difference or the so-called "gap" between the intention to buy organic products and the specific consumption of these products. Vehapi [18], in his research of organic consumers in Serbia, states that health is the most crucial motive for buying organic food. What still characterizes consumers in this market, according to him, is the insufficient presence of ethical and environmental reasons when purchasing organic food, which distinguishes them from consumers in developed countries. Consumers who do not buy organic food products or buy them very rarely and in limited quantities do so because of the higher price of these products compared to similar products from conventional production. Grubor and Djokić [19] conducted a conjoint and cluster analysis to data about Serbian consumers in the organic sector. The mentioned authors state that the market of organic products in Serbia is insufficiently researched and that it is generally characterized by scarce food consumer research.

In an attempt to address these shortcomings, this research was conducted to determine the characteristics of the Serbian organic market, discover attitudes, practices, and identify factors that influence the purchase of organic products, that is, to define the organic consumer profile in Serbia. The uniqueness and originality of this research stems from the fact (mentioned in the previous paragraph) that this type of research is covered relatively little in scientific research on the territory of the Republic of Serbia. This study is among few conducted on this territory, organized to analyze the consumer behavior towards organic products, contributing to enrich that part of the economic literature which states that consumers have a positive attitude towards organic products. Given the fact that the very concept of organic agriculture in Serbia has been current for some twenty years (since 1990, when the first organizations in this field appeared) [20], it is necessary to make additional efforts in researching the market and consumers of these products. Especially if one keeps in mind the fact that organic agriculture emphasizes the importance of the local market and local product purchases. Relying on both original and derived data, using the appropriate methods, relevant elements will be identified, which play a significant role in improving the consumption of organic products in the Republic of Serbia. The findings of this study could potentially apply to other countries with a similar socio-economic level of development.

The paper's structure is as follows: In the first part of the article (Introduction), the fundamental determinants (green marketing, organic agriculture) are presented together with the initial hypotheses. The literature review shortly presents past research related to the subject area. After that, the material and used methods and sources are presented. The results ofthe discussion present the obtained results with references from other research related to the same topic. The last part of the paper refers to the authors' conclusions based on the presented results.

2. Literature Review

Early studies focused on ad hoc research into sustainable food production systems, eco-marketing, and factors influencing food purchasing decisions. Eco-marketing is a part of economic theory which is focused on researching the market itself and its components. For example, Shabbir et al. [21] define green or eco-marketing as marketing that encompasses all activities to promote any change to satisfy a human need or desire that has a minimal detrimental impact on the natural environment. This definition contains the traditional components of the definition of marketing (consumer satisfaction), but also contains environmental protection, in a way that minimizes the harmful impact on the environment (to be accurate, the eco-claim on the product should read "less harmful to the environment", not "environmentally friendly").

A certain part of previous research was related to determining the factors that influence consumer decisions when buying environmentally friendly products. One of the first studies that brought to the forefront the attitude of consumers towards the price of products from organic agriculture is a study conducted in Italy in 1992 [22]. The basic assumption that guided it is that information on the market is insufficient, and that when making a purchase decision, the currently valid (objective) price is not considered, but the consumer's choice to buy is based on the fact of what consumers believe that the price should be at that moment (subjective price). In addition, one of the earlier studies was conducted by Margetts et al. [23]. In their study, when asked to assess the choice of diet in the future, 32% of respondents put future efforts to use "healthy" food in the first place. The decision on the choice of food is most often made by women, and the price of food has proven to be a decisivefactor only for pensioners and the unemployed. Special emphasis during the research was given to examining the influence of the level of education of the respondents on the choice of "healthy" food, and the result indicates that this is the factor of the strongest influence. Torjusen et al. [24] explored the potential of organic agriculture, starting from the view that it is very important to know both consumers and producers thinking about food quality and systemic issues. Multivariate analysis has shown that traditional aspects of food quality, such as freshness and taste, are especially important for consumers. Additionally, those who bought organic food were more concerned about ethical, environmental, and health issues and consumer orientations in the food market were identified. Wier et al. [25] explored organic food markets in the UK and Denmark, identifying major differences and similarities in consumer perceptions and priorities, labelling schemes, and sales channels as a basis for assessing market stability and future growth prospects. The authors found that decisions to buy organic food were primarily motivated by "private" attributes such as freshness, taste, and health. Gracia and de Magistris [26] analyzed the demands of organic food consumers in southern Italy. The authors hypothesized that consumer behavior depends on the characteristics of the product instead of the product itself. In this way, consumers will choose a product (organic vs. conventional) that has a combination of attributes that increase utility. Garg et al. [27] state in research that consumers, regardless of their differences by segments, in all developed countries, are not satisfied with having enough food but they want to be able to choose foods that will maximally protect and improve their health. Nie and Zepeda [28] with the so-called food-related lifestyle model, widely used in European research, research the consumption of organic and local food. The authors also segmented consumers and compared segments with respect to organic and local food consumption variables. They correlated with environmental concerns, knowledge and practice, practices, and health concerns, as well as some demographic characteristics (race, gender, age, education), income, and variables that measured access to these foods. Van Huy et al. [29] applied the unique food-related lifestyles (FRL) approach to segment organic food consumers in Vietnam. Data were obtained from 203 organic food consumers, and a two-step cluster analysis established three identifiable market segments which we named "Conservatives", "Trendsetters", and "Unengaged". The Conservatives were interested in the health aspects of food and preferred natural products. The Trendsetters were interested in healthy food, liked to cook, and held a positive attitude toward organic food and local food products. The Unengaged consumers were not concerned about food-related issues, and they reported the least consumption of organic food. In their research, certain authors turn to the market of individual products determining what motivates consumers to buy specific products. In their research, Migliore et al. [30] tried to understand which wine quality characteristics, consumers' attitudes, and socio-demographic characteristics affect the consumers' willingness to pay (WTP) a premium price for a bottle of natural wine. The research results reveal that drink frequency and occasion, organic production method, the content of sulfites, income, and the attitudes towards healthy eating and the environment are positively associated with a higher WTP for natural wine. Likewise, Drugova et al. [31] examined determinants of consumer interest in organic versions of wheat products by analyzing differences in selected factors among groups of consumers, distinguished by their likelihood of purchasing organic wheat products. They conclude that consumer preferences and willingness to pay (WTP) for organic foods depend on product type. Additionally, significant differences are found across consumer groups—regardless of product type—in the importance they place on labels and product characteristics, WTP, reasons for (not) purchasing organic products, and consumption limitations.

Novel studies are also focused on consumer's behavior in the organic market, as well as their intentions and decision-making factors. Unlike previous studies, some modern research focuses on developing countries and the market for organic products. Pacho [32] empirically examined the impact of attitude, subjective norms, and perceived behavior control on the consumer intention to buy organic food. The author stated that the behavior compelling the consumer's intention to purchase organic food has received little attention in developing countries. There is limited knowledge concerning the factors that impact consumer's intention to purchase organic food in these countries. The findings in this research showed that subjective norms and attitudes were positively correlated, and they significantly impacted the purchase intention of organic food. The findings also showed that knowledge about organic food and health consciousness has an indirect effect on the relationship between attitude, subjective norms, and intention to buy. Troudi and Bouyoucef [33] proposed a model based on reasoned action theory that combines two types of variables, the green marketing type and personal type, in order to predict purchasing behavior of green food. The model was confirmed, and the results showed how green marketing and personal factors influence the green food purchasing behavior in a direct and indirect way, in presence of the mediating variables' attitude toward green food and intention to buy green food. Nair and Shams [34] examined whether food and grocery (F&G) shoppers in India are strongly influenced by store-attributes. The findings show that store-attributes—atmosphere, promotion, convenience, facilities, merchandise, store personnel interaction, and services affect store choice decisions. Wang et al. [35] associated intention to buy organic food with knowledge. Findings show that knowledge positively moderates the relationship among subjective norms, personal attitude, health consciousness, and organic food purchase intention.

Frewer et al. [36] concluded that the research of the requirements and attitudes of citizens—potential consumers—in the countries of the European Union, which was conducted between 1989 and 2000, can be summarized as follows:

- Citizens generally accept the use of mechanized tillage methods better than the use of chemicals;
- Old, known tillage and livestock technologies are more accepted than new, automated ones;
- Conventional reproduction in plant and livestock production is much more desirable than reproduction with the use of genetic modifications;
- Natural methods in food processing are more accepted than modernized, so-called hi-tech food industry.

On the other hand, certain authors had a different approach, analyzing the influence of the state, its regulations, and regulations in the field of organic production on the very decisions of consumers to purchase organic products. Vadnal [37], in the mid-1990s, researched the influence of the state in the field of "healthy" food in accordance with the increasingly pronounced demands of citizens to define the difference between health and food produced in the usual (conventional) way. Efforts seem to be made to adequately regulate this issue, which is obviously very interesting for citizens, as well as for potential producers. During this period, the first legal regulations in the field of production, which began to be called organic agriculture, began to be passed [20]. Regulations in the field of the right to certification and penal provisions are also adopted in cases of non-compliance with the provisions of the law. Over time, national legislation becomes uniform for all EU member states, with each country still striving to develop its own national logo. At the same time, there are major changes in the field of marketing, which begins to respect the desire of consumers to be accurately informed, and marketing efforts are focused on highlighting the benefits of products based on the certificate, and adopting a number of new features of health food marketing. In their work, Grunert and Juhl [38] presented the first regulations regarding the labeling of health-safe food in the European Union dating back to 1990. As the public's interest in the link between diet and health increases, efforts are intensifying to improve regulations in this area and to harmonize as much as possible in all member states. The same authors [38] presented the results of research with Danish school teachers regarding their values, attitudes on environmental protection, and the purchase of organic food. The aim was to examine the applicability of value theory and approach measurement in explaining specific aspects of consumer behavior. They conducted two large studies which indicate that health-safe food has acquired a niche market dimension, and the most important role in consumers' decision to buy is availability, trust, and an adequate price level [39]. Yang et al. [40] stated that general confusion is exacerbated by the fact that consumers (as confirmed by a large number of studies) are largely insufficiently acquainted with the essence of environmentally friendly food, with the methods of production, processing, certification, and control. On the other hand, even when they are offered additional information that should indicate, for example, the proper use and treatment of food, the vast majority of consumers do not notice such instructions on the product at all (as many as 95%), and from those who see additional instructions, 79% of them did their best to read and reread the notice. Therefore, in most European Union countries, as part of programs for the affirmation of environmental suitability and eco-marketing, the process of educating the population is given special attention.

3. Materials and Methods

The empirical research of the organic food market consisted of gathering data through a survey in the form of face-to-face interviews, with interviewers recording respondents' answers to a list of questions in a questionnaire. The largest amount of data was collected directly through conversations with respondents, partly in health food stores, farmers' markets, and partly in front of large supermarket chains. The authors oversaw conducting interviews at the mentioned locations. Respondents who showed interest in organic products in stores were approached by the authors with the question of whether they could set aside a few minutes of their time to help research the organic market. In front of large markets, the authors approached a larger number of respondents with the question of whether they had heard or whether they used organic products. If the answer was yes, they asked the above question and approached the interview in case of a positive answer.

The research was conducted in Novi Sad and Belgrade in late 2015 and early 2016. The sample included 496 respondents over 18 years of age, of varying levels of education, and different marital status and other sociodemographic characteristics. The sampling was non-probabilistic, occasional, and volunteer which is by convenience procedure.

The questionnaire created for this research relies significantly on experiences of researchers from other countries such as Great Britain, Denmark, Germany, USA; Croatia, Italy, Switzerland, Sweden, Australia, and others [5,24,25,28,41–47]. Papers used in this study come predominantly from reputable international scholarly journals.

The questionnaire consisted of 18 (groups of) questions about respondents' attitudes towards factors which influence their decision of whether to purchase food products as well as their opinion about most relevant characteristics of food products. Their knowledge of organic products, satisfaction with them, and satisfaction with their offer on the domestic market were also tested with individual questions. When it comes to the type of question in the questionnaire, semi-open-ended questions and (mostly) closed-ended questions are combined with more than two offered answers, and only one open-ended question. In addition, five questions were formulated in the form of the Likert scale, where number one expressed complete disagreement and number five completely agreed with the stated statement, i.e., attitude. As already mentioned, when compiling the questionnaire, previous research was used, i.e., the results obtained in the same, as well as certain attitudes used in these surveys in order to find out the attitudes of respondents when it comes to the market and consumption of organic food products in Serbia. In addition to the 18 questions, the questionnaire included questions about the respondents' sociodemographic characteristics such as: sex, age, marital status, education, employment status, the number of household members, and the total monthly income of the household. The complete questionnaire is included in the Supplementary.

One of the methods used to analyze data collected in the empirical research of the organic market in Serbia was cluster analysis, which segmented respondents into three mutually heterogeneous clusters according to following factors:

- Factors which influence respondents' food purchase in general;
- Factors pertaining to products' characteristics;

- Factors pertaining to respondents' food consumption habits; and
- Sociodemographic characteristics.

Keeping in mind the volume of data, the paper presents only an excerpt of the second question, which served as a basis for conducting cluster analysis and, as already mentioned, a complete survey is available in the Supplementary. Cluster analysis was applied to the second question in the questionnaire. The second group of questions was about discovering consumers' habits and their attitude towards certain statements (Table 1).

Table 1. An extract from the questionnaire from the second group of questions.

On the Scale from 1 to 5, Indicate to What Extent You Agree with the Following Statements *						
1	I consume healthy food products; I care about what gets into my body	1	2	3	4	5
2	Imported food products are of better quality	1	2	3	4	5
3	I am careful about the energy value (calories) and fat which I consume	1	2	3	4	5
4	I prefer buying food products produced in Serbia	1	2	3	4	5
5	Packaging plays a significant role when I buy food	1	2	3	4	5
6	My diet consists mainly of fresh produce	1	2	3	4	5
7	I try to be informed about healthy lifestyles	1	2	3	4	5
8	I habitually consume fast food	1	2	3	4	5
9	Conservation of nature and living in accordance with nature are important to me	1	2	3	4	5

* 1—strongly disagree; 2—disagree; 3—I do not know; 4—agree; 5—strongly agree

The results of processing these questions are first presented (in the Results section) through descriptive statistics (simple frequency distributions). In the later presentation of the results, the results of the cluster analysis are presented, which, with the help of these questions, gave answers about the basic clusters, i.e., they could be called "consumer profiles" in relation to organic food products. F statistics for cluster analysis purposes, mean values for variables within a cluster, analysis of the variance of the first question, i.e., the factors that generally influence the decision of the respondents when buying food products and distribution of respondents by clusters based on socio- demographic characteristics, are attached in the Supplementary.

4. Results and Discussion

In this part of the paper, the research results will be presented through descriptive statistics (simple frequency distributions). The next step was factor analysis, to determine the components that most influence the decision of the respondents when buying food products. For this purpose, it was first checked whether it was possible to perform a factor analysis on the collected data, and therefore the Kaiser–Meyer–Olkin test (KMO test) and the Bartlett test were performed. Given that the value of the KMO coefficient was greater than 0.6, and the significance of Bartlett was less than 0.05, it was concluded that it is possible to perform a factor analysis. In the later presentation of the results, the result of the cluster analysis will be presented, which, with the help of the second block of questions, gave answers about the consumers tendencies when buying organic produce. The second block of questions was aimed at discovering consumers' buying habits as well as their opinions on certain issues (as evidenced by the given extract from the questionnaire). In addition to conducting cluster analysis, it is important to recognize what governs consumers' purchasing behavior and could therefore be relevant for organic products and their manufacturers.

4.1. Consumers Buying Habits

Replying to the question to what extent they agree with the statement that they consume "healthy" food products and that they care about what gets into their body, over 68% of the respondents replied either that they agree or strongly agree with the statement.

Due to the fact that there is a great number of food products imported into Serbia from abroad, it is important to discover to what extent the respondents from the analyzed sample agree with the statement that foreign products are of higher quality. The results show that 38.67% of respondents either strongly disagree or disagree with that statement, which is somewhat greater than the percentage of those who either agree or strongly agree (34.82%). There is a significant portion of those respondents who do not know, that is, those who are undecided. These percentages testify of a certain degree of trust that Serbian consumers have in local food producers [1]. Further analysis of the obtained results indicates that there is a statistically significant dependence (p = 0.049) of the marital status of the respondents and the attitude that foreign, imported food products are of better quality. Unmarried respondents mostly agree that foreign, imported food products are of better quality, while widows agree the least with this attitude. Regarding the quality of food products, the analysis showed that the students most agree with the attitude that foreign, imported food products are of better quality than domestic ones. Correlation analysis of these obtained answers showed that older respondents are less likely to believe that foreign, imported food products are of better quality (correlation coefficient r = −0.144, p = 0.001).

Regarding respondents' eating habits, the aim was to discover to what degree they look after their health by consuming the optimal number of calories and amount of fat. The results show that the number of respondents who do not have such habits is larger than the number of respondents who do (Figure 1).

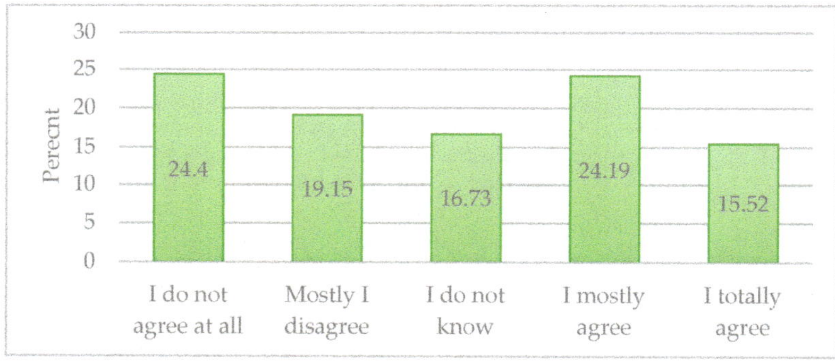

Figure 1. Results showing agreement with the statement "I am careful about the energy value (calories) and fat which I consume".

As in the previously discussed result, a significant percentage of respondents from the sample either agree or strongly agree with the statement that they prefer buying food products produced in Serbia (57.78%). This result, together with the result of the question about buying imported food products, speaks to the fact that the respondents prefer buying products made in Serbia to products imported from foreign countries, which is particularly important for organic food products which are the subject of this research (Figure 2). Correlation analysis shows that as the education level of respondents increases, the likelihood of them buying food products from Serbia decreases (the correlation coefficient r= −0.144). Furthermore, the older the respondents, the more likely they are to buy Serbian food products [48] (the correlation coefficient r = 0.164). Thus, the analysis shows that retirees are more likely to buy Serbian products. However, the higher the income of the respondents, the less significant this factor becomes, meaning that people with higher income are less likely to buy Serbian food products (the correlation coefficient r= −0.125).

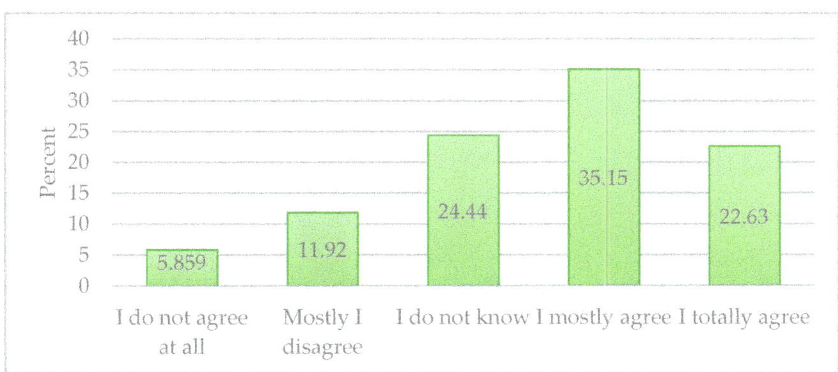

Figure 2. Results showing agreement with the statement "I prefer buying food products from Serbia".

Significant findings came from the analysis of the respondents' responsesto the question about the role packaging plays in decisions governing food purchase (Figure 3). For different reasons, packaging has turned out to be an important factor which influences the purchase of food products. As shown in Figure 3, only 33.87% of the respondents either strongly disagree or disagree with the statement while almost half of the respondents think packaging is an important factor in buying food products (48.59%). No statistically significant correlations were shown between this factor and other variables.

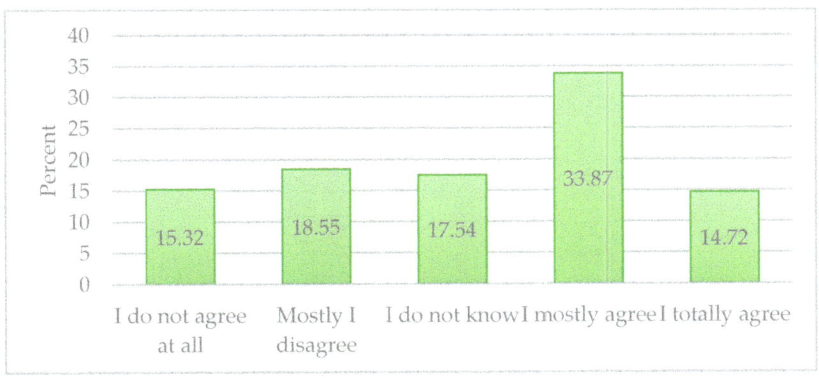

Figure 3. Results showing agreement with the statement "Packaging plays a significant role when I buy food".

The majority of respondents mainly eat fresh produce, which is important when it comes to food consumption, that is, consumption of food products available in the Serbian market where both supply and demand include predominantly fruits and vegetables and much less other types of organic food products (Figure 4). There is a statistically significant correlation between education and the intention to eat fresh produce (the correlation coefficient r = 0.104). At the same time, as the respondents' age increases, their habit of consuming fresh produce decreases, but that correlation is weak (the correlation coefficient r= −0.090).

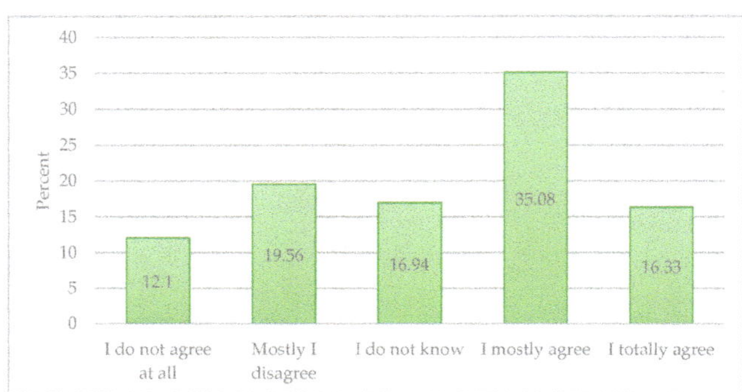

Figure 4. Results showing agreement with the statement "My diet consists mainly of fresh produce".

The research into the respondents' purchase of food products, their eating habits, and lifestyles needs to discover to what extent the respondents are informed about healthy lifestyles and to what extent they eat the so-called "fast food", the consumption of which is noticeably increasing (Figures 5 and 6). The results show a correlation between responses to these two questions, meaning that the majority of respondents from the analyzed sample makes efforts to be informed about healthy lifestyles (61%) while a small percentage habitually eat so-called "fast food" (27.6%). Correlation analysis shows that the higher the income, the more likely the respondents are to be informed about healthy lifestyles, but the correlation is weak (the correlation coefficient r = 0.075). The same is true of education, meaning that the higher the education, the more likely the respondents are to pay attention to healthy lifestyles (the correlation coefficient r = 0.110). Regarding "fast food" consumption, the analysis shows that single respondents are more likely to consume "fast food", and that the older the respondents, the less likely they are to consume "fast food" (the correlation coefficient r = −243). Accordingly, retirees are the least likely and students are the most likely to consume "fast food." Considering the respondents' sex, the analysis shows that men are more prone to this habit than women.

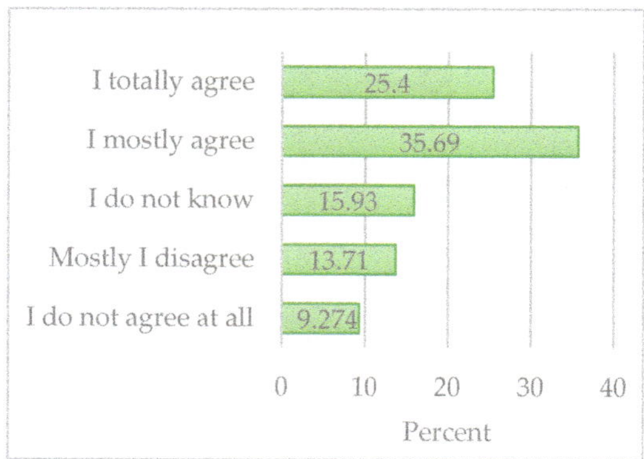

Figure 5. How informed the respondents are about healthy lifestyles and healthy eating.

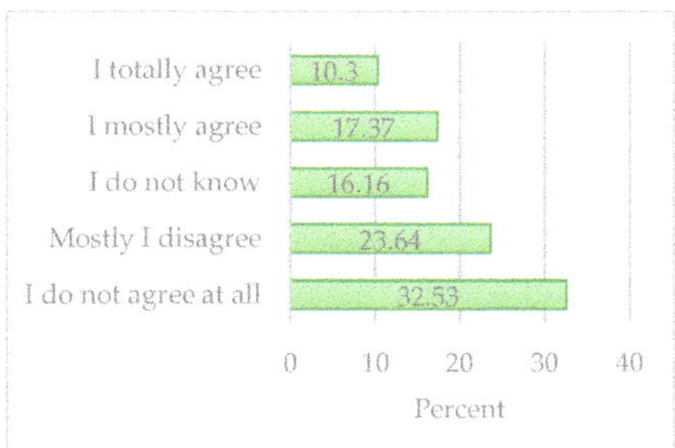

Figure 6. How informed the respondents are about healthy lifestyles and healthy eating.

It was important in this study to discover the respondents' opinion about the significance of conservation of nature and living in accordance with nature (Figure 7). A small percentage of respondents in the sample either strongly disagree or disagree with this statement (about 26%) while the percentage of those who agree with the statement is almost twice as large (53.43%). Correlation analysis shows that people with a higher level of education are more likely to hold this view, but the correlation is weak (the correlation coefficient r = 0.079). In addition, older people are more likely to hold this view (the correlation coefficient r = 0.090) as well as people with higher income, but the correlation is weak (the correlation coefficient r = 0.075).

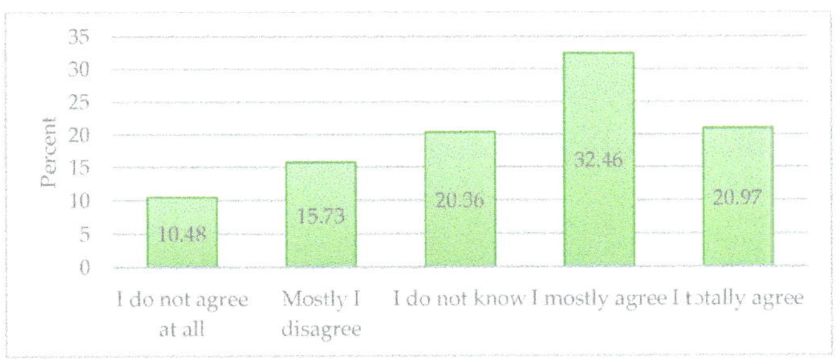

Figure 7. Results showing agreement with the statement "Conservation of nature and living in accordance with nature are very important to me".

4.2. Factor Analysis

In the next step, a factor analysis was performed to determine the components that most influence the decision of the respondents when buying organic food products. First, it is necessary to check whether it is possible to do a factor analysis on these data, and therefore, a KMO and Bartlett test were performed. Given that the value of the KMO coefficient is greater than 0.6, and the significance of Bartlett is less than 0.05, it is possible to perform a factor analysis. Then, the factor analysis of the first question was done. The column contains the factors, 14 of them, because analysis was done on the first question (first question and complete survey is in Supplementary). For a factor to be significant, it must have a value greater than or equal to 1, and these values are read from the column Initial value—Total. Based on that, it is concluded that 4 components/factors were obtained that explain 56% of the variations of the observed variables. This percentage can be seen in the column Initial value—Cumulative%. This analysis shows the statistical equality of the observed values based on the following assumptions: (1) The choice of respondents in groups should be random and independent, (2) The variability of results in the populations of the analyzed groups should be statistically equal, (3) The results of groups should be normal distributed, i.e., not to deviate statistically significantly from the normal distribution. Data are shown in Table 2.

Table 2. KMO (Kaiser–Meyer–Olkin) and Bartlett's test, data related to analysis of variance (software excerpt), and data related to pattern matrix analysis *.

Description	Data
Kaiser–Meyer–Olkin Measure of Sampling Adequacy	0.756

Bartlett's Test of Sphericity		Data
	Approx. Chi-Square	1425.032
	df	0.91
	Sig.	0

Component	Initial value			The sum of squares loading			Rotation sums of squares loadings
	Total	Variance	Cumulative	Total	Variance	Cumulative	
1	3.455	24.676	24.676	3.455	24.676	24.676	2.444
2	1.792	12.801	37.477	1.792	12.801	37.477	2.509
3	1.54	11.003	48.48	1.54	11.003	48.48	1.487
4	1.058	7.559	56.039	1.058	7.559	56.039	2.157
5	0.946	6.761	62.8				
6	0.783	5.596	68.396				
7	0.735	5.247	73.644				
8	0.682	4.873	78.516				
9	0.661	4.722	83.238				
10	0.577	4.125	87.363				
11	0.518	3.702	91.065				
12	0.472	3.373	94.438				
13	0.42	2.997	97.435				
14	0.359	2.565	100				

Question	Component			
	1	2	3	4
1.8 On a scale of 1 to 5, rate how much the product advertisement affects your decision when buying food products.	0.814			
1.9 On a scale of 1 to 5, rate the fact that the manufacturer is knowninfluences your decision when buying food products.	0.697			
1.7 On a scale of 1 to 5, rate how much recommendations (friends, experts, nutritionist, doctor) influence your decision when buying food products.	0.636			
1.10 On a scale of 1 to 5, evaluate how clearly written the composition of the packaging influences your decision when buying food product.		0.77		
1.4 On a scale of 1 to 5, evaluate how much the product does not contain additives and harmful substances affects your purchase decision food products.		0.706		
1.5 On a scale of 1 to 5, assess how clearly indicated the shelf life of the product influences your decision when buying food products.		0.663		
1.13 On a scale of 1 to 5, rate the fact that the product is environmentally friendly (bio) influences your decision when buying food products.		0.505		
1.6 On a scale of 1 to 5, evaluate how much the favorable price of the product affects your decision when buying food.			0.827	
1.14 On a scale of 1 to 5, rate the best value for money the product influences your decision when buying food products.			0.786	
1.12 On a scale of 1 to 5, rate the fact that the product looks nice influences your decision when buying food products.				0.788
1.11 On a scale of 1 to 5, rate how pleasant the environment in which the product is sold influences your decision when buying food products.				0.63

* author's research.

A problem arose here, because each component should contain at least 3 factors, and here that is not the case, i.e., component three contains only two factors. Based on that, it was concluded that the number of components should be reduced to 3. When the number of components was changed, a new table was obtained (Table 3).

Table 3. Data related to analysis of variance II *.

Component	Initial Value			The Sum of Squares Loading			Rotation Sums of Squares Loadings
	Total	Variance	Cumulative	Total	Variance	Cumulative	
1	3.455	24.676	24.676	3.455	24.676	24.676	2.978
2	1.792	12.801	37.477	1.792	12.801	37.477	2.648
3	1.540	11.003	48.480	1.540	11.003	48.480	1.606
4	1.058	7.559	56.039				
5	0.946	6.761	62.800				
6	0.783	5.596	68.396				
7	0.735	5.247	73.644				
8	0.682	4.873	78.516				
9	0.661	4.722	83.238				
10	0.577	4.125	87.363				
11	0.518	3.702	91.065				
12	0.472	3.373	94.438				
13	0.420	2.997	97.435				
14	0.359	2.565	100.000				

* author's research.

Based on the following table (Table 4), three components were obtained that explain 48.5% of the variations of the observed variables. Based on the second (new) pattern matrix table, it is possible to conclude that each component has more than two factors.

Table 4. Data related to the analysis of the II pattern matrix (excerpt from the software) *.

Question	Component			
	1	2	3	4
1.8 On a scale of 1 to 5, rate how much the product advertisement affects your decision when buying food products.	0.780			
1.3 On a scale of 1 to 5, evaluate how attractive the product packaging is on your decision when buying food products.	0.737			
1.9 On a scale of 1 to 5, rate the fact that the manufacturer is known influences your decision when buying food products.	0.710			
1.11 On a scale of 1 to 5, rate how pleasant the environment in which the product is sold influences your decision when buying food products.	0.657			
1.12 On a scale of 1 to 5, rate the fact that the product looks nice influences your decision when buying food products.	0.594			
1.10 On a scale of 1 to 5, evaluate how clearly written the composition of the packaging influences your decision when buying food product.		0.764		
1.4 On a scale of 1 to 5, evaluate how much the product does not contain additives and harmful substances affects your purchase decision food products.		0.754		
1.5 On a scale of 1 to 5, assess how clearly indicated the shelf life of the product influences your decision when buying food products.		0.536		
1.2 On a scale of 1 to 5, assess how much the origin of the product affects your decision when buying food products.	0.302	0.521		
1.13 On a scale of 1 to 5, rate the fact that the product is environmentally friendly (bio) influences your decision when buying food products.		0.511		
1.1 On a scale of 1 to 5, evaluate how much the quality of the product affects your decision when buying food products.		0.464		
1.14 On a scale of 1 to 5, rate the best value for money the product influences your decision when buying food products.			0.757	

Table 4. *Cont.*

Question	Component			
	1	**2**	**3**	**4**
1.6 On a scale of 1 to 5, evaluate how much the favorable price of the product affects your decision when buying food.			0.755	
1.7 On a scale of 1 to 5, rate how much recommendations (friends, experts, nutritionist, doctor) influence your decision when buying food products.	0.319		0.351	

* author's research.

Each component is characterized by certain factors, based on which components are defined, with the factors that have the highest value in each component, best reflect the characteristics of a given component. Based on the same table, it is possible to determine the following components:

Component 1: AESTHETICALLY PLEASING AND EASILY RECOGNIZABLE PRODUCT: The product is advertised, of attractive packaging. The manufacturer is well known, it is sold in a pleasant ambience and it looks nice.

Component 2: HEALTHY AND SAFE PRODUCT: The composition of the product and shelf life are clearly indicated, it does not contain additives, it is environmentally friendly and high quality.

Component 3: CHEAP PRODUCT: The product is affordable, characterized by the best price-quality ratio and is recommended by friends, experts, nutritionists, doctors.

4.3. Cluster Analysis

After factor analysis cluster analysis, dividing the respondents into heterogeneous clusters according to three criteria followed: factors which generally influence their food purchasing decisions, their opinions regarding characteristics of food products and their eating habits, and their sociodemographic characteristics (All values of this calculation are given in the Supplementary).

The sample was divided into 3 clusters which were internally homogenous yet mutually heterogeneous. The clusters were formed after 13 iterations (Table 5). This method resulted in clusters with the values as shown in Table 5.

Table 5. Iterative partitioning method of the first cluster analysis.

Iterations	Variations between Cluster Means *		
	1	**2**	**3**
1	2.948	3.282	3.266
2	0.295	0.303	0.315
3	0.101	0.115	0.167
4	0.054	0.077	0.088
5	0.053	0.036	0.095
6	0.031	0.079	0.096
7	0.023	0.108	0.107
8	0.032	0.119	0.154
9	0.065	0.110	0.160
10	0.060	0.080	0.116
11	0.039	0.033	0.073
12	0.026	0.000	0.033
13	0.000	.0000	0.000

* author's calculation.

As shown in Table 6, the first cluster is the largest while the second is the smallest. To describe the clusters, the F-statistic was used to determine which variables were the most relevant for partitioning. First, the F-statistic was applied to the second question. Values of the F-statistic show that three variables which have the highest values are at the same time the variables which are the most significant for partitioning, and they are in order of significance:

2.7. I try to be informed about healthy lifestyles.
2.1. I consume healthy food products; I care about what gets into my body.
2.3. I am careful about the energy value (calories) and fat which I consume.

Table 6. Clusters after the first cluster analysis *.

Description	Number of Clusters	Number of Respondents
	1	204.000
Cluster	2	134.000
	3	155.000
Valid		493.000
Not valid		3.000

* author's calculation.

Variables with the lowest F-value are those which have the least influence on the formation of the clusters, and they are as follows (2.4 variable has the least influence of the three):

2.5. Packaging plays a significant role when I buy food.
2.8. I habitually consume fast food.
2.4. I prefer buying food products produced in Serbia.

To confirm the results, comparisons were made between mean values of all variables within each cluster through LSD (Least Significant Difference) test (in Supplementary).

The analysis showed that differences are largest between the arithmetic means of variables 1, 3, and 7, while differences are smallest between variables 4, 5, and 8, confirming the results of our study. This led to the exclusion of variables 2.4, 2.5, and 2.8 from the following stage, because they were proven to have the smallest influence. After that, another cluster analysis was carried out with the aim of reaching more homogenous clusters.

The second cluster analysis had 9 iterations (Table 7). Based on this procedure, clusters with the following values were formed (Table 8).

Table 7. Iteration procedure for the purpose of the second cluster analysis *.

Iterations	Variations between Cluster Means		
	1	2	3
1	2.388	2.200	2.331
2	0.267	0.232	0.399
3	0.121	0.135	0.243
4	0.076	0.093	0.133
5	0.092	0.061	0.130
6	0.032	0.049	0.077
7	0.017	0.034	0.055
8	0.017	0.000	0.011
9	0.000	0.000	0.000

* author's calculation.

Table 8. Clusters after the second cluster analysis *.

Description	Number of Clusters	Number of Respondents
	1	105.000
Cluster	2	222.000
	3	167.000
Valid		494.000
Not valid		2.000

* author's calculation.

Cluster structure is much more homogenous in this instance. However, the size of clusters has changed, too. The largest cluster now is the second one, and the smallest is the first one. In order to compare arithmetic means of different clusters, analysis of variance was used.

The results show that there is a statistically significant difference in every variable between arithmetic means of the clusters. This can be seen in the *Sig* column in the table, where all values are lower than 0.05, bearing in mind that variables 2.4, 2.5, and 2.8 were excluded from the analysis because they lower homogeneity of the cluster.

Every statistically significant result is marked by an asterisk (*) next to the number in the mean difference (I–J). The cluster in column I is compared with each cluster in column J individually, with respect to every variable, and the result is statistically relevant if the Sig in the row of cluster J is lower than 0.05

In the next step, an analysis of variance was applied to the first question, that is, the question regarding factors which generally influence food purchasing decisions of respondents. As it turned out, the factor of "Good quality and value for money" did not play a part in the distribution of respondents into clusters, meaning that there was no statistically significant difference between clusters with regard to this factor.

After that, the respondents were distributed into clusters according to their sociodemographic characteristics and it turned out that there was no statistically significant difference between clusters with regard to: sex, number of children, employment status, and total monthly income of the household. That means that there was a statistically significant difference between clusters regarding: age, marital status, number of household members, education, and occupation [49,50].

To define clusters more easily, mean plots were used in this analysis. Based on all analyses conducted in this study, the following groups (i.e., clusters) and their characteristics have emerged:

Cluster 1:

This group consists mainly of respondents who are over 35 years of age, single or married, have one household member and a lower level of education, and are tradesmen or blue-collar workers. What they consider of least importance is whether they consume "healthy" products and what the energy value of those products is. When purchasing food items, they pay least attention to whether the products are organic or not. Nor is a clearly marked expiration date of much importance to this group, and this constitutes the largest difference between this cluster and the other two clusters. In addition, they do not assign any importance to the origin of products, packaging, or pleasantness of setting in which the product is sold. When buying food items, respondents in this cluster pay attention to getting value for money, and the most important factor which influences their decision is product price.

Cluster 2:

This group consists of respondents who are under 45 years of age, with most of them between 25 and 34, single or married, (but this cluster has a higher number of divorced people than either Cluster 1 or Cluster 3), and have 2–4 household members. Regarding the level of education, most respondents in this cluster have high school or university education. Their occupations include tradesmen, engineers, and professionals, with the largest number of professionals, that is, people with a university degree. When compared with other two clusters, respondents in this cluster pay most attention to what they consume, and they want to be informed about healthy lifestyles. Factors which they find important when purchasing food items include quality, recommendation, marketing, renowned manufacturer, product aesthetics, and a pleasant setting in which the product is sold. The importance they assign to these factors puts this group in stark contrast with the other two groups. The least important factor to this group is product origin, but they still assign more importance to it than Clusters 1 and 3. Two most important factors to this group are product quality and a clearly marked expiration date.

Cluster 3:

This group consists of respondents between 25 and 54 years of age, predominantly single or married, with 3 or 4 household members. The level of education of this group is like that of Cluster 2, that is, most respondents in this group have high school or university education. This group is like the previous one regarding occupations, too, but engineers with university degrees are prevalent in this cluster. Respondents in this cluster think that it is important to eat "healthy" food, but they pay less attention to product origin than either Cluster 1 or 2. Regarding factors which influence purchase of food items, it is important to them that the product comes from an organic system of production and has a clearly marked expiration date, and the most important factor is product quality. Respondents in this group do not assign importance either to recommendations and product aesthetics or to the way it is marketed and packaged.

Based on this, it should be noted that combining cluster and factor analysis leads to the following conclusion:

If we look at the relationship between clusters and components that affect food purchases, it can be concluded that:

Cluster 1 corresponds to component 3 (CHEAP PRODUCT).

Cluster 2 corresponds to components 1 and 2 (AESTHETICALLY PLEASING AND EASILY RECOGNISABLE PRODUCT and HEALTHY AND SAFE PRODUCT.

Cluster 3 corresponds to component 2 (HEALTHY AND SAFE PRODUCT).

The abovementioned conclusion is presented in Table 9.

Table 9. Results as divided into clusters *.

Cluster 1:	
This group consists mainly of respondents who are over 35 years of age, single or married, have one household member and a lower level of education, and are tradesmen or blue-collar workers. Of least importance to them is whether they consume healthy products and the number of calories those products have. When purchasing products, it is of least importance to them whether the products are environmentallyfriendly or not. A clearly marked expiration date is an equally insignificant factor, and that constitutes the largest difference between this cluster and the other two clusters. In addition, they do not assign any importance to the origin of products, packaging, or a pleasant setting in which the product is sold. When buying food items, respondents in this cluster pay attention to getting value for money, and the most important factor which influences their decision is product price.	**Component 3:** CHEAP PRODUCT: The product is reasonably priced or sold at a bargain price; it is good value for money and it is recommended by a friend, expert, nutritionist, doctor.

Table 9. *Cont.*

Cluster 2: This group consists of respondents who are under 45 years of age, with most of them between 25 and 34, single or married, (but this cluster has a higher number of divorced people than either Cluster 1 or Cluster 3), and have 2–4 household members. Regarding the level of education, most respondents in this cluster have high school or university education. They are mostly tradesmen, engineers, and professionals, with the largest number of professionals. When compared with the other two clusters, respondents in this cluster pay most attention to what they consume and they want to be informed about healthy lifestyles. Factors which they find important when purchasing food items include: quality, recommendation, marketing, renowned manufacturer, aesthetic, and a pleasant setting in which the products is sold. The importance they assign to these factors sets this group starkly apart from Cluster 1 and Cluster 3. The least important factor to this group is product origin, but they still assign more importance to it than Clusters 1 and 3. The two most important factors to this group are product quality and a clearly marked expiration date.	**Component 1:** AESTHETICALLY PLEASING AND EASILY RECOGNISABLE PRODUCT: The product is well-marketed, attractively packaged. It is produced by a renowned manufacturer, sold in a pleasant setting, and looks nice. **Component 2:** HEALTHY AND SAFE PRODUCT: Ingredients and the expiration date are clearly stated; the product does not contain additives, it is eco-friendly, and of good quality.
Cluster 3: This group consists of respondents between 25 and 54 years of age, predominantly single or married, with 3 or 4 household members. Level of education of this group is similar to that of Cluster 2, that is, most respondents in this group have high school or university education. This group is similar to the previous one with regard to occupations, too, but engineers are prevalent in this cluster. Respondents in this cluster think that it is important to eat "healthy" food, but they pay less attention to product origin than either Cluster 1 or 2. Regarding factors which influence purchase of food items, it is important to them that the product is eco-friendly and has a clearly marked expiration date, and the most important factor is product quality. Respondents in this group do not assign importance to recommendations and product aesthetics. This sets them apart from Clusters 1 and 2. In addition, respondents in this cluster do not assign importance to marketing and packaging.	**Component 2:** HEALTHY AND SAFE PRODUCT: Ingredients and the expiration date are clearly stated; the product does not contain additives; it is eco-friendly and of good quality.

* author's research.

4.4. Discussion

As can be seen from the presented research results, the analysis on a random sample from the two largest cities in Serbia, Belgrade, and Novi Sad, showed that consumers of organic products from the analyzed sample can be grouped into three clusters corresponding to certain components related to characteristics products that can be found in the domestic market.

In the research, the education factor showed that most of those with secondary and higher education are in the group of those who care about eating "healthy" food. In addition, the research showed that there is a statistically significant difference between clusters based on: age, marital status, household members, education, and occupation [49].

The mentioned research [28] showed that there is a segmentation of consumers in relation to the variable of organic food consumption in the USA. They were correlated with environmental concerns, knowledge, and health concerns, as well as some demographic characteristics (race, gender, age, education). In contrast, in the research segment marked as "segment 3 -" negligent individuals "were younger respondents with a higher level of education, but who did not show much concern for nutrition, quality, and nutritional value. However, a similar study was shown in the study by Magetts et al. [23], that there is a special emphasis on the influence of the level of education of the respondents on the choice of "healthy" food, where the result showed that it is the factor of the strongest influence. Other research shows that among the surveyed consumers, health care is in the first place, 44% of respondents stated that it is the most important reason, 15% emphasize health as an important factor, only 2% believe that health is not an important factor when buying organic agri-food products [51]. The results of research in several cities in Serbia

(Belgrade, Novi Sad, Nis, Kragujevac, and Novi Pazar), given by Vehapi [18], also show that health is the most important motive for buying organic food in Serbia. As many as 81.6% of consumers rated health as the leading motive when ranking. Slightly different influencing factors are given in two studies from Italy that indicate that health-safe food has acquired the dimension of a niche market, and the most important role in consumers' decision to buy it have availability (availability), trust, and an adequate price level [52]. On the other hand, in their research, Canio Francesca and Elisa [53] showed that regional and locally produced food, as well as organic products, are experiencing an increasing success amongst consumers as perceived as authentic high-quality food products, able to contribute to sustainable methods of production and consumption. Applying the theory of reasoned action, this work compares consumers' intention to buy EU quality labels and organic food products. Results revealed different motives at the roots of the two products buying choices. On one hand, shoppers are willing to pay a premium price for sustainable EU quality label foods. On the other hand, recyclable packaging is mandatory to shape the intention to buy organic foods. Ditlevsen et al. [54] elaborated on organic consumers features and established that they were more likely to live in the capital and have a higher education.

5. Conclusions

The results of the research showed that organic consumers in Serbia can be grouped into 3 clusters based on certain characteristics and components.

As Table 9 shows, Cluster 1 is rather specific and is characterized by a category of respondents who assign least importance to whether they eat "healthy" products and to the number of calories in those products. When buying food items, they assign least importance to whether the products are organic. Similarly, a clearly marked expiration date is not considered important, which sets this group apart from the other two clusters. When buying food products, this category of respondents pays attention mainly to the price, which leads to the conclusion that these buyers are highly unlikely to consume organic food in the future, partly because those products cost more and partly because of inherent characteristics of these specific products whose eco-friendly features, emphasis on "healthiness" and similar factors clearly have no influence on buyers in this cluster. Respondents in Cluster 2 differ from respondents in other two clusters because they care most about what they consume, and they try to be informed about healthy lifestyles. Regarding factors which relates to purchase of food, they care about: quality, recommendation, marketing, renowned manufacturer, product aesthetics, and pleasantness of setting in which the product is sold. Respondents in Cluster 3 assign importance to eating "healthy" food, and they care less about product origin than either respondents in Cluster 1 or Cluster 2. Regarding factors that are associated with purchase of food products, it is important to them that the product comes from an organic system of production and has a clearly marked expiration date, and the most important factor is product quality.

In such a case, eco-marketing should be more directed towards those consumers who are already "more environmentally and health conscious", which are those in Cluster 2, because evidence demonstrates that respondents who buy organic products are largely attracted to product provenance, an attractive packaging design, an absence of additives and harmful substances, a clearly marked expiration date, a pleasant setting in which the product is sold, and the fact that the product is environmentally friendly. This indicates that it is important for eco-marketing to emphasize these product characteristics to increase their sales and influence the consumers.

Keeping in mind the results of the research, certain measures of economic and agrarian policy should go in the direction of encouraging this form of agricultural production, especially to consumers that can be subsumed under Cluster 1. First, it is necessary to improve consumer knowledge about what organic production is and how products from this production system are properly labeled. This can be achieved through media and other forms of campaigns (in retail outlets) where the promotion of organic products would be carried out continuously and at the same time with education of consumers. In addition, education within primary and secondary schools can improve the state of organic demand in the domestic market in the long run. Promotions of organic products at certain events (Organic Street of Novi Sad, etc.) contribute to raising the visibility of this system of production and local self-government, by supporting such events; contributes to raising environmental awareness among consumers.

The main limitation of this study is related to both the convenient sample and the limited geographical area. It is reflected in the fact that only the markets of Belgrade and Novi Sad are covered. Although these are currently the largest and most significant domestic markets for organic products, further research should go in the direction of determining the possibilities of developing this market in smaller urban and rural areas. In addition, it would be important to determine (and possibly expand) the cluster of consumers of organic products based on the opinions of consumers from smaller places. In this way, local development policies could be harmonized.

The insights gained from this study suggest several important theoretical and managerial implications aimed at increasing consumer demand for organic food. Based on the demographic and psychographic profiles identified for the three clusters of organic food consumers, key stakeholders such as organic food marketers and associations, policymakers, and socio-environmental organizations need to segregate their target policies, adjusting them for different organic consumer profiles identified in every cluster. In the long run, this should contribute to development of the local organic market and organic sector in the Republic of Serbia. Likewise, this study will enrich scientific, economic, and social literature when it comes to understanding consumers behavior at "green" or "organic" markets in developing countries.

Supplementary Materials: The following are available online at https://www.mdpi.com/2071-1050/13/1/131/s1, Table S1: F statistics for cluster analysis purposes, Table S2: Mean values for variables within a cluster, Survey: Organic Food Product Market Research.

Author Contributions: Conceptualization, V.R. and M.T.S.; methodology, M.T.S.; validation, V.R., D.M. and D.G.T.; formal analysis, V.R.; investigation, D.G.T. and M.T.S.; resources, V.R. and D.M.; data curation, V.R.; writing—original draft preparation, V.R. and M.T.S.; writing—review and editing, D.M.; supervision, D.G.T.; project administration, M.T.S. and D.G.T.; funding acquisition, V.R. All authors have read and agreed to the published version of the manuscript.

Funding: This research was funded by the Ministry for Education, Science and Technological Development of the Republic of Serbia, contract No. 451-03-68/2020-14/200117 and by Provincial Secretariat for Education, Regulations, Administration and National Minorities—National Communities contract No. 142-451-3158/2020-02.

Conflicts of Interest: The authors declare no conflict of interest.

References

1. Vapa-Tankosić, J.; Ignjatijević, S.; Kiurski, J.; Milenković, J.; Milojević, I. Analysis of Consumers' Willingness to Pay for Organic and Local Honey in Serbia. *Sustainability* **2020**, *12*, 4686. [CrossRef]
2. Tomaš-Simin, M.; Glavaš-Trbić, D.; Petrović, M. Organic production in the Republic of Serbia: Economic aspects. *Ekon. Teor. Praksa* **2019**, *12*, 88–101. [CrossRef]
3. Aarset, B.; Beckmann, S.; Bigne, E.; Beveridge, M.; Bjorndal, T.; Bunting, J.; McDonagh, P.; Mariojouls, C.; Muir, J.; Prothero, A.; et al. The European consumers' understanding and perceptions of the "organic" food regime: The case of aquaculture. *Br. Food J.* **2004**, *106*, 93–105. [CrossRef]
4. Venkat, K. Comparison of Twelve Organic and Conventional Farming Systems: A Life Cycle Greenhouse Gas Emissions Perspective. *J. Sustain. Agric.* **2012**, *36*, 620–649. [CrossRef]
5. Prodanović, R.; Ignjatijević, S.; Bošković, J. Innovative potential of beekeeping production in AP Vojvodina. *J. Agron. Technol. Eng. Manag.* **2019**, *2*, 268–277.
6. Gomiero, T. Food quality assessment in organic vs. conventional agricultural produce: Findings and issues. *Appl. Soil Ecol.* **2018**, *123*, 714–728. [CrossRef]
7. Milošević, G.; Kulić, M.; Đurić, Z.; Đurić, O. The Taxation of Agriculture in the Republic of Serbia as a Factor of Development of Organic Agriculture. *Sustainability* **2020**, *12*, 3261. [CrossRef]

8. Puvača, N.; Lika, E.; Cocoli, S.; ShtyllaKika, T.; Bursić, V.; Vuković, G.; TomašSimin, M.; Petrović, A.; Cara, M. Use of Tea Tree Essential Oil (Melaleucaalternifolia) in Laying Hen's Nutrition on Performance and Egg Fatty Acid Profile as a Promising Sustainable Organic Agricultural Tool. *Sustainability* **2020**, *12*, 3420. [CrossRef]

9. Sarkar, A.N. Green Branding and Eco-innovations for Evolving a Sustainable Green Marketing Strategy. *Asia-Pac. J. Manag. Res. Innov.* **2012**, *8*, 39–58. [CrossRef]

10. Fernando, Y.; ChiappettaJabbour, C.J.; Wah, W.-X. Pursuing green growth in technology firms through the connections between environmental innovation and sustainable business performance: Does service capability matter? *Resour. Conserv. Recycl.* **2019**, *141*, 8–20. [CrossRef]

11. Peattie, K. Towards Sustainability: The Third Age of Green Marketing. *Mark. Rev.* **2001**, *2*, 129–146. [CrossRef]

12. Simula, H.; Lehtimäki, T. Managing greenness in technology marketing. *J. Syst. Inf. Tech.* **2009**, *11*, 331–346. [CrossRef]

13. Feil, A.A.; da Cyrne, C.C.S.; Sindelar, F.C.W.; Barden, J.E.; Dalmoro, M. Profiles of sustainable food consumption: Consumer behavior toward organic food in southern region of Brazil. *J. Clean. Prod.* **2020**, *258*, 120690. [CrossRef]

14. Falguera, V.; Aliguer, N.; Falguera, M. An integrated approach to current trends in food consumption: Moving toward functional and organic products? *Food Control* **2012**, *26*, 274–281. [CrossRef]

15. Hansmann, R.; Baur, I.; Binder, C.R. Increasing organic food consumption: An integrating model of drivers and barriers. *J. Clean. Prod.* **2020**, *275*, 123058. [CrossRef]

16. Chekima, B.; Oswald, A.I.; Wafa, S.A.W.S.K.; Chekima, K. Narrowing the gap: Factors driving organic food consumption. *J. Clean. Prod.* **2017**, *166*, 1438–1447. [CrossRef]

17. Chekima, B.; Chekima, K.; Chekima, K. Understanding factors underlying actual consumption of organic food: The moderating effect of future orientation. *Food Qual. Prefer.* **2019**, *74*, 49–58. [CrossRef]

18. Vehapi, S. A study of the consumer Motives which influence the purchase of organic food in serbia. *Ekon. Teme* **2015**, *53*, 105–121. [CrossRef]

19. Grubor, A.; Djokić, N. Organic food consumer profile in the Republic of Serbia. *Br. Food J.* **2016**, *118*, 164–182. [CrossRef]

20. Tomaš-Simin, M.; Glavaš-Trbić, D. Historical development of organic production. *Ekon. Polj.* **2016**, *63*, 1083–1099. [CrossRef]

21. Shabbir, M.S.; Bait Ali Sulaiman, M.A.; Hasan Al-Kumaim, N.; Mahmood, A.; Abbas, M. Green Marketing Approaches and Their Impact on Consumer Behavior towards the Environment—A Study from the UAE. *Sustainability* **2020**, *12*, 8977. [CrossRef]

22. Lotter, D.W. Organic Agriculture. *J. Sustain. Agric.* **2003**, *21*, 59–128. [CrossRef]

23. Margetts, B.M.; Martinez, J.A.; Saba, A.; Holm, L.; Kearney, M.; Moles, A. Definitions of "healthy" eating: A pan-EU survey of consumer attitudes to food, nutrition and health. *Eur. J. Clin. Nutr.* **1997**, *51* (Suppl. 2), S23. [PubMed]

24. Torjusen, H.; Lieblein, G.; Wandel, M.; Francis, C.A. Food system orientation and quality perception among consumers and producers of organic food in Hedmark County, Norway. *Food Qual. Prefer.* **2001**, *12*, 207–216. [CrossRef]

25. Wier, M.; O'Doherty Jensen, K.; Andersen, L.M.; Millock, K. The character of demand in mature organic food markets: Great Britain and Denmark compared. *Food Policy* **2008**, *33*, 406–421. [CrossRef]

26. Gracia, A.; de Magistris, T. The demand for organic foods in the South of Italy: A discrete choice model. *Food Policy* **2008**, *33*, 386–396. [CrossRef]

27. Garg, N.; Wansink, B.; Inman, J.J. The Influence of Incidental Affect on Consumers' Food Intake. *J. Mark.* **2007**, *71*, 194–206. [CrossRef]

28. Nie, C.; Zepeda, L. Lifestyle segmentation of US food shoppers to examine organic and local food consumption. *Appetite* **2011**, *57*, 28–37. [CrossRef]

29. Van Huy, L.; Chi, M.; Lobo, A.; Nguyen, N.; Long, P. Effective Segmentation of Organic Food Consumers in Vietnam Using Food-Related Lifestyles. *Sustainability* **2019**, *11*, 1237. [CrossRef]

30. Migliore, G.; Thrassou, A.; Crescimanno, M.; Schifani, G.; Galati, A. Factors affecting consumer preferences for "natural wine": An exploratory study in the Italian market. *BFJ* **2020**, *122*, 2463–2479. [CrossRef]

31. Drugova, T.; Curtis, K.R.; Akhundjanov, S.B. Organic wheat products and consumer choice: A market segmentation analysis. *BFJ* **2020**, *122*, 2341–2358. [CrossRef]

32. Pacho, F. What influences consumers to purchase organic food in developing countries? *BFJ* **2020**, *122*, 3695–3709. [CrossRef]

33. Troudi, H.; Bouyoucef, D. Predicting purchasing behavior of green food in Algerian context. *EMJB* **2020**, *15*, 1–21. [CrossRef]

34. Nair, S.R.; Shams, S.M.R. Impact of store-attributes on food and grocery shopping behavior: Insights from an emerging market context. *EMJB* **2020**. [CrossRef]

35. Wang, X.; Pacho, F.; Liu, J.; Kajungiro, R. Factors Influencing Organic Food Purchase Intention in Developing Countries and the Moderating Role of Knowledge. *Sustainability* **2019**, *11*, 209. [CrossRef]

36. Frewer, L.J.; Risvik, E.; Schifferstein, H. (Eds.) *Food, People and Society*; Springer: Berlin/Heidelberg, Germany, 2001; ISBN 978-3-642-07477-6.

37. Vandal, K. Trženje s sonarovnimikmetijskimipridelki. *Sodobno Kmetijstvo* **1997**, *30*, 363–369.

38. Grunert, S.C.; Juhl, H.J. Values, environmental attitudes, and buying of organic foods. *J. Econ. Psychol.* **1995**, *16*, 39–62. [CrossRef]

39. Weiner, M. Cradle to Cradle@_ An analysis of the market potential in the German outdoor apparel industry. Bachelor's Thesis, Hochschule Hannover University of Applied Sciences and Arts, Hanover, Germany, 2016.

40. Yang, S.; Angulo, F.J.; Altekruse, S.F. Evaluation of Safe Food-Handling Instructions on Raw Meat and Poultry Products. *J. Food Prot.* **2000**, *63*, 1321–1325. [CrossRef]

41. Fotopoulos, C.; Chryssochoidis, G.M.; Pantzios, C.J. Critical factors affecting the future of the Greek market of organic produce. *Mediterr. J. Econ.Agric. Environ.* **1999**, *10*, 30–35.
42. MacRae, R.J.; Lynch, D.; Martin, R.C. Improving Energy Efficiency and GHG Mitigation Potentials in Canadian Organic Farming Systems. *J. Sustain. Agric.* **2010**, *34*, 549–580. [CrossRef]
43. Magnusson, M.K.; Arvola, A.; Hursti, U.-K.K.; Åberg, L.; Sjödén, P.-O. Choice of organic foods is related to perceived consequences for human health and to environmentally friendly behaviour. *Appetite* **2003**, *40*, 109–117. [CrossRef]
44. Lockie, S.; Lyons, K.; Lawrence, G.; Grice, J. Choosing organics: A path analysis of factors underlying the selection of organic food among Australian consumers. *Appetite* **2004**, *43*, 135–146. [CrossRef] [PubMed]
45. Tsakiridou, E.; Boutsouki, C.; Zotos, Y.; Mattas, K. Attitudes and behaviour towards organic products: An exploratory study. *Int. J. Retail. Distrib. Manag.* **2008**, *36*, 158–175. [CrossRef]
46. Ness, M.R.; Ness, M.; Brennan, M.; Oughton, E.; Ritson, C.; Ruto, E. Modelling consumer behavioural intentions towards food with implications for marketing quality low-input and organic food. *Food Qual. Prefer.* **2010**, *21*, 100–111. [CrossRef]
47. Stolz, H.; Stolze, M.; Hamm, U.; Janssen, M.; Ruto, E. Consumer attitudes towards organic versus conventional food with specific quality attributes. *NJAS Wagening. J. Life Sci.* **2011**, *58*, 67–72. [CrossRef]
48. Ignjatijević, S.; Prodanović, R.; Bošković, J.; Puvača, N.; Tomaš-Simin, M.; Peulić, T.; Đuragić, O. Comparative analysis of honey consumption in Romania, Italy and Serbia. *Food Feed Res.* **2019**, *46*, 125–136. [CrossRef]
49. Brkanlić, S.; Sánchez-García, J.; Esteve, E.B.; Brkić, I.; Ćirić, M.; Tatarski, J.; Gardašević, J.; Petrović, M. Marketing Mix Instruments as Factors of Improvement of Students' Satisfaction in Higher Education Institutions in Republic of Serbia and Spain. *Sustainability* **2020**, *12*, 7802. [CrossRef]
50. Tatarski, J.; Brkanlić, S.; Sanchez Garcia, J.; Esteve, E.B.; Brkić, I.; Petrović, M.; Okanović, A. Measuring Entrepreneurial Orientation of University Employees in Developing Countries Using the ENTRE-U Scale. *Sustainability* **2020**, *12*, 8911. [CrossRef]
51. National Research Council (US); Institute of Medicine (US); Woolf, S.H.; Aron, L. *Public Health and Medical Care Systems*; National Academies Press: Washington, DC, USA, 2013.
52. Scaglioni, S.; De Cosmi, V.; Ciappolino, V.; Parazzini, F.; Brambilla, P.; Agostoni, C. Factors Influencing Children's Eating Behaviours. *Nutrients* **2018**, *10*, 706. [CrossRef]
53. Canio Francesca, D.; Elisa, M. EU quality label vs Organic food products: A multigroup structural equation modeling to assess consumers' intention to buy in light of sustainable motives. *Food Res. Int.* **2020**, 109846, In Press, Corrected Proof. [CrossRef]
54. Ditlevsen, K.; Denver, S.; Christensen, T.; Lassen, J. A taste for locally produced food—Values, opinions and sociodemographic differences among 'organic' and 'conventional' consumers. *Appetite* **2020**, *147*, 104544. [CrossRef] [PubMed]

 *sustainability*

Article

Analysis of Consumers' Willingness to Pay for Organic and Local Honey in Serbia

Jelena Vapa-Tankosić [1], Svetlana Ignjatijević [1,*], Jelena Kiurski [1], Jovana Milenković [2] and Irena Milojević [3]

1   Faculty of Economics and Engineering Management in Novi Sad, Cvećarska 2, 21000 Novi Sad, Serbia; jvapa@fimek.edu.rs (J.V.-T.); jelena.kiurski7@gmail.com (J.K.)
2   Faculty of Pharmacy in Belgrade, Vojvode Stepe 450, 11221 Belgrade, Serbia; infinity.feed.rs@gmail.com
3   Institute of Applied Sciences Belgrade, Lomina 2, 11000 Belgrade, Serbia; drimilojevic@gmail.com
*   Correspondence: svetlana.ignjatijevic@gmail.com

Received: 6 May 2020; Accepted: 2 June 2020; Published: 8 June 2020

Abstract: In times of increased concern for human health and care for the environment, it is important to investigate the consumer behavior models in order to better manage the product supply. From the perspective of our research, it is important to learn about consumer attitudes, of a specific product, such as honey, so as to potentially strengthen the economic position of honey producers. The purpose of this article was to explore the consumers' perceptions of organic and local honey in the Republic of Serbia and identify factors that contribute to predicting consumers' willingness to pay (WTP) for organic and local honey. The ordinal regression was used to determine which factors influence the WTP for organic and local honey, and the findings show that the consumers were WTP more for organic honey than for local honey. Socio-economic characteristics of respondents and honey attributes affect consumers' WTP. The higher monthly household income positively influences the WTP for organic honey, while on the other hand, the higher level of education has a positive influence on the WTP for local honey. The WTP for organic honey positively affected by the perceived importance of honey attributes, such as food safety and support for the local community. The frequency of the purchase of the local honey, the recommendations for the local honey purchase and the perceived importance of attributes, such as the care for the environment and nutritional properties, have a positive influence on the consumers' WTP for local honey. From this, we can conclude that our understanding of all the factors that influence a consumer's decision to allocate budget expenses for honey can help all stakeholders in creating an adequate pricing and promotional strategy for honey products.

Keywords: willingness to pay; honey; organic; local; Serbia

---

1. Introduction

Natural resources provide the Republic of Serbia with the possibility of more intensive development of organic production. Simić [1] points out that the country has favorable ecological, climatic, and technical conditions to produce traditional berries and fruits, as well as organic vegetables, fruits, cereals, and oilseeds. According to the Directorate for National Reference Laboratories, Organic Production Group, in the Republic of Serbia the share of areas under organic production in previous years has had a substantial rise (with a registered increase of 204% in organic production from the year 2012 till 2018) [2]. Lazić [3] classified organic production by farm size and type. Berenji, Milenković, Kalentić, and Stefanović [4] indicate that characteristics of organic farms differ by the production region and that in the Province of Vojvodina the average organic farm is larger than 10 ha with a specialization in a smaller number of plant species (cereals, industrial plants and vegetables).

The food market in the Republic of Serbia is dominated by locally produced foods [5–8], both organic or conventionally grown. They are positioned in the market, either as complements or substitutes. The consumers in the Republic of Serbia prefer the attributes of quality, freshness and taste, but are also increasingly concerned about food safety [9–12]. Consumers tend to prefer certified products, because they consider that the certificate indicates the notion of food safety [13]. They believe that local products contribute to halting biodiversity loss [5], and improving ecosystems. The consumers who are concerned about the environment consider the organic products safe from the pesticide and GMO aspects. There is still some skepticism surrounding the organizations that are responsible for monitoring food safety and pesticide levels, which can potentially influence the consumers' willingness to pay (WTP) levels. On the other hand, there is a belief that the consumption of local products has a positive impact on the development of the local economy, as it supports the local agricultural producers and employment. Because more than 80% of Serbia is rural, and the dominant agricultural production has low productivity, it is important to conserve and sustainably use the existing resources. Previous research by Cvijanović and Ignjatijević [14] points out that it is necessary to conduct further analysis of the impact of the honey sector on rural development. In recent years, the Republic of Serbia has seen an increase both in honey production and export of honey [15–17]. Local honey from the territory of Fruška Gora, Homolja, and Šumadija has the potential for branding and recognition in the domestic and international markets [15]. Consequently, an increase in honey consumption would lead to an increase in production, and would have a positive impact on the economic development of the honey sector, local areas, and the country itself.

In the Serbian market, the consumers can differentiate between organically produced and locally produced honey (also with a geographical origin). The first aspect of honey refers to organic honey production, therefore the consumers are aware that organic honey, as well as all organically produced foods, are controlled during the production process and that they do not contain harmful substances, GMO residues, drug residues, etc. Another aspect relates to the authenticity of geographical and botanical origins. Thus, honey from certain geographical regions or from the local area is associated with a well-known local producer's brand. The largest number of consumers is interested in consuming high-quality products, such as honey, that is new and has specific flavors, also of organic origin expressing a willingness to pay a higher price [18,19]. The WTP is also determined by socio-economic factors (gender, age, income, and education), so lower-income consumers in the Republic of Serbia have pointed out that the organic products are too expensive and income levels are still the main limiting factors affecting demand for organic food products [20].

In Serbia, the WTP has been researched in recent years by a small number of authors, and on the other hand, such an analysis is focused on a limited number of food products. Such a situation imposes the need to shed light on the factors influencing consumer behavior with regard to WTP for specific food products, especially honey. The identification and understanding of consumer behavior, that is, the perceptions of local and organic products and the factors that influence the purchase and consumption decision-making processes have so far not, to our best knowledge, been investigated. Many authors have been investigating problems in the production and consumption of honey from various aspects. Some authors point to oscillations in honey production [15–17], others point to differences in consumption motives and preferences [6,7,12,21,22] interconnectedness of the honey production and the environment [8,23–25], the impact of the honey production on the employment [26–30], and the local regional development [31,32]. Their findings show that honey is considered to be a healthy and safe product [33], which consumers eat if they nurture healthy lifestyles and are committed to preserving the environment and local community progress.

The subject of the research has been to evaluate the consumer's willingness to pay for local and organic honey, while assessing whether the socio-demographic characteristics and attributes of organic and local honey affect the consumers' WTP. The current study contributes to the relevant literature in two important ways. The authors have investigated honey, which is considered extremely important for human health and for the environment. The choice of this product is motivated by the fact that Serbia is one of the largest producers of honey and that the production is steadily increasing registering an increase in the production of the organic honey and the geographical origin honey [15,17]. Second, the consumers' perceptions and willingness to pay for the local and organic honey in the Republic of Serbia have not been investigated so far. The specific objectives of the research paper are to: (1) Investigate the current organic and local honey perceptions in the Republic of Serbia; (2) assess Serbian consumers' willingness to pay (WTP) for the organic and local honey; and (3) identify factors that contribute to predicting the consumers' WTP for the organic and local honey. In the light of above mentioned, this research paper seeks to answer several important questions that are presented as formal hypotheses:

Hypothesis 1 (H1). *The consumers are willing to pay more for organic honey than for local honey.*

Hypothesis 2 (H2). *Consumers' education and monthly household income positively affect the WTP for organic and local honey.*

Hypothesis 3 (H3). *Consumers' WTP for organic and local honey is positively affected by the perceived importance of honey attributes.*

Due to the importance of honey and the modest results of previous research on honey consumption in Serbia, the obtained results fill the gap in knowledge about the preferences of honey consumers. As there are no previous comparative studies on the consumption of local and organic honey in Serbia, the findings of this study will be of interest to authors and researchers, on the one hand, the producers and the distributors of honey on the other, and potentially to the representatives of state authorities. The results of the research indicate the reasons behind the choice of honey and highlight the diversity of motives for buying and consuming organically produced or local honey. Understanding the preferences of honey consumers, especially WTP, can serve in creating a marketing strategy for beekeepers, policies to encourage organic production incentives and/or to create strategies for local development and define incentive measures for honey production. Given the favorable conditions and long tradition in honey production [27,34,35], the cooperation between science, the real sector and state representatives is especially important. Finally, the arguments presented speak of the need to adequately manage the market supply in response to increased consumer demands. The paper is structured as follows: The introduction points out the need, importance and aspects of research and sets out hypotheses. Then, the authors review the current literature and available research on WTP for organic and local products. In the following section, the authors present the methodology, the research findings, and a discussion of the results. The final section provides concluding remarks.

2. Literature Review

2.1. WTP for Organic Products

The first study on consumer attitudes regarding certified pesticide-free fresh produce in the United States, by Ott [36], showed that the majority of respondents were willing to pay 5% to 10% more than the standard price, in order to buy certified fresh pesticide-free products. The identified consumer target were highly educated consumers, who earned an average or above-average income. WTP research on radiation-treated food products in the United States, by Malone [37], used a probit regression model, and identified that three variables were significantly associated with the consumption of the irradiation-treated food products—education, income, and gender. Further research on WTP for organic food products in the United States was conducted by Jolly [38], who analyzed the views of organic food consumers in California, using an analysis of variance (ANOVA). They found that the WTP varied by product type, and depended on the conventional price of the product. Misra, Huang, and Ott [39] have used a probit regression model to analyze Georgia consumers' attitudes toward organic food consumption, and their findings show that socio-economic factors (race, age, income, and education) are among the factors that have a significant impact on the WTP for organic products. The findings of Buzby and Skees [40] (who used a national survey in the US), point out that female, younger respondents, and those less educated, had a higher WTP for the organic produce (while the household income, race, and size did not influence the WTP). The results of a survey of organic consumers in Norway by Wandel and Bugge [41] showed that over 70% of Norwegian organic consumers were willing to pay a price premium of 5%, while less than 10% of respondents were willing to pay 25% and more for the organic products. Gil, Gracia, and Sanchez [42] studied organic consumers in Madrid and Navarre, and found a WTP of 15% to 25% for organic fruits and vegetables. In Ireland, O'Donovan and McCarthy [43] found that 73% of Irish consumers were not prepared to pay more than 10% for organic meat, compared to conventionally produced meat. Millock, Hansen, Wier, and Andersen, [44] in Denmark, found that 51% of respondents indicated a different WTP for different types of organic foods, and their WTP ranged from 18.5% for organic minced meat, up to the 40% for the organic potatoes.

Radman [45] found that the majority of Croatian consumers have expressed the view that organic products are too expensive and their WTP was in the range of 11–20% over the price of conventionally produced food. The findings of Sakagami, Sato, and Ueta [46], in Japan, show that consumers who are concerned about fresh foods and prefer certified vegetables to conventionally grown vegetables, expressed a WTP ranging from 8–22%. Krystallis, Fotopoulos, and Zotosal [47] showed that Greek consumers WTP levels ranged from 55% for organic grape wine to 100% for organic oranges. Rodriguez, Lacaze, and Lupinal [48] point out that the Argentine consumer' WTP depends on the type of organic product. Haghjou, Hayati, Pishbahar, Mohammadrezaei, and Dashti [49] showed that 95% of Iranian respondents expressed a WTP between 5 and 24% on the price of conventional products, while the identified target market showed that the consumers were female, married, and with children under the age of 10. The findings of Attanasio, Carelli, Cappelli, and Papetti [50] showed that 50.6% of Italian respondents were willing to pay a lower price for the organic produce (in regard to the market price for a conventional product). Vietoris et al. [51] pointed out that consumers in Romania are willing to pay 5% to 10% more for organic food compared to the price of conventional food.

One of the first studies in the Balkans region included samples from Macedonia and Serbia [52]. It showed that the average consumer of organic products was 50 years old, had a university degree, earned an above-average income, and were living in a family (up to a maximum of 3 family members). They also had a WTP for organic products up to 30%. Vlahović, Puškarić, and Jeločnik [53] findings show that a small number of consumers in Serbia are regular buyers of organic products, due to the low purchasing power and high retail prices—which are the main factors limiting the demand and the consumption of organic products. According to a study by Vehapi [54], the majority of organic consumers in Serbia were WTP up to 20% over the cost of conventional products. Vlahović and Šojić [55] findings show that there is an increased interest in organic agricultural products in the Republic of Serbia, but organic food prices and respondents' income levels are still major limiting factors affecting demand for organic food products. Vehapi and Dolićanin's study [56] identified the potential organic consumer in Serbia: They had university or college degree, a monthly net household income that exceeded RSD 100.000,00, and a WTP up to 30% over the price of conventional products. Jovanović, Joksimović, Kašćelan, and Despotović [57] findings indicate that about 81% of Montenegro consumers are still not ready to pay a higher price for organically produced products. The consumers in the Republic of Serbia had an increased interest for value-added food products, and were willing to pay a premium price for organic products up to 20% [19].

2.2. WTP for Local Products

On the other hand, the term "local food" can be linked to a concept of natural goods or services produced or provided by different enterprises in rural areas with an established socio-economic identity [58]. However, the consumers show great variation in the definition of the "local food". It is interesting that the majority of food consumers have associated the term "local" food with the term "foods grown locally" [59]. As far as the distance for the local produce is concerned, it may differ in case of fresh and processed products [60]. In the UK, La Trobe [61] points out that local food products are regarded as produced and sold within a 30 to 40-mile radius of the market. In Europe, Karner [62] found that alternative local food networks, as an emerging European sector, differ from the conventional food system and large-scale agro-food enterprises in terms of their organizational structures, farming systems, territorial setting, food supply chains, policy support, focus on 'quality' of food and social, cultural, ethical, economic and environmental aspects. Guided by the idea of sustainable local development, the consumers are willing to pay more for the local product than for the organic product [63]. Local food consumers using direct channels (farmers markets, community-supported agriculture outlets, and roadside stands) reported a significantly higher WTP for local produce [64]. Nganje, Hughner, and Lee [65] findings show that local produce bearing the Arizona Grown label had a higher WTP than local produce labeled USDA-certified pointing out the brand association between local food and safe food. Gracia, De Magistris, and Nayga [66] elicited consumers' WTP for local lamb, confirming that social influence affects WTP values. Grebitus, Lusk, and Nayga [67] findings show that the belief to support the local economy, when buying food that traveled fewer miles, affects positively both the consumers' WTP and the consumers' perceptions that fresh local food has superior attributes. As females are the main household shoppers, their attitudes towards organic, local, GM-free, and U.S. grown are also stronger [68].

The studies that have focused on analyzing consumers' attitudes towards organic, local or other types of food have focused on different kinds of fresh produce (apples, tomatoes, blueberries, potatoes, corn, etc.). The findings of Costanigro, McFadden, Kroll, and Nurse [69] have shown that the consumers' WTP for local apples is higher than their WTP for organic apples. The findings on German consumers show that the WTP for the local organic food label is higher than the WTP for EU organic declaration [70]. The findings of Onken, Bernard, and Pesek [71] on the influence of purchasing venue on WTP for strawberry preserves at US farmers market, in five states, show that consumers have expressed a higher WTP for natural preserves than for the organic produce. Gracia, Barreiro-Hurlé and López-Galán [72] in their research on whether local and organic claims are complements or substitutes show that consumers are willing to pay a positive premium price for an enhanced method of production and the proximity of production, while the consumers with a higher WTP for origin related attributes valued higher the local claim (but when combined with other claims, the most valued combination is local plus organic). The findings on WTP of conventional and organic potatoes and sweet corn and its' versions with two individual organic parts (such as no use of pesticides and non-GM) show that all versions of each food were viewed as substitutes for one another and the consumers were willing to pay significant premiums for organic and its' parts in regard to conventional versions [73]. Onozaka and McFadden [74] analysis of differential values and interactive effects of sustainable production claims and location claims (local, domestic, not local and imported), elicited through a conjoint choice experiment, have shown that if the Gala apples are produced both organically and locally the WTP values range from 9% to 15% price premium. USA consumers have expressed positive WTPs for the attributes of 'organic' and 'locally produced' blueberries, although a higher WTP for organic than for the locally produced blueberries has been expressed [75]. Dominican consumers are willing to pay 17.5% more for organic and 12% more for locally grown produce [76]. The authors' findings on elicited consumer WTP for local and organic attributes for fresh tomatoes show that the average premiums the consumers were willing to pay for organic tomatoes and locally grown tomatoes were about the same [77]. The findings of Tempesta and Vecchiato [78] on WTP for milk have shown that the higher amount of milk consumed brings about a reduction of WTP of 26%.

Cicia and Colantuoni [79] by meta-analysis on 23 studies have shown that "on-farm traceability" is important for consumers and that they are willing to pay a premium of 16.71% over the base price to be fully informed on the "meat's production path". The findings on WTP for a country-of-origin labeling program show for the consumers WTP for the U.S. certified steak and hamburger is equivalent to 38% and 58% [80]. The findings on the importance of the country of origin in food consumption in a developing country have shown that the majority of consumers consume imported foods because of the lower price or good price/quality ratio and that the origin was more significant than either the price or the packaging in the decision to purchase beef [81]. The principal-component analysis reveals that the strong association of local and organic apples labels with the desirable environmental and food safety outcomes, combined with the distrust for the government agencies responsible for monitoring food safety and pesticide levels, is the most important predictor of consumers' WTP [82].

3. Materials and Methods

According to Breidert, Hahsler, and Reutterer [83], the concept of consumer willingness to pay for a product or service is used when formulating competitive strategies and developing new products. Many methods for measuring WTP have been presented in the scientific literature. There are two main groups of non-market valuation techniques in the scientific literature: Revealed preferences (RP) techniques that observe consumer behavior and their choices in the real market, and stated preferences (SP) techniques used to elicit individual reported preferences over hypothetical alternatives. Boccaletti and Nardella [84] point out that the contingent valuation method allows a direct evaluation of the WTP. The consumers should indicate their WTP without purchasing a hypothetical product. With this method, the consumer is directly asked to state their WTP for a particular good or service. Carson and Hanemann [85] point out that this method has become known as "conditional valuation", as the "valuation" estimate obtained from preferential information is a "conditional" valuation of an environmental good within the "built market for research purposes". The most commonly used questionnaire formats for measuring WTP are direct (open) questions, discrete choice experiments [86–89], bidding games, payment card system and referendum question format [90]. The authors have tested four WTP elicitation methods the Becker–DeGroot–Marschak mechanism, multiple price lists, multiple price lists with stated quantities, and real-choice experiments, and their findings were closely related [91]. In the present survey, the authors have opted for a payment card system as the consumers were asked in the survey to indicate predesigned price premiums (nothing more (1); up to 10% more (2); 10–20% (3); 20–30% (4); more than 30% (5)) expressing the willingness to pay for local and organic honey.

The total number of consumers investigated was 1000. Of the distributed questionnaires, 788 consumers of organic and local honey consumer questionnaires were returned complete (79% response rate). Before the data collection, the questionnaire was tested in cooperation with the Association of Beekeeping Organizations of Serbia and Vojvodina to improve its validity and reliability. Data collection began in September 2019 and finished in January 2020. In the consumers' survey, the authors have used the snowball method as a random sampling technique. The questionnaire, or link of the questionnaire, was sent to the initial seed informants within the researchers' professional and personal network, in order to be further distributed [92,93]. The previous research [94] has shown that the snowball method is suitable for exploring under-researched topics, where the knowledge and awareness of the product is not sufficiently explored. The questionnaire was created according to questionnaires from the relevant researchers on the topic of examining consumers' WTP [19,76,95–101]. The first part of the questionnaire is based on the collection of socio-demographic data of the respondents, including characteristics such as gender, age, qualifications, level of monthly income, and presence of children and parents in the household. The second part of the questionnaire is focused on the general characteristics of consumer behavior when buying organic and local honey: frequency of buying the honey, the place of the honey purchase, and the recommendations of other people that influence the honey purchase. The third part involves eliciting consumers' willingness to pay more for the organic and local honey using a five-point scale (nothing more, up to 10%, 10–20%, 20–30% and more than 30%). The fourth part deals with the respondents' perceptions of intrinsic and extrinsic attributes of honey, such as taste, health, environmental care, food safety, nutritional properties, price, and support to the local community (interval level from 1–5).

Model explanatory variables include socio-demographic variables, such as gender (male—1, female—2), age (five intervals: 1—lowest and 5—oldest), monthly household income (four intervals:

1—lowest, 4—highest), education level (three intervals: 1—lowest, 3—highest), and the presence of children and parents in the household (four intervals: 1—no children and parents, 2—children in the household, 3—parents in the household 4—both children and parents in the household). The model explanatory variables are also general honey purchase variables, such as the place of shopping of organic honey (1—manufacturer, 2—health food stores, 3—specialized stores, 4—large supermarkets, 5—markets), the frequency of purchase of organic honey (1—once in six months, 2—once in three months, 3—once a month, 4—once in every two weeks), the recommendations for the organic honey purchase (1—doctors, 2—friends, 3—mass media, 4—family, 5—I decide by myself), as well as the respondents' opinions concerning general honey attributes (5 intervals: 1—lowest, 5—highest): 'I eat honey because it tastes good', 'I eat honey because it is good for health', 'I think that honey consumption helps the environment', 'I am very concerned about my food safety', 'I eat honey because of its nutritional properties', 'the price of honey is important to me', 'honey consumption is beneficial to the local community'. The dependent variable in the models is the WTP for organic and local honey (five intervals: 1—0%, 2—<10%, 3—10–20%, 4—20–30%, 5—>30%).

The statistical package SPSS (Statistical Package for Social Sciences) was used to analyze the data of this study. In all statistical tests, the significance threshold (α) was set at 5%. Descriptive statistics (frequencies and percentages) have addressed the socio-demographic and consumer characteristics and their perceptions of local and organic honey. To test the hypotheses, we used ordinal regression analysis to determine which factors influence the WTP for organic and local honey. The ordinal regression is a statistical method that examines the influence of multiple independent factors on a single dependent factor [102]. The ordinal regression equation is defined by (1):

$$\ln\left(\frac{p}{1-p}\right) = \beta_0 + \beta_1 X_1 + \beta_2 X_2 + \dots \beta_n X_n \tag{1}$$

where the independent factors are X_i, the regression coefficients are β_i and p the probability that the event has occurred. In this case, the dependent factor is binary.

However, when the ordinal type variable is dependent with more than two categories, then we define the final cumulative logit model as (2):

$$\ln\left(\frac{p(Y \leq j)}{1-p\,(Y \leq j)}\right) = \beta_0 + \beta_1 X_1 + \beta_2 X_2 + \dots \beta_n X_n \tag{2}$$

where: X_i = independent (explanatory) variables or predictors; β_i = regression coefficients or parameters; p = the probability of an event occurring; Y = dependent variable divided into j categories.

We have decided to apply the Ordinal Regression [103,104] with the cumulative logistic regression model (with the increasing outcome), that is, the "proportional odds model". This model converts the ordinal scale into a series of binary cut-off values (the number of these cut-off values is always one less than the number of categories of the dependent variable). Each cut-off value, i.e., critical points of separation or classification criteria represent a threshold that must be crossed to move from one category of criteria to another. Proportional odds models assume that the true regression coefficients (beta) are the same in all models and that the only difference between the models is the cut-off values. The standard interpretation of ordinal logit coefficients is that the regression parameter expressed in ordinal logit (odds logarithm) with each independent variable shows the expected degree of change of the ordinal dependent variable, when the observed independent variable increases by one unit, while the other independent variables are constant. The positive values of beta indicate higher odds of moving to the next higher ordered category for higher values of the independent variable. The final model incorporates certain predictor variables and has undergone iterative processes that account for the maximum likelihood function and parameter estimates. The requirements that are required to complete in order to achieve Multidimensional Ordinary Logistic Regression are:

The dependent variable must be an ordinal scale, which means it should be displayed in encoded categories, which must be ranked: Independent variables are interval, nominal or ordinal scales; Multidimensionality should not exist between independent variables; Proportional Odds—each independent variable must have the same effect.

The authors have checked the assumptions of the validity of ordinal regression by applying the models with proportional odds. The following models have all been satisfied: model fitting; goodness-of-fit, including the Pearson and deviance goodness-of-fit tests; the Cox and Snell, Nagelkerke, and McFadden measures of R2; the likelihood-ratio test, and the assumption of parallel lines.

4. Results

The results of the descriptive statistical analysis indicate that in the sample of 788 respondents, the majority are female (58.4%). The female respondents are more willing to participate in the research—this can be due to the fact that they are more involved in purchasing food for the whole family. The average consumer belongs to the age group of 20–30 years (39.7%) with a significant percentage (39.8%) of respondents in the age group of 31–50 years. Most consumers have a high school diploma (41.6%), followed by a professional college degree and higher education degree (Ma, PhD) (32.6%). The highest percentage of respondents has an average income of less than 500 euros a month and lives with their children in the household (36.5%) (Table 1).

Table 1. Socio-demographic characteristics of respondents.

Socio-Demographic Characteristics	Category	Number of Respondents	Percentage (%)
Gender	Male	328	41.6%
	Female	460	58.4%
Age	<20	57	7.2
	20–30	313	39.7
	31–40	184	23.4
	41–50	129	16.4
	More than 50	105	13.3
Level of Education	High school	328	41.6
	Professional degrees	203	25.8
	Bachelor's degrees and other (Masters, PhD)	257	32.6
Monthly Income	<500 euros	334	42.4
	501–1000 euros	168	21.3
	1001–2000	152	19.3
	>2000	134	17.0
Presence of Children and Parents in the Household	I have children and parents living in the household	56	7.1
	I have children living in the household	288	36.5
	I have no children and parents in the household	214	27.2
	I have parents living in the household	230	29.2

In the continuation of the descriptive analysis (Table 2) the WTP for organic and local honey, place and frequency of purchase, as well as the recommendations for purchase for organic and local honey, have been presented. The results indicate that the majority of the respondents (44.9%) are willing to pay premium prices of 20–30% for the organic honey over the price of conventional honey. For local honey consumers, the percentages of respondents that are willing to pay 20–30% more is significantly lower (17.5%), while the majority is willing to pay 10–20% more for the local honey. The consumers buy organic honey in speciality stores and large supermarkets, while the local honey is most commonly bought at local markets. The organic honey is usually purchased once every three months, while the local honey is purchased at least once a month. The majority of respondents decide on the purchase of organic and local honey by themselves (53.9% and 43%, respectively), although recommendations for the organic honey purchase are also accepted from the family (24.5%) and for the local honey purchase from friends (34.3%).

Table 2. Questions related to general characteristics of honey consumption.

Questions Related to General Characteristics of Honey Consumption		Organic		Local	
		Fr *	%	Fr *	%
Willingness to Pay (WTP) more	0%	55	7.0	93	11.8
	<10%	92	11.7	235	29.8
	10–20%	243	30.8	293	37.2
	20–30%	354	44.9	138	17.5
	>30%	44	5.6	29	3.7
Place of Purchase of Honey	Markets	73	9.3	283	35.9
	Large supermarkets	261	33.1	203	25.8
	Specialized stores	387	49.1	175	22.2
	Health food stores	31	3.9	98	12.4
	Directly from manufacturer	36	4.6	29	3.7
Frequency of Purchase of Honey	Once in 2 weeks	54	6.9	245	31.1
	At least once a month	221	28.0	361	45.8
	Once in 3 months	330	41.9	112	14.2
	Once in 6 months	183	23.2	70	8.9
Recommendations for the Honey Purchase	From doctors	33	4.2	38	4.8
	From friends	116	14.7	270	34.3
	From mass media	21	2.7	33	4.2
	From my family	193	24.5	108	13.7
	I decide by myself	425	53.9	339	43.0

* Frequency.

The consumers find honey to be good for their health (4.40) and of good taste (4.34). In the recent years, the consumer awareness that the honey production and consumption are linked to the support of the environmental conditions (4.14) and local development (4.09), has been substantially increased (Table 3). The consumers value honey in terms of food safety (3.99) and nutritional properties (3.90). The consumers also consider the price of honey to be a relevant factor in their purchase (3.96).

Table 3. Consumer-reported perceptions of honey attributes.

	It Tastes Good	It Is Good for Health	I Think That Honey Consumption Helps the Environment	I Am Very Concerned about My Food Safety	I Eat Honey Because of Its' Nutritional Properties	Price of Honey Is Important to me	Honey Consumption Is Beneficial to the Local Community
Mean	4.34	4.40	4.14	3.99	3.90	3.96	4.09
Std.dev.	0.902	0.823	0.891	0.948	1.026	1.104	0.940

4.1. WTP for Organic and Local Honey

Chi-square test for an association has been used to determine if there is an association between the variables of WTP for organic/local honey and the consumers' socio-demographic characteristics (Table 4).

Table 4. Chi-square test for association.

	WTP Organic—Chi-Square Value	WTP Local—Chi-Square Value
Gender	26.661; df (4); $p = 0.000$ *	5.327; df (4); $p = 0.255$
Age	15.597; df (16); $p = 0.481$	21.399; df (16); $p = 0.164$
Level of Education	14.457; df (8); $p = 0.041$ *	10.121; df (8); $p = 0.257$
Monthly income	14.013; df (12); $p = 0.300$	61.034; df (12); $p = 0.000$ *
Presence of children and parents in the household	9,046; df (12); $p = 0.699$	11.785; df (12); $p = 0.463$
Place of purchase of organic/local honey	34.808; df (16); $p = 0.004$ *	53.680; df (16); $p = 0.000$ *
Frequency of purchase of organic/local honey	6.604; df (12); $p = 0.883$	207.079; df (12); $p = 0.000$ *
Recommendations for the organic/local honey purchase	11.848; df (16); $p = 0.754$	20.138; df (16); $p = 0.214$

* statistically significant at $p < 0.05$ level.

Significant associations were determined between the WTP for the organic honey and gender ($p = 0.000$) and the WTP for the organic honey and the level of education ($p = 0.041$). As for the local honey, the significant associations were determined between the WTP for the local honey and the respondents' monthly income ($p = 0.000$). As a next step, the chi-square test for association has been used to determine if there is an association between the WTP for the organic and the local honey and the general characteristics of honey purchase (Table 4). A significant association was determined between the WTP for the organic honey and the place of purchase of organic honey ($p = 0.004$). The significant associations was found between the WTP for the local honey and the place of purchase of local honey ($p = 0.000$), as well as between the WTP for the local honey and the frequency of purchase of local honey ($p = 0.000$). The significant association was also tested on the variables of the WTP for the organic and the local honey and the perceived honey attributes. The significant association was determined between the WTP for the organic honey and the perceived honey attribute taste ($p = 0.042$) and the WTP for local honey and the perceived honey attribute taste ($p = 0.008$), health (p = 0.000), concern for the environment ($p = 0.000$) and the food safety ($p = 0.001$), that are presented in Table 5.

Table 5. Chi-square test for association.

	WTP Organic Honey	WTP Local Honey
I eat honey because it tastes good	14.854; df (8); $p = 0.042$ *	32.787; df (8); $p = 0.008$ *
I eat honey because it is good for health	5.372; df (8); $p = 0.717$	47.510; df (8); $p = 0.000$ *
I think that honey consumption helps the environment	9.318; df (8); $p = 0.316$	51.482; df (8); $p = 0.000$ *
I am very concerned about my food safety	9.352; df (8); $p = 0.313$	40.555; df (8); $p = 0.001$ *
I eat honey because of its' nutritional properties	5.010; df (8); $p = 0.757$	19.103; df (8); $p = 0.263$
The price of honey is important to me	12.880; df (8); $p = 0.116$	22.315; df (8); $p = 0.133$
Honey consumption is beneficial to the local community	13.317; df (8); $p = 0.101$	19.528; df (8); $p = 0.242$

* statistically significant at $p < 0.05$ level.

4.2. Ordinal Regression Models

By using the ordinal logistic regression with the logit link function, we assume that the effect of the independent variables shall be the same for each level of the dependent variable (WTP). Before interpreting the model's regression coefficients, the assumptions regarding model adequacy must be examined. Using the PLUM procedure in SPSS—first, we have checked whether the final models improve the outcome prediction. To confirm this, we have analyzed the model fitting information model. In Table 6, we have received information about whether the models of the logit link function improve the ability to predict the resulting variable. Based on the results obtained from model fitting we can conclude that the statistical significance of the models is present and that the models with predictor variables make a significant contribution to the prediction of the dependent variable consumers' willingness to pay for organic honey (Models 1 and 2) and consumers' willingness to pay for local honey (Models 3 and 4).

Table 6. Model fitting information for the WTP for organic and local honey models.

Dependent Variable	Model Explanatory Variables	Model	-2 Log Likelihood	Chi-Square	Df	Sig.
1. WTP organic honey	Socio-demographic variables	Intercept only Final	1453.608 1431.896	21.712	13	0.020
2. WTP organic honey	Organic honey purchase and honey attributes variables	Intercept only Final	1901.113 1864.887	36.226	39	0.05
3. WTP local honey	Socio-demographic variables	Intercept only Final	1901.113 1864.887	36.226	39	0.040
4. WTP local honey	Local honey purchase and honey attributes variables	Intercept only Final	2109.694 1982.423	127.270	39	0.00

Link function: Logit.

4.2.1. WTP for Organic Honey

Model explanatory variables include gender (male—1, female—2), respondent age (five intervals: 1—lowest and 5—oldest), monthly household income (four intervals: 1—lowest, 4—highest), education level (three intervals: 1—lowest, 3—highest) and presence of children and parents in the household (four intervals: 1—no children and parents, 2—children in the household, 3—parents in the household 4—both children and parents in the household) and the dependent variable in the model is the WTP for organic honey (five intervals: 1—0%, 2—<10%, 3—10–20%, 4—20-30%, 5—>30%). The estimation results of the Ordered Logit model has been presented in Table 7. We have first analyzed the influence of independent socio-demographic predictors on the dependent variable (WTP for organic products). Multicollinearity detection was then performed using VIF (Variance Inflation Factor). Multicollinearity was not present in the model. Therefore, we have decided to keep all the predictor variables in our ordinal logistic regression model. We have further tested whether the final model, which included all explanatory variables improves the outcome of the base model. We started from the null hypothesis that the fit is good. As we did not reject this hypothesis ($p > 0.05$), we have concluded that the data and the model predictions are similar and that we have a good model (χ^2 (1067) = 1102.13; $p = 0.22$). The values of the coefficient of determination (Pseudo R2) have indicated a 17.4% Nagelkerke variance of the model was explained by the explanatory variables. In addition to the results of assumptions regarding the model adequacy, we have performed a test of parallel lines. If the general model gave a significantly better fit to the data than the ordinal (proportional odds) model ($v < 0.05$), then we could be led to reject the assumption of proportional odds. The assumption of good fit of the ordinal model was confirmed by the result ($\chi^2 = 66.70$, df = 39, $p = 0.06$). The process of verifying the adequacy of the model has been fully completed. When analyzing the final Model in Table 7, the statistical significance was found for only one variable of the five explanatory variables. The monthly household income has proven to be a significant explanatory variable (in the second category of monthly household income of 501–1000 euros). Given that the obtained regression coefficient is less than 1, we come to the conclusion that the monthly household income and the willingness to pay for organic honey are negatively correlated. Based on the B value for the explanatory variable the monthly household income, provided that all other predictors in the model are kept constant, we can say that the likelihood of the willingness to pay for the organic honey of the respondents in the second category with a monthly income of 501–1000 euros decreases by 0.10 compared to those with a monthly income of over 2000 euros.

Table 7. Results of the final model obtained by ordinal regression (Model 1–4).

Regression Models 1–4		B	Std. Error	Wald	df	Sig.	Lower Bound	Upper Bound	Exp(B)
(Model 1)	Monthly Household Income = 3	−0.099	0.232	1.182	1	0.046 *	−0.356	0.554	0.104
(Model 2)	I am very concerned about my food safety = 3	−0.700	0.241	8.466	1	0.004 *	−1.172	−.229	0.496
	Honey consumption is beneficial to the local community = 4	−0.383	0.183	4.367	1	0.003 *	0.024	0.743	0.467
(Model 3)	Education = 2	−0.298	0.179	2.781	1	0.040 *	−0.649	0.052	0.742
(Model 4)	Once in 3 months = 2	−0.621	0.254	5.971	1	0.015 *	−1.119	−0.123	0.537
	From my family = 4	0.544	0.207	6.879	1	0.009 *	0.137	0.950	1.723
	I think that honey consumption helps the environment = 1	−2.228	1.194	3.482	1	0.042*	−4.568	0.112	0.108
	I eat honey because of its nutritional properties = 4	−0.325	0.187	3.027	1	0.042 *	−0.691	0.041	0.723

* statistically significant at $p < 0.05$ level.

In a second model, model explanatory variables are the variables reflect the respondents opinions concerning the following: Place of shopping of organic honey (1—manufacturer, 2—health food stores, 3—specialized stores, 4—large supermarkets, 5—markets); frequency of purchase of organic honey (1—once in six months, 2—once in three months, 3—once a month, 4—once in every two weeks); recommendations for the organic honey purchase (1—doctors, 2—friends, 3—mass media, 4—family, 5—I decide by myself), as well as the respondents opinions concerning general honey attributes (5 intervals: 1—lowest, 5—highest): 'I eat honey because it tastes good', 'I eat honey because it is good for health', 'I think that honey consumption helps the environment', 'I am very concerned about my food safety', 'I eat honey because of its nutritional properties', 'the price of honey is important to me', 'honey consumption is beneficial to the local community', and the dependent variable in the model is WTP for organic honey (five intervals: 1—0%, 2—<10%, 3—10–20%, 4—20–30%, 5—>30%). Multicollinearity detection by VIF (Variance Inflation Factor) has shown that no multicollinearity was present in the model. Therefore, we have further tested the final model starting from the null hypothesis that the fit is good. Goodness-of-Fit with χ^2 test ($p > 0.05$) has shown that we have a good model (χ^2 (2621) = 2593.32; $p = 0.65$). The value of the coefficient of determination (Pseudo R2) has indicated 48% Nagelkerke variance of the model was explained by the independent variables. In addition to the results of assumptions regarding the model adequacy, we have performed a test of parallel lines (χ^2 = 1636.83, df = 117, $p = 0.07$). When analyzing the final Model 2 in Table 7, the statistical significance was found for only two variables of the ten predictor variables. Respondents' perceived importance of Food Safety and Support of the Local Community in honey consumption have proven to be significant explanatory variables. Given that the obtained regression coefficient for the perceived food safety importance is less than 1, and based on its' Exp (B) value, provided that all other predictors in the model are kept constant, we can say that the likelihood of the willingness to pay for organic honey of respondents who have rated the claim "I am very concerned about my food safety" with the mark 3 decreases by 0.49 compared to those respondents who have rated food safety of honey with the mark 5. Having in mind, that the obtained regression coefficient for the claim "Honey consumption is beneficial to the community" is also less than 1, and based on its' Exp (B) value, provided that all other predictors in the model are kept constant, we can say that the likelihood of the willingness to pay for the organic honey of the respondents who have rated the claim "Honey consumption is beneficial to the community" with the mark 4 decreases by 0.46 compared to those respondents who have rated the same claim with the mark 5.

4.2.2. WTP for Local Honey

In the third model, we have started from the analysis of the influence of independent socio-demographic predictors on the dependent variable (WTP for local honey). We started from the null hypothesis that the fit is good, and the goodness-of-fit with χ^2 test ($p > 0.05$) has shown that we have a good model (χ^2 (1067) = 1194.32; $p = 0.078$). The values of the coefficient of determination (Pseudo R_2) have indicated that 29% of Nagelkerke variance of the model was explained by the independent variables. In addition to the results of assumptions regarding the model adequacy, we have performed a test of parallel lines ($\chi^2 = 101.82$, df = 39, $p = 0.08$). When analyzing the final Model 3 in Table 7, the statistical significance was found for only one variable of the five explanatory variables. The level of education has proven to be a significant explanatory variable in Model 3. Given that the obtained regression coefficient is less than 1, we have come to the conclusion that education and the WTP for local honey are negatively correlated. Based on the Exp (B) value for the explanatory variable 'Education', provided that all other predictors in the model are kept constant, we can say that the likelihood of the willingness to pay for local honey of the respondents with the professional college degree decreases by 0.74 compared to those respondents with the bachelor's degrees and higher degrees (Masters, PhD).

As a next step, we tested by means of Ordered Logit model the WTP for the local honey and the variables that reflected the respondents' perceptions on the general purchase of local honey and the honey attributes. As no multicollinearity was present in the model, we have further tested the final model. The Goodness-of-Fit with χ^2 test ($p > 0.05$) has shown that we have a good model (χ^2 (2685) = 1983.24; $p = 0.98$) and the values of the coefficient of determination (Pseudo R_2) have indicated 36% Nagelkerke variance of the model was explained by the independent variables. We have performed a test of parallel lines ($\chi^2 = 1736.31$, df = 117, $p = 0.55$). When analyzing the final Model 4 in Table 7, the statistical significance was found for four variables of the ten predictor variables. The frequency of the purchase of the local honey, the recommendations for the local honey purchase and the perceived importance of preserving the environment and the nutritional properties in the honey consumption have proven to be significant explanatory variables. Given that the obtained regression coefficient for the frequency of purchase of local honey is less than 1, and based on its' Exp (B) value, provided that all other predictors in the model are kept constant, we can say that the likelihood of the willingness to pay for the local honey of respondents who purchase honey once in three months decreases by 0.53 compared to those respondents who purchase honey once in every two weeks. Based on Exp (B) value for the recommendations for the local honey purchase, provided that all other predictors in the model are kept constant, we can say that the likelihood of the willingness to pay for the local honey of the respondents who accept recommendations from the family to purchase honey once a month increases by 1.72 compared to those respondents who decide by themselves. The likelihood of the willingness to pay for the local honey of the respondents who have rated the claim "I think that honey consumption helps the environment" with the mark 1 decreases by 0.10 compared to those respondents who have rated the same claim with the mark 5. The likelihood of the willingness to pay for the local honey of respondents who have rated the claim "I eat honey because of its nutritional properties" with the mark 4 decreases by 0.72 compared to those respondents who have rated the same claim with the mark 5.

5. Discussion

The results show that Serbian consumers were willing to pay more for organic honey than for local honey, which confirmed our first hypothesis. Hypothesis 1 has also been confirmed by the findings of other authors [75,76,105]. The significant associations were found between the WTP for organic honey and gender, as well as the WTP for organic honey and level of education. The concern of mothers for their family [16,106–110] has determined that women are the most important honey buyers [111]. Our results are confirmed by Davies et al. [112] in Northern Ireland who have shown that women, ages 30–45, who have children and higher income are regular customers of organic products. The results obtained were confirmed by Storstad and Bjorkhaug [113] as a higher percentage of women in Norway was found to have positive attitudes towards organic food consumption. Padel and Foster [97] also identified that younger women in employment, as well as middle-aged women, were the major buyers of organic foods, as they are better informed of all aspects related to organic products, and reflect the standard gender divide related to the division of household chores, where women are still more involved in food preparation and family care. The research in the US has also concluded that younger and more educated women are more likely to be regular buyers of organic food [114]. In our research, the WTP for local honey has been determined to be 10–20% above the conventional price of honey. When the products are promoted as locally grown the consumers are willing to pay a price premium [115], therefore the WTP level, as well as the perception of the product quality, is influenced by the provided locally grown information. Therefore, when the consumers are given additional information, their willingness to pay a premium for local produce can increase.

The findings of Model 1 obtained by ordinal regression have shown that the higher monthly household income is a significant explanatory variable of the WTP for organic honey. This is also confirmed by the findings of other authors [112,116–119]. As expected, the frequency of organic purchases increases significantly with higher household incomes. Findings in Croatia [120], and in Serbia of Vlahović and Šojić [55], also confirm this hypothesis. Some authors believe that demographic variables, such as age, income, and education may more closely define organic consumers, but not significantly, as price continues to block organic food consumption [121]. We can conclude that consumers with higher income levels are more willing to buy (and pay more for) organic products. Moreover, the price of organic products and the level of respondents' income are still the main limiting factors in the Republic of Serbia, which influence the demand for organic products and households with higher income levels have shown a higher tendency to buy organic products. This is also confirmed by findings on the interconnectedness of the honey consumption and the income levels [122–124]. The significant association was found between the WTP for local honey and the monthly income. Thus, the price plays a significant role for consumers and is often linked to quality, the existence of certificates [125–130]. The findings of Model 3 show that only the higher level of education has proven to be a significant explanatory variable for predicting the WTP for local honey. This finding is confirmed by the results of Carpio and Isengildina-Massa [131] and Andam, Ragasa, Asante, and Amewu [132]. As expected, the respondents with higher education [124], value more the local honey consumption, and therefore, are willing to contribute to the development of the local honey community. Therefore, we can confirm the second hypothesis that higher education and higher monthly household income of consumers positively affect the WTP for organic and local honey.

The significant association was also found between WTP for local honey and the perceived honey attribute of taste, health, concern for environment and food safety. The perceived honey attribute of taste has been determined to be significantly associated with both WTP for organic honey and WTP for local honey. Good taste has been a motive for purchasing organic products as confirmed by Saba and Messina [133], Özcelik and Ucar [134], Hamzaoui, Essoussi, and Zahaf [135], Padel and Foster [97], Lea and Worsley [136], Brčić-Stipčević and Petljak [137], Vlahović and Šojić [55]. Taste has been an important feature of honey consumption [110]. The local honey consumers are concerned with the state of natural resources and the potential contamination of honey with GMO residues, pesticide residues and antibiotics [138,139]. They perceive honey as a healthy and safe product [140–142], and environmentally friendly [67,143]. More often they associate honey with the attributes of taste, color and appearance and consume honey because of its nutritional properties [144,145], which in this study was found to be associated with the WTP for local honey. The perceived attributes of food safety and support for the local community have proven to be significant explanatory variables of the WTP for organic honey (Model 2). This confirms the fact that WTP for organic honey is positively affected by the perceived importance of honey attributes as the consumers are increasingly concerned with the food safety, as honey is a natural product [124,146–152]. The consumers of honey value the local community support [109,153,154]. The relative influence of the general purchase variables and honey attributes variables on consumers' WTP for local honey (Model 4) has determined the positive influence of the frequency of the purchase of the local honey, the recommendations for the local honey purchase [100,153–157], and the perceived attributes of the care for the environment and nutritional properties [141,158,159]. The third hypothesis is that the consumers' WTP for organic and local honey is positively affected by the perceived importance of honey attributes can be confirmed. On the other hand, the local community support has proven to influence the WTP of organic honey. Thus, the findings signal an increased environmental awareness of consumers of organic products in Serbia. The findings of Hamzaoui-Essoussi and Zahaf [135], as well as Padel and Foster [97], also show that the organic products consumers value the social aspect, such as support for local agriculture, fair trade, and environmental protection. Therefore, identifying other important factors of consumer behavior that accompany the decision to buy organic and local honey can help honey producers gain price premiums.

6. Conclusions

This research aimed to understand the preferences for the organic and local honey, and the respective willingness to pay, in a selected sample of consumers in the Republic of Serbia. The results of the statistical analysis according to the defined hypotheses have been duly presented, and the findings have shown which factors in the proposed model have significantly influenced the WTP for organic and local honey. We found that consumers perceived different factors as important for their purchase. The findings in the research in the Republic of Serbia can be potentially different from the research conducted in countries where the organic products market is in a mature stage of development. In line with the literature review, we have seen that the willingness to pay for agricultural products varies within the developed and underdeveloped markets, according to the consumer segments and different food products categories. The findings indicate that the respondents in Serbia are willing to pay 20–30% more for the organic honey, while their willingness to pay for the local honey is slightly lower (10–20% more for the local honey). The profile of honey consumers shows that they are predominantly female, having a high school diploma, an average income of less than 500 euros a month with the presence of children in the household. The consumers of honey value highly honey attributes and the perceived attribute of health, taste and the care for the environment were rated the highest. The higher monthly household income positively influences the WTP for the organic honey. The higher level of education has a positive influence on the WTP for the local honey. The WTP for organic honey is positively affected by the perceived importance of honey attributes, such as food safety and support for the local community. The frequency of the purchase of the local honey, the recommendations for the local honey purchase and the perceived importance of attributes, such as the care for the environment and nutritional properties, have a positive influence on the consumers' WTP for local honey. From this, we can conclude that understanding of all factors that influence a consumer's decision to allocate budget expenses for honey products can shed additional light and help all stakeholders in the honey production, processing and sales sectors. The above socio-demographic characteristics of the consumers that are willing to pay more for the organic or local honey could be used as a starting point for the creation of an adequate organic or local honey marketing strategy. Furthermore, it is necessary to further investigate the extent to which Serbian consumers of the organic and local honey products value the regional provenance. The labeling of different types of honey, and the additional information on organic and local honey, as well as the preference of direct contact with the honey producers, can potentially contribute to the higher WTP levels and should be further examined. As a direction of future research—due to the quite high acceptance among Serbian consumers for higher prices, it is worth conducting similar analyzes in other both developing and developed countries.

One of the limitations of this study is the study sample. This survey was limited to consumers in the Republic of Serbia. Another limitation might be the overstated willingness to pay more as consumers do not necessarily take into account all the factors they consider when making an actual purchase. The consumers' purchase decision at the point of sale takes into account the essential attributes of honey, but also the product availability on the point of sale, the latest recommendations, or their budget allocations at the time of purchase. The elicited willingness to pay is a conditional elicitation for research purposes. Thus, the elicited willingness to pay may not necessarily coincide with the consumers' willingness to actually pay a higher price for organic and local honey products at the point of sale.

Author Contributions: All authors contributed equally in the writing and reading of this paper. Conceptualization, methodology, formal analysis, and original draft preparation, S.I., J.V.-T. and J.K.; investigation and resources, J.M. and I.M.; supervision and project administration, J.K. All authors have read and agreed to the published version of the manuscript.

Funding: The paper is part of the research on the project: 142-451-2505/2019-02 "Improving the competitiveness of traditional food products in the function of sustainable development of AP Vojvodina" financed by the Provincial Secretariat for Higher Education and Research, AP Vojvodina.

Conflicts of Interest: The authors declare no conflict of interest.

References

1. Simić, I. Organic Agriculture in Serbia at a Glance 2017. National Association Serbia Organica. 2016. Available online: http://www.serbiaorganica.info/wp-content/uploads/2012/07/Organic-Agriculture-in-Serbia-At-a-glance-2017.pdf (accessed on 12 April 2018).
2. Directorate for National Reference Laboratories. Organic Production Group. Available online: http://www.dnrl.minpolj.gov.rs/lat/o_nama/o_nama.html (accessed on 15 May 2019).
3. Lazić, B. Organska poljoprivreda—Zalog za budućnost. *Org. News* **2010**, *1*, 8–9.
4. Berenji, J.; Milenković, S.; Kalentić, M.; Stefanović, E. Nacionalna istraživačka agenda za sektor organske proizvodnje. In *ACCESS-Program for Development of Private Sector in Serbia*; GIZ: Belgrade, Serbia, 2013.
5. Cvijanović, D.; Ignjatijević, S.; Vapa Tankosić, J.; Cvijanović, V. Do Local Food Products Contribute to Sustainable Economic Development? *Sustainability* **2020**, *12*, 2847. [CrossRef]
6. Vučurević, T.; Ignjatijević, S.; Milojević, I. Ekonomska analiza konkurentnosti potrošnje meda u Srbiji i zemljama okruženja. *Ecologica* **2016**, *23*, 599–606.
7. Ćirić, M.; Ignjatijević, S. The influence of demographic characteristics of consumers on honey buying behaviour. In *Sustainable Agriculture and Rural Development in Terms of the Republic of Serbia Strategic Goals Realization with the Danube Region-Development and Application of Clean Technologies in Agriculture-Thematic Proceedings*; Institute of Agricultural Economics: Belgrade, Serbia, 2017; pp. 374–392.
8. Simeunović, T. Menadžerski aspekti međuzavisnosti zaštite životne sredine i budžeta. *Oditor* **2016**, *2*, 25–30. [CrossRef]
9. Vehapi, S. Istraživanje motiva potrošača koji utiču na kupovinu organske hrane u Srbiji. *Econ. Themes* **2015**, *53*, 105–121.
10. Đokić, N.; Grubor, A.; Milićević, N.; Petrov, V. New Market Segmentation Knowledge in the Function of Bioeconomy Development in Serbia. *Amfiteatru Econ.* **2018**, *20*, 700. [CrossRef]
11. Grubor, A.; Đokić, N. Organic food consumer profile in the Republic of Serbia. *Br. Food J.* **2016**, *118*, 164–182. [CrossRef]
12. Ćirić, M.; Ignjatijević, S.; Cvijanović, D. Istraživanje ponašanja potrošača meda u Vojvodini. *Ekon. Poljopr.* **2015**, *62*, 627–644.
13. Driouech, N.; Bilali, H.; Berjan, S.; Radović, M.; Despotović, A. Exploring the Serbian consumer attitude towards agro-food products with ethical values: Organic, fair-trade and typical/traditional products. In Proceedings of the 5th International Scientific Conference "Rural Development 2011", Kaunas, Lithuania, 24–25 November 2011; Volume 5, pp. 37–43. Available online: https://orgprints.org/20162/1/Driouech_et_al_Rural_development_conference_in______Lithuania_Rev_H26-08.pdf (accessed on 10 January 2020).
14. Cvijanović, D.; Ignjatijević, S. *Exploring the Global Competitiveness of Agri-Food Sectors and Serbia's Dominant Presence: Emerging Research and Opportunities*; IGI Global: Hershey, PA, USA, 2017. [CrossRef]

15. Pocoli, C.; Ignjatijević, S.; Cavicchioli, D. *Production and Trade of Honey in Selected European Countries: Serbia, Romania and Italy, Honey Analysis*; De Toledo, V.A., Ed.; InTech: Rijeka, Croatia, 2017; Available online: http://www.intechopen.com/books/honey-analysis/production-and-trade-of-honey-in-selected-european-countries-serbia-romania-and-italy (accessed on 24 October 2019). [CrossRef]

16. Ignjatijević, S.; Ćirić, M.; Čavlin, M. Analysis of honey production in Serbia aimed at improving the international competitiveness. *Custos Agronegocio On Line* **2015**, *11*, 194–213.

17. Ignjatijević, S.; Milojević, I.; Andžić, R. Economic analysis of exporting serbian honey. *Int. Food Agribus. Manag. Rev.* **2018**, *21*, 929–944. [CrossRef]

18. Centre for the Promotion of Imports from Developing Countries. CBI Product Factsheet: Organic Honey in Germany. Available online: https://www.cbi.eu/market-information/honey-sweeteners/trade-statistics/ (accessed on 10 December 2016).

19. Vapa-Tankosić, J.; Ignjatijević, S.; Kranjac, M.; Lekić, S.; Prodanović, R. Willingness to pay for organic products on the Serbian market. *Int. Food Agribus. Manag. Rev.* **2018**, *21*, 791–801. [CrossRef]

20. Kranjac, M.; Vapa-Tankosić, J.; Knežević, M. Profile of organic food consumers. *Econ. Agric.* **2017**, *64*, 497–514. [CrossRef]

21. Murphy, M.; Cowan, C.; Henchion, M.; O'reilly, S. Irish consumer preferences for honey: A conjoint approach. *Br. Food J.* **2000**, *102*, 585–598. [CrossRef]

22. Brščić, K.; Šugar, T.; Poljuha, D. An empirical examination of consumer preferences for honey in Croatia. *Appl. Econ.* **2017**, *49*, 5877–5889. [CrossRef]

23. Prodanović, R.; Bošković, J.; Ignjatijević, S. Organic honey production in function of enviromental protection. *ECOLOGICA* **2016**, *23*, 315–321.

24. Gibbs, D.; Muirhead, I. The Economic Value and Environmental Impact of the Australian Beekeeping Industry. 1998, pp. 9–21. Available online: https://www.honeybee.com.au/Library/gibsmuir.html (accessed on 15 May 2019).

25. Prdić, N. Konkurentska prednost preduzeća na osnovu benčmarking analize poslovanja. *Oditor* **2017**, *3*, 107–117. [CrossRef]

26. Pocol, C.B.; Moldovan-Teselios, C.; Arion, F.H. Beekeepers' association: Motivations and expectations. *Bull. Univ. Agric. Sci. Vet. Med. Hortic.* **2014**, *1*, 141–147. [CrossRef]

27. Bekić, B.; Jeločnik, M.; Subić, J. Honey and Honey Products. In Proceedings of the Works with XXVI Consulting Cgronomists, Ceterinarians, Cechnologists and Cconomists, Belgrade, Serbia, 20–21 February 2013; Volume 18, pp. 3–4.

28. Marinković, S.; Nedić, N. Analysis of production and competitiveness on small beekeeping farms in selected districts of Serbia. In *Applied Studies in Agribusiness and Commerce*; Agroinform Publishing House: Budapest, Hungary, 2010; pp. 1–5.

29. Popa, A.A.; Mărghitaş, L.A.; Pocol, C.B. Economic and socio-demographic factors that influence beekeepers' entrepreneurial behavior. *Agronomy Series of Scientific Research/Lucrari Stiintifice Seria Agronomie* **2011**, *54*, 450–455. Available online: http://www.uaiasi.ro/revagrois/PDF/2011-2/paper/pagini_450-455_Popa.pdf (accessed on 27 December 2019).

30. Qaiser, T.; Ali, M.; Taj, S.; Akmal, N. Impact Assessment of Beekeeping in Sustainable Rural Livelihood. *J. Soc. Sci. (COES&RJ-JSS)* **2013**, *2*, 82–90.

31. Pocol, C.B.; Ilea, M. The development of local products in Romania: A case study of honey. *Agric. Manag. Lucr. Stiintifice Ser. I Manag. Agric.* **2011**, *13*, 153–160.

32. Ahmad, T.; Shah, G.M.; Ahmad, F.; Partap, U.; Ahmad, S. Impact of Apiculture on the Household Income of Rural Poor in Mountains of Chitral District in Pakistan. *J. Soc. Sci. (COESRJ-JSS)* **2017**, *6*, 518–531. [CrossRef]

33. Kowalczuk, I.; Jeżewska-Zychowicz, M.; Trafiałek, J. Conditions of honey consumption in selected regions of Poland. *Acta Sci. Pol. Technol. Aliment* **2017**, *16*, 101–112. [CrossRef] [PubMed]

34. Tasić, J. Budući trendovi i pravci razvoja ruralnog turizma u Srbiji i u svetu. *Oditor* **2018**, *4*, 7–19. [CrossRef]

35. Keča, L.; Marčeta, M.; Bogojević, M. Commercialisation of non-wood forest products on the territory of AP Vojvodina. *Bull. Fac. For.* **2012**, *105*, 99–116. [CrossRef]

36. Ott, S.L. Supermarkets shoppers' pesticide concerns and willingness to purchase certified pesticide residue-free fresh produce. *Agribusiness* **1990**, *6*, 593–602. [CrossRef]

37. Malone, J.W., Jr. Consumer Willingness to Purchase and to Pay More for Potential Benefits of Irradiated Fresh Food Products. *Agribus Int. J.* **1990**, *6*, 163–178. [CrossRef]

38. Jolly, D. Differences between buyers and nonbuyers of organic produce and willingness to pay organic price premiums. *J. Agribus.* **1991**, *9*, 97–111. [CrossRef]
39. Misra, S.; Huang, C.L.; Ott, S.L. Consumer willingness to pay for pesticide-free fresh produce. *West. J. Agric. Econ.* **1991**, *16*, 218–227. [CrossRef]
40. Buzby, J.C.; Skees, J. Consumers want reduced exposure to pesticides in food. *Food Rev.* **1994**, *17*, 19–22. [CrossRef]
41. Wandel, M.; Bugge, A. Environmental concern in consumer evaluation of food quality. *Food Qual. Prefer.* **1997**, *8*, 19–26. [CrossRef]
42. Gil, J.M.; Gracia, A.; Sanchez, M. Market segmentation and willingness to pay for organic products in Spain. *Int. Food Agribus. Manag. Rev.* **2000**, *3*, 207–226. [CrossRef]
43. O'Donovan, P.; McCarthy, M. Irish consumer preference for organic meat. *Br. Food J.* **2002**, *104*, 353–370. [CrossRef]
44. Millock, K.; Hansen, L.G.; Wier, M.; Andersen, L.M. Willingness to pay for organic foods: A comparison between survey data and panel data from Denmark. In Proceedings of the 12th Annual EAERE Conference, Monterey, CA, USA, 24–27 June 2002; Available online: https://mpra.ub.uni-muenchen.de/id/eprint/47588 (accessed on 24 December 2019).
45. Radman, M. Consumer consumption and perception of organic products in Croatia. *Br. Food J.* **2005**, *107*, 263–273. [CrossRef]
46. Sakagami, M.; Sato, M.; Ueta, K. Measuring consumer preferences regarding organic labelling and the JAS label in particular. *N. Z. J. Agric. Res.* **2006**, *49*, 247–254. [CrossRef]
47. Krystallis, A.; Fotopoulos, C.; Zotos, Y. Organic Consumers' Profile and Their Willingness to Pay (WTP) for Selected Organic Food Products in Greece. *J. Int. Consum. Mark.* **2006**, *19*, 81–106. [CrossRef]
48. Rodriguez, E.; Lacaze, V.; Lupin, B. Willingness to pay for organic food in Argentina: Evidence from a consumer survey. In *International Marketing and International Trade of Quality Food Products, Proceedings of the 105th Seminar of the European Association of Agricultural Economists, Bologna, Italy, 8–10 March 2007*; Canavari, M., Regazzi, M.D., Spadoni, R., Eds.; 2007; pp. 187–213. Available online: https://www.researchgate.net/publication/235625001_Willingness_to_Pay_for_organic_food_in_Argentina_Evidence_from_a_consumer_survey (accessed on 5 December 2019).
49. Haghjou, M.; Hayati, B.; Pishbahar, E.; Mohammadrezaei, R.; Dashti, G. Factors affecting consumers' potential willingness to pay for organic food products in Iran: Case study of Tabriz. *J. Agric. Sci. Technol.* **2013**, *15*, 191–202.
50. Attanasio, S.; Carelli, A.; Cappelli, L.; Papetti, P. Organic Food: A Study on Demographic Characteristics and Factors Influencing Purchase Intentions Among Consumers in Pontina Province. *Int. J. Latest Res. Sci. Technol.* **2013**, *2*, 128–132.
51. Vietoris, V.; Kozelova, D.; Mellen, M.; Chrenekova, M.; Potclan, J.E.; Fikselova, M.; Kopkas, P.; Horska, E. Analysis of Consumer Preferences at Organic Food Purchase in Romania. *Pol. J. Food Nutr. Sci.* **2015**, *66*, 139–146. [CrossRef]
52. Sekovska, B.; Vlahović, B.; Bunevski, G. Consumption of Organic Food in Macedonia and Serbia: Similarities and Differences. In *Consumer Attitudes to Food Quality Products, European Federation of Animal Science (EAAP) 133*; Klopčić, M., Kuipers, A., Hocquette, J.F., Eds.; Wageningen Academic Publishers: Wageningen, The Netherlands, 2012; pp. 239–247. [CrossRef]
53. Vlahović, B.; Puškarić, A.; Jeločnik, M. Consumer attitude to Organic Food Consumption in Serbia. *Pet. Gas Univ. Ploiesti Bull.* **2011**, *18*, 45–52. [CrossRef]
54. Vehapi, S. Marketing Strategija Proizvođača Organske Hrane. Ph.D. Thesis, Ekonomski Fakultet, Niš, Serbia, 2014.
55. Vlahović, B.; Šojić, S. Istraživanje stavova potrošača o organskim poljoprivredno-prehrambenim proizvodima i njihovim brendovima. *Agroekonomika* **2016**, *45*, 33–46.
56. Vehapi, S.; Dolićanin, E. Consumers Behavior on organic food: Evidence from the Republic of Serbia. *Econ. Agric.* **2016**, *63*, 871–889. [CrossRef]
57. Jovanović, M.; Joksimović, M.; Kašćelan, L.J.; Despotović, A. Consumer attitudes to organic foods: Evidence from Montenegrin market. *Agric. For.* **2017**, *63*, 223–234. [CrossRef]
58. European Committee of the Regions. *Promoting and Protecting Local Products: A Trumpcard for the Regions*; Committee of the Regions: Brussels, Belgium, 1996.

59. Wilkins, J.L.; Bowdish, E.; Sobal, J. Consumer perceptions of seasonal and local foods: A study in a U.S. community. *Ecol. Food Nutr.* **2002**, *41*, 415–439. [CrossRef]

60. Durham, C.A.; King, R.P.; Roheim, C.A. Consumer definitions of 'locally grown' for fresh fruits and vegetables. *J. Food Distrib. Res.* **2009**, *40*, 56–62. [CrossRef]

61. La Trobe, H. Farmers' markets: Consuming local rural produce. *Int. J. Consum. Stud.* **2001**, *25*, 181–192. [CrossRef]

62. Karner, S. *Local Food Systems in Europe: Case Studies from Five Countries and What They Imply for Policy and Practice*; FAAN Report; IFZ: Graz, Austria, 2010; ISBN 978-3-9502678-2-2. Available online: http://www.genewatch.org/uploads/f03c6d66a9b354535738483c1c3d49e4/FAAN_Booklet_PRINT.pdf (accessed on 31 January 2020).

63. Brown, C. Consumers' preferences for locally produced food, a study in southeast Missouri. *Am. J. Altern. Agric.* **2003**, *18*, 213–224. [CrossRef]

64. Onozaka, Y.; Nurse, G.; Thilmany McFadden, D. Defining Sustainable Food Market Segments: Do Motivations and Values Vary by Shopping Locale? *Am. J. Agric. Econ.* **2011**, *93*, 583–589. [CrossRef]

65. Nganje, W.E.; Hughner, R.S.; Lee, N.E. State-Branded Programs and Consumer Preference for Locally Grown Produce. *Agric. Resour. Econ. Rev.* **2011**, *40*, 20–31. [CrossRef]

66. Gracia, A.; De Magistris, T.; Nayga, R.M. Importance of Social Influence in Consumers' Willingness to Pay for Local Food: Are There Gender Differences? *Agribusiness* **2012**, *28*, 361–371. [CrossRef]

67. Grebitus, C.; Lusk, J.L.; Nayga, R.M. Effect of distance of transportation on willingness to pay for food. *Ecol. Econ.* **2013**, *88*, 67–75. [CrossRef]

68. Bellows, A.C.; Alcaraz, V.G.; Hallman, W.K. Gender and food, a study of attitudes in the USA towards organic, local, U.S. grown, and GM-free foods. *Appetite* **2010**, *55*, 540–550. [CrossRef] [PubMed]

69. Costanigro, M.; McFadden, D.T.; Kroll, S.; Nurse, G. An in-store valuation of local and organic apples: The role of social desirability. *Agribusiness* **2011**, *27*, 465–477. [CrossRef]

70. Roosen, J.; Köttl, B.; Hasselbach, J. Can Local Be the New Organic? Food Choice Motives and Willingness to Pay. Selected Paper for Presentation at the AAEA/EAAE Food Environment Symposium. 2012, pp. 1–12. Available online: http://ageconsearch.umn.edu/bitstream/123512/2/Roosen_CanLocalBeTheNewOrganic.pdf (accessed on 10 December 2019).

71. Onken, K.A.; Bernard, J.C.; Pesek, J.D. Comparing willingness to pay for organic, natural, locally grown, and state marketing program promoted foods in the mid-Atlantic region. *Agric. Resour. Econ. Rev.* **2011**, *40*, 33–47. [CrossRef]

72. Gracia, A.; Barreiro-Hurlé, J.; López-Galán, B. Are local and organic complement or substitutes labels? A Consumer Preferences Study for Eggs. *J. Agric. Econ.* **2013**, *65*, 49–67. [CrossRef]

73. Bernard, J.C.; Bernard, D.J. Comparing parts with the whole: Willingness to pay for pesticide-free, non-GM, and organic potatoes and sweet corn. *J. Agric. Resour. Econ.* **2010**, *35*, 457–475.

74. Onozaka, Y.; McFadden, D.T. Does local labeling complement or compete with other sustainable labels? A conjoint analysis of direct and joint values for fresh produce claim. *Am. J. Agric. Econ.* **2011**, *93*, 689–702. [CrossRef]

75. Shi, L.; House, L.A.; Gao, Z. Impact of purchase intentions on full and partial bids in BDM auctions: Willingness-to-pay for organic and local blueberries. *J. Agric. Econ.* **2013**, *64*, 707–718. [CrossRef]

76. Boys, K.A.; Willis, D.B.; Carpio, C.E. Consumer willingness to pay for organic and locally grown produce on Dominica: Insights into the potential for an "Organic Island". *Environ. Dev. Sustain.* **2014**, *16*, 595–617. [CrossRef]

77. Yue, C.; Tong, C. Organic or local? Investigating consumer preference for fresh produce using a choice experiment with real economic incentives. *HortScience* **2009**, *44*, 366–371. [CrossRef]

78. Tempesta, T.; Vecchiato, D. An analysis of the territorial factors affecting milk purchase in Italy. *Food Qual. Prefer.* **2013**, *27*, 35–43. [CrossRef]

79. Cicia, G.; Colantuoni, F. Willingness to pay for traceable meat attributes: A meta-analysis. *Int. J. Food Syst. Dyn.* **2010**, *1*, 252–263. [CrossRef]

80. Loureiro, M.L.; Umberger, W.J. Estimating consumer willingness to pay for country-of-origin labeling. *J. Agric. Resour. Econ.* **2003**, *28*, 287–301. [CrossRef]

81. Schnettler, B.; Ruiz, D.; Sepúlveda, O.; Sepúlveda, N. Importance of the country of origin in food consumption in a developing country. *Food Qual. Prefer.* **2008**, *19*, 372–382. [CrossRef]

82. Costanigro, M.; Kroll, S.; Thilmany, D.; Bunning, M. Is it love for local/organic or hate for conventional? Asymmetric effects of information and taste on label preferences in an experimental auction. *Food Qual. Prefer.* **2014**, *31*, 94–105. [CrossRef]

83. Briedert, C.; Hahsler, M.; Reutterer, T. A. Review of Methods for Measuring Willingness-to-Pay. *Innov. Mark.* **2006**, *2*, 8–32.

84. Boccaletti, S.; Nardella, M. Consumer Willingness to Pay for Pesticide-Free Fresh Fruit and Vegetables in Italy. *Int. Food Agribus. Manag. Rev.* **2000**, *3*, 297–310. [CrossRef]

85. Carson, R.T.; Hanemann, W.M. Contingent valuation. In *Handbook of Environmental Economics 2*; Maler, K.G., Vincent, J.R., Eds.; Elsevier: Amsterdam, The Netherlands; New York, NY, USA, 2005; pp. 517–1103.

86. Balogh, P.; Békési, D.; Gorton, M.; Popp, J.; Lengyel, P. Consumer willingness to pay for traditional food products. *Food Policy* **2016**, *61*, 176–184. [CrossRef]

87. Ceschi, S.; Canavari, M.; Castellini, A. Consumer's Preference and Willingness to Pay for Apple Attributes: A Choice Experiment in Large Retail Outlets in Bologna (Italy). *J. Int. Food Agribus. Mark.* **2017**, 1–18. [CrossRef]

88. Wongprawmas, R.; Canavari, M. Consumers' willingness-to-pay for food safety labels in an emerging market: The case of fresh produce in Thailand. *Food Policy* **2017**, *69*, 25–34. [CrossRef]

89. Xie, J.; Gao, Z.; Swisher, M.; Zhao, X. Consumers' preferences for fresh broccolis: Interactive effects between country of origin and organic labels. *Agric. Econ.* **2015**, *47*, 181–191. [CrossRef]

90. Hoyos, D.; Mariel, P. Contingent valuation: Past, present and future. *Prague Econ. Pap.* **2010**, *4*, 329–343. [CrossRef]

91. Alphonce, R.; Alfnes, F. Eliciting Consumer WTP for Food Characteristics in a Developing Context: Application of Four Valuation Methods in an African Market. *J. Agric. Econ.* **2016**, *68*, 123–142. [CrossRef]

92. Henryks, J.; Pearson, D. Attitude behavior gaps: Investigating switching amongst organic consumers. *Int. Food Mark. Res. Symp. Conf. Proc.* **2013**, *2*, 3–19.

93. Palys, T. Purposive sampling. In *The Sage Encyclopedia of Qualitative Research Methods*, 2nd ed.; Given, L., Ed.; Sage: Los Angeles, CA, USA, 2008; Volume 2, pp. 697–698.

94. Venter, K.; Van der Merwe, D.; De Beer, H.; Kempen, E.; Bosman, M. Consumers' perceptions of food packaging: An exploratory investigation in Potchefstroom, South Africa. *Int. J. Consum. Stud.* **2011**, *35*, 273–281. [CrossRef]

95. Llorens, M.; Puelles, M.; Manzano, R. Consumer behavior and brand preferences in organic grocery products. Store brands vs. manufacturer brands. *Innov. Mark.* **2011**, *7*, 109–115.

96. Mukul, A.Z.; Afrin, A.S.; Hassan, M.M. Factors affecting consumers' perception about organic food and their prevalence in Bangladeshi organic preference. *J. Bus. Manag. Sci.* **2013**, *1*, 12–118. [CrossRef]

97. Padel, S.; Foster, C. Exploring the gap between attitudes and behavior: Understanding why consumers buy or do not buy organic food. *Br. Food J.* **2005**, *107*, 606–625. [CrossRef]

98. Aryal, K.P.; Chaudhary, P.; Pandit, S.; Sharma, G. Consumers' willingness to pay for organic products: A case from Kathmandu valley. *J. Agric. Environ.* **2009**, *10*, 12–22. [CrossRef]

99. Nasır, V.A.; Karakaya, F. Underlying Motivations of Organic Food Purchase Intentions. *Agribus. Int. J.* **2014**, *30*, 290–308. [CrossRef]

100. Shang, W.; Fooks, J.R.; Messer, K.D.; Delaney, D. Consumer demand for local honey. *Appl. Econ.* **2015**, *47*, 4377–4394. [CrossRef]

101. Wekeza, S.V.; Sibanda, M. Factors Influencing Consumer Purchase Intentions of Organically Grown Products in Shelly Centre, Port Shepstone, South Africa. *Int. J. Environ. Res. Public Health* **2019**, *16*, 956. [CrossRef] [PubMed]

102. Rosner, B. *Fundamentals of Biostatistics*, 7th ed.; Brooks/Cole: Boston, MA, USA, 2011.

103. Greene, W. *Econometric Analysis*; Macmillan: New York, NY, USA, 2006.

104. Ryan, T. *Modern Regression Methods*, 2nd ed.; Wiley: New York, NY, USA, 2009.

105. Cosmina, M.; Gallenti, G.; Maragon, F.; Troiano, S. Attitudes towards honey among Italian consumers: A choice experiment approach. *Appetite* **2016**, *99*, 52–58. [CrossRef] [PubMed]

106. National Honey Board. Honey Market Research. 2013. Available online: https://www.researchgate.net/publication/263733375_Honey_Demand_and_the_Impact_of_the_National_Honey_Board\T1\textquoterights_Generic_Promotion_Programs (accessed on 16 May 2019).

107. Yeow, S.H.C.; Chin, S.T.S.; Yeow, J.A.; Tan, K.S. Consumer purchase intentions and honey related products. *J. Mark. Res. Case Stud.* **2013**, *1*, 1–12. [CrossRef]

108. Pocol, C.B.; Bolboacă, S.D. Perceptions and trends related to the consumption of honey: A case study of North-West Romania. *Int. J. Consum. Stud.* **2013**, *37*, 642–649. [CrossRef]

109. Pocol, C.B. Consumer preferences for different honey varieties in the North West Region of Romania. *Lucr. Stiintifice Ser. Agron.* **2012**, *55*, 263–266.

110. Schifferstein, H.N.J.; Oude Ophius, P.A.M. Health-related determinants of organic food consumption in the Netherlands. *Food Qual. Prefer.* **1998**, *9*, 119–133. [CrossRef]

111. Tregear, A.; Dent, J.B.; McGregor, M.J. The demand for organically-grown produce. *Br. Food J.* **1994**, *96*, 21–26. [CrossRef]

112. Davis, A.; Titterington, A.J.; Cochrane, A. Who buys organic food? A profile of the purchasers of organic food in Northern Ireland. *Br. Food J.* **1995**, *97*, 17–23. [CrossRef]

113. Storstad, O.; Bjorkhaug, H. Foundations of production and consumption of organic food in Norway: Common attitudes among farmers and consumers. *Agric. Hum. Values* **2003**, *20*, 151–163. [CrossRef]

114. Onyango, B.; Hallman, W.; Bellows, A. Purchasing Organic Food in US Food Systems: A Study of Attitudes and Practice. In Proceedings of the Selected Paper Prepared for Presentation at the American Agricultural Economics Association Annual Meeting, Long Beach, CA, USA, 23–26 July 2006; pp 23–26. Available online: https://core.ac.uk/download/pdf/6779438.pdf (accessed on 21 January 2020).

115. Fan, X.; Gómez, M.; Coles, P. Willingness to Pay, Quality Perception, and Local Foods: The Case of Broccoli. *Agric. Resour. Econ. Rev.* **2019**, *48*, 414–432. [CrossRef]

116. Kuhar, A.; Juvancic, L. Modelling consumer's preferences towards organic and integrated fruits and vegetables in Slovenia. In Proceedings of the Contributed paper at the 97rd EAAE Seminar "The economics and policy of diet and health", Reading, UK, 21–22 April 2005; pp. 21–22.

117. Tsakiridou, E.; Konstantinos, M.; Tzimitra-Kalogianni, I. The influence of consumer's characteristics and attitudes on the demand for organic olive oil. *J. Int. Food Agribus. Mark.* **2006**, *18*, 23–31. [CrossRef]

118. Gracia, A.; De Magistris, T. The demand for organic foods in the South of Italy: A discrete choice model. *Food Policy* **2008**, *33*, 386–396. [CrossRef]

119. Wier, M.; O'Doherty, J.K.; Andersen, L.M.; Millock, K. The character of demand in mature organic food markets: Great Britain and Denmark compared. *Food Policy* **2008**, *33*, 406–421. [CrossRef]

120. Anić, I.D.; Jelenc, L.; Šebetić, N. Istraživanje demografskih obilježja i ponašanja kupaca ekoloških prehrambenih proizvoda u Karlovačkoj Županiji. *Ekon. Misao Praksa Časopis Sveučilista Dubrov.* **2015**, *24*, 367–388.

121. Shafie, F.A.; Rennie, D. Consumer perceptions towards organic food. *Procedia Soc. Behav. Sci.* **2012**, *49*, 360–367. [CrossRef]

122. Ikeme, A.I. *Meat Science and Technology*; Africans FEP Publishers: Ibadan, Nigeria, 1990; pp. 67–72.

123. Ekine, D.I.; Albert, C.O.; Peregba, T.A. Expenditure Pattern for Beef Consumption in Selected Households in Southern Nigeria. *Dev. Ctry. Stud.* **2012**, *2*, 1–6. Available online: https://www.iiste.org/Journals/index.php/DCS/article/view/2442/2464 (accessed on 19 July 2019).

124. Pocol, C.B.; Al Mărghitaş, L. *National and International Trends Regarding Production and Consumption of Honey*; Academic Press: Cluj Napoca, Romania, 2010; pp. 41–79.

125. Árváné, V.G.; Csapó, Z.; Kárpáti, L. Evaluation of consumers' honey purchase habits in Hungary. *J. Food Prod. Mark.* **2011**, *17*, 227–240. [CrossRef]

126. Sanzo, M.J.; Del Río, A.B.; Iglesias, V.; Vázquez, R. Attitude and satisfaction in a traditional food product. *Br. Food J.* **2003**, *105*, 771–790. [CrossRef]

127. Sagan, A. Modele Zachowań Konsumenta. CEM Instytut Badań Rynku i Opinii Publicznej. 2005. Available online: http://www.cem.pl/?a=page-s&id=42 (accessed on 15 July 2019).

128. Semkiw, P.; Ochal, J. *Sektor Pszczelarski w Polsce w 2012 Roku. Instytut Ogrodnictwa*; Oddział Pszczelnictwa: Puławy, Poland, 2012; Available online: http://www.opisik.pulawy.pl/pdf/SP2012.pdf (accessed on 19 July 2019).

129. Mieczkowski, M. Krajowy rynek miodu 2000–2005. Analiza i prognozy. *Biul. Inf. Agencji Rynk. Rolnego* **2005**, *12*, 42–62.

130. Mieczkowski, M. Rynek miodu—Uwarunkowania rozwoju. *Biul. Inf. Agencji Rynk. Rolnego* **2007**, *2*, 25–32.

131. Carpio, C.E.; Isengildina-Massa, O. Consumer willingness to pay for locally grown products: The case of South Carolina. *Agribusiness* **2009**, *25*, 412–426. [CrossRef]
132. Andam, K.S.; Ragasa, C.; Asante, S.; Amewu, S. *Can Local Products Compete against Imports in West Africa? Supply-and Demand-Side Perspectives on Chicken, Rice, and Tilapia in Accra, Ghana (Vol. 1821)*; Discussion Paper No. 01821; International Food Policy Research Institute: Washington, DC, USA, 2019; Available online: https://ssrn.com/abstract=3366406 (accessed on 12 January 2020).
133. Saba, A.; Messina, F. Attitudes towards organic foods and risk/benefit perception associated with pesticides. *Food Qual. Prefer.* **2003**, *14*, 637–645. [CrossRef]
134. Özcelik, A.Ö.; Ucar, A. Turkish academic staffs' perception of organic foods. *Br. Food J.* **2008**, *110*, 948–960. [CrossRef]
135. Hamzaoui Essoussi, L.; Zahaf, M. Exploring the decision-making process of Canadian organic food consumers. Motivations and trust issues. *Qual. Mark. Res.* **2009**, *12*, 443–459. [CrossRef]
136. Lea, E.; Worsley, T. Australians' organic food beliefs, demographics and values. *Br. Food J.* **2005**, *107*, 855–869. [CrossRef]
137. Brčić-Stipčević, V.; Petljak, K. Research on organic food purchase in Croatia. *Tržište* **2011**, *23*, 189–207.
138. Prodanović, R.; Bošković, J.; Ignjatijević, S. *Organska Proizvodnja Meda u Funkciji Zaštite Životne Sredine, Međunarodna Konferencija*; Ekološka Kriza: Tehnogeneza I Klimatske Promene: Beograd, Serbia, 2016; pp. 96–97.
139. Zielińska, H.; Zieliński, K. Spożycie żywności w Polsce. Tendencje i determinanty zmian. *Zesz. Naukowe Uniw. Rzeszowskiego* **2004**, *5*, 105–118.
140. Vanyi, G.A.; Csapo, Z. Evaluation of Honey Consumption in the Main Cities of the North-Great Plain Region. In Proceedings of the International Congress on the Aspects and Visions of Applied, Chania, Greece, 3–6 September 2009.
141. Arvanitoyannis, I.; Krystallis, A. An empirical examination of the determinants of honey consumption in Romania. *Int. J. Food Sci. Technol.* **2006**, *41*, 1164–1176. [CrossRef]
142. Jakšić, S.; Mihaljev, Ž.; Kartalović, B.; Babić, J.; Vidaković, S.; Živkov Baloš, M. Evaluation of Elisa tests as screening methods for determination of antibiotics and sulfonamides in honey. *Food Feed Res.* **2018**, *45*, 11–18. [CrossRef]
143. Radtke, J.; Etzold, E.; Iilies, I.; Siede, R.; Büchler, R.; Wegener, J.; Huang, Z.; Bienefeld, K.; Kleinhenz, M.; Bujok, B.; et al. Association of Institutes for Bee Research Report of the 54th seminar in Veitshöchheim 27–29 March 2007. *Apidologie* **2007**, *38*, 482–504. [CrossRef]
144. Ghorbani, M.; Khajehroshanaee, N. The Study of Qualitative Factors Influencing on Honey Consumers Demand: Application of Hedonic Pricing Model in Khorasan Razavi Province. *J. Appl. Sci.* **2009**, *9*, 1597–1600. [CrossRef]
145. Hooker, N.H.; Caswell, J.A. Regulatory Targets and Regimes for Safety: A Comparison of North American and European Approaches. In *The Economics of Reducing Health Risk from Food*; Caswell, J., Ed.; Food Marketing Policy Center: Storrs, CT, USA, 1996; pp. 1–17.
146. Zamudio, F.; Kujawska, M.; Hilgert, N.I. Honey as Medicinal and Food Resource. Comparison between Polish and Multiethnic Settlements of the Atlantic Forest, Misiones, Argentina. *Open Complement. Med. J.* **2010**, *2*, 58–73. [CrossRef]
147. Muli, E.; Munguti, A.; Raina, S.K. Quality of honey harvested and processed using traditional methods in rural areas of Kenya. *Acta Vet. Brno* **2007**, *76*, 315–320. [CrossRef]
148. Parvanov, P.; Dinkov, D. More insight into organic bee honey processing, storage and shelf life. *Bulg. J. Vet. Med.* **2012**, *15*, 206–210.
149. Majewska, E.; Kowalska, J.; Łapińska, M. Analiza czynników dotyczących miodów naturalnych kształtujących preferencje konsumenckie studentów. *Towarozn. Probl. Jakości* **2012**, *2*, 78–86.
150. Tornuk, F.; Karaman, S.; Ozturk, I.; Toker, O.S.; Tastemur, B.; Sagdic, O.; Dogan, M.; Kayacier, A. Quality characterization of artisanal and retail Turkish blossom honeys: Determination of physicochemical, microbiological, bioactive properties and aroma profile. *Ind. Crop. Prod.* **2013**, *46*, 124–131. [CrossRef]
151. Mizrahi, A.; Lensky, Y. *Bee Products: Properties, Applications, and Apitherapy*; Plenum Press: New York, NY, USA, 1997; pp. 83–87.
152. Cooper, R.; Gray, D. Is manuka honey a credible alternative to silver in wound care? *Wounds UK* **2012**, *8*, 54–64.

153. Desrochers, P.; Shimizu, H. *Yes, We Have No Bananas: A Critique of the Food-Miles Perspective*; Mercatus Policy Series Policy Primer, No. 8; Mercatus Center at George Mason University: Arlington, VA, USA, 2008; Available online: http://mercatus.org/ploadedFiles/Mercatus/Publications/Yes%20We%20Have%20No%20Bananas_%20A%20Critique%20of%20the%20Food%20Mile%20Perspective.pdf (accessed on 30 March 2019).

154. Lusk, J.L.; Norwood, F.B. The Locavore's Dilemma. Featured Article at the Library of Economics and Liberty (EconLib). 2011. Available online: http://www.econlib.org/library/Columns/y2011/LuskNorwoodlocavore.html (accessed on 21 March 2019).

155. Allsopp, M.H.; De Lange, W.J.; Veldtman, R. Valuing insect pollination services with cost of replacement. *PLoS ONE* **2008**, *3*, e3128. [CrossRef]

156. Häagen-Dazs. *Häagen-Dazs—Help the Honey Bees*; Häagen-Dazs: Wilkes-Barre, PA, USA, 2013; Available online: http://www.haagendazs.us/Learn/HoneyBees/ (accessed on 21 February 2020).

157. Adams, D.C.; Salois, M.J. Local versus organic: A turn in consumer preferences and willingness-to-pay. *Renew. Agric. Food Syst.* **2010**, *25*, 331–341. [CrossRef]

158. Magnusson, M.K.; Avrola, A.; Hursti Koivisto, U.K.; Aberg, L.; Sjoden, P.O. Attitudes towards organic foods among Swedish consumers. *Br. Food J.* **2001**, *103*, 209–226. [CrossRef]

159. Bogoeva, I. Production of safety food by phytoremediation methods. *J. Agron. Technol. Eng. Manag.* **2018**, *1*, 39–44.

 *sustainability*

Article

Analysis of Production and Sales of Organic Products in Ukrainian Agricultural Enterprises

Roman Ostapenko [1], Yuliia Herasymenko [1], Vitalii Nitsenko [2], Svitlana Koliadenko [3], Tomas Balezentis [4,*] and Dalia Streimikiene [4]

[1] Department of Statistics and Economic Analysis, Kharkiv National Agrarian University Named after V.V. Dokuchaiev, 62483 Kharkiv, Ukraine; rm_ostap@knau.kharkov.ua (R.O.); interstudent@knau.kharkov.ua (Y.H.)
[2] Department of Accounting and Taxation, Interregional Academy of Personnel Management, 03039 Kyiv, Ukraine; vitaliinitsenko@onu.edu.ua
[3] Department of Computer Science and Economic Cybernetics, Vinnytskiy National Agricultural University, 21008 Vinnytsia, Ukraine; koladenko@vsau.vin.ua
[4] Division of Farm and Enterprise Economics, Lithuanian Institute of Agrarian Economics, 03220 Vilnius, Lithuania; dalia@mail.lei.lt
* Correspondence: tomas@laei.lt

Received: 25 January 2020; Accepted: 20 April 2020; Published: 22 April 2020

Abstract: As organic farming gains more popularity across the world, it is important to discuss the underlying trends of its development in Ukraine, who is an important agricultural producer. Organic farming may have lower environmental pressures—therefore, we seek to identify the major trends in the production and sales of the organic agricultural products in Ukraine. In this study, data on the production structure, costs, and selling prices from Ukrainian enterprises are analyzed. Conventional and organic enterprises are contrasted in order to identify the possibilities for the development of organic agriculture in Ukraine. Our results suggest that enterprises that use organic farming in Ukraine tend to produce higher output per hectare, as opposed to those engaged in conventional farming. However, labor profitability remains low in labor-intensive organic farming, especially in larger companies, and organic products remain a low percentage of Ukraine's agricultural exports. This calls for further study into the development of organic production and consumption in the domestic market, as well as the implementation of appropriate certification practices in order to ensure the growth of organic exports.

Keywords: organic farming; production structure; production intensity; Ukraine

---

1. Introduction

Organic agricultural production has gained more popularity recently due to changes in customer tastes and income [1–3]. In 2017, organic agriculture was licensed in 181 countries, and approximately 2.4 million farmers were engaged in organic production [4,5]. In 2017, global organic farming increased from 11 million hectares in 1999 to 69.8 million hectares. The largest areas used for organic farming, as of 2017, were in Australia (35.6 million hectares), Argentina (3.4 million hectares), and China (3 million hectares). The largest numbers of organic enterprises were in India (835,000), Uganda (210,300), and Mexico (210,000). According to the research of the Forschungsinstitut fur biologischen Landbau (FiBL), organic production has been expanding rapidly in the world [6]. In Europe, agricultural land area for organic production has increased tenfold over the last 10 years. The area of organic farming is growing at a slower pace in countries where the process began a relatively long time ago (Germany, the Netherlands, and France). In Austria, the share of the land area under organic production reached 24.0% [7].

Ukraine is a developing country with underdeveloped organic agriculture sector. The major reason for this is the limited purchasing power of the domestic population. However, it is important to identify Ukraine's major trends in the production of organic agriculture in order to ensure the balanced development of the sector. Among the determinants of customers' decisions to purchase organic products, willingness to avoid contamination of food with pesticides and other chemical compounds is an important factor. A survey of respondents in Thailand indicated that the reasons for purchasing organic products were due to consumers' desires to get healthier products. At the same time, respondents who bought organic vegetables are usually older, have higher education and higher family income than those who did not buy them [8]. A similar study was also conducted in Ukraine. It was found that consumers were willing to pay no more than 25% for eco-friendly products [9]. As organic products are more expensive in Ukraine, the demand remains low if compared to developed countries of the world [10].

The Ministry of Agrarian Policy of Ukraine in addition to Germany, Switzerland, and some companies from the European Union has sought to improve the regulation of the organic market in Ukraine. An example of this is the long-term cooperation agreement concluded in 2005 between the Ukrainian company Ukragrofin and the German ECOLAND Grains & Legumes for organic soybean cultivation [11]. Since the end of the twentieth century, there has been a steady increase in global production, sales, and consumption of organic products, which is up to 20% annually [12]. At the same time, growth in conventional food sales was only up 3% per year [13].

This paper seeks to identify major trends in the production and sales of the organic agricultural products in Ukraine over the past two nationwide agricultural censuses. Data on the production structure, selling prices, labor use, and production costs from Ukrainian agricultural enterprises were used to calculate profits for organic and conventional farms (grouped by the cost of production intervals). Economic indicators for organic farms were compared to those for conventional farms in order to identify possible areas for improvement for further growth and development of Ukrainian organic agriculture.

2. Literature Review

There are different views as to what the "organic farming" category includes. In particular, one of the most common definitions considers it as a production system that supports the health of soils, existing ecosystems, and humans. It depends on ecological processes, biodiversity, and natural cycles specific to local conditions, avoiding the use of non-renewable resources [14].

Ayuya stressed the need for certification of organic farming at the state level within the home country of such enterprises. His analysis confirmed the positive economic impact of developed and adopted programs at the state level for farmers engaged in organic farming [15]. He also emphasized that there is no deliberate policy on the part of supermarkets and manufacturers themselves regarding the clear positioning of their products and their promotion in the markets. Today, organic products are sold by specialized stores, but the vast majority of these products are not organic—they are other so-called eco-friendly products [16]. The importance of the factor of trust in the quality of products is also highlighted by a study in Denmark and the United Kingdom [17]. Some authors even emphasize that this problem has an ethical aspect [18].

Klitna and Bryzhan highlighted the following difficulties of developing the organic sector: Innovative passivity of manufacturers and management structures; lack of institutional support and lack of state financial support; poor awareness of producers about the specifics of organic production and the population regarding organic products; predominance of exports of organic raw materials; processing, production, wholesale, and retail sale of organic products are still underdeveloped; deficiency of grain and other crops of organic origin [19,20]. Løes and Adler noted the need to solve problems between stakeholders involved in organic farming, such as making better use of resources [21]. Wallenbeck et al. showed the heterogeneity of the results obtained from a study of the main types of organic dairy farms in seven EU countries. They noted that businesses have significant differences in both farm size, technological features, animal care, and management strategy. Other limiting factors, such as the natural and climatic features of the geographical location of farms, resource availability, land size and state regulation, are also important [22]. Furthermore, Muller et al. noted that organic farming requires larger land areas to produce a similar amount of production compared to the conventional farming system [23]. Kobets argued that the transition to organic farming requires not only suitable land, but also a long conversion period, which lasts two to five years [24].

Kulish emphasized that organic production can be used as an alternative model of management that does not use any chemicals [25]. Lotter [12] and Smith [26] identified the positive environmental impact of organic farming. They note that organic farming contributes to the conservation of biological diversity, nutrients in the soil, and reduces the level of wind flow and water erosion. Other research indicates a negative influence from the use of chemicals, which damages soil ecosystems and compromises soil fertility—thus, reducing crop yields [27]. Caradonna stressed the negative impact of conventional agriculture, such as reduced agroecosystem diversity, due to state support for agricultural monocultures. Adherents of conventional farming practices are introducing organic farming practices to save costs and conserve biodiversity [28]. Porodina argued that organic products that are produced in Ukraine are of high quality and can satisfy the requirements of the foreign markets [29].

In addition to marketing policy, other factors can lead to farmers' decision to switch to organic production. A study in Finland concluded that lower prices for produce, and an increase in direct subsidies, could encourage the transition to organic farming. Here, a transition to organic production was also more likely on farms with large land areas and low yields, where intensive livestock production and labor-intensive requirements discourage the switch to organic methods [30]. Therefore, it is important to identify the key challenges to organic agriculture within different contexts.

3. The Market of Organic Products in Ukraine

In 2007, at the initiative of public organizations of the organic movement, with the support of the Ministry of Agrarian Policy of Ukraine, the certification body, Organic Standard, was established to deal with the review and issuance of licenses to enterprises and farmers who want to produce and sell organic products [31]. This certification body is recognized in the European Union (EU) and Switzerland [32]. According to official data of the Ministry of Agrarian Policy of Ukraine, there were 504 certified functioning agricultural enterprises in 2017. The largest number of organic producers is concentrated in the following areas: Vinnytsia, Zhytomyr, Kyiv, Odesa, Kharkiv, and Kherson regions [33].

Ukrainian organic agricultural area, producers, and sales volume and value constitute a small percentage of Ukraine's total agricultural production, as well as European and global organic agriculture (Table 1). Ukrainian organic agricultural exports include corn, wheat, barley, sunflower, soybeans, spelt, apples/juice, peas, millet, and rapeseed. Top importing countries are The Netherlands, Germany and United Kingdom.

Table 1. Main indicators of organic market development in Ukraine, 2017.

Indicator	Value	Ukraine's Total Agriculture	Share (%) in European Organic	Global Organic
Organic agricultural land (ha)	381,000	0.911	2.617	0.546
Including in-conversion areas (ha)	92,000	0.220	0.632	0.132
Just certified organic (ha)	289,000	0.691	1.985	0.414
Organic agricultural businesses	304	0.667	0.076	0.011
Organic product producers	200	0.439	0.050	0.007
Organic exporters/importers	169	0.371	2 053	1.255
Organic processors	87	0.191	0 122	0.099
Organic producers/farmers	504	1.106	0.127	0.018
Total revenue from exports (thousand €)	99,000	0.658	0.265	0.108

Source: Calculated by Reference [34] and State Statistics Service of Ukraine (Agriculture of Ukraine: Statistical Yearbook 2018).

The Ukrainian organic market is still developing. As of 2017, the total share of organic farmland was negligible (less than 1% of the total agricultural area in Ukraine). Meanwhile, Ukraine ranks 11th in Europe in terms of agricultural land for organic production. During 2013–2017, the area for organic production increased 1.5 times. Approximately 45.5% of all organic area in 2017 was sown under cereal crops [35].

The growth in organic production in Ukraine is one of the highest in the world: The growth rate exceeds the European one by 5.5 times and the global one—by 4.9 times [36]. About 90% of the organic products produced in Ukraine are exported. The most popular organic products exported from Ukraine include corn, wheat, barley, sunflower, soybeans, spelt, apples (juice), peas, millet and rapeseed. The total volume of the exported organic products was 264 metric tons, and the total value was €99 million in 2017. These products were exported to the Netherlands, Germany, the United Kingdom, among others. Such a situation improves the balance of payments of Ukraine, yet the domestic market faces increasing pressure in terms of the prices of the organic products. The lack of the local market may induce a lack in the resilience of organic farming in Ukraine in the case of economic turmoil resulting in a decline of international trade.

The low penetration of the organic products in the Ukrainian market can be illustrated by a comparison to a developed economy—the United States (US). In Ukraine, the domestic retail sales of organic products total €29 million, which corresponds to €1 per inhabitant. As for the US, the domestic market is €40 billion—corresponding to per capita consumption of €122. Thus, the Ukrainian organic products market comprises 0.072% of that of the US, and the rate of consumption per capita is 0.82% of that in the US. Cernansky argues that the US, as one of the global leaders in the production of corn, imports substantial quantities of organic corn for the needs of the feed industry and for the production of organic livestock products [37].

The Ukrainian enterprises are engaged in production and supply of different types of organic products. We use the data on the certificates issued by Organic Standard in order to quantify the different activities undertaken by Ukrainian enterprises. The highest number of certificates were issued for organic crop producers (mostly cereal growing). Beekeeping is the second most frequent type of certified production type, with five times fewer certificates issued if compared to crop farming (Figure 1). The least popular farming type is aquaculture. As regards the components of the supply chain, the foreign trade operations appear as the most important activity after production in terms of the number of certificates issued. Retail trade of the organic products was certified for 69 enterprises in Ukraine.

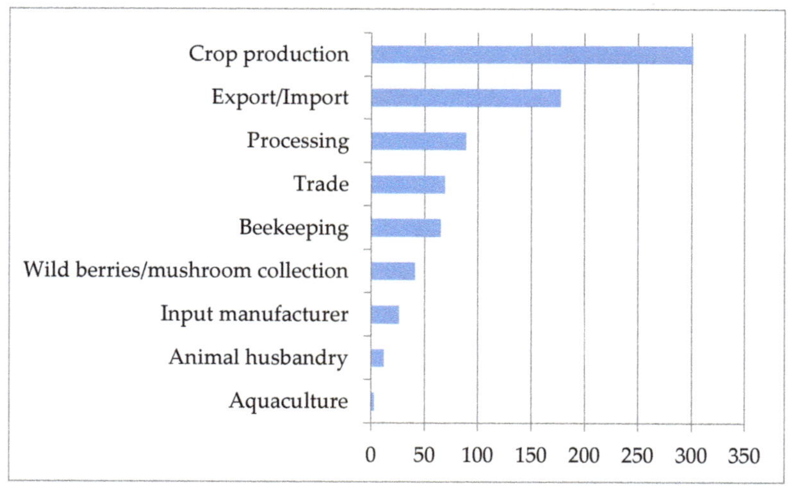

Figure 1. The number of enterprises certified for certain activities as of 2017.

As it was mentioned above, Ukraine's organic agricultural production is mainly oriented towards foreign markets. The EU is a market with favorable geographical location and developed logistics network. Therefore, it has become the most attractive market for export of Ukraine's organic producers. In Ukraine, 489 organic producers have implemented standards equivalent to European Commission Regulations No 834/07 and 889/08.

Among the exported products, most of the total sales came from cereals (maize, wheat, barley, millet) and industrial crops (sunflower, soybean, and rapeseed). Table A1 presents detailed data on the export structure. The shares of crops in the national export structure fluctuate from 0.02% for sunflower to 5.5% for millet. These values are rather low and should increase in case a successful policy for promotion of organic farming is implemented.

4. Data

In Ukraine, no official body keeps official statistics on the operation of the organic producers at the national level. Information on certified enterprises and products certified by 16 internationally accredited certification bodies can be found in the Organic Business Directory of Ukraine [38]. The data on organic producers were collected from the website of Organic Standard Ltd [39]. The data from report F-50 of agricultural enterprises are used. The comparison of organic and conventional enterprises is based on indicators of enterprise size, output size and structure, the structure of crop area, input intensity, and profitability. We were able to find data on 82 enterprises in 2012, and 75 enterprises in 2017, in the websites mentioned above. The changes in the number of enterprises may be due to the changes in the form of ownership of enterprises, certification expiry, merger or acquisition, changes in their type of activity or liquidation. The data for conventional agricultural companies come from the State Statistical Reporting of Ukrainian Enterprises. The number of conventional enterprises selected for the analysis is equal to that of organic enterprises (75 enterprises in 2017). The random sampling was applied to select conventional enterprises from the database of enterprises with similar characteristics (compared to organic enterprises).

5. Results

5.1. Production Structure in Ukrainian Agricultural Enterprises

The volume of production in organic enterprises is higher than it is the case for conventional enterprises in Ukraine (Table 2). In 2012, on average, an organic enterprise generated revenue of €2104.8 thousand from the sales, whereas an average conventional enterprise generated just €1269.6 thousand. In 2017, these figures went down to €1776.5 thousand and €1075.7 thousand respectively. Thus, the ratio of these indicators virtually did not change with time. The share of the livestock output is much lower than that of the crop output in organic enterprises of Ukraine.

Table 2. Average farm and crop area, livestock inventory, employees, and total farm revenues for organic and conventional enterprises in Ukraine, 2012 and 2017.

Indicator	Organic Enterprises			Conventional Enterprises		
	2012	2017	% 2017 of 2012	2012	2017	% 2017 of 2012
Land area (ha)						
Agricultural land area (ha)	3237.8	3368.7	104.0	2,018.7	2,080.6	103.1
Arable land (ha)	3098.7	3423.9	110.5	1923.5	1987.6	103.3
Crops (ha)						
Winter Wheat	654.0	980.1	149.8	395.2	478.8	121.2
Corn for grain	635.9	587.4	92.4	324.6	323.3	99.6
Sunflower	516.0	637.7	123.6	369.7	483.7	130.8
Soybean	300.6	427.3	142.2	118.4	159.1	134.4

Table 2. *Cont.*

Indicator	Organic Enterprises			Conventional Enterprises		
	2012	2017	% 2017 of 2012	2012	2017	% 2017 of 2012
Livestock (animals)						
Milk cows	99.4	126.1	126.8	57.0	87.1	152.8
Cattle	148.3	215.8	145.5	97.7	362.2	370.7
Pigs	249.5	417.5	167.3	298.7	55.8	18.7
Employees						
Crop farming	75.4	67.6	89.7	36.8	34.5	93.8
Animal husbandry	19.0	20.2	106.3	16.1	12.8	79.1
Total	94.4	85.4	90.4	53.0	47.3	89.3
Farm total revenue (thousand €)						
Crop farming	1857.5	1602.8	86.3	970.3	894.5	92.1
Animal husbandry	247.3	173.7	70.2	299.6	181.2	60.1
Total	2104.8	1776.5	84.4	1269.6	1075.7	84.7

Source: Calculated by the authors using the State Statistical Reporting of Ukrainian Enterprises.

In regards to the agricultural land area, organic enterprises are approximately 1.5 times larger than conventional enterprises in Ukraine. The areas sown under organic soybean, winter wheat and sunflower increased during 2012–2017, whereas that under organic corn for grain fell. The number of workers remained rather stable, although the declining trend persists for both conventional and organic enterprises. The organic farming showed growth in the number of livestock, yet the rate of growth was lower for milk cows and cattle. The structure of the agricultural output of conventional and organic enterprises of Ukraine is presented in Table 3. Even though the share of the livestock output is rather low in organic enterprises, it shows an increase during 2012–2017, whereas the opposite trend is observed for conventional farms. In the output of the organic sector, the share of grain and legume crops shrunk by 2.5 p.p. and that of soybeans increased by 5.1% p.p. during 2012–2017.

Table 3. The structure of the agricultural output of organic and conventional enterprises in Ukraine (%), 2012 and 2017.

Product	Organic Enterprises			Conventional Enterprises		
	2012	2017	Change, p.p.	2012	2017	Change, p.p.
Cereals and legumes	52.5	50.0	−2.5	41.5	43.9	2.3
Winter wheat	15.1	18.6	3.5	15.3	18.7	3.4
Corn for grain	30.7	28.3	−2.4	20.2	19.1	−1.1
Sunflower seeds	19.5	22.2	2.7	20.6	23.6	3.0
Soybean	7.8	12.9	5.1	5.3	9.0	3.7
Winter rape	3.7	2.9	−0.8	3.5	3.1	−0.4
Other crop production	7.4	2.6	−4.9	7.8	5.5	−2.3
Total crop output	**90.9**	**86.4**	**−4.5**	**78.7**	**85.0**	**6.3**
Milk	5.2	6.5	1.4	5.1	4.8	−0.2
Growth of live weight of cattle	1.5	1.4	−0.1	1.7	1.2	−0.5
Growth of live weight of pigs	1.7	2.8	1.1	4.8	3.5	−1.3
Bird	0.0	0.0	0.0	1.0	0.5	−0.5
Eggs	0.0	0.0	0.0	5.8	2.8	−3.0
Other livestock products	0.8	0.2	−0.5	9.8	5.4	−4.4
Total livestock output	**9.1**	**10.7**	**1.6**	**21.3**	**15.0**	**−6.3**
Agricultural output	**100.0**	**100.0**		**100.0**	**100.0**	

Source: Calculated by the authors using the State Statistical Reporting of Ukrainian Enterprises.

There has been an increase in the share of areas sown under winter wheat, sunflowers and soybeans (Table 4). At the same time, the share of corn for grain and cereals and legumes decreased. The largest difference between organic and conventional enterprises in terms of land use structure was noted for soybean. In 2017, its share in conventional enterprises amounted to 8.0%, while there was an 11.6% increase for organic products. These changes are related to the relative prices of the agricultural products and the changing patterns in the international markets. Particularly, the transition towards winter wheat is evident. Indeed, Ukraine and countries in the neighboring region may exploit their competitive advantage in wheat production, due to favorable geoclimatic conditions.

Table 4. Crop structure of organic and conventional enterprises in Ukraine, 2012 and 2017.

Crop	Organic Enterprises			Conventional Enterprises		
	2012	2017	%, 2017 of 2012	2012	2017	%, 2017 of 2012
Cereals and legumes, including:	57.0	58.6	1.6	51.7	51.6	−0.1
- Winter Wheat	21.1	29.2	8.1	20.5	24.1	3.6
- Summer Wheat	0.5	0.7	0.2	0.8	0.5	−0.3
- Buckwheat	0.9	1.3	0.4	1.0	0.5	−0.5
- Corn	20.5	15.2	−5.3	16.9	16.3	−0.6
- Other cereals and legumes	14.0	12.2	−1.8	12.5	10.2	−2.3
Sunflower seeds	16.7	16.8	0.1	19.2	24.3	5.1
Soybean	9.7	11.6	1.9	6.2	8.0	1.8
Winter Rape	2.2	3.1	0.9	2.3	2.2	−0.1
Other crops	14.4	9.9	−4.5	20.6	13.9	−6.7
Total	100	100		100	100	

Source: Prepared by the authors using data from the State Statistical Reporting of Ukrainian Enterprises.

The differences in the production structure of the Ukrainian organic agricultural companies can be illustrated by considering particular cases. Table A2 summarizes data on some of the Ukrainian organic producers. In general, the increase in the livestock output would increase diversification and ensure the generation of higher agricultural value-added.

5.2. Profitability in Ukrainian Agricultural Enterprises

The differences in revenue can be due to land intensity (per labor unit) and labor productivity. In this sub-section, we further analyze the differences in the input use intensity and the profitability of both conventional and organic enterprises in Ukraine. Table 5 presents the cost, revenue and profit indicators per hectare of agricultural land for conventional enterprises.

Table 5. Profitability of conventional agricultural enterprises in Ukraine in 2017.

Cost per 1 Hectare (€)	Revenue per 1 ha (€)	Profit per 1 Hectare (€)	Profit of Crop Farming per 1 Hectare (€)	Cost Profitability (%)	Cost Profitability of Crop Farming (%)	Agricultural Land Area (ha)	Number of Employees per 1 ha
<70.0	24.6	10.3	10.1	20.7	84.7	3592.5	0.5
70.1–176.0	220.0	83.6	84.6	34.5	66.4	1558.1	1.5
176.1–282.0	363.2	131.1	132.4	42.4	62.4	1987.3	1.8
282.1–387.0	502.0	169.0	168.9	42.2	55.3	2084.5	2.0
387.1–493.0	608.5	173.7	171.7	34.2	44.5	2579.4	2.4
493.1–704.0	774.8	204.2	200.8	32.9	40.2	3068.4	2.5
>704.1	1031.6	164.3	159.2	20.4	22.0	2267.6	3.4
Average	519.6	148.9	147.9	34.6	45.1	2266.5	2.1

Note: Profit of Crop Farming refers to the results of the crop farming excluding livestock farming. Cost Profitability = Profit per 1 hectare/Cost per 1 hectare · 100%. Source: Calculated by the authors using the State Statistical Reporting of Ukrainian Enterprises

The increasing input intensity corresponds to the increasing cost per hectare. As a result, there exists an almost linear dependence between the cost and revenue per hectare: At the cost level of 70 €/ha, the revenue is 24.6 €/ha, whereas, at the cost level of more than 704.1 €/ha, the revenue approaches 1,031.6 €/ha. The relationship between costs and profit per hectare is not that straightforward. Indeed, the profit is maximized at the cost level of 493–704 €/ha and declines thereafter. The largest share of the profit per hectare is generated from crop farming (99.3% on average in Ukrainian enterprises).

Cost profitability follows an inverted U-shape curve with cost per hectare. Therefore, the lowest profitability is observed for the lowest and highest cost intensity. The land area varies with the cost level, whereas the number of employees keeps increasing. This suggests that labor costs comprise a substantial share of the total costs. Crop farming profitability exceeds that of the animal husbandry for conventional enterprises in Ukraine.

The same kind of analysis is carried out for the organic enterprises (Table 6). The organic enterprises show higher levels of revenue per 1 ha of agricultural land (658.5 €/ha against 519.6 €/ha in conventional enterprises). The same applies to the profits: Organic enterprises show the average value of 218.8 €/ha, whereas 148.9 €/ha is observed for conventional enterprises. The profit per hectare fluctuates with the cost level. Again, the influence of the labor intensity is evident.

Table 6. The profitability of organic agricultural enterprises in Ukraine, 2017.

Cost per 1 Hectare (€)	Revenue per 1 ha (€)	Profit per 1 Hectare (€)	Profit of Plant Growing on 1 Hectare (€)	Cost Profitability (%)	Cost Profitability of Crop Farming (%)	Agricultural Land Area (ha)	Number of Employees per 1 ha
<176.0	247.6	89.2	93.7	43.5	59.4	2027.9	2.1
176.1–282.0	489.4	228.2	211.3	61.8	128.1	2029.7	2.7
282.1–387.0	525.9	159.0	153.3	38.8	53.8	3206.1	2.6
387.1–598.0	862.1	327.1	308.6	70.2	71.6	3894.1	3.2
>598.1	1270.7	290.7	308.2	40.0	32.2	1823.5	3.9
Average	658.5	218.8	213.3	52.0	57.6	2550.4	2.8

Note: Profit of Crop Farming refers to the results of the crop farming excluding livestock farming. Cost Profitability = Profit per 1 hectare/Cost per 1 hectare · 100%. Source: Calculated by the authors using the State Statistical Reporting of Ukrainian Enterprises.

In the case of Ukraine, we can note the declining returns to labor if comparing different groups of conventional and organic farms once the most optimal scale size is exceeded (Figure 2a). The results suggest that the economies of scale exist in the production of both organic and conventional products. Indeed, the largest farms (in terms of the labor force) show decreasing returns to scale. Nevertheless, another observation is that organic farms operate at an inferior technology compared to conventional farms, in the sense that the same amount of labor can produce a lower profit, in general. A deeper analysis of data in Table 6 suggests that there are two options for organic farms in terms of the optimal scale size as the declining profit per hectare is observed for medium- and high-intensity organic farms. Furthermore, organic farms are more labor-intensive than conventional farms, which further reduces the profitability.

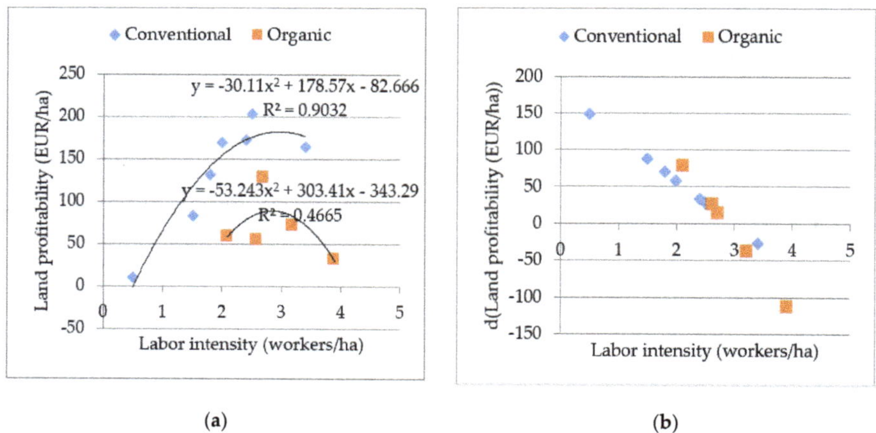

Figure 2. Labor—profit relationship in Ukrainian agricultural enterprises, 2017. (**a**) Levels; (**b**) marginal responsiveness (change in land profitability due to change in labor intensity).

The previous findings indicate that there exist differences in the levels of input use intensity (as indicated by the cost per hectare) and profitability. Therefore, Ukrainian organic enterprises are further analyzed in the regional aspect as different regions may be associated with differences in soil quality. Table 7 presents the distribution of the organic enterprises across different levels of cost per hectare and regions they operate in.

Table 7. Grouping of organic agricultural enterprises by cost per 1 hectare (€) in Ukraine, 2017.

Cost per 1 Hectare (€)	Number of Enterprises and Their Locations
<176.0	Zhytomyr—2, Zaporizhzhia—1, Odesa—3, Poltava—1, Ternopil—1, Kherson—1, Khmelnitskyi—1, Chernihiv—3
176.1–282.0	Ivano-Frankivsk—1, Vinnytsia—1, Dnipropetrovsk—1, Zhytomyr—2, Zaporizhzhia—1, Kirovohrad—3, Kyiv—3, Odesa—4, Ternopil—1, Kharkiv—1
282.1–387.0	Vinnytsia—1, Zhytomyr—1, Zaporizhzhia—1, Kirovohrad—1, Kyiv—1, Odesa—2, Poltava—2, Rivne—1, Kherson—1, Khmelnitskyi—1, Chernihiv—1
387.1–598.0	Ivano-Frankivsk—1, Volyn—1, Zhytomyr—1, Zaporizhzhia—1, Kyiv—1, Odesa—1, Poltava—1, Sumy—2, Cherkasy—2, Chernihiv—3
>598.1	Ivano-Frankivsk—1, Volyn—1, Dnipropetrovsk—1, Kirovohrad—2, Kyiv—1, Odesa—2, Poltava—3, Sumy—2, Kharkiv—1, Khmelnitskyi—2, Chernihiv—1

Source: Calculated by the authors using the State Statistical Reporting of Ukrainian Enterprises.

The data on costs levels per hectare in organic enterprises are further summarized in Figure 3. The regions in northern and central Ukraine show the highest cost per ha. In addition, the largest areas of agricultural land under organic farming are also found in these regions.

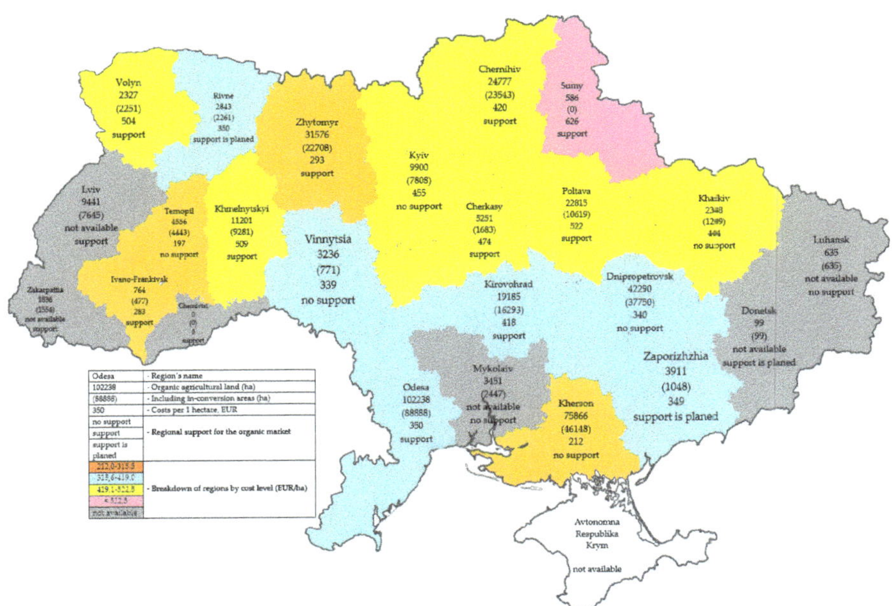

Figure 3. Distribution of regions by cost level per hectare of organic farmland, €/ha. Source: Data from the Research Institute of Organic Agriculture (FiBL) and own calculation.

In order to further explain the differences in profitability of conventional and organic enterprises in Ukraine, we compare the crop yields across these two types of enterprises. The yields for 2017 are presented in Table 8. The organic enterprises how higher yields (with the exception of wheat) if compared to conventional farms. This counterintuitive result can be explained by (i) higher level of development of enterprises embarking on organic farming, and (ii) generally low crop yields in Ukraine.

Table 8. Yield levels in organic and conventional enterprises in Ukraine in 2017.

Crop	Organic Enterprises (100 kg/ha)	Conventional Enterprises (100 kg/ha)	% Organic of Conventional
Wheat	41.3	41.5	99.4
Corn	80.3	60.1	133.6
Sunflower	27.1	24.1	112.3

The results indicate that organic farming should be promoted in regions with favorable climatic conditions and soils. The crop structure also needs to be adjusted to increase the profitability of farming. The scale of operation also plays an important role in terms of cost intensity.

Besides the input intensity and productivity, price level also determines the profitability of farming. The organic production features high rates of output growth (10–20% per year) and higher prices for products if compared to conventional enterprises (30–50% on average) [40]. At the same time, production costs increase, due to energy consumption corresponding to a large number of mechanical treatments of soil and high costs of the organic fertilizers. In European countries, the level of prices for organic products is often 1.5–2 times higher than usual, which compensates for additional costs [41]. The supply chain is also longer for the organic products, which further adds to the prices of the organic products [42,43].

In Ukraine, organic enterprises receive higher prices than conventional enterprises (Table 9). For example, the price for organic corn is 29% higher than for a conventional one—moreover, it is 26% higher for winter wheat, and 45% higher for pork. The smallest difference of 7% is observed for milk. The difference in milk price is significant at the 10% level, whereas the other prices are significantly different at the 1% level. As the previously discussed differences in the profitability suggest, the price differences are still not high enough to support the development of organic farming.

Table 9. Statistical parameters for estimating prices for agricultural products in conventional and organic enterprises in Ukraine in 2017.

Product	Mean Price (€/100kg)		St. Dev.		t Statistic	*p*-Value	Sig.
	Organic Enterprises	Conventional Enterprises	Organic Enterprises	Conventional Enterprises			
Winter Wheat	13.88	11.03	7.81	1.67	2.75	0.007	***
Corn	14.28	11.03	8.35	2.18	2.56	0.012	**
Pork	108.43	74.93	16.86	8.25	6.11	0.000	***
Milk	19.56	18.22	2.36	2.44	1.73	0.092	*

Note: *, **, *** represent significance at the levels of 10%, 5% and 1% respectively. Source: Calculated by the authors using the State Statistical Reporting of Ukrainian Enterprises.

The low difference in milk prices can be explained by the fact that livestock production is primarily directed at the domestic market, whereas the largest share of crop production is oriented towards the external markets. Successful integration into foreign markets requires direct contracts with European companies. The dairy sector of Ukraine has not yet managed to increase its productivity or provide products to foreign markets [44].

6. Discussion

Lotter [12] and Smith et al. [26] noted a lower level of economic efficiency of organic farms associated with lower crop yields. Moreover, the level of operating expenses is lower than that of conventional enterprises, with a higher price level for such products, which increases gross profit per hectare of organic land. State substantiated the need to provide technical support and financial incentives for the widespread dissemination of advanced methods of conducting organic farming, increasing the effectiveness of management decisions [45]. Adamchak corroborated the aforementioned trends, also notes the need to solve a complex of interrelated complex problems related to increasing the yield (productivity) of organic farms, lowering prices and maintaining the environmental benefits of organic farming aimed at implementing the principles of sustainable development of organic farming [46]. A study, conducted by Freyer et al., identified seven myths related to organic agriculture and food research. The organic production still needs to ensure that both higher performance (yield, productivity, efficiency and others), and reduction in the environmental pressures are achieved [47]. However, this requires a systematic approach towards the regulation of the economic activities [48,49].

Freedman noted that development of the organic production system requires (1) access to organic products; and (2) systematic studies with the participation of state institutions, businesses and research institutions aimed at mitigating damage from farming [50]. Our research has also shown that the Ukrainian market is not yet ready for a large-scale introduction of organic produce as this market is associated with a high level of poverty and low income. Thus, the most acceptable scenario for the development of organic production is its export. Taking into account transportation costs, product prices remain higher than in the Ukrainian market, which will stimulate further growth in the number of manufacturers in Ukraine.

Regarding the microeconomics of organic production in Ukraine, the labor input appears as a limiting factor. Specifically, the marginal profit (per land area) declines with the labor input (Figure 2b) for both conventional and organic enterprises. The land profitability starts declining when 2.85 workers/ha is reached in organic farms, whereas the turning point for conventional farms is 2.97 workers/ha. However, the organic enterprises are located in the high labor intensity region where this issue becomes even more acute. This indicates that, besides economic and technological considerations, the Ukrainian organic producers may face a bottleneck in terms of the labor supply. Due to recently increasing depopulation of the rural regions in Ukraine (as it is the case in Eastern Europe), the labor-intensive organic farming faces increasing labor costs and pressure.

7. Conclusions

The organic products market in Ukraine has been growing rapidly during 2002–2017. There is a significant discrepancy between conventional and organic enterprises from the viewpoint of the size of the sown area, the level of economic efficiency and the intensity of production. Further development of the organic products market is highly dependent on growth in the income of the domestic population and state support for producers.

The low difference in prices for conventional and organic products (e.g., milk) indicates the absence of developed markets for organic production in Ukraine. In order to further develop organic farming specialized in production of the aforementioned products, the support schemes are required. In particular, the direct sales promotion may help to supply the organic production to the domestic markets, whereas large scales measures are needed to support exports of the organic products.

The results suggest that organic farms in Ukraine utilize a larger amount of agricultural area when compared to conventional farms. This indicates that organic farming in Ukraine is still in its early development stage, where only well-established agricultural enterprises can embark on organic farming. Indeed, conversion to organic farming induces transitional losses in revenue and additional costs which can only be borne by large enterprises which are able to diversify without undermining their cash flows.

Certain products (e.g., pig meat) face significant differences in prices under organic and conventional farming. However, the results also show that organic farms face lower land profitability (profitability) than conventional farms in Ukraine. In this instance, the promotion of organic farming can rely on market signals. The further development of these sectors should rely on improvement of certification and labelling practices. In any case, the state-wide programs aimed at increasing the quality of organic products should be supported in order to increase the prevalence of organic farming in Ukraine. The promotion of the organic animal husbandry should be a topical issue as the livestock output currently corresponds to a meagre share of the total agricultural output in Ukraine.

The results showed that organic farms are also more labor intensive than conventional farms. This adds to the land profitability (productivity) gap between the two farming systems. Accordingly, training and advisory services are important to improve the labor productivity.

The results suggest that when enterprises embarked on organic farming in Ukraine, they tended to produce lower output per hectare, as opposed to enterprises engaged in conventional farming. The profitability (per hectare) is also lower for the organic companies. It should also be noted that organic enterprises in Ukraine require more labor force per land unit. This, of course, improves the level of qualification of employees in the activities of organic enterprises. The problem, in this case, is caused not so much by the additional costs, but by the scarcity of skilled staff in a rural area. Moreover, in turn, this can lead to disruption of the work schedule and significant losses. However, cost productivity is higher in organic farms, due to the different levels of intermediate consumption. Therefore, organic farming can be developed in Ukraine, given the economic arguments. Moreover, further research is needed regarding the environmental aspects of organic farming in Ukraine. Furthermore, another important direction for further research is the analysis of the homogeneity of organic farms—in order to propose effective support measures, one should identify the groups of organic farms exhibiting similar performance and output-mix.

This study is limited by the random sampling used to establish the sample of conventional farms, compared to the organic farms. In future studies, propensity score matching and bootstrapping can be applied to ensure more accurate inference. In addition, this is an exploratory study seeking to present the stylized facts about the development of Ukrainian organic agriculture. Future research should analyze the enterprise-level data, in order to obtain the estimates of microeconomic indicators.

Author Contributions: Conceptualization, R.O. and V.N.; Methodology, T.B.; Validation, Y.H., V.N. and S.K.; Formal Analysis, Y.H.; Investigation, S.K.; Data Curation, R.O.; Writing—Original Draft Preparation, R.O.; Writing—Review & Editing, T.B. and D.S.; Visualization, T.B. and D.S. All authors have read and agreed to the published version of the manuscript.

Funding: The research has not received any funding.

Conflicts of Interest: The authors declare no conflicts of interest.

Appendix A. Context Data for Organic Farming in Ukraine

Table A1. The most export-oriented types of organic produce, 2017.

Indicator	Corn	Wheat	Barley	Sunflower	Soybeans	Peas *	Millet *	Rapeseed
Export of organic products, thousand tons	100	58	23	12	11	5	4	3
Total exports of products, million tons	19.4	17.3	4.9	73.2	2.9	873.5	72.7	2.1
Share (%)	0.52	0.34	0.47	0.02	0.38	0.57	5.50	0.14

* For these crops, total exports are measured in thousands of tons (1 Ton = 1000 kg).

Table A2. General characteristics of some organic producers.

Title	Operation Area	Certified Activity	Number and Validity of the Certificate	Products	Main Retailers
Svit Bio ("Lybid-K")	Khmelnytsky region	Plant growing; Trade	19-0110-09-01, 2020-12-31	Chicken eggs, Walnut	Ashan, Silpo, Mehamarket, Varus
Zolotyi Parmen	Chernihiv region	Plant growing; Processing; Export/Import	19-0186-08-01, 2020-12-31	Cereals, fodder crops, apples, pumpkins, currants, juice production	Ok Wine, Silpo
Orhanik milk ("Haleks-ahro")	Zhytomyr region	Livestock; Plant growing; Processing; Export/Import; Apiculture	19-0038-08-03, 2020-12-31	Honey, milk, meat, fodder crops, vegetables, cereals, cereals,	Large trading networks
Skvyrskyi kombinat khliboproduktiv	Kyiv region	Processing; Export/Import	19-0294-06-01, 2020-12-31	Buckwheat, oatmeal, corn grits, flour and flakes	Large trading networks
Etnoprodukt	Chernihiv region	Plant growing; Livestock; Processing; Export/Import; Trade	19-0196-12-03, 2020-12-31	Meat, milk, milk products, cereals	Not available
Kasper	Odesa region	Trade; Export/Import; Processing	19-0182-08-01, 2020-12-31	Production of oil, cake, sunflower, flax, rapeseed, soybean, wheat, corn, spelt, barley, oats, rye	Fozzy, Ashan, Mehamarket
Orhanik Oryhinal	Kyiv region	Export/Import; Trade; Processing	19-0135-09-01, 2020-12-31	Flour, flakes, bran, cereals, honey, sunflower oil, beans, peas, lentils	Novus, Ashan, MehaMarket

Source: References [51,52].

References

1. Mariyono, J.; Kuntariningsih, A.; Suswati, E.; Kompas, T. Quantity and monetary value of agrochemical pollution from intensive farming in Indonesia. *Manag. Environ. Qual.* **2018**, *29*, 759–779. [CrossRef]
2. Czyżewski, B.; Matuszczak, A.; Miśkiewicz, R. Public goods versus the farm price-cost squeeze: Shaping the sustainability of the EU's common agricultural policy. *Technol. Econ. Dev. Eco.* **2019**, *25*, 82–102. [CrossRef]
3. Trukhachev, V.; Sklyarov, I.; Sklyarova, Y.; Gorlov, S.; Volkogonova, A. Monitoring of Efficiency of Russian Agricultural Enterprises Functioning and Reserves for Their Sustainable Development. *Montenegrin J. Econ.* **2018**, *14*, 95–108. [CrossRef]
4. The World of Organic Agriculture 2019. Available online: https://www.organic-world.net/yearbook/yearbook-2019/data-tables.html (accessed on 19 September 2019).
5. Kovalchuk, S.Y.; Muliar, L.V. Sustainable development of the world economy: The role of organic production. *Agrosvit* **2014**, *23*, 61–66. (In Ukrainian)
6. Forschungsinstitut für Biologischen Landbau (FiBL). Available online: http://www.fibl.org/en/homepage.htm (accessed on 6 September 2019).
7. Novak, N.P. Principles and competitive advantages of the development of organic agricultural production in Ukraine. *Agrosvit* **2016**, *9*, 23–28. (In Ukrainian)
8. Roitner-Schobesberger, B.; Darnhofer, I.; Somsook, S.; Vogl, C.R. Consumer Perceptions of Organic Foods in Bangkok, Thailand. *Food Policy* **2008**, *33*, 112–121. [CrossRef]
9. Kucher, A.; Heldak, M.; Kucher, L.; Fedorchenko, O.; Yurchenko, Y. Consumer willingness to pay a price premium for ecological goods: A case study from Ukraine. *Environ. Socio-Econ. Stud.* **2019**, *7*, 38–49. [CrossRef]
10. Kosark, N.S.; Kuzio, N.I. Research of the organic Food Market of Ukraine and Directions of Increasing its Competitiveness. *Electron. Sci. Prof. Ed. Eff. Econ.* **2016**, *3*. Available online: http://www.economy.nayka.com.ua/?op=1&z=4777&p=1 (accessed on 10 February 2020). (In Ukrainian)

11. Federation of Organic Movement of Ukraine. Organic Soybean Cultivation in Ukraine. Available online: http://organic.com.ua/en/organic-soybean-cultivation-in-ukraine/ (accessed on 3 November 2019). (In Ukrainian)

12. Lotter, D.W. Organic Agriculture. *J. Sustain. Agric.* **2003**, *21*, 59–128. [CrossRef]

13. Bhavsar, H. The rise of organic food and farming practices. *J. Agric. Sci. Bot.* **2017**, *1*, 1. [CrossRef]

14. Act of Sweden on Organic Production Control (SFS 2013:363) from 23 May 2013 H Wcb-sitc Eco-Lcx. Available online: https://www.ecolex.org/details/legislation/act-on-organic-production-control-sfs-2013363-lex-faoc125965/ (accessed on 3 November 2019).

15. Ayuya, O.I. Organic certified production systems and household income: Micro level evidence of heterogeneous treatment effects. *Org. Agric.* **2019**, *9*, 417–433. [CrossRef]

16. Voskobiinyk, Y.P.; Havaza, I.V. Capacity of the organic produce market in Ukraine. *Agroinkom* **2013**, *4*, 7–10. (In Ukrainian)

17. Wiera, M.; Jensen, K.O.; Andersena, L.A.; Millock, K. The character of demand in mature organic food markets: Great Britain and Denmark compared. *Food Policy* **2008**, *33*, 406–421. [CrossRef]

18. Browne, A.W.; Harris, P.J.C.; Hofny-Collins, A.H.; Pasiecznik, N.; Wallace, R.R. Organic production and ethical trade: Definition, practice and links. *Food Policy* **2000**, *25*, 69–89. [CrossRef]

19. Klitna, M.; Bryzhan, I. State and Development of Organic Production and Market of Organic Products in Ukraine. *Eff. Econ.* **2013**, *10*. Available online: http://www.economy.nayka.com.ua/?op=1&z=2525 (accessed on 26 November 2019).

20. Dabija, D.C.; Bejan, B.M.; Dinu, V. How sustainability oriented is generation Z in retail? A literature review. *Transform. Bus. Econ.* **2019**, *18*, 140–155.

21. Løes, A.; Adler, S. Increased utilisation of renewable resources: Dilemmas for organic agriculture. *Org. Agric.* **2019**, *9*, 459–469. [CrossRef]

22. Wallenbeck, A.; Rousing, T.; Sørensen, J.T.; Bieber, A.; Neff, A.S.; Fuerst-Waltl, B.; March, S. Characteristics of organic dairy major farm types in seven European countries. *Org. Agric.* **2019**, *9*, 275–291. [CrossRef]

23. Muller, A.; Schader, C.; El-Hage Scialabba, N.; Brüggemann, J.; Isensee, A.; Erb, K.H.; Niggli, U. Strategies for feeding the world more sustainably with organic agriculture. *Nat. Commun.* **2017**, *8*, 1290. [CrossRef]

24. Kobets, M.I. Organic farming in the context of sustainable development. In *Agrarian Policy for Human Development Project*; UNDP: New York, NY, USA, 2004. (In Ukrainian)

25. Kulish, L. Development of Competitive Organic Production in Ukraine. *Invest. Prac. Exp.* **2019**, *1*, 42–46. (In Ukrainian) [CrossRef]

26. Smith, O.M.; Cohen, A.L.; Rieser, C.J.; Davis, A.G.; Taylor, J.M.; Adesanya, A.W.; Jones, M.S.; Meier, A.R.; Reganold, J.P.; Orpet, R.J.; et al. Organic Farming Provides Reliable Environmental Benefits but Increases Variability in Crop Yields: A Global Meta-Analysis. *Front. Sustain. Food Syst.* **2019**, *3*, 82. [CrossRef]

27. Madhusudhan, L. Organic Farming-Ecofriendly Agriculture. *J. Ecosys Ecograph* **2016**, *6*, 209. [CrossRef]

28. Caradonna, J.L. Organic Agriculture is Going Mainstream, but not the Way You Think it is. Available online: https://theconversation.com/organic-agriculture-is-going-mainstream-but-not-the-way-you-think-it-is-92156 (accessed on 5 December 2019).

29. Porodina, L.V. The current state of regulation of the safe food market: World experience. *Commod. Sci. Innov.* **2013**, *5*, 188–197. (In Ukrainian)

30. Pietola, K.S.; Lansink, A.O. Farmer response to policies promoting organic farming technologies in Finland. *Eur. Rev. Agric. Econ.* **2001**, *28*, 1–15. [CrossRef]

31. Wikipedia. Organic Farming. Available online: https://en.wikipedia.org/wiki/Organic_farming (accessed on 18 December 2019).

32. Organic Standard—More than 10 Years of Leadership and Reliability. Available online: https://organicstandard.ua/en (accessed on 16 December 2019).

33. Ministry of Agrarian Policy and Food of Ukraine. Organic Production in Ukraine. 4 February 2019. Available online: https://agro.me.gov.ua/en/napryamki/organichne-virobnictvo-v-ukrayini (accessed on 17 December 2019).

34. Willer, H.; Lernoud, J. (Eds.) The World of Organic Agriculture. In *Statistics & Emerging Trends 2019*; Research Institute of Organic Agriculture FiBL; IFOAM—Organics International: Bonn, Germany, 2019.

35. AgroPolit.com. Organic Production in Ukraine is Growing 5 Times Faster than in the EU. 17 June 2019. Available online: https://agropolit.com/news/12556-organichne-virobnitstvo-v-ukrayini-zrostaye-u-5-raziv-shvidshe-nij-v-yes (accessed on 20 October 2019).

36. AgroPortal. Organic Ukraine in Infographic. 13 March 2019. Available online: https://agroportal.ua/en/publishing/infografika/organicheskaya-ukraina-v-infografike/ (accessed on 29 October 2019).

37. Cernansky, R. We don't Have Enough Organic Farms. Why Not? Available online: https://www.nationalgeographic.com/environment/future-of-food/organic-farming-crops-consumers/ (accessed on 2 February 2020).

38. Prokopchuk, N.; Zigg, T.; Vlasyuk, J. Organizational Business Owner of Ukraine. Available online: http://www.ukraine.fibl.org/fileadmin/documentsukraine/UKRAINE_ORGANIC_BUSINESS_DIRECTORY_part2.pdf (accessed on 15 November 2019).

39. Organic Standard Ltd. Available online: https://www.organicstandard.ua/ua (accessed on 16 December 2019).

40. Makarenko, N.A.; Bondar, V.I.; Nikityuk, Y.A. Ecological inspection of technologies of growing the grain crops (on the example of technologies of growing summer wheat in the area of northern forest-steppe). *Agroecol. J.* **2009**, *1*, 24–30. (In Ukrainian)

41. Kovalenko, N.P. Ecologically balanced crop rotation of alternative agriculture: Historical aspects. *Agroecol. J* **2012**, *4*, 95–99. (In Ukrainian)

42. Feshchenko, N.M. Problematic moments of the organic agricultural market. *Innov. Econ.* **2013**, *7*, 141–149. (In Ukrainian)

43. Yatsenko, O.M.; Yatsenko, O.V.; Nitsenko, V.S.; Butova, D.V.; Reva, O.V. Asymmetry of the development of the world agricultural market. *Financ. Credit Act. Probl. Theory Prac.* **2019**, *30*, 423–434. [CrossRef]

44. Kruhliak, O.V. Organic production in dairy cattle breeding in Ukraine. *Econ. Agric.* **2017**, *5*, 33–38. (In Ukrainian)

45. Penn State. Conservation Dairy Farming Could Help Pa. Meet Chesapeake Target. Available online: http://www.sciencedaily.com/releases/2018/09/180913134548.htm (accessed on 16 February 2020).

46. Adamchak, R. *Organic Farming*; Encyclopædia Britannica, Inc.: Chicago, IL, USA, 2018; Available online: https://www.britannica.com/topic/organic-farming (accessed on 19 February 2020).

47. Freyer, B.; Bingen, J.; Fiala, V. Seven myths of organic agriculture and food research. *Org. Agric.* **2019**, *9*, 263–273. [CrossRef]

48. Leu, A.; Regeneration International. Organic Agriculture Can Feed the World. Available online: https://regenerationinternational.org/2018/10/22/organic-agriculture-can-feed-the-world/ (accessed on 15 February 2020).

49. Zhang, L.; Li, X. Changing institutions for environmental policy and politics in New Era China. *Chin. J. Popul. Resour. Environ.* **2018**, *16*, 242–251. [CrossRef]

50. Freedman, B. Organic Farming. Available online: https://science.jrank.org/pages/4904/Organic-Farming-popularity-organic-culture.html (accessed on 22 February 2020).

51. Organic Standard. Available online: https://organicstandard.ua/en/clients (accessed on 8 February 2020).

52. Ekonomichna Pravda. Who Produces Real Organic Produce in Ukraine. Available online: https://www.epravda.com.ua/rus/publications/2019/04/3/646613/ (accessed on 10 February 2020).

Article

Investment in Research and Development and New Technological Adoption for the Sustainable Beekeeping Sector

Jelena Vapa-Tankosić [1,*], Vera Miler-Jerković [2], Dejan Jeremić [3], Slobodan Stanojević [4] and Gordana Radović [5]

[1] Department of Business Economics and Finance, Faculty of Economics and Engineering Management in Novi Sad, University Business Academy in Novi Sad, Cvećarska 2, 21000 Novi Sad, Serbia

[2] School of Electrical Engineering, University of Belgrade, Bulevar kralja Aleksandra 73, 11000 Belgrade, Serbia; veramiler@gmail.com

[3] Faculty of Business and Financial Studies, University of Business Studies in Banja Luka, Jovana Dučića 23a, 78000 Banja Luka, Bosnia and Herzegovina; djeremic@sequestergroup.com

[4] Faculty of Applied Management, Economics and Finance, University Business Academy in Novi Sad, Jevrejska 24/1, 11000 Belgrade, Serbia; sloba_leo@yahoo.com

[5] Dnevnik-Poljoprivrednik AD Novi Sad, Bulevar Oslobođenja 81, 21000 Novi Sad, Serbia; gordana.radovic09@gmail.com

* Correspondence: jvapa@libero.it

Received: 22 June 2020; Accepted: 15 July 2020; Published: 20 July 2020

Abstract: The purpose of this paper is to understand how members of beekeeping associations, with long-standing sustainable traditions and products with registered geographical origins, perceive the investments in research and development (R&D) and new technological adoptions. By means of a binary logistic regression, the socio-demographic factors of the members of beekeeping associations predicting the investments in R&D and new technological adoptions were analyzed. Our findings point out that higher level of education and professional beekeeping experience predicts the willingness of investing in research and development. The higher level of education positively influences the willingness to hire professional consultants or bodies for the research and development of beekeeping practices. Serbian female beekeepers, beekeepers aged more than 41 years and professionally engaged beekeepers are more likely to admit that they need support of scientific and research institutions in the further development of beekeeping practices. A higher education has been shown to significantly predict the value added hive products due to new technology adoption. There is also a positive influence of the education level on new technology adoption.

Keywords: apiculture investment; research and development; sustainability; beekeepers; honey; geographical origin; food; label

---

1. Introduction

Beekeeping is a very important sector in European agriculture. In recent years, the honey market in the EU has been showing a constant increase in demand. In the period of 2017–2019, 216 million EUR will be spent on national beekeeping programs in 28 EU member states, an increase of 9% from the 2014–2016 budget period with an increase in the number of beekeeping incentive measures from six to eight [1–3]. According to the data of the ANSES Sophia Antipolis laboratory, designated as the European Union Reference Laboratory for bee health [4], the total number of beekeepers in Europe was estimated at 620,000 in 2010. The European honey production was evaluated at around 220,000 tons in 2010. The price of honey varied from 1.5 to 40 EUR/kg and the estimated colony winter mortality varied from 7% to 28% depending on the country and the origin of the data. The beekeeping industry has a specific role and is different from other food productions, having in mind that the size of the apiary is generally small (22.4 colonies/beekeeper) while most beekeepers were still hobby beekeepers or "non-professional" beekeepers [4]. One of the current proposals at the EU level, in the scope of beekeeping research is focused on promotion of initiatives to boost European beekeeping research projects [5].

The European integration agenda has an aim to promote a higher level of awareness for health and environmental sustainability in countries that are EU candidates, including the Republic of Serbia. The authors findings [2] point out that Serbian consumers of organic honey are willing to pay (WTP) up to 20% over the price of conventional honey. Understanding customer needs is a key factor in retaining them, as well as acquiring new customers in the process of achieving greater market share [6]. It is necessary to timely allocate efficient and sustainable use of ecological goods [7]. Ignjatijević and Cvijanović [8] point out that the current economic and political position of Serbia imposes the need to analyze the comparative advantages of the agro business sector, especially if it is known that Serbia has a significant potential that is not sufficiently exploited. The honey market in the Republic of Serbia is gradually developing since, in the period from 2004 to 2013, the average value of the exported honey from Serbia amounted to 4.79 million dollars, and imports to 62.4 thousand dollars. The exported quantities of honey indicate a significant increase at the rate of 61.74% per annum. Germany and Italy are the leading countries for import of honey from the Republic of Serbia. Monitoring the variations in comparative advantages in export in the Republic of Serbia transition period and accession to the EU is important for the analysis of the effects of trade liberalization and integration in international flows of the honey sector [9].Western Balkan exporters are gradually positioning themselves on the markets of the EU and countries of ex-Yugoslavia (CEFTA) [10]. The Republic of Serbia has achieved a positive comparative advantage in exports of the processed food sector [11]. High export and minimal import values contributed to Serbia's positive comparative advantage (RCA) in the export of agricultural and food products, sugar industry, molasses [12] and especially honey [13,14]. Research shows that EU countries, especially Germany and Italy, are the target markets for the export of Serbian honey [6]. Ignjatijević et al. [15] suggest that "companies must continually work on the sustainable development of trade by applying new technologies in management".

Bearing in mind the developing honey market, the aim of the present study is to analyze how the beekeeping association members of honey with a registered geographical origin "Fruškogorski lipov med" (Fruška Gora linden honey) perceive investments in R&D and new technological adoptions. The specific aims of this paper are: (a) to investigate the attitudes of beekeeping association members towards investments in R&D and new technological adoptions, and (b) to identify socio-demographic factors that contribute to predicting the investment in R&D and the new technological adoption. In light of the aim of the paper, several hypotheses have been defined:

Hypothesis 1. *A beekeeping association member's education and professional engagement in beekeeping positively predicts their willingness to invest in the investment in research and development and engage professional bodies for the research and development of beekeeping production.*

Hypothesis 2. *A beekeeping association member's willingness to rely on support of scientific and research institutions in the further development of beekeeping practices is positively predicted by their gender, age and professional engagement in the beekeeping.*

Hypothesis 3. *The need for significant funds and further research in the development of beekeeping practices is positively predicted by the beekeeping association member's age, family size, number of beehives and professional engagement in beekeeping.*

Hypothesis 4. *Beekeepers with larger families are more prepared to use new technologies, which is positively predicted by the beekeeping association member's number of beehives, level of education and family size.*

Hypothesis 5. *A higher education positively predicts the perception of new technology adoption and the subsequent increase in the value added hive products.*

For this research, the authors have chosen beekeeping association with a registered label of geographic origin, "Fruška Gora linden honey", which has received a label of geographic origin in 2015, thanks to the support of EU funds in the process of certification. The concept of honey quality integrates the features of the honey production process with the social and environmental impacts, animal welfare and the link of the food with a certain agricultural area. The later aspect at European Union level can be achieved by obtaining two designations: Protected Designation of Origin and Protected Geographical Indication [5]. Authenticity and traceability are the main aspects in the case of the Protected Designation of Origin or the Protected Geographical Indication recognition process [13]. According to Bertozzi [16], the use of geographical name for an agricultural product dates from ancient times: "honey from Sicily" is a good example in this sense. The research findings of Pocol et al. [17] have shown the significance of supplying honey with a protected geographic origin status, while emphasizing several European Countries where honey is protected by the protected designation of origin/protected geographical indication status: Greece, Spain, France, Italy, Luxembourg, Malta, Poland, Portugal and Ukraine. In Italy those are "Miele della Lunigiana", registered from 2004 [18], "Miele delle Dolomiti Bellunesi" registered from 2011 [19], and "Miele Varesino", registered in 2014 [20]. In Romania, the Ministry of Agriculture and Rural Development encourages the protected designation of origin/protected geographical indication honey certification [21]. On the other hand, while the United Kingdom is pursuing withdrawal from the European Union, the Protected designations of origin, the Protected geographical indications, the Geographical indications and Traditional specialties, in force in the EU-27 Member States, may be subject to securing alternative ways of protection of the relevant geographical names in the United Kingdom, as per United Kingdom law [18].

Recent findings on consumer attitudes, shows that majority of consumers, as many as 83%, are willing to try Fruška Gora's linden honey that is of above average quality and is certified, regardless of the fact that linden honey does not belong to the type of honey which consumers usually buy. The consumers are willing to pay even a 30% higher price than the average market price of noncertified linden honey [22]. It should also be mentioned that this study is among the first to empirically analyze the current issues in the Serbian beekeeping sector. Thus, field research was conducted to collect the technical and economic data necessary for this study. More specifically, a structured questionnaire was completed by 250 Serbian beekeepers, from March to December 2019.

2. Literature Review

A plethora of various studies has pointed out the significance of beekeeping associations. Ferreira et al. [23] pointed out the economic effects of the associations in terms of the impact on employment regardless of nationality [24], while Androulidakis and Harizanis [25] emphasized the importance of the associations in organized beekeeping education, technical support, providing relevant statistical data [26], clarifying or interpreting legislation [27], strengthening entrepreneurial orientations [28] and marketing the performance of honey and honey products [29], market positioning [30] and more. Pocol et al. [31] point out the importance of beekeepers' associations in establishing cooperation with other beekeepers, while Popa et al. [27] point out that associations should cooperate with other beekeeping associations, but also with other companies, in order to modernize beekeeping. According to Mushonga et al. [32], local authorities should promote the establishment of beekeeping associations as the findings show that the beekeepers' operations should be supported by education and training.

The issue of the efficiency of investing in beekeeping has been analyzed by Ismail and Ismail [33], indicating that assistance in the form of cheaper loans for the purchase of bees is the most effective initial investment. The authors conclude that the final effects of investment in beekeeping shall be the reduction of rural and urban poverty. Grgić et al. [34] have investigated the beekeeping sector of the Mediterranean area by conducting an analysis of investments, yields and income and have concluded that investments per hive are around 316 to 395 EUR, where the largest investments are in the categories of small and large beekeepers and in the procurement of vehicles and trailers for migration beekeeping. Grgić et al. [35] have conducted an assessment of the main factors of economic performance, the justification of the investment in the bee colony migration system and the increase in the range of hive products. They have presented two investment models: in the first model the stationary way of beekeeping was retained, while migration beekeeping was presented in the other model. The authors conclude that the migration of bee colonies affects the pasture conditions and thus produces an average of 60 kg of honey per hive.

The increased collaboration of beekeepers with science and real sector stakeholders [36], development of social entrepreneurship [37] and development of entrepreneurial behavior, would contribute to a decrease in the total of (especially rural) unemployment [2,17,28,38–40]. In fact, Pocol et al. [41], as well as other researchers [42], have concluded that establishing social beekeeping enterprises may represent an innovative answer to minimizing social problems and can preserve local specificity, promote traditional agriculture and create additional local branded products. New technological adoption and investment in R&D in the apiculture sector should be crucial in order to add value to the product and improve beekeeping practices, including bee health and apiary management. Developing new beekeeping methods and raising awareness of good practices should be done with the help of professional bodies and research institutes. "Investment in knowledge infrastructure and R&D is an important component of any science, technology and innovation policy as well as building up of a set of linkages between main actors and the encouragement of productive interactions and learning processes among them in the context of national, sectoral, regional and in fact transnational systems of innovation" [43].

Keiyoro et al. [44] have revealed that sociocultural factors have a positive and negative influence on the adoption of beekeeping technologies. The study, on the comparison of the new technology adoption rate and income through beekeeping at the farmers' level, has shown that the transfer of improved technology to farmers could help generate income and alleviate poverty [45]. The findings of a study by Popescu and Siceanu [46] in Romania argues that the use of the new reproduction techniques based on the instrumental insemination of queen bees is not only a valuable way for improving bee-breeding programs, but is also a possibility to change the actual apiaries into sustainable and modernized beekeeping farms. The findings of a study on the perception of farmers towards the use of modern beehive technology in Ethiopia showed that education, the off-farm income, availability of credit, beekeeping training and perception in the price of box hives were important factors influencing the adoption of modern beehive technology, recommending that credit should also be given as a part of the package for the proper adoption of this technology [47]. The findings also revealed that beekeepers in Nigeria have increased the quantity of honey produced with a significant difference in the level of the use of modern beekeeping technologies before and after the training [48]. A study of the determinants of technical efficiency of beekeeping farms in Turkey [49] and the association between the beekeeping subsidies and farm efficiency, using a stochastic frontier analysis on beekeeping farming, showed that the beekeepers were generally found to be fairly inefficient thus "increasing of education level of farmer was found as one of the important determinants of efficiency due to access to information, good farm management and adaptation of new production methods". The findings of a study on the factors affecting the adoption of beekeeping and associated technologies in India [50] underline the most important constraints in beekeeping, such as a lack of equipment, pest and predator attacks, bad weather, lack of credit, inadequate skill and knowledge, fear of bees, lack of starting capital to buy hives and equipment, level of income, information on technology and the other technicalities involved. The findings of Muya et al. [51] show that the following economic factors influenced the adoption of beekeeping technologies: product prices, substitute product prices, consumer income, beekeeper's income and government policies.

3. Materials and Methods

A product of the society of beekeepers "Jovan Živanović" from Novi Sad, Region of Vojvodina that lies in the north of the Republic of Serbia, established in 1973, was used. Their product "Fruška Gora linden honey" has recently received a product quality certificate, which represents a chance to contribute to a greater visibility of the Fruška Gora linden honey in the world market and further promote beekeeping as a potentially profitable and environmentally-friendly business. The entire product is traditionally produced in the specific region of Fruška Gora. For these reasons, the beekeepers "Jovan Živanović" from Novi Sad have been chosen for participation in the study.

The research was carefully prepared. The survey was conducted in the territory of Vojvodina, from March to December 2019, and the questionnaire was tested in cooperation with the Association of Beekeeping Organizations of Vojvodina to improve its validity and reliability. The research was directed towards honey with a geographical indication, therefore it was decided that the target group was the members of the Society of beekeepers "Jovan Živanović" from Novi Sad, with whom the Faculty of Economics and Engineering Management in Novi Sad has already collaborated on several projects. The initial questionnaire was improved based on the suggestions of the management of the focus group. The survey was electronically sent to all of the members (in total, 294), while 250 questionnaires were returned in full. The response rate was 85%. The questionnaire was created according to previous research on the application of innovations in the beekeeping industry [3] and the perceived factors towards the use of modern technologies in beekeeping [3,44,47]. All instrument items, except for the socio-demographical characteristics, were answered on a five-point psychometric Likert scale (anchored on 1–"strongly disagree" through 5–"strongly agree"). Likert-type scales usually contain either five or seven response categories [52]. The literature suggests that a five-point scale appears to be less confusing and increases response rates [53], so the authors opted for the five-point Likert scale. A questionnaire that consisted of three sets of questions was prepared. The first group of questions assessed the socio-demographic characteristics of beekeepers (e.g., sex, number of hives, education, family size, age and engagement in beekeeping). The second group consisted of statements regarding the characteristics of investment in R&D in beekeeping (e.g., I am willing to invest funds in the research and development of beekeeping production; I am willing to engage professionals bodies for the research; I need support of scientific and research institutions in the further development of beekeeping practices; Significant funds and further research in the development of beekeeping practices is needed). The third group consisted of statements which indicate the characteristics of the new technological adoptions in beekeeping (e.g., beekeepers with larger families are more willing to use new technologies; new technology adoption can increase the value added hive products; education levels can positively influence new technology adoption).

Thus, field research was conducted to collect the empirical data necessary for this study. The answers that were received in full totaled 250 respondents (response rate 83%) and were further elaborated. The qualitative and quantitative data were analyzed and the appropriate statistical tools and techniques were employed. Several different models have been tested for each dependent variable, IRD 1, 2, 3, 4 and NTA 1, 2, 3, in order to find the best model. At the start, the ordinal logistic regression was conducted with the complementary clog–log link function, with the following set of predictors: sex, number of beehives, education, family size, age, and engagement in beekeeping. Additionally, for each model, a backward step analysis was conducted to find the optimal model. This analysis was used to determine if any of the independent variables can be excluded. However, the accuracy of the original model was small (less than 50%), and this is the reason why the alternative model was used. Improvement was achieved by a reduction of the categories of the independent variables and the models were tested again. Further on, because the accuracy of the new models was small (less than 60%), the alternative model has been produced, in which a dependent variable was gradually collapsed. Finally, for the purpose of a higher accuracy, the dependent variables have been collapsed into two levels and we ran a binary logistic regression. In order to determine the optimum threshold value for a binary logistic regression, we used the receiver operator characteristic curve (the ROC curve). The optimization method which has been used is the iteratively reweighted least squares (IRLS). For the imbalanced data, the SMOTE technique was used. For all models, the ratio of the train and the test has been set on a 80–20 ratio [54]. The authors wanted to find the best fitting model. The best model was set as the model with the highest accuracy. The data was processed in the R Studio software (Version 0.98.976).

4. Results

4.1. Demographic Analysis of the Respondents

The final sample consisted of 250 respondents, of whom 16 (6.4%) were female and 234 (93.6%) were male. On average, they were 38.7 years of age. The 73 (29.2%) beekeepers were 26–40 years old. The 96 (38.4%) of beekeepers were 41–55 years old. The 81 (32.4%) beekeepers were over 55 years old. In the entire sample, 161 (64.4%) beekeepers completed high school, while 89 (35.6%) of the beekeepers completed college or had a higher education. One hundred and five (42%) of the beekeepers have a small family (up to three members) while 145 (58%) have a bigger family (four or more members). The 32 (12.8%) beekeepers have a maximum of 10 beehives, 104 (41.6%) of them have 10 to 25 bee hives, while 114 (45.6%) have more than 25 bee hives.

Table 1 shows the characteristics of investment in the R&D of beekeeping. The beekeepers do invest funds in the research and development of beekeeping (mean average of 3.74) and they are willing to hire experts in the research and development of beekeeping practices. They agree that additional funds and further research in the field of beekeeping is needed. The beekeepers believe they need the support of scientific and research institutions in the further development of beekeeping practices (mean average of 4.64).

Table 1. Characteristics of the investment in the research and development (R&D) of beekeeping.

	Characteristics of Investment in R&D	Mean	SD
IRD1	I am willing to invest funds for research and development of beekeeping production	3.74	1.41
IRD2	I am willing to engage professionals bodies for the research	3.32	1.18
IRD3	I need support of scientific and research institutions in the further development of beekeeping practices	3.50	1.43
IRD4	Significant funds and further research in the development of beekeeping practices is needed	4.64	0.65

Table 2 shows the characteristics of the new technological adoptions in beekeeping. The beekeepers are well aware that the education level positively influences the adoption of new technologies (mean average of 4.19) and that this may increase their production and introduction of new value-added products (mean average of 4.19).

Table 2. Characteristics of the new technological adoptions in beekeeping.

Characteristics of New Technological Adoption		Mean	SD
NTA1	Beekeepers with larger families are more prepared to use new technologies	3.64	1.09
NTA2	New technology adoption can increase the value-added hive products (such as royal jelly, propolis, bee pollen and beeswax)	4.19	0.97
NTA3	Education level can positively influence new technology adoption	4.19	0.93

4.2. Main Results of the Investigation

The aim of the research was to detect the interdependence between the nominal outcomes, which were: new technological adoption and investment in R&D, and the socio-demographic characteristics of beekeepers (e.g., sex, number of beehives, education, family size, age and engagement in beekeeping).

For IRD1, the accuracy of an ordinal model with the modified independent variables was 49.01%, whereas the accuracy of a logistic model was higher, at 63.26%. The backward procedure did not see any significant improvement. For the model based on a logistic regression, the optimal threshold was set at 0.75. Table 3 represents the results given from a logistic regression with the dependent variable IRD1 and the set of predictors.

Table 3. Impact of the socio-demographic predictors on IRD1.

	Estimate Coefficient	Odds Ratio	Test Statistics	Significance Level
(Intercept)	−3.11	0.04	−3.27	<0.001
Sex (female)	−18.23	0.01	−0.02	0.99
Number of beehives (10–25)	0.56	1.75	1.04	0.31
Number of beehives (>25)	−0.66	0.52	−1.29	0.21
Education (higher)	1.47	4.36	3.45	<0.001 **
Family (bigger)	0.68	1.97	1.82	0.07
Age (41–55)	0.25	1.29	0.61	0.55
Age (>55)	−0.02	0.98	−0.05	0.96
Professional (Yes)	2.57	13.05	3.47	<0.001 **

Significant at ** $p < 0.01$.

Only two independent variables from the set of predictors (higher education and professional engagement in beekeeping) made a statistically significant contribution to the model with regard to the dependent variable IRD1 (I am willing to invest funds for the research and development of beekeeping production). The strongest predictor of the above-mentioned statement that respondents invest in research and development was their professional engagement in beekeeping, recording an odds ratio of 13.05. This indicates that the respondents who engage in professional beekeeping are 13.05 times more likely to agree to invest in research and development, than those who engage in beekeeping as an additional job, controlling for all other factors in the model. The beekeepers that are more educated (having finished a college or a university) are 4.36 times more likely to agree with the statement that they invest in research and development than beekeepers that have only finished high school, controlling for other factors in the model.

For IRD2, the accuracy of an ordinal model with the modified independent variables was 44.12%, whereas the accuracy of a logistic model was higher, at 89.80%. The backward procedure did not see any significant improvement. For the model based on a logistic regression, the optimal threshold was set at 0.50. Table 4 represents the results given from a logistic regression with the dependent variable IRD2 and the set of predictors. The independent variable (higher education) made a unique, statistically significant contribution to the model with regard to the dependent variable IRD2 (I am willing to engage professional bodies for the research and development of beekeeping practices). The only significant predictor of the above-mentioned statement, that respondents should engage professional bodies for the research and development of beekeeping practices, was higher education, recording an odds ratio of 13.27. This indicates that the respondents that are more educated (having finished a college or a university) are 13.27 times more likely to agree with the statement that they should engage professional bodies for the research and development of beekeeping practices, than those who only finished high school, controlling for other factors in the model.

Table 4. Impact of the socio-demographics predictors on IRD2.

	Estimate Coefficient	Odds Ratio	Test Statistics	Significance Level
(Intercept)	−19.13	0.01	−0.01	0.99
Sex (female)	−18.51	0.01	−0.01	0.99
Number of beehives (10–25)	−17.86	0.01	0.01	0.99
Number of beehives (>25)	−18.25	0.01	0.01	0.99
Education (higher)	2.58	13.27	4.96	<0.001 **
Family (bigger)	−0.15	0.86	−0.34	0.74
Age (41–55)	0.59	1.81	1.23	0.22
Age (>55)	0.48	1.61	0.97	0.33
Professional (Yes)	0.42	1.53	0.66	0.51

Significant at ** $p < 0.01$.

For IRD3, the accuracy of an ordinal model with the modified independent variables was 59.11%, whereas the accuracy of a logistic model with the SMOTE technique was 90.61%. The backward procedure did not see any significant improvement. For the model based on a logistic regression, the optimal threshold was set at 0.25. Table 5 represents the results obtained from a logistic regression with the dependent variable IRD3 and the set of predictors.

Table 5. Impact of the socio-demographics predictors on IRD3.

	Estimate Coefficient	Odds Ratio	Test Statistics	Significance Level
(Intercept)	−2.69	0.07	−24.27	<0.001
Sex (female)	0.24	1.27	2.14	0.03
Number of beehives (10–25)	−2.31	0.10	−31.29	<0.001 **
Number of beehives (>25)	−3.01	0.05	−38.95	<0.001 **
Education (higher)	0.06	1.07	1.54	0.12
Family (bigger)	−1.24	0.29	−30.75	<0.001 **
Age (41–55)	2.41	10.97	29.82	<0.001 **
Age (>55)	2.29	9.88	30.79	<0.001 **
Professional (Yes)	1.71	5.45	18.61	<0.001 **

Significant at ** $p < 0.01$.

Five of the independent variables (all except higher education) made a statistically significant contribution to the model with regard to the dependent variable IRD3 (I need the support of scientific and research institutions in the further development of beekeeping practices). The strongest significant predictor of the above-mentioned statement, that the respondents need the support of scientific and research institutions in the further development of beekeeping practices, was age (41–55), recording an odds ratio of 10.97. This indicates that the respondents aged 41–55 are 10.97 times more likely to agree with the statement than those aged 26–40, controlling for other factors in the model. The beekeepers older than 50 are 9.8 times more likely to agree with the statement than those aged 26–40, controlling for other factors in the model. The respondents who are engaged in professional beekeeping are 5.5 times more likely to agree with the previous statement, that they need the support of scientific and research institutions in the further development of beekeeping practices, than those who engage in beekeeping as an additional job, controlling for all other factors in the model. Regarding the female beekeepers, they are 1.27 times more likely to agree with the statement than men, controlling for other factors in the model. The respondents that have 10 to 25 beehives are 0.1 times less likely to agree with the statement than those who have less than 10 beehives. The respondents that have more than 25 beehives are 0.05 times less likely to agree that they need the support of scientific and research institutions in the further development of beekeeping than those who have less than 10 beehives. The beekeepers that have a bigger family are 0.29 times less likely to agree with the statement than those with a smaller family.

For IRD4, the accuracy of an ordinal model with the modified independent variables was 59.19%, whereas the accuracy of a logistic model with the SMOTE technique was 90.6%. The backward procedure did not get any significant improvement. For the model based on a logistic regression, the optimal threshold was set at 0.25. Table 6 represents the results obtained from a logistic regression with the dependent variable IRD4 and the set of predictors.

Table 6. Impact of the socio-demographics predictors on IRD4.

	Estimate Coefficient	Odds Ratio	Test Statistics	Significance Level
(Intercept)	−42.7	<0.001	−0.14	0.89
Sex (female)	1.12	3.08	<0.001	0.99
Number of beehives (10–25)	1.05	2.85	11.9	<0.001 **
Number of beehives (>25)	−0.99	0.37	−11.88	<0.001 **
Education (higher)	−1.51	0.22	−21.12	<0.001 **
Family (bigger)	2.55	12.86	32.23	<0.001 **
Age (41–55)	−19.13	<0.001	0.12	0.91
Age (>55)	−21.39	<0.001	0.13	0.90
Professional (Yes)	−20.04	<0.001	0.08	0.94

Significant at ** $p < 0.01$.

As shown in Table 6, four of the independent variables (number of beehives 10–25 and >25, and higher education and family size) made a statistically significant contribution to the model with regard to the dependent variable IRD4 (significant funds and further research in the development of beekeeping practices are needed). The strongest significant predictor of this statement was having a bigger family, recording an odds ratio of 12.88. This indicates that the respondents with bigger families are 12.8 times more likely to agree with the statement than those with smaller families (up to four members), controlling for other factors in the model. The beekeepers that have between 10 and 25 beehives are 2.85 times more likely to agree that significant funds and further research in the development of beekeeping practices are needed (compared to those who have fewer than 10 beehives). The beekeepers that have more than 25 beehives are 0.37 times less likely to agree that significant investments and further research in the development of beekeeping are needed than those who have less than 10 beehives. The beekeepers with a higher education are 0.22 times less likely to agree that significant investments and further research in the development of beekeeping are needed.

For NTA1, the accuracy of an ordinal model with the modified independent variables was 41.87% whereas the accuracy of a logistic model was 70.39%. The backward procedure did not get any significant improvement. For the model based on a logistic regression, the optimal threshold was set at 0.50. Table 7 represents the results given from a logistic regression with dependent variable NTA1 and the set of predictors.

Table 7. Impact of the socio-demographics predictors on NTA1.

	Estimate Coefficient	Odds Ratio	Test Statistics	Significance Level
(Intercept)	−1.21	0.30	−1.48	0.14
Sex (female)	0.33	1.39	0.40	0.69
Number of beehives (10–25)	0.81	2.24	1.30	0.20
Number of beehives (>25)	1.52	4.59	2.59	0.01 *
Education (higher)	1.33	3.80	3.47	<0.001 **
Family (bigger)	0.91	2.49	2.39	0.02 *
Age (41–55)	−1.19	0.30	−2.70	0.01 *
Age (>55)	0.60	1.82	1.42	0.16
Professional (Yes)	−0.88	0.41	−1.51	0.13

Significant at * $p < 0.05$, ** $p < 0.01$.

Four of the independent variables (number of beehives >25, higher education, bigger family, age 41–55) made a statistically significant contribution to the model with the dependent variable NTA1 (beekeepers having large families are more prepared to use new technologies). The strongest significant predictor of the previous statement is beekeepers having more than 25 beehives, recording an odds ratio of 4.59. This indicates that the beekeepers with more than 25 beehives are 4.59 times more likely to agree with this statement than those beekeepers with fewer than 10 beehives, controlling for other factors in the model. The beekeepers that have a higher education (having a college or university degree) are 3.8 times more likely to agree that beekeepers with large families are more prepared to use new technologies than those with finished high school. The beekeepers with a larger family (more than four members) are 2.49 times more likely to agree with the statement that beekeepers with large families are more prepared to use new technologies, than those with a smaller family. The beekeepers aged 41–55 are 0.3 times less likely to agree with the statement than the beekeepers aged 26–40, controlling for other factors in the model.

For NTA2, the accuracy of an ordinal model with the modified independent variables was 51.92% whereas the accuracy of a logistic model with the SMOTE technique was 92.23%. The backward procedure did not see any significant improvement. For the model based on a logistic regression, the optimal threshold was set at 0.20. Table 8 represents the results given from a logistic regression with the dependent variable NTA2 and the set of predictors. In Table 8, we can see that the strongest significant predictor of the statement indicating that that the adoption of new technologies increases the value added hive products was higher education, recording an odds ratio of 18.81. The beekeepers with a higher education were 18.81 times more likely to agree that the adoption of new technologies increases their production of the value added hive products.

Table 8. Impact of the socio-demographics predictors on NTA2.

	Estimate Coefficient	Odds Ratio	Test Statistics	Significance Level
(Intercept)	−0.48	0.62	−3.56	<0.001
Sex (female)	−18.56	0.001	−0.13	0.89
Number of beehives (10–25)	−0.10	0.91	−0.97	0.33
Number of beehives (>25)	−0.33	0.72	−3.21	<0.001 **
Education (higher)	2.94	18.81	39.73	<0.001 **
Family (bigger)	−2.95	0.05	−44.95	<0.001 **
Age (41–55)	−0.99	0.37	−12.95	<0.001 **
Age (>55)	−1.99	0.14	−27.26	<0.001 **
Professional (Yes)	−0.07	0.94	−0.74	0.46

Significant at ** $p < 0.01$.

On the other hand, the beekeepers with larger families (more than four members), compared to beekeepers with fewer family members, are 0.05 less likely to agree that the adoption of new technologies increases their production of value added hive products. The beekeepers who have more than 25 beehives are 0.72 less likely to agree that the adoption of new technologies increases their production offer compared to the ones who have fewer than 10 beehives. The beekeepers aged 41–55 years are 0.37 times less likely to agree with the statement than those aged 26–40, controlling for other factors in the model. The beekeepers older than 50 are 0.14 times less likely to agree with the statement than those who are aged 26–40, controlling for other factors in the model.

For NTA3, the accuracy of an ordinal model with the modified independent variables was 58.13%, whereas the accuracy of a logistic model was 81.63%. The backward procedure did not see any significant improvement. For the model based on a logistic regression, the optimal threshold was set at 0.60. Table 9 shows the results given from a logistic regression with the dependent variable NTA3 and the set of predictors. In Table 9 we can see that the strongest significant predictor of the statement indicating that the education level can positively influence new technology adoption was higher education, recording an odds ratio of 4.3. The beekeepers with a higher education are 4.3 times more likely to agree that the education level can positively influence new technology adoption, compared to beekeepers with lower education levels. The beekeepers with larger families compared to beekeepers with fewer family members are 0.36 times less likely to agree that the educational level in beekeeping can positively influence new technology adoption. The beekeepers aged 41–55 years are 0.14 times less likely to agree than those aged 26–40, controlling for other factors in the model.

Table 9. Impact of the socio-demographics predictors on NTA3.

	Estimate Coefficient	Odds Ratio	Test Statistics	Significance Level
(Intercept)	−18.18	0.001	−0.02	0.99
Sex (female)	−16.7	0.001	−0.01	0.99
Number of beehives (10–25)	−0.52	0.59	−0.78	0.43
Number of beehives (>25)	0.01	1.01	0.02	0.99
Education (higher)	1.46	4.29	3.03	<0.001 **
Family (bigger)	−1.01	0.36	−2.46	0.01 *
Age (41–55)	−1.96	0.14	−3.07	<0.001 **
Age (>55)	−0.55	0.57	−1.08	0.28
Professional (Yes)	0.08	1.05	0.01	0.99

Significant at * $p < 0.05$, ** $p < 0.01$.

5. Discussion

The findings of this research show that socio-demographic factors influence the adoption of new technologies and investments in R&D in the Serbian beekeepers association. Our findings point out that the higher the level of education and professional beekeeping influenced the willingness to invest in research and development. A higher level of education has predicts the willingness to hire professional consultants or bodies for the research and development of beekeeping practices, therefore our first hypothesis has been confirmed. Female Serbian beekeepers, beekeepers aged over 41 and professionally engaged beekeepers are more likely to admit that they need the support of scientific and research institutions in the further development of beekeeping practices. Thus, the second hypothesis has also been confirmed. The respondents with larger families and those who have between 10 and 25 beehives were more likely to agree that significant funds and further research in the development of beekeeping practices is needed. It is interesting to note that beekeepers that have more than 25 beehives and those with a higher education were less likely to agree that significant investments and further research in the development of beekeeping are needed. The beekeeping association members' age and professional engagement in beekeeping did not prove to be significant predictors, so the third hypothesis has been partially confirmed. The following findings point out that the Serbian beekeepers who are more educated and who have more than 25 beehives are possibly well aware of current research, and already use the incentives available to the beekeepers by the state or province and thus consider that may consider that no additional significant investments and further research are needed. A higher number of beehives, level of education and family members (more than four members) predicted the easier adoption of new technologies in large families, as they may consider that their productivity as well as income can be increased which is significant, especially for the household income of large families, therefore these findings confirm our forth hypothesis. The findings are in line with the study findings [45] that show that the improved technology adoption was higher in households involving both genders in beekeeping activities, while the findings have shown the relationship between the number of honeybee colonies, extent of improved technology adoption, honey yield and farm income from beekeeping. The finding of a study [46] on the economic performance of the new beekeeping technology, in comparison with the traditional one, has shown that the modern technology assures a 48% higher profit than in case of classic technology. A higher education has been shown to be the strongest significant predictor of a new technology increase of value added hive products (such as royal jelly, propolis, bee pollen and beeswax). The higher education exerted the highest positive influence on new technology adoption therefore the fifth hypothesis has been confirmed. The findings are in line with a study on the use of modern beehive technology, which revealed that the education was an important factor [47]. The findings are in line with Keiyoro et al. [44], who concluded that in female beekeeping groups, the education levels (among several other factors) can positively influence the adoption of new technologies. The findings are also confirmed by the authors of a study on the innovative potential of beekeeping production in AP Vojvodina [3], which has concluded that "the basic recommendations for encouraging the introduction of innovations in beekeeping are providing adequate and relevant agricultural advisory services, provision of credit services to beekeepers for the purchase of modern equipment; incentives for the introduction of innovations" (p. 276).

6. Conclusions

This research paper focuses on how members of the society of beekeepers, that have developed a product that is traditionally manufactured in the specific region of Fruška Gora, perceive the investment in the R&D and the adoption of new technologies in beekeeping. The socio-demographic characteristics of beekeepers (sex, number of hives, education, family size, age and engagement in beekeeping) have been analyzed to see whether they influence the adoption of new technologies and the investment in R&D. The findings show that the majority of Serbian beekeepers have a very high level of awareness of the level of education positively influencing the adoption of new technologies. A positive influence of the education level on new technology adoption has also been proven. Serbian beekeepers are well aware that further research in the field of beekeeping is needed and would be willing to hire experts in the research and development of beekeeping practices. Higher education has had an influence on the adoption of new technology with the aim of increasing the production towards the value added hive products (such as royal jelly, propolis, bee pollen and beeswax). Therefore, several conclusions could be drawn. Firstly, the authors can conclude that the education on sustainable beekeeping practices and new technology adoption should be supported by national beekeeping education programs, in accordance with the best practices of EU beekeeping programs. Secondly, the further promotion of beekeeping as a potentially profitable and environmentally-friendly business, of honey with the protected geographical indication, is an important marketing tool and the consumers, on the other hand, should recognize the product as having a higher quality than other similar products. Because of the low consumer purchasing power in the Serbian market, the cheaper substitutes for honey, which are largely present on the market, may appeal more to consumers. This constitutes a major threat. The findings on the Polish food-processing industry after their EU accession has shown the higher importance of quality guarantees and successful branding in international activities, whereas the importance of taste and price is higher in domestic market [55]. In this regard the quality of the honey represents a chance to contribute to a greater visibility of the "Fruška Gora linden honey" in the market. Thirdly, the initiatives aimed at encouraging cooperation and R&D with different faculties and research institutes should nurture further apiculture research [56]. This implies that the potential for R&D in the apiculture sector, and the adoption of new technologies in the apiculture communities, is still not fully exploited with a need to be further investigated.

The issue of raising the efficiency in honey production and enhancing the export potential of the apiculture sector was traditionally based on increasing the number of hives. However, the authors point out that the modernization of production, the education of beekeepers and the organizational improvements are essential parts of an increased competitiveness of the sector. The authors stress the necessity of improving all the factors of competitiveness in the apiculture sector, in relation to their economic strength and the level of the entrepreneurial approach. Beekeepers do need increased institutional support, but the expansion of the beekeeper network and the efficient use of production resources represent a significant factor. By defining targeted measures to encourage beekeeping production, the state has an important role in ensuring the simulative conditions for the apiculture sector in the country. This would be a step forward in the development of beekeeping, which is in line with the beekeepers stated willingness to invest in R&D activities and to cooperate with professional consultants. By adopting new technologies and directing activities and investments in R&D, the potential of the apiculture can be further exploited, thus ensuring the positive impact of beekeeping on the overall development of agriculture.

On the basis of the study findings, the following practical recommendations can be outlined. Awareness creation can be raised through intensive training and workshops for the beekeepers. Experts of different faculties and research institutes, willing to work with the society of beekeepers, need to assure service provisions to beekeepers with the aim of assuring help and support in the introduction of the new technologies and promotion of investment in the R&D. The management staff of the society of beekeepers can be trained to improve their management skills in terms of offering members a better placement of honey on the local market, as it can have a significant effect on promoting honey production and honey quality. In view of the foregoing, all the stakeholders should work towards creating an enabling environment in which beekeeping entrepreneurs can exploit opportunities to provide economic security to local communities as a basic or additional income generating activity. A measure on cooperation with the specialized bodies for the implementation of applied research programs in the field of beekeeping and apiculture products, is already included in the national programs developed by member states, provides co-financing of 50% by the European Union. For example, in the Croatian National Beekeeping Program, for the period from 2020 to 2022, the same measure (financing mutual cooperation in applied research programs) that is relevant for improving beekeeping conditions through supplementation and building new knowledge was introduced in the following thematic areas: bee diseases and pests and their interactions, the conservation of bee biodiversity and the impact of the environment on bee communities, the confirmation of the authenticity of the species, geographical origin and the manner of production of bee products (honey, royal jelly, pollen, propolis, wax) [57]. The National Rural Development Program of the Republic of Serbia from 2018 to 2020 offers financial support to beekeepers, only in terms of subsidies per hive and the procurement of beekeeping equipment [58]. In the future, also prior to EU accession, the Serbian National Association of Beekeepers should initiate the adoption of the national program for the development of beekeeping that also includes a measure on increasing the mutual cooperation in applied research programs. By harmonizing domestic legislation with European legislation in the field of honey production, the domestic producers should have more support (that in the future could be equivalent to the support provided by European honey producers), which would in turn contribute to improving the competitiveness of Serbian honey producers and a better recognition of honey from the Republic of Serbia on the European market.

Since this survey targeted its resources exclusively among the members of the society of beekeepers in the specific region of "Fruška Gora", it is questionable if the conclusions could be generalized to be taken as representative. The researchers suggest that similar research be carried out in different locations of the country and in different beekeeping societies/associations, to establish other possible influences. Then the findings of the following study could be compared to the sociocultural factors in different regions and the beekeepers' willingness to adopt new beekeeping technologies or professional advice from specialized bodies. The originality of this paper stems from the need to investigate further sources of uniting traditional beekeeping with the perceived investment in R&D and the adoption of new technologies in order to investigate their future performance. The perceived willingness of beekeepers to invest in the R&D and the adoption of new technologies, especially for the honey products of protected geographical indications, has not been exploited enough and deserves further attention.

Author Contributions: Conceptualization, J.V.-T.; methodology, J.V.-T.; software, V.M.-J.; validation, S.S., G.R. and J.V.-T.; formal analysis, D.J.; investigation, S.S.; resources, G.R.; data curation, V.M.-J.; writing—original draft preparation, J.V.-T.; writing—review and editing, V.M.-J.; visualization, S.S.; supervision, G.R.; project administration, S.S.; funding acquisition, D.J. All authors have read and agreed to the published version of the manuscript.

Funding: This research received no external funding.

Conflicts of Interest: The authors declare no conflict of interest.

References

1. Vapa-Tankosić, J.; Mirković, Z.; Ignjatijević, S. Economic incentive measures and legal framework in the field of beekeeping in the European Union. *Ecologica* **2018**, *25*, 29–35.
2. Vapa-Tankosić, J.; Ignjatijević, S.; Kiurski, J.; Milenković, J.; Milojević, I. Analysis of Consumers' Willingness to Pay for Organic and Local Honey in Serbia. *Sustainability* **2020**, *12*, 4686. [CrossRef]
3. Prodanović, R.; Ignjatijević, S.; Bošković, J. Innovative potential of beekeeping production in AP Vojvodina. *J. Agron. Technol. Eng. Manag.* **2019**, *2*, 268–277.
4. Chauzat, M.-P.; Cauquil, L.; Roy, L.; Franco, S.; Hendrikx, P.; Ribière-Chabert, M. Demographics of the European Apicultural Industry. *PLoS ONE* **2013**, *8*, e79018. [CrossRef] [PubMed]
5. EU. Agricultural Product Quality Policy. Available online: https://ec.europa.eu/info/food-farming-fisheries/food-safety-and-quality/certification/quality-labels/quality-schemes-explained_en (accessed on 21 June 2020).
6. Jovanović, D.; Milenković, N.; Damjanović, R. Ocenjivanje i predviđanje potreba potrošača. *Oditor* **2017**, *3*, 71–79.
7. Milosavljević, S.; Pantelejić, Đ.; Međedović, D. Primena i mogućnost unapređenja ekonomskih činilaca u realizaciji održivog razvoja. *Održiv. Razvoj* **2019**, *1*, 7–16. [CrossRef]
8. Ignjatijevic, S.; Cvijanović, D. Exploring the Global Competitiveness of Agri-Food Sectors and Serbia's Dominant Presence: Emerging Research and Opportunities. In *Advances in Business Strategy and Competitive Advantage*; IGI Global: Hershey, PA, USA, 2017.
9. Ignjatijević, S.; Milojević, I.; Andžić, R. Economic analysis of exporting Serbian honey. *Int. Food Agribus. Manag. Rev.* **2018**, *21*, 929–944. [CrossRef]
10. Vapa-Tankosić, J.; Redžepagić, S.; Stojsavljević, M. Trade, Regional Integration and Economic Growth: MEDA Region and the Western Balkan Countries. In *Financial Integration*; Springer: Berlin/Heidelberg, Germany, 2013; Volume 36, pp. 215–229. ISBN 978-3-642-35696-4.
11. Ignjatijević, S.; Matijašević, J.; Milojević, I. Revealed comparative advantages and competitiveness of the processed food sector for the Danube countries. *Custos E Agronegocio Line* **2014**, *10*, 256–281.
12. Raičević, V.; Ignjatijević, S.; Matijašević, J. Economic and legal determinants of export competitiveness of the food industry of Serbia. *Industrija* **2012**, *40*, 201–226.
13. Ignjatijević, S.; Čavlin, M.; Djordjevic, D. Measurement of comparative advantages of processed food sector of Serbia in the increasing the export. *Econ. Agric.* **2014**, *61*, 677–693. [CrossRef]

14. Ignjatijević, S.; Prodanović, R.; Bošković, J.; Puvača, N.; Tomaš-Simin, M.; Peulić, T.; Đuragić, O. Comparative analysis of honey consumption in Romania, Italy and Serbia. *Food Feed Res.* **2019**, *46*, 125–136. [CrossRef]

15. Ignjatijević, S.; Milojević, I.; Cvijanović, G.; Jandrić, M. Balance of Comparative Advantages in the Processed Food Sector of the Danube Countries. *Sustainability* **2015**, *7*, 6976–6993. [CrossRef]

16. Bertozzi, L. Designation of origin: Quality and specification. *Food Qual. Prefer.* **1995**, *6*, 143–147. [CrossRef]

17. Pocol, C.B.; Ignjatijević, S.; Cavicchioli, D. Production and Trade of Honey in Selected European Countries: Serbia, Romania and Italy. In *Honey Analysis*; InTech: Rijeka, Croatia, 2017; pp. 1–20.

18. Commission Regulation (EC) No 1845/2004 of 22 October 2004 Supplementing the Annex to Regulation (EC) No 2400/96 on the Entry of Certain Names in the Register of Protected Designations of Origin and Protected Geographical Indications (Tergeste, Lucca, Miele della Lunigiana and Άγιος Ματθαίος Κέρκυρας (Agios Mathaios Kerkyras)). Available online: http://www.legislation.gov.uk/eur/2004/1845/annex/division/1 (accessed on 21 June 2020).

19. Union, P.O. of the E. Commission Implementing Regulation (EU) No 243/2011 of 11 March 2011 Entering a Name in the Register of Protected Designations of Origin and Protected Geographical Indications (Oravský korbáčik (PGI). CELEX1. Available online: http://op.europa.eu/en/publication-detail/-/publication/4c0e5038-3947-4c84-98be-f8158a6c4e2e/language-en (accessed on 21 June 2020).

20. Commission Implementing Regulation (EU) No 328/2014 of 26 March 2014 Entering a Name in the Register of Protected Designations of Origin and Protected Geographical Indications (Miele Varesino (PDO)) EUR-Lex-32014R0328-EN-EUR-Lex. Available online: https://eur-lex.europa.eu/legal-content/EN/ALL/?uri=CELEX%3A32014R0328 (accessed on 21 June 2020).

21. Ministry of Agriculture and Rural Development and Romanian Beekeepers Association. Romanian Presidency of the Council of the European Union: Romania's Environmental Mission and Priorities. *EIRP Proc.* **2019**, *14*. Available online: https://ec.europa.eu/environment/eir/pdf/report_ro_en.pdf (accessed on 20 July 2020).

22. Ćirić, M.; Ignjatijević, S.; Cvijanović, D. Research of honey consumers' behavior in province of Vojvodina. *Ekon. Poljoprivred.* **2015**, *62*, 627–644. [CrossRef]

23. Ferreira, M.L.B.; Arnaud, E.D.R.; Leite, D.T.; Sousa, L.C.F.S.; Schmidt, F.R. Social environmental and economic study of family production of the beekeepers cooperative of Catole do Rocha. *Rev. Verde De Agroecol. E Desenvolv. Sustent.* **2012**, *7*, 34–44.

24. Ramadani, V.; Hisrich, R.D.; Dana, L.-P.; Palalic, R.; Panthi, L. Beekeeping as a family artisan entrepreneurship business. *Int. J. Entrep. Behav. Res.* **2019**, *25*, 717–730. [CrossRef]

25. Androulidakis, S.; Harizanis, P. Identification of Greek beekeepers' educational needs, a case study. *Eur. J. Agric. Educ. Ext.* **1996**, *3*, 47–53. [CrossRef]

26. Semkiw, P.; Skubida, P. Evaluation of the economical aspects of Polish Beekeeping. *J. Apic. Sci.* **2010**, *54*, 10–12.

27. Pocol, C.B. Sustainable policies for the development of beekeeping in Romania. *Probl. Rol. Świat.* **2011**, *11*, 3.

28. Popa, A.A.; Mărghitaş, L.A.; Pocol, C.B. Economic and socio-demographic factors that influence beekeepers' entrepreneurial behavior. *Agron. Ser. Sci. Res.* **2011**, *54*, 450–455.

29. Gomes, M.; Casaca, J.; Cabo, P.; Dias, L.G.; Vilas Boas, M. Trade barriers and economic impact of organic beekeeping in Portugal. In *Book of Abstracts of the IInd International Symposium on Bee Products, Annual Meeting of the International Honey Commission*; School of Agriculture, Polytechnic Institute of Bragança: Braganca, Portugal, 2012; pp. 38–39.

30. Lengler, L.; Lago, A.; Coronel, D.A. A organização associativa no setor apícola: Contribuições e potencialidades. *Org. Rurais Agroind.* **2011**, *9*, 152–163.

31. Pocol, C.; Moldovan-Teselios, C.; Arion, F.H. Beekeepersâ Association: Motivations and Expectations. *Bull. Univ. Agric. Sci. Vet. Med. Cluj Napoca Hortic.* **2014**, *71*, 141–147.

32. Mushonga, B.; Hategekimana, L.; Habarugira, G.; Kandiwa, E.; Samkange, A.; Ernest Segwagwe, B.V. Characterization of the Beekeeping Value Chain: Challenges, Perceptions, Limitations, and Opportunities for Beekeepers in Kayonza District, Rwanda. *Adv. Agric.* **2019**, *2019*, 1–9. [CrossRef]

33. Ismail, M.M.; Ismail, W.I.W. Development of stingless beekeeping projects in Malaysia. *E3S Web. Conf.* **2018**, *52*, 00028. [CrossRef]

34. Grgić, Z.; Filipi, J.; Bićanić, D.; Bobić, B.Š. Opportunities for developing a business model of Mediterranean beekeeping. *J. Cent. Eur. Agric.* **2017**, *19*, 206–216. Available online: https://hrcak.srce.hr/507719 (accessed on 10 May 2020).

35. Grgić, Z.; Filipi, J.; Bićanić, D.; Šakić Bobić, B. Opportunities for developing a business model of Mediterranean beekeeping. *J. Cent. Eur. Agric.* **2018**, *19*, 287–296. [CrossRef]

36. Tesser, F.; Cavicchioli, D. *Economic Aspects of Beekeeping and Honey Production in Italy and in Lombardy Region*; Museo Lombardo di Storia dell'Agricoltura, Sant'Angelo Lodigiano: Lodi, Italy, 2014; Volume 1, pp. 41–52.

37. Pocol, C.B.; Marghitas, L.A.; Popa, A.A. Evaluation of sustainability of the beekeeping sector in the North West Region of Romania. *J. Food Agric. Environ.* **2012**, *10*, 132–138.

38. Praća, N.; Paspalj, M.; Paspalj, D. Ekonomska analiza uticaja savremene poljoprivrede na održivi ravoj. *Oditor* **2017**, *3*, 37–51. [CrossRef]

39. Plavša, N.; Stojanović, D.; Stojanov, I.; Puvača, N.; Stanaćev, V.; Đuričić, B. Evaluation of oxytetracycline in the prevention of American foulbrood in bee colonies. *AJAR* **2011**, *6*, 1621–1626. [CrossRef]

40. Puvača, N.; Lika, E.; Cocoli, S.; Shtylla Kika, T.; Bursić, V.; Vuković, G.; Tomaš Simin, M.; Petrović, A.; Cara, M. Use of Tea Tree Essential Oil (Melaleuca alternifolia) in Laying Hen's Nutrition on Performance and Egg Fatty Acid Profile as a Promising Sustainable Organic Agricultural Tool. *Sustainability* **2020**, *12*, 3420. [CrossRef]

41. Pocol, C.B.; Bârsan, A.; Popa, A.A. A model of social entrepreneurship developed in Barču Valley, Slaj County. *Analele Univ. Din Oradea Fasc. Ecotoxicol. Zooteh. Si Tehnol. De Ind. Aliment.* **2012**, *11*, 183–190.

42. Cvijanović, D.; Ignjatijević, S.; Vapa Tankosić, J.; Cvijanović, V. Do Local Food Products Contribute to Sustainable Economic Development? *Sustainability* **2020**, *12*, 2847. [CrossRef]

43. Schot, J.; Steinmueller, W.E. Three frames for innovation policy: R&D, systems of innovation and transformative change. *Res. Policy* **2018**, *47*, 1554–1567. [CrossRef]

44. Keiyoro, P.N.; Muya, B.I.; Gakuo, C.M.; Mugo, K. Impact of Sociocultural factors on adoption of modern technologies in beekeeping projects among women groups in Kajiado County- Kenya. *IJIER* **2016**, *4*, 55–64. [CrossRef]

45. Bhusal, S.; Thapa, R. Comparative Study on the Adoption of Improved Beekeeping Technology for Poverty Alleviation. *J. Inst. Agric. Anim. Sci.* **2005**, *26*, 117–125. [CrossRef]

46. Popescu, A.; Siceanu, A. Economic efficiency of various queen bees maintenance systems. *J. Cent. Eur. Agric.* **2003**, *4*, 153–160.

47. Yehuala, S.; Birhan, M.; Melak, D. Perception of Farmers towards the Use of Modern Beehives Technology in Amhara Region, Ethiopia. *Engineering* **2013**, *5*, 1–8.

48. Eforuoky, F.; Thomas, K.A.; Eforuoky, F.; Thomas, K.A. Effect of training on the use of modern beekeeping technologies in Oyo State, Nigeria. *Niger. J. Rural Sociol.* **2015**, *15*, 61–66. [CrossRef]

49. Gürer, B.; Akyol, E. The determinants of technical efficiency in beekeeping farms and the role of agricultural subsidies: The case of Nigde, Turkey. *J. Agric. Environ. Int. Dev.* **2018**, *112*, 343–360. [CrossRef]

50. Suraj, S.; Das, D. Factors affecting adoption of beekeeping and associated technologies in Kamrup (rural) district, Assam state, India. *BIJ* **2018**, *2*, 253–258. [CrossRef]

51. Muya, B.I.; Gakuu, C.M.; Keiyoro, P.N. Economic factors influencing adoption of modern beekeeping technologies among women beekeeping projects in Kajiado county- Kenya. *Int. J. Curr. Res.* **2018**, *10*, 72779–72782.

52. Walford, G.; Tucker, E.; Viswanathan, M. *The SAGE Handbook of Measurement*; SAGE: London, UK, 2010; ISBN 978-1-4462-0688-1.

53. Hayes, B.E. *Measuring Customer Satisfaction: Development and Use of Questionnaires*; Irwin Professional Pub: Milwaukee, WI, USA, 1992; ISBN 978-0-87389-131-8.

54. Harrell, F.E. *Regression Modeling Strategies: With Applications to Linear Models, Logistic and Ordinal Regression, and Survival Analysis*; Springer International Publishing: Cham, Switzerland, 2015; ISBN 978-3-319-19424-0.

55. Bryla, P. The impact of EU accession on the marketing strategies of Polish food companies. *Br. Food J.* **2012**, *114*, 1196–1209. [CrossRef]

56. Cristina, R.; Kovačević, Z.; Cincović, M.; Dumitrescu, E.; Muselin, F.; Imre, K.; Militaru, D.; Mederle, N.; Radulov, I.; Hădărugă, N.; et al. Composition of a Natural Phytotherapeutic Association Beneficially Used to Combat Nosemosis in Honey Bees. *Sustainability* **2020**, in press.

57. Ministry of Agriculture, National Beekeeping Program for the Period from 2020 to 2022. Available online: https://poljoprivreda.gov.hr/UserDocsImages/dokumenti/poljoprivreda/pcelarstvo/Nacionalni%20p%C4%8DerIarski%20program%202020.-2022.pdf (accessed on 12 July 2020).
58. Nacionalni Program Ruralnog Razvoja od 2018. do 2020. godine Available online: /subvencije_regulative_u_poljoprivredi_detalji/ministarstvo_poljoprivrede_donelo_je_nacionalni_programruralnog_razvoja_od_2018__do_2020__godine-1533643094 (accessed on 13 July 2020).

Article

Sustainable Rearing for Kid Meat Production in Southern Italy Marginal Areas: A Comparison among Three Genotypes

Maria Antonietta Colonna, Pasqua Rotondi, Maria Selvaggi *, Anna Caputi Jambrenghi, Marco Ragni and Simona Tarricone

Department of Agricultural and Environmental Science, University of Bari Aldo Moro, 70125 Bari, Italy; mariaantonietta.colonna@uniba.it (M.A.C.); pasqua.rotondi@uniba.it (P.R.); anna.caputijambrenghi@uniba.it (A.C.J.); marco.ragni@uniba.it (M.R.); simona.tarricone@uniba.it (S.T.)
* Correspondence: maria.selvaggi@uniba.it; Tel.: +39-080-544-2236

Received: 2 July 2020; Accepted: 25 August 2020; Published: 26 August 2020

Abstract: Sustainable goat breeding plays an important role in the economy of marginal areas. The present study aimed to compare performances and meat quality traits in kids of a native Apulian genotype (Garganica) in comparison with two Mediterranean breeds (Maltese and Derivata di Siria). Kids suckled dam milk until they were 21 (±2) days old, hence three groups of 12 male kids per each genotype were made. The kids received a pelleted feed ad libitum in addition to dam milk and were slaughtered at 60 days of age. The Maltese kids showed the lowest net cold-dressing percentage, with statistical differences compared to Garganica and Derivata di Siria. Meat obtained from Garganica kids showed a rosy color due to a significantly lower a* index and were also more tender since a lower WBS was recorded in comparison with the other two genotypes. As for the nutritional value of meat, the best n-6/n-3 ratio was found for the Derivata di Siria breed. In conclusion, Garganica kid meat showed the lowest content of SFA and atherogenic index, with potential beneficial effects for human health.

Keywords: kids; Garganica; Maltese; Derivata di Siria; meat quality; fatty acids; biodiversity; sustainability

1. Introduction

Goat breeding plays an important role in the economy of several countries because of their ability to adapt to different climates, management conditions and to regulate their feeding regimens according to the availability of food resources, which may be very poor in harsh geographical areas, where it is difficult to rear other animal species [1]. In the Basilicata region, the pedo-climatic conditions and the rocky and hostile territory of the inland mountain areas prevent the cultivation of food resources for livestock animals reared under semi-intensive or intensive conditions. Moreover, the lack of infrastructures is a severe limitation for farmers. Traditionally, sheep and goat grazing systems are the only possible source of income from livestock in this region. This has led rural populations to preserve the historical and cultural heritage related to local small ruminant breeds, also to avoid biodiversity loss.

To date, there is a growing interest in the nutritional properties and high quality of goat products [2]. Consumers are oriented towards foods of animal origin that are perceived as being healthy and wholesome. Goat-meat acceptability varies widely depending on cultural traditions, age, socioeconomic conditions and family habits that influence the consumer's preference [3,4]. In Italy "capretto" is the main kind of goat meat appreciated by the consumer. This meat has a light or rosy color, with a delicate and wild taste due to kids suckling milk from goats that graze on spontaneous pastures and scrubs typical of the Mediterranean area [5,6]. The traditional "capretto" is consumed during Easter and Christmas festivities that correspond to the two seasonal kidding times. Kids are typically slaughtered at about 4–7 weeks of age with an average weight that ranges from 10 to 12 kg.

The EU Regulation N. 2018/848 concerning organic production strongly suggests taking into account the genetic value of animal breeds along with longevity, vitality, ability of the animals to adapt to the environment and disease resistance. Preference should be given to native breeds and genetic lines, privileging genetic diversity. As the only form of animal husbandry that is gaining growing interest in the last decade is organic livestock farming, the exploitation of autochthonous breeds has a renewed significance—especially for small ruminant breeders who apply multifunctional agriculture. The concept of integrated agroforestry and silvopastoral systems, based on connections between conventional agriculture with other on-farm services (e.g., tourism, educational activities, social agriculture, etc.) offers rural populations the chance to diversify and increase incomes from their multifunctional farms involved in food production, environmental conservation and on-farm services. In this system, there is the integration of forests, tree plantations and herbaceous crops with grazing small ruminant species.

Genetic and environmental factors are also known to affect meat production and quality in goats [7,8]. As reported by several authors, the feeding system, breed, age and gender are able to influence growth, muscle and fat deposition and, therefore, may affect meat quality in small ruminants [2,3,6,8–13]. In the Basilicata region, the Garganica, Maltese and Derivata di Siria goat breeds are usually reared for milk production, which is processed into traditional cheeses.

The Garganica breed originated in the Gargano promontory in the Apulia region by crossing the autochthonous population of goat with west European goats. This breed shows an exceptional ability to utilize poor pasture that would not otherwise be used. Animals are medium-sized and have black glossy hair with some reddish shade and long twisted horns in both sexes. The average milk production ranges from 200 to 250 L in 210 days; the milk is high in protein (3.5%) and fat (4.8%) [14,15]. Nowadays, the Garganica goat breed is included in the list of Italian endangered breeds maintained by the Italian Department for Environment, Food and Rural Affairs [16]. The Maltese goat has white body with long hair, black head and large drooping ears; this breed has no horns. Milk production is about 350 L with high fat (3.8%) and protein (3.3%) content; prolificacy is high (180%) [17]. The Derivata di Siria is a Sicilian domestic dairy goat which derives from the Damascus goat of Syria. This breed has a peculiar reddish-brown coat and a milk production of ~570 L per lactation. The milk contains 4.11% fat and 3.53% protein [18]. Garganica, Maltese and Derivata di Siria goats are reared also in other regions in Southern Italy (Basilicata, Campania, Calabria).

The aim of the research was to carry out a comparative study on growth performances and meat physical, chemical and fatty acid composition in Garganica kids, an Apulian autochthonous breed, in comparison with two Mediterranean breeds (Maltese and Derivata di Siria) reared by low-input traditional farming system in marginal areas of the Basilicata region.

2. Materials and Methods

2.1. Animal Management and Diet

The trial was carried out during March–May 2017 on a total of 36 unrelated male kids—all born as twins—of the Garganica, Maltese and Derivata di Siria breeds in Muro Lucano (Basilicata region, Southern Italy, latitude: 40°45′13″64 N, longitude: 15°29′17.1″ E, 650 m above sea level). Kids were reared according to the traditional farming system: they were milk-fed, suckling from their dams until they were about 21 days old. Afterwards, three groups of 12 kids per each genotype—the progeny of six dams for each breed—were made, homogeneous for age (21 ± 2 d). The kids were housed in individual pens (1 m²/head) in an open-sided barn that complies with welfare standards. Each kid had free access to water and received a pelleted feed ad libitum (Table 1) that was formulated in order to meet nutritional requirements [19]. Feed was offered daily at 08:00 a.m. at a rate of 110% of ad libitum intake, calculated by weighing-back refusals once per week. Feed samples were taken weekly and stored at −20 °C until analyses were performed. In addition to pelleted feed, kids had access to maternal milk throughout the trial period. Suckling occurred twice daily, in the morning at about 07:00 a.m., before the dams were led to pasture and in the evening at 07:00 p.m., when the dams came back from grazing. During suckling time, twin kids were put in the same pen together with their dam. Feed refusals were recorded weekly for each kid in order to calculate the average daily gain (ADG), the average daily feed intake (ADFI) and the feed-conversion ratio (FCR).

Table 1. Feed ingredients (% as fed), chemical (% dry matter basis) and fatty acid composition (% of total fatty acid methyl esters) of the pelleted feed administered to kids.

Ingredient Composition (g/kg as Fed Basis)	
Dehulled soybeans (37% crude protein)	6.00
Corn	31.00
Barley	9.00
Wheat flour middlings	9.00
Faba bean	10.00
Bran	10.00
Dehydrated beet pulp	6.00
Soybean oil	1.00
Sunflower meal	8.00
Molasses	3.00
Soybean hulls (12% crude protein)	4.00
Vitamin mineral premix	3.00
Total	100
Chemical composition (% dry matter)	
Moisture (% as fed)	12.00
Crude protein	16.80
Ether extract	4.60
Ash	9.10
Crude fiber	15.18
NDF (neutral detergent fiber)	33.85
ADF (acid detergent fiber)	10.94
ADL (acid detergent lignin)	2.64
Meat forage units (n/kg dry matter)	1.03
Fatty acid composition (% of total fatty acids methyl esters)	
$C_{12:0}$ (lauric)	0.95
$C_{14:0}$ (myristic)	0.95
$C_{16:0}$ (palmitic)	9.17
$C_{18:0}$ (stearic)	1.15
$C_{20:0}$	0.73
$C_{18:1}$ n9 c9 (oleic)	17.91
$C_{18:2}$ n6 (linoleic)	39.17
$C_{18:3}$ n3 (α-linolenic)	4.55
$C_{18:3}$ n6	0.36
$C_{20:3}$ n3	0.65
$C_{20:4}$ n6 (arachidonic)	0.21
$C_{22:2}$ n6	1.17
$C_{22:5}$ n3 (DPA)	0.54
$C_{22:6}$ n3 (DHA)	0.30

The dams grazed during the day on a spontaneous vegetation characterized by several shrubs and grass species such as *Spartium* sp., *Rosa canina*, *Prunum spinosa*, *Quercus pubescens*, *Lolium perenne*, *Festuca* sp., *Trifolium pratense*, *Cichorium sativus*, *Avena fatua*, *Avena sterilis*, *Feniculurn* sp., *Vicia sativa*, *Onobrychis viciifolia*, *Lotus corniculatus* and *Thymus serpyllum*. At housing, in the evening, the dams received hay ad libitum and a commercial feed (500 g/head/day).

2.2. Feed Chemical Composition

Samples of the pelleted feed were ground in a hammer mill with a 1 mm screen and analyzed using the following procedures [20]: dry matter (DM; Method 934.01), ether extract (EE; Method 920.39), ash (Method 942.05), crude protein (CP; Method 954.01), crude fiber (CF; Method 945.18), acid detergent fiber (ADF), acid detergent lignin (ADL) (Method 973.18) and amylase-treated NDF (Method 2002.04).

2.3. Slaughtering and Carcass Traits

Kids were slaughtered at 60 days of age by exsanguination, according to the veterinary police rules. The animals were deprived of feed 12 h before slaughter, with free access to water. Kids were weighed immediately prior to slaughter (live weight at slaughter, LWS). The hot carcass, skin and fleece, pluck, full and empty gastrointestinal tract (GIT) were weighed according to the Italian ASPA procedures [21]. Empty body weight (EBW) was calculated by deducting the weight of the full GIT from the LWS. Net hot-dressing percentage was calculated as hot carcass weight/EBW * 100. The carcasses were hung by the Achilles tendon, chilled at 4 °C (80–82% relative humidity) for 24 h and then weighed again. The net cold-dressing percentage was calculated as cold-carcass weight/EBW * 100. The refrigerated carcasses were split into two halves by the midline; the right side was divided into cuts (neck, steaks, brisket, shoulder, abdominal region, loin, leg) according to ASPA methods [21]. The leg and loin were transported from the slaughterhouse to the laboratory using a portable refrigerated box. The two meat cuts were stored at 4 °C for further 24 h and then dissected into tissue components (lean, separable fat and bone) [15].

2.4. Physical Analysis

The pH value was measured on the *longissimus lumborum* muscle of the right half carcass at the time of slaughter (pH 0) and after 24 h of refrigeration at 4 °C (pH 24), using a portable instrument (Hanna Instruments HI 9025, Woonsocket, RI) equipped with a penetrating electrode (FC 23 °C; Hanna Instruments) calibrated at two pH standards points (7.01 and 4.01).

Samples of the *longissimus lumborum* muscle were taken in order to evaluate meat quality characteristics. Meat color (L* = lightness; a* = redness; b* = yellowness) was assessed using a HunterLab MiniScanTM XE Spectrophotometer (4500/L, 45/0 LAV, 3.20-cm-diameter aperture, 10° standard observer, focusing at 25 mm, illuminant D65/10; Hunter Associates Laboratory, Inc.; Reston, VA, USA) by taking three readings for each sample. The instrument was normalized to a standard white tile supplied with the instrument before performing analysis (Y = 92.8; x = 0.3162 and y = 0.3322). The reflectance measurements were performed after the sample had oxygenated in air for at least 30 min, in order to allow the measurements to get stable [22].

Meat tenderness was assessed on raw samples of the *longissimus lumborum* muscle by the Warner–Bratzler shear (WBS) force system using an Instron 5544 universal testing machine (Instron Corp., Canton, MA, USA). The meat samples had a cylindrical shape with a 12.5-mm-diameter; they were assessed in triplicate and sheared perpendicularly to the direction of muscle fibers (load cell: 50 kg; shearing speed: 200 mm/min). Peak force was expressed as kg/cm^2.

2.5. Chemical and Fatty Acid Analyses and Lipid Oxidation

Chemical analysis and fatty acid profile were performed on raw meat of the *longissimus lumborum* muscle using samples devoid of external fat, epimysium and parts with visible metmyoglobin.

The AOAC procedures [20] were used to assess moisture, crude fat, protein and ash. Total lipids were extracted from the homogenized *longissimus lumborum* samples (100 g) according to the chloroform/methanol method [23]. Fatty acids (FA) were methylated using BF3-methanol solution (12% *v/v*) [24]. The fatty acid profile was assessed using a Chrompack CP 9000 gas chromatograph, with a silicate glass capillary column (70% cyanopropyl polysilphenylene-siloxane BPX 70 of SGE Analytical

Science, length = 50 m, internal diameter = 0.22 mm, film thickness = 0.25 µm). The temperature program was as follows: 135 °C for 7 min followed by increases of 4 °C per minute up to 210 °C.

The $\Delta 9$ desaturase and elongase enzymatic activities were mathematically determined [25] as follows:

$\Delta 9$ desaturase 16 index = 100 [(C16:1cis9)/(C16:1cis9 + C16:0)];
$\Delta 9$ desaturase 18 index = 100 [(C18:1cis9)/(C18:1cis9 + C18:0)];
Elongase index = 100 [(C18:0 + C18:1cis9)/(C16:0 + C16:1cis9 + C18:0 + C18:1cis9)].

The food risk factors of meat were determined by calculating the atherogenic (AI) and thrombogenic (TI) indices [26]:

AI = [(C12:0 + 4 × C14:0 + C16:0)] ÷ [ΣMUFA + Σn−6 + Σn−3];
TI = [(C14:0 + C16:0 + C18:0)] ÷ [(0.5 × ΣMUFA + 0.5 × Σn−6 + 3 × Σn−3 + Σn−3)/Σn−6];

where MUFA are monounsaturated fatty acids.

Fatty acids were expressed as percentage (*wt/wt*) of total methylated fatty acids.

Lipid oxidation was evaluated in *longissimus lumborum* muscle samples stored at 4 °C for 48 h after slaughtering by measuring the concentration of 2-thiobarbituric acid reactive substances (TBARS) [27] and expressed as mg malondialdehyde (MDA)/kg meat.

2.6. Collagen Analysis

The *longissimus* muscle was excised from the right half-carcass (chilled for 24 h at 4 °C) between the 4th–5th lumbar vertebrae; samples of approximately 100 g of muscle (wet weight) were removed from the cranial end of the muscle and dry-frozen. The total and soluble collagen content were analyzed according to the method based on the spectrophotometric determination of the hydroxyproline content [28]. Assuming that collagen weighed 7.25 times the measured hydroxyproline weight, the amounts of collagen contents were calculated as follows:

Total collagen = hydroxyproline * 7.25/1000/(weight/250)
Soluble collagen = hydroxyproline * 7.25/1000/(weight/400)

The results were expressed as mg/g of frozen dry matter. Insoluble collagen was calculated from total collagen minus soluble collagen and expressed as µg/mg of frozen dry matter.

2.7. Statistical Analysis

Data were analyzed for variance (ANOVA) using the GLM procedure of SAS software [29]. The statistical model included genotype as the main effect and experimental error. When the genotype effect was significant ($p \leq 0.05$), means were compared by the Bonferroni post hoc test. Results are reported as least squares mean and pooled standard error of the mean (SEM). Significance was declared at $p \leq 0.05$.

3. Results and Discussion

3.1. In Vivo Performance and Slaughtering Data

No differences due to genotype were observed for kids' birth weight, growth performance and slaughter weights (Table 2). The birth weight of kids may depend on the conformation and size of the adults which are typical for each genotype [30].

Table 2. In vivo performance and slaughtering data of kids.

	Genotype			SEM [1]	*p*-Value
	Garganica	**Maltese**	**Derivata di Siria**		
Initial body weight (kg)	3.16	2.88	3.06	0.350	0.463
Live weight at slaughter (kg)	11.00	12.44	12.18	2.320	0.592
Average daily gain (kg/d)	0.13	0.16	0.15	0.036	0.459
Average daily feed intake (kg)	0.62	0.70	0.64	0.052	0.549
Feed-conversion ratio (kg/kg)	4.77	4.37	4.27	2.309	0.356
Empty body weight (kg)	9.27	10.16	10.06	1.978	0.742
Skin + fleece (%)	10.18 [b]	11.18 [a]	10.08 [b]	0.637	0.033
Omentum (%)	0.52	0.98	1.61	0.691	0.081
Head (%)	5.99	6.06	6.48	0.415	0.175
Pluck (%)	5.90	6.56	5.77	0.788	0.275
Net hot-dressing percentage	67.72 [a]	63.48 [b]	67.27 [a]	2.553	0.042
Net cold-dressing percentage	64.22 [Aa]	56.42 [Bb]	62.51 [ABa]	3.723	0.015

[1] SEM—standard error of the mean; a, b: $p < 0.05$; A, B: $p < 0.01$.

Comparable results in terms of live weights at slaughter have been reported for suckling kids of the Criollo Cordobes and Anglonubian breeds reared in Argentina according to the traditional system and slaughtered at the same age, i.e., 60 days [31]. Kids of the Maltese breed showed a significantly greater incidence of skin and fleece in comparison with both Garganica and Derivata di Siria ($p < 0.05$). Furthermore, the Maltese kids showed a lower net hot-dressing percentage in comparison with the other two breeds ($p < 0.05$). In the present study, the net hot dressing values recorded were higher than those reported in other experiments [5,31,32], but similar to those observed by other authors in Girgentana [6] and Garganica [15] kids slaughtered at the same age. The lowest net cold-dressing percentage was recorded for the Maltese kids, which was significantly lower in comparison with both the Garganica ($p < 0.01$) and Derivata di Siria ($p < 0.05$) breeds. Several studies have reported that genotype, gender and feeding are the major factors able to affect the slaughter yield which is positively correlated to live weight at slaughter [31,33]. As for the effect of genotype, these differences could be attributed to a higher milk production from their mothers along to differences in size and conformation among breeds [31].

Table 3 shows the section data expressed in terms of percentage of the half-carcass weight. The differences between breeds with regards to carcass conformation and to the incidence of the single meat cuts reflects genetic differences in muscle growth: Maltese kids showed a significantly lower percentage of the loin and shoulder in comparison with the other two breeds ($p < 0.05$) and of the leg than Garganica kids ($p < 0.05$).

The Derivata di Siria and Maltese breeds provided a similar percentage of brisket that was greater than that observed for Garganica kids ($p < 0.01$). Ekiz et al. 2010 [34] reported a significant influence of breed or genotype on certain carcass measurements and indices used as indicators of carcass conformation and size which increase proportionally to slaughter weight.

Genotype did not show any influence on the dissection data of the loin and leg (Table 4). In small ruminants, tissue composition of meat cuts is influenced by nutrition, gender, litter size and body weight [3,6,12]. Moreover, fat depots may develop in a different way in relation to the animal genotype and age; goat carcasses have usually more than 60% dissectible lean and about 5–14% fat [3]. Meat breeds are selected due to an efficient utilization of feed for maximum muscle deposition, while local and indigenous breeds have evolved in response to their ability to use fibrous feeds typical of the production systems in marginal areas, besides their disease resistance [10].

Table 3. Section data (% of half-carcass weight) of kids.

	Genotype			SEM [1]	*p*-Value
	Garganica	**Maltese**	**Derivata di Siria**		
Half carcass (kg)	4.66	4.80	4.94	1.192	0.937
		Meat cuts (%)			
Neck	8.04	8.30	7.16	1.133	0.286
Steaks	16.01	16.84	16.49	1.541	0.703
Brisket	9.12 [B]	10.85 [A]	10.89 [A]	0.517	0.001
Loin	7.06 [a]	5.78 [b]	6.86 [a]	0.711	0.031
Abdominal region	4.77	5.86	5.39	0.669	0.070
Leg	31.21 [Aa]	30.15 [ABb]	29.69 [B]	0.388	0.001
Shoulder	19.43 [a]	17.80 [b]	18.61 [a]	0.683	0.045
Shins	2.24	2.35	2.45	0.242	0.409
Perirenal fat	1.53	1.36	1.88	0.786	0.574
Kidney	0.59	0.71	0.58	0.129	0.278

[1] SEM—standard error of the mean; a, b: $p < 0.05$; A, B: $p < 0.01$.

Table 4. Dissection data of the loin and leg (% on weight).

	Genotype			SEM [1]	*p*-Value
	Garganica	**Maltese**	**Derivata di Siria**		
Loin weight (kg)	0.332	0.280	0.326	0.076	0.510
Lean (%)	49.98	49.92	43.91	4.759	0.109
Fat (%)	7.68	6.41	6.51	1.900	0.521
Bone (%)	42.34	43.67	49.57	4.853	0.080
Leg weight (kg)	1.44	1.58	1.46	0.394	0.840
Lean (%)	66.29	64.57	61.62	4.783	0.330
Fat (%)	3.39	4.66	5.67	1.818	0.180
Bone (%)	30.32	30.77	32.70	4.783	0.712

[1] SEM—standard error of the mean.

3.2. Physical and Chemical Parameters of Meat

The results of the physical and chemical analyses of meat are reported in Table 5. No significant differences were found between the three breeds for the pH values of meat. The final pH values varied from 5.41 to 5.68 that fall within the acceptable pH range [35] and are optimal for high quality goat meat [36,37].

Table 5. Physical and chemical characteristics and lipid oxidation of meat from the *longissimus lumborum* muscle.

	Genotype			SEM [1]	*p*-Value
	Garganica	**Maltese**	**Derivata di Siria**		
pH_0	6.69	6.69	6.78	0.186	0.702
pH_{24}	5.45	5.41	5.68	0.173	0.067
L*	47.82	48.36	46.49	3.043	0.618
a*	6.21 [Bb]	7.84 [A]	7.64 [a]	0.781	0.012
b*	12.01	12.91	12.42	0.893	0.312
WBS (kg/cm^2)	5.15 [b]	7.95 [a]	7.27 [a]	1.553	0.036
Moisture	76.63	74.95	73.94	2.246	0.203
Crude protein	19.36	19.62	19.19	0.994	0.794
Ether extract	2.34 [b]	2.89 [b]	4.55 [a]	1.407	0.045
Ash	1.39	1.63	1.44	0.299	0.451
MDA (mg/kg meat)	0.77 [A]	0.38 [B]	0.28 [B]	0.206	0.005

[1] SEM—standard error of the mean; a, b: $p < 0.05$; A, B: $p < 0.01$.

The pH values found in the present study are comparable to those reported in a previous research carried out in Garganica kids [38] as well as in other goat breeds [6], while they were lower than those recorded for the Criollo Cordobes and Anglonubian breeds [31], thus supporting the hypothesis that kids were not subjected to pre-slaughter stress, that is responsible for high pH values which negatively affect meat visual appearance and quality [39].

Meat color features were quite similar among the three genotypes studied, except for the a* index that was significantly lower in Garganica kids in comparison with Derivata di Siria ($p < 0.05$) and Maltese ($p < 0.01$). Meat obtained in the present study may be classified as light and pale red in comparison with the results reported by other authors [31]. Since the kids in this trial did not differ for gender, litter size, slaughter age and muscle type, the only factor which may have affected meat color is genotype [31,40,41]. Whether these differences may be due to dams' individual eating preferences during grazing which have potential effects on milk must be ascertained.

"Capretto" must show a pink color in order to be appreciated by consumers and in the present study the overall colorimetric features may be judged as satisfactory.

The tenderness varied widely among the three genotypes examined: a significantly lower WBS value was recorded for Garganica kids in comparison with Maltese and Derivata di Siria ($p < 0.05$). The WBS values found in the present study ranged from 5.15 to 7.95 kg/cm^2 which are comparable to those reported by several authors for other goat breeds [6,42,43], but higher in comparison to Garganica kids slaughtered at the same age [15] and to Jonica kids slaughtered at 45 days [44]. The study of the factors influencing meat tenderness is particularly relevant for goat meat due to its lower tenderness compared to lamb/mutton and beef [42]. Among these factors, undoubtedly genotype [10,31,45] plays an important role, along with gender, age at slaughter and animal management system [5,30,42,46,47].

In our study, the chemical composition was not affected by genotype except for the intramuscular fat content, in agreement with previous research [31].

Lipid oxidation of meat is one of the major causes of its qualitative degradation during storage due to the formation of aldehydes responsible for the development of rancid off-flavors and meat color darkening [31]. In the present study, the TBARS values (mg of malondialdehyde-MDA/kg meat) recorded were similar for Maltese and Derivata di Siria kids and significantly lower than those found for the Garganica breed ($p < 0.01$). Although there is evidence on the relationship between meat lipid oxidation and its iron and myoglobin content, in this study we found a higher MDA value in the Garganica breed notwithstanding a similar L value than the other two genotypes and a significantly lower a* value of meat.

The TBARS values recorded for the three genotypes, however, are below the concentration of 2 mg MDA/kg meat, which is considered to be the limit above which rancidity could be revealed by consumers [48].

3.3. Collagen Analysis

Table 6 shows the collagen fractions (total, soluble and insoluble) of the *longissimus lumborum* muscle of kids from the three genotypes. The assessment of the total amount and solubility of collagen is considered to be a simple way to evaluate meat eating quality, with regards to tenderness. No significant effect of genotype was found on neither of the collagen fractions, confirming the results reported by other studies [36]. This may be attributable to the fact that in our study kids had the same age at slaughter. In fact, the distribution of collagen fractions seems to vary widely in relation to animal age [3]. Although the three genotypes did not differ significantly for the soluble and insoluble collagen concentrations; it is possible to suppose that the slightly lower total concentration found for the Garganica breed may have determined a lower shear force.

Table 6. Collagen content of the *longissimus lumborum* muscle (µg/mg lyophilized muscle).

	Genotype			SEM [1]	*p*-Value
	Garganica	Maltese	Derivata di Siria		
Total collagen (µg/mg)	44.96	55.96	58.41	11.554	0.177
Soluble collagen (µg/mg) (%)	22.09	27.66	28.84	4.838	0.090
	49.49	49.76	49.38	12.651	0.885
Insoluble collagen (µg/mg) (%)	22.87	28.30	29.57	12 163	0.647
	50.51	50.23	50.62	12 651	0.885

[1] SEM—standard error of the mean.

3.4. Fatty Acid Profile of Meat

The fatty acid composition of the *longissimus lumborum* intramuscular fat is shown in Table 7. Among the SFAs, the main fatty acids identified were palmitic, stearic and myristic, in agreement with several studies carried out on kid and goat meat [15,31,49,50]. In particular, meat from Garganica kids showed a significantly lower ($p < 0.05$) amount of total SFAs than the Derivata di Siria breed and a lower concentration of several individual fatty acids such as $C_{10:0}$ ($p < 0.05$), $C_{14:0}$ ($p < 0.01$) and $C_{17:0}$ ($p < 0.05$). Meat from Maltese kids had overall intermediate values, in particular as for myristic acid ($C_{14:0}$) that was significantly lower ($p < 0.05$) in comparison with the Derivata di Siria breed while higher ($p < 0.05$) respect to Garganica kids. No effect of genotype was recorded for the oleic acid concentration ($C_{18:1\ n9\ c9}$), which is the most representative fatty acid among MUFAs, in agreement with previous studies [15,31,50], while the trans-oleic fatty acid concentration, i.e., elaidic acid ($C_{18:1n9t9}$), was highest in meat from Garganica kids, in comparison with both Derivata di Siria ($p < 0.05$) and Maltese ($p < 0.01$). The main trans-monoenoic acids in ruminants are elaidic and vaccenic acid. In adult ruminants, these trans-fatty acids are produced by microbial hydrogenation of linoleic acid and linolenic acid in the rumen; as a consequence, a variety of positional and stereoisomers of both cis and trans-fatty acids may appear in both meat and milk [51]. This finding in Garganica kid meat may by hypothesized as a combination of factors that depend on the fatty acid profile of dam's milk and on a different maturity of the kid's gut at this age.

Meat from Garganica kids showed a higher concentration of both isomers of conjugated linoleic acid (CLA) than the other two breeds ($p < 0.01$). The CLA content in ruminant milk and meat depends on many factors, such as genotype, age and diet [52]. Although CLA accounts for a relatively small amount of the total fatty acid composition of foods, it is very important for human health since it shows positive effects with regard to cancer, atherosclerosis, growth, obesity, osteoporosis and immune responses [53,54].

Garganica meat showed the lowest concentration of α-linolenic acid, with a significant difference than Maltese ($p < 0.05$) and Derivata di Siria ($p < 0.01$) breeds. In the present study, the average $C_{18:3}$ n3 (α-linolenic) concentration was similar [15,49], lower [31] or greater [50] compared to the findings reported by other authors, thus confirming the wide variability of the fatty acid profile of meat due to animal genotype and slaughter age [55]. α-linolenic acid is converted into its longer chain homologs of which the most important for their nutritional interest are $C_{20:5}$ n3 (eicosapentaenoic acid, EPA) and $C_{22:6}$ n3 (docosahexaenoic acid, DHA) since they are important components of animal cell membranes [56]. In turn of a lower concentration of α-linolenic in Garganica meat, we found a significantly higher concentration of EPA ($p < 0.01$) in comparison with the other two breeds. Adversely, Garganica kid meat showed a significantly lower ($p < 0.05$) content of total fatty acids of the n-3 series in comparison with the Derivata di Siria breed.

The n-6/n-3 ratio ranged from a minimum of 5.26 in Derivata di Siria kid meat to 9.22–10.73, respectively for the Maltese and Garganica breeds ($p < 0.01$). A ratio close to one is considered ideal for human health while values less than five are acceptable [56–58]. The n-6/n-3 ratios found in our study were higher than those found by several authors [47,59,60], but comparable to those reported by Todaro et al. 2000 [61]. The controversial findings indicate that in the pre-ruminant stage this ratio is more influenced by the composition of mother's milk rather than by genotype [62,63].

Table 7. Mean (± SE) fatty acid composition (% of total fatty acid methyl esters) of meat from the *longissimus lumborum* muscle.

	Genotype			SEM [1]	*p*-Value
	Garganica	**Maltese**	**Derivata di Siria**		
Total Fatty acids (g/100 g muscle)	2.07	2.31	3.64	0.650	0.802
$C_{10:0}$ (capric)	0.29 [b]	0.26 [b]	0.59 [a]	0.196	0.042
$C_{12:0}$ (lauric)	0.97	0.73	1.03	0.364	0.413
$C_{14:0}$ (myristic)	2.65 [Bc]	5.37 [b]	7.38 [Aa]	1.636	0.002
$C_{16:0}$ (palmitic)	24.72	21.98	24.79	2.580	0.187
$C_{17:0}$	0.72 [b]	1.10 [a]	1.04 [a]	0.199	0.021
$C_{18:0}$ (stearic)	14.86	14.09	11.86	2.565	0.199
Total SFA [2]	44.92 [b]	46.24 [ab]	48.97 [a]	2.204	0.037
$C_{16:1}$ n7 (palmitoleic)	1.22	1.74	1.99	0.780	0.317
$C_{18:1}$ n9 t9 (elaidic)	2.99 [Aa]	0.47 [B]	1.23 [b]	1.288	0.026
$C_{18:1}$ n9 c9 (oleic)	31.84	34.83	31.06	3.633	0.261
Total MUFA [3]	40.28	40.61	39.10	3.804	0.628
$C_{18:2}$ n6 c9 c12 (linoleic)	5.18	6.23	5.09	1.782	0.548
CLA c9, t11	0.09 [B]	0.01 [A]	0.02 [A]	0.032	0.003
CLA t10, c12	0.13 [A]	0.01 [B]	0.02 [B]	0.023	0.001
$C_{18:3}$ n3 (α-linolenic)	0.21 [Bb]	0.59 [a]	0.81 [A]	0.202	0.002
$C_{20:5}$ n3 (EPA)	0.13 [A]	0.01 [B]	0.01 [B]	0.027	0.001
$C_{22:6}$ n3 (DHA)	0.14	0.21	0.33	0.260	0.507
Total n-6 [4]	6.71	7.28	6.10	2.066	0.676
Total n-3 [5]	0.66 [b]	0.80 [ab]	1.15 [a]	0.258	0.030
Total PUFA [6]	7.96	8.08	7.25	2.260	0.824
Unidentified fatty acids	6.84	5.07	4.68	1.832	0.579
n-6/n-3	10.73 [A]	9.22 [A]	5.26 [B]	1.925	0.002
PUFA/SFA	0.16	0.18	0.15	0.060	0.785
Δ^9 desaturase 16 index	4.76	7.08	7.12	2.268	0.210
Δ^9 desaturase 18 index	68.12	71.21	72.40	5.172	0.428
Elongase index	64.10	67.45	61.69	3.799	0.094
Atherogenic index	1.11 [b]	1.18 [b]	1.43 [a]	0.171	0.032
Thrombogenic index	1.74	1.57	1.68	0.195	0.377

[1] SEM: standard error of the mean; [2] SFA—saturated fatty acids (sum of $C_{10:0}$ + $C_{12:0}$ + $C_{14:0}$ + $C_{15:0}$ + $C_{16:0}$ + $C_{17:0}$ + $C_{18:0}$ + $C_{21:0}$ + $C_{22:0}$ + $C_{24:0}$); [3] MUFA—monounsaturated fatty acids (sum of $C_{14:1}$ + $C_{15:1}$ + $C_{16:1}$ c9 + $C_{17:1}$ c10 + $C_{18:1}$ t11 + $C_{18:1}$ t9 + $C_{18:1}$ t10 + $C_{18:1}$ c9 + $C_{20:1}$ + $C_{24:1}$); [4] Total n-6 (sum of $C_{18:2}$ c9;c12 + $C_{18:2}$ c9;t11 + $C_{18:3}$ + $C_{20:3}$ + $C_{20:4}$); [5] Total n-3 (sum of $C_{18:3}$ + $C_{20:3}$ + $C_{20:4}$ + $C_{20:5}$ + $C_{22:6}$); [6] PUFA—polyunsaturated fatty acids (sum of n-6 + n-3); a, b, c: $p < 0.05$; A, B: $p < 0.01$.

No influence of breed was found on the PUFA/SFA ratio that ranged from 0.15 to 0.18. These values are far below the limit of 0.45 recommended for human health [15]. Similarly, no effect of genotype was observed for the indices of Δ9-desaturase, neither 16 nor 18 and elongase enzyme activities.

In this study, the meat from Derivata di Siria kids showed a markedly greater atherogenic index in comparison with the other two goat breeds ($p < 0.05$). The indices of atherogenicity and of thrombogenicity are indicators assessing the level and the interrelation of some fatty acids that have effects on occurrence of coronary heart diseases [26].

4. Conclusions

This comparative study of growth performances, carcass and meat quality traits in Garganica, Maltese and Derivata di Siria kids reared by low-input farming systems and slaughtered at 60 days of age showed that the Maltese breed provided a better meat yield, whereas meat from the Garganica kids was light-rose and more tender. In terms of the nutritional value of meat with regards to the fatty acid composition of intramuscular fat, genotype had no effect on the PUFA/SFA ratio, although Garganica kid meat had a lower content of SFA and atherogenic index. These positive results in terms of potential benefits for human health may represent an opportunity of valorization and promotion of this breed in order to improve the profitability of low environmental impact rearing systems in marginal areas.

Author Contributions: Conceptualization, M.A.C. and S.T.; methodology, P.R., M.A.C. and S.T.; software, M.S.; validation, M.A.C., A.C.J. and M.S.; formal analysis, P.R., M.S.; investigation, M.A.C., A.C.J., M.S.; resources, A.C.J and M.R.; data curation, M.A.C., M.S. and S.T.; writing—original draft preparation, M.A.C.; writing—Review & Editing, M.A.C., M.S. and S.T.; visualization, A.C.J.; supervision, A.C.J. and M.S.; project administration, S.T.; funding acquisition, A.C.J. and M.R. All authors have read and agreed to the published version of the manuscript.

Funding: This research was funded by the Basilicata Region, Grant number DGR No. 1096 08/08/2012, "Valorization of sheep and goat genotypes production bred in Basilicata for the biodiversity preservation and conservation" (Basilicata 2014–2020 PSR-Mis. 10 Environmental agroclimatic payments, Sub-measure 10.2.).

Acknowledgments: The authors would like to thank Dott. Salvatore Claps of the Council for Agricultural Research and Economics, Research Center for Animal Production and Aquaculture (CREA-ZA) in Muro Lucano (PZ) for management and care of the experimental animals. The authors are grateful to the technicians Massimo Lacitignola, Nicolò Devito and Domenico Gerardi for their laboratory assistance.

Conflicts of Interest: The authors declare no potential conflict of interest.

References

1. Boyazoglu, J.; Hatziminaoglou, I.; Morand-Fehr, P. The role of the goat in society: Past, present and perspectives for the future. *Small Rumin. Res.* **2005**, *60*, 13–23. [CrossRef]
2. Zervas, G.; Tsiplakou, E. The effect of feeding systems on the characteristics of products from small ruminants. *Small Rumin. Res.* **2011**, *101*, 140–149. [CrossRef]
3. Webb, E.C.; Casey, N.H.; Simela, L. Goat meat quality. *Small Rumin. Res.* **2005**, *60*, 153–166. [CrossRef]
4. Borgogno, M.; Corazzin, M.; Saccà, E.; Bovolenta, S.; Piasentier, E. Influence of familiarity with goat meat on liking and preference for capretto and chevon. *Meat Sci.* **2015**, *106*, 69–77. [CrossRef]
5. Dhanda, J.S.; Taylor, D.G.; McCosker, J.E.; Murray, P.J. The influence of goat genotype on the production of capretto and chevon carcass. 1. Growth and carcass characteristics. *Meat Sci.* **1999**, *52*, 355–361. [CrossRef]
6. Todaro, M.; Corrao, A.; Alicara, M.L.; Schinelli, R.; Giaccone, P.; Priolo, A. Effects of litter size and sex on meat quality traits of kid meat. *Small Rumin. Res.* **2004**, *54*, 191–196. [CrossRef]
7. Warmington, B.G.; Kirton, A.H. Genetic and non-genetic influences on growth and carcass traits of goats. *Small Rumin. Res.* **1990**, *3*, 147–165. [CrossRef]
8. Casey, N.H.; Webb, E.C. Managing goat production for meat quality. *Small Rumin. Res.* **2010**, *89*, 218–224. [CrossRef]
9. Goetsch, A.L.; Merkel, R.C.; Gipson, T.A. Factors affecting goat meat production and quality. *Small Rumin. Res.* **2011**, *101*, 173–181. [CrossRef]
10. Xazela, N.M.; Chimonyo, M.; Muchenje, V.; Marume, U. Effect of sunflower cake supplementation on meat quality of indigenous goat genotypes of South Africa. *Meat Sci.* **2012**, *90*, 204–208. [CrossRef]
11. Ozcan, M.; Yalcintan, H.; Tölü, C.; Ekiz, B.; Yilmaz, A.; Savas, T. Carcass and meat quality of Gokceada Goat kids reared under extensive and semi-intensive production systems. *Meat Sci.* **2014**, *96*, 496–502. [CrossRef]
12. Facciolongo, A.M.; Lestingi, A.; Colonna, M.A.; De Marzo, D.; Toteda, F. Effect of diet lipid source (linseed vs. soybean) and gender on performance, meat quality and intramuscular fatty acid composition in fattening lambs. *Small Rumin. Res.* **2018**, *159*, 11–17. [CrossRef]
13. Facciolongo, A.M.; De Marzo, D.; Ragni, M.; Lestingi, A.; Toteda, F. Use of alternative protein sources for finishing lambs. 2. Effects on chemical and physical characteristics and fatty acid composition of meat. *Progr. Nutr.* **2015**, *17*, 165–173.
14. Selvaggi, M.; Fedrica, I.; Pinto, F.; Dario, C. Analysis of A Sequence Nucleotide Polymorphism of STAT5A Gene in Garganica Goat Breed. *Int. J. Adv. Sci. Eng. Inf. Technol.* **2015**, *5*, 323. [CrossRef]
15. Rotondi, P.; Colonna, M.A.; Marsico, G.; Giannico, F.; Ragni, M.; Facciolongo, A.M. Dietary supplementation with Oregano and Linseed in Garganica Suckling Kids: Effects on Growth Performances and Meat Quality. *Pak. J. Zool.* **2018**, *50*, 1421–1433. [CrossRef]
16. MiPAAF. 2007. Available online: https://www.anmvioggi.it/media/files/ELENCO%20RAZZE%20MINACCIATE.pdf (accessed on 1 August 2020).

17. Pesce Delfino, A.R.; Selvaggi, M.; Celano, G.V.; Dario, C. Heritability Estimates of Lactation Traits in Maltese Goat. *Int. J. Biol. Biomol. Agric. Food Biotech. Eng.* **2011**, *5*, 362–364.

18. Noè, L.; Gaviraghi, A.; D'Angelo, A.; Bonanno, A.; Di Trana, A.; Sepe, L. Le razze caprine d'Italia. In *L'alimentazione Della Capra da Latte*; Pulina, G., Ed.; Avenue Media: Bologna, Italy, 2005; pp. 381–435.

19. INRA. *Ruminant Nutrition, Recommended Allowances and Feed Tables*; Jarrige, R., Ed.; Institut National de la Recherche Agronomique, INRA: Paris, France, 1989.

20. AOAC. *Official Methods of Analysis of Association of Official Analytical Chemists*, 18th ed.; Association of Official Analytical Chemists: Arlington, VA, USA, 2004.

21. ASPA. *Metodiche Per la Determinazione Delle Caratteristiche Qualitative Della Carne (Procedures for Meat Quality Evaluation)*; Scientific Association of Animal Production, Università di Perugia: Perugia, Italy, 1996. (In Italian)

22. Šicklep, M.; Candek-Potokar, M. Pork color measurements as affected by bloom time and measurement location. *J. Muscle Foods* **2007**, *18*, 78–87. [CrossRef]

23. Folch, A.J.; Lees, M.; Stanley, G.H. A simple method for the isolation and purification of total lipids from animal tissues. *J. Biol. Chem.* **1957**, *226*, 497–509. [PubMed]

24. Christie, W.W. *Lipid Analysis-Isolation, Separation, Identification and Structural Analysis of Lipids*; Pergamon Press: Oxford, UK, 1982; p. 270.

25. Malau-Aduli, A.E.O.; Siebert, B.D.; Bottema, C.D.K.; Pitchford, W.S. A comparison of the fatty acid composition of triacylglycerols in adipose tissue from Limousin and Jersey cattle. *Aust. J. Agric. Res.* **1997**, *48*, 715–722. [CrossRef]

26. Ulbricht, T.L.V.; Southgate, D.A.T. Coronary heart disease: Seven dietary factors. *Lancet* **1991**, *338*, 985–992. [CrossRef]

27. Salih, A.M.; Smith, D.M.; Price, J.F.; Dawson, L.E. Modified extraction 2-tiobarbituricacid method for measuring lipid oxidation in poultry. *Poult. Sci.* **1987**, *66*, 1483–1488. [CrossRef] [PubMed]

28. Starkey, C.P.; Geesink, G.H.; Oddyc, V.H.; Hopkins, D.L. Explaining the variation in lamb longissimus shear force across and within ageing periods using protein degradation, sarcomere length and collagen characteristics. *Meat Sci.* **2015**, *105*, 32–37. [CrossRef]

29. SAS. *SAS/STAT User's Guide: Statistics*; SAS Institute Inc.: Cary, NC, USA, 2000.

30. Dhanda, J.S.; Taylor, D.G.; Murray, P.J. Carcass composition and fatty acid profiles of adipose tissue of male goats: Effects of genotype and live weight at slaughter. *Small Rumin. Res.* **2003**, *50*, 67–74. [CrossRef]

31. Peña, F.; Bonvillani, A.; Freireb, B.; Juárezc, M.; Perea, J.; Gómez, G. Effects of genotype and slaughter weight on the meat quality of Criollo Cordobes and Anglo Nubian kids produced under extensive feeding conditions. *Meat Sci.* **2009**, *83*, 417–422. [CrossRef] [PubMed]

32. Peña, F.; Juárez, M.; Bonvillani, A.; García, P.; Polvillo, O.; Domenech, V. Muscle and genotype effects on fatty acid composition of goat kid intramuscular fat. *Ital. J. Anim. Sci.* **2011**, *10*, 212–216. [CrossRef]

33. Marichal, A.; Castro, N.; Capote, J.; Zamorano, M.J.; Argüello, A. Effects of live weight at slaughter (6, 10, 25 kg) on kid carcass and meat quality. *Livest. Prod. Sci.* **2003**, *83*, 247–256. [CrossRef]

34. Ekiz, B.; Ozcan, M.; Yilmaz, A.; Tölü, C.; Savaş, T. Carcass measurements and meat quality characteristics of dairy suckling kids compared to an indigenous genotype. *Meat Sci.* **2010**, *85*, 245–249. [CrossRef]

35. Peña, F.; Perea, J.; Garcia, A.; Acero, R. Effects of weight at slaughter and sex on the carcass characteristics of Florida suckling kids. *Meat Sci.* **2007**, *75*, 543–550. [CrossRef]

36. Santos, V.A.C.; Silva, A.O.; Cardoso, J.V.F.; Silvestre, A.J.D.; Silva, S.R.; Martins, C.; Azevedo, J.M.T. Genotype and sex effects on carcass and meat quality of suckling kid protected by the PGI "Cabrito de Barroso". *Meat Sci.* **2007**, *75*, 725–736. [CrossRef]

37. Bonvillani, A.; Peña, F.; Domenech, V.; Polvillo, O.; García, P.T.; Casal, J.J. Meat quality of Criollo Cordobes goat kids produced under extensive feeding conditions. Effects of sex and age/weight at slaughter. *Span. J. Agric. Res.* **2010**, *8*, 116–125. [CrossRef]

38. Todaro, M.; Corrao, A.; Barone, C.M.A.; Alicata, M.L.; Schinelli, R.; Giaccone, P. Use of weaning concentrate in the feeding of suckling kids: Effects on meat quality. *Small Rumin. Res.* **2006**, *66*, 44–50. [CrossRef]

39. Lowe, T.E.; Gregory, N.G.; Fisher, A.D.; Payne, S.R. The effects of temperature elevation and water deprivation on lamb physiology welfare and meat quality. *Aust. J. Agric. Res.* **2002**, *53*, 707–714. [CrossRef]

40. Herold, P.; Snell, H.; Tawfik, E.S. Growth, carcass and meat quality parameters of purebred and crossbred goat kids in extensive pasture. *Arch. Anim. Breed.* **2007**, *50*, 186–196. [CrossRef]

41. Madruga, M.S.; Torres, T.S.; Carvalho, F.F.; Queiroga, R.C.; Narain, N.; Gaarutti, D.; Souza Neto, M.A.; Mattos, C.W.; Costa, R.G. Meat quality of Moxotó and Canindé goats as affected by two levels of feeding. *Meat Sci.* **2008**, *80*, 1019–1023. [CrossRef]

42. Dhanda, J.S.; Taylor, D.G.; Murray, P.J.; McCosker, J.E. The influence of goat genotype on the production of Capretto and Chevon carcasses. 2. Meat quality. *Meat Sci.* **1999**, *52*, 363–367. [CrossRef]

43. Ragni, M.; Tufarelli, V.; Pinto, F.; Giannico, F.; Laudadio, V.; Vicenti, A.; Colonna, M.A. Effect of Dietary Safflower Cake (Carthamus tinctorius L.) on growth performances carcass composition and meat quality traits in Garganica breed kids. *Pak. J. Zool.* **2015**, *47*, 193–199.

44. Caputi Jambrenghi, A.; Colonna, M.A.; Giannico, F.; Coluccia, A.; Crocco, D.; Vonghia, G. Meat quality in suckling kids reared by different production systems. *Progr. Nutr.* **2009**, *11*, 36–46

45. Simela, L.; Webb, E.C.; Frylinck, L. Effect of sex, age, and pre-slaughter conditioning on pH, temperature, tenderness and colour of indigenous South African goats. *S. Afr. J. Anim. Sci.* **2004**, *34*, 208–211.

46. Dhanda, J.S.; Taylor, D.G.; Mccosker, J.E.; Murray, P.J. The influence of goat genotype on the production of capretto and chevon carcass. 3. Dissected carcass composition. *Meat Sci.* **1999**, *52*, 369–374. [CrossRef]

47. Solaiman, S.; Kerth, C.; Willian, K.; Min, B.R.; Shoemaker, C.; Jones, W.; Bransby, D. Growth performance, carcass characteristics and meat quality of boer-cross wether and buck goats grazing marshall ryegrass. *Asian-Australasian J. Anim. Sci.* **2011**, *24*, 351–357. [CrossRef]

48. Guillen-Sans, R.; Guzman-Chozas, M. The Thiobarbituric Acid (TBA) Reaction in Foods: A Review. *Crit. Rev. Food Sci.* **1998**, *38*, 315–350. [CrossRef]

49. Longobardi, F.; Sacco, D.; Casiello, G.; Ventrella, A.; Contessa, A.; Sacco, A. Garganica kid goat meat: Physico-chemical characterization and nutritional impacts. *J. Food Comp. Anal.* **2012**, *28*, 107–113. [CrossRef]

50. Zhou, X.; Yan, Q.; Yang, H.; Ren, A.; Kong, Z.; Tang, S.; Han, X.; He, Z.; Bamikcle, M.A.; Tan, Z. Effects of maternal undernutrition during mid-gestation on the yield, quality and composition of kid meat under an extensive management system. *Animals* **2019**, *9*, 173. [CrossRef]

51. Sommerfeld, M. Trans unsaturated fatty acids in natural products and processed foods. *Prog. Lipid Res.* **1983**, *22*, 221–233. [CrossRef]

52. Dhiman, T.R.; Nam, S.H.; Ure, A.L. Factors affecting conjugated linoleic acid content in milk and meat. *Crit. Rev. Food Sci. Nutr.* **2005**, *45*, 463–482. [CrossRef]

53. Pariza, M.W. Perspective on the safety and effectiveness of conjugated linoleic acid. *Am. J. Clin. Nutr.* **2004**, *79*, 1132S–1136S. [CrossRef]

54. Park, Y. Conjugated linoleic acid (CLA): Good or bad trans fat? *J. Food Comp. Anal.* **2009**, *22S*, S4–S12. [CrossRef]

55. Besserra, F.J.; Madruga, M.S.; Leite, A.M.; da Silva, E.M.C.; Maia, E.L. Effect of age at slaughter on chemical composition of meat of Moxotó goats and their crosses. *Small Rumin. Res.* **2004**, *55*, 177–181. [CrossRef]

56. Simopoulos, A.P. The importance of omega-6/omega-3 fatty acid ratio in cardiovascular disease and other chronic diseases. *Exp. Biol. Med.* **2008**, *233*, 64–688. [CrossRef]

57. Harnack, K.; Andersen, G.; Somoza, V. Quantitation of alpha-linolenic acid elongation to eicosapentaenoic and docosahexaenoic acid as affected by the ratio of n6/n3 fatty acids. *Nutr. Metab.* **2009**, *6*, 8. [CrossRef]

58. Werdi Pratiwi, N.M.; Murray, P.J.; Taylor, D.G.; Zhang, D. Comparison of breed, slaughter weight and castration on fatty acid profiles in the longissimus thoracic muscle from male Boer and Australian feral goats. *Small Rumin. Res.* **2006**, *64*, 94–100. [CrossRef]

59. Banskalieva, V.; Sahlub, T.; Goetschc, A.L. Fatty acid composition of goat muscles and fat depots: A review. *Small Rumin. Res.* **2000**, *37*, 255–268. [CrossRef]

60. Lee, J.H.; Kannan, G.; Eega, K.R.; Kouakou, B.; Getz, W.R. Nutritional and quality characteristics of meat from goats and lambs finished under identical dietary regime. *Small Rumin. Res.* **2008**, *74*, 255–259. [CrossRef]

61. Todaro, M.; Console, A.; Giaccone, P.; Genna, G. Indagine preliminare sulla carne del capretto di razza Girgentana: Rilievi in vita e post mortem (Preliminary investigation of Girgentana meat kid yield: In vita and post mortem data). In Proceedings of the XIV SIPAOC Congress, Vietri sul Mare, Italy, 18–21 October 2000; pp. 233–236.

62. Hedrick, H.B.; Aberle, E.D.; Forrest, J.C.; Judge, M.D.; Merkel, R.A. *Principles of Meat Science*, 3rd ed.; Kendall and Hunt: Dubuque, IA, USA, 1994.
63. Mateo, L.; Delgado, P.; Ortuño, J.; Bañón, S. Maternal grazing on stubble and Mediterranean shrubland improves meat lipid profile in light lambs fed on concentrates. *Anim. Int. J. Anim. Biosci.* **2018**, *12*, 1547–1554. [CrossRef] [PubMed]

Article

The Effect of Using Natural or Biotic Dietary Supplements in Poultry Nutrition on the Effectiveness of Meat Production

Nikola Puvača [1,*], Ivana Brkić [2], Miralem Jahić [1], Svetlana Roljević Nikolić [3], Gordana Radović [4], Dragan Ivanišević [1], Milorad Đokić [5], Dragana Bošković [1], Dragan Ilić [2], Sandra Brkanlić [2] and Radivoj Prodanović [1]

[1] Department of Engineering Management in Biotechnology, Faculty of Economics and Engineering Management in Novi Sad, University Business Academy in Novi Sad, Cvećarska 2, 21000 Novi Sad, Serbia; jahic.miralem@fimek.edu.rs (M.J.); dragan.ivanisevic@fimek.edu.rs (D.I.); pgboskovic@gmail.com (D.B.); rprodanovic@fimek.edu.rs (R.P.)

[2] Department of Business Economics and Finance, Faculty of Economics and Engineering Management in Novi Sad, University Business Academy in Novi Sad, Cvećarska 2, 21000 Novi Sad, Serbia; ivana.j.milosevic@fimek.edu.rs (I.B.); dragan.ilic@fimek.edu.rs (D.I.); sandrab@fimek.edu.rs (S.B.)

[3] Institute of Agricultural Economics Belgrade, Volgina 15, 11060 Belgrade, Serbia; svetlana_r@iep.bg.ac.rs

[4] Dnevnik–Poljoprivrednik AD Novi Sad, Bulevar Oslobođenja 81, 21000 Novi Sad, Serbia; gordana.radovic09@gmail.com

[5] Faculty of Biofarming Bačka Topola, University Megatrend, Bulevar maršala Tolbuhina 8, 11070 Novi Beograd, Serbia; milorad59@yahoo.com

[*] Correspondence: nikola.puvaca@fimek.edu.rs; Tel.: +381-65-219-1284

Received: 11 May 2020; Accepted: 22 May 2020; Published: 26 May 2020

Abstract: The goal of the research was to investigate the effect of dietary natural or biotic additives such as garlic, black pepper, and chili pepper powder in poultry nutrition on sustainable and economic efficiency of this type of production. A total of eight dietary treatments with 1200 broiler chickens of hybrid line Hubbard were formed, with four replicates. During the experimental period, chickens were fed with three period mixtures diets of different average costs: Starter compound mixture two weeks (0.38 €/kg in all treatments), grower compound mixture next three weeks (0.36, 0.38, 0.40, 0.41, 0.46, 0.39, 0.42, and 0.39 €/kg, respectively), and finisher compound mixture for the final week (0.34, 0.36, 0.38, 0.39, 0.44, 0.37, 0.40, and 0.37 €/kg, respectively). The experiment lasted a total of 42 days. Upon finishing the experiment, results have shown statistically significant ($p < 0.05$) differences regarding the European broiler index (EBI) as one of the indicators of economic efficacy. The EBI was lowest in the control treatment (220.4) and significantly higher in experimental treatments (298.6), respectively. In cost, a calculation included the cost of feed and used natural or biotic supplements in chicken nutrition. The findings of the study of economic efficiency revealed that the cost per treatment rises depends on the natural additive used. Economic efficiency analysis showed that the most economical natural additive with the lowest cost is garlic (0.68 €/kg), while the most uneconomical is treatment with black pepper with the highest cost of body weight gain (0.82 €/kg). This higher cost of the gained meat is minimal as a consequence of a much healthier and more nutritious food meant for human use, which often promotes sustainable aspects, compared to conventional and industrialized poultry production.

Keywords: meat; organic; biotic; natural; agriculture; poultry; economic efficiency; costs

---

1. Introduction

Animal nutrition plays an especially important role and feed accounts for up to almost eighty percent of total livestock production costs [1]. Each segment in the process has to have a strong goal on how feed quality can be optimized and feed cost controlled in animal production, and work towards achieving these goals daily [2]. The constant increase of feed ingredient costs, particularly major protein feedstuff such as soybean meal, forces producers to refocus how they will use their resources in feeding to increase economic efficiency to make the extra mile more efficient to convert feed protein into high-quality meat products [3,4]. Another problem comes with the prohibition of antibiotics used in livestock growth promotion due to antimicrobial tolerance development [5], therefore the natural alternative for growth promoting in animal nutrition has been required [6]. When excluded from daily nutrition, antibiotics as growth promoters were not replaced with natural alternatives, which has led to numerous problems in production, such as the increase of feed conversion ratio (FCR) and increased incidence and outbreaks of animal diseases [7–9]. There are now numerous alternatives to antibiotics as growth stimulators [10]. In addition to herbs' and spices' essential role in human nutrition, these natural and biotic additives have been used to improve the health and general wellbeing of animals as well, especially in poultry [11–16]. Biotic additives originate from plants that have been used in animal nutrition to improve performance. The exact mode of action of these biotic additives and their derivates is not yet quite clear. Some studies have shown that this additive contributes to the balance of gastrointestinal microbiota through controlling pathogens [17]. A significant number of bioactive substances present in essential oils leads to a reduction of the *Clostridium* sp. population in the digestive tract and poultry feces [18,19]. In industrial poultry production, many different forms of oregano [20], rosemary [21], sage [22], thyme [23], garlic [24], black pepper [3], and chili [25] have been used separately or in a mixture as feed additives [26–28]. In recent years, the natural additive cost has been decreased mainly because of a large number of competitors present in the market [29]; however, it is expected that the cost of protein and energy feedstuffs will remain high and continue increasing in the long term [30]. There is no question that extra efforts will be necessary to optimize the use of feed and natural additives to promote animal growth in the long term [11]. The European industry is now under tremendous pressure to reduce the usage of antibiotics and is faced with a growing desire from customers for good quality and safe products and meat [3] and eggs [5,31,32] from sustainable and welfare production systems. Such problems include the return of each component in the supply chain in terms of feeding and management while remaining centered on customer demand [33,34]. This complicated condition cannot be fixed easily. The productivity and economic and social dimensions of a method of production are major problems [35,36]. In the European economy, though, the biggest challenge to development is the failure to produce economic outcomes, so the producers need solutions for long-term survival [37]. There is no single change that can optimize dietary performance; only multi-level changes can be the possible solution [6]. It is therefore necessary to re-evaluate and fine-tune established nutritional concepts.

The goal of the research was to investigate the effect of dietary natural or biotic additives such as garlic, black pepper, and chili pepper powder in poultry nutrition on the sustainable and economic efficiency of this type of production.

2. Materials and Methods

Ethical Approval: Biological experiment was performed following the EU legislation and principle of the Three Rs within Directive 2010/63/EU.

Animals and Experimental Design: At the start of the trial, in four iterations, eight treatments on a total of 1200 chickens with 150 one-day-old chickens per treatment were formed. For the nutrition of chicks, three mixtures were used. Starter, grower, and finisher compound feed mixtures, respectively. In the first two weeks, chicks were fed with a starter compound mixture of an average cost of 0.38 €/kg. Over the next three weeks, chickens were fed with grower compound mixtures of average cost of 0.36, 0.38, 0.40, 0.41, 0.46, 0.39, 0.42, and 0.39 €/kg, respectively. Afterwards, in the last week of the experiment, chickens were fed with finisher compound mixtures of average cost of 0.34, 0.36, 0.38, 0.39, 0.44, 0.37, 0.40, and 0.37 €/kg, respectively. The experimental design of the experiment is given in Figure 1. Chicks were provided with fed and water ad libitum during the whole trial period, while microclimate conditions were regularly monitored and maintained following specific hybrid requirements provided by the chickens' producer. Chickens were reared on the floor wheat straw bedding system. To monitoring the productive performance of chickens, body weight and feed consumption were recorded every week.

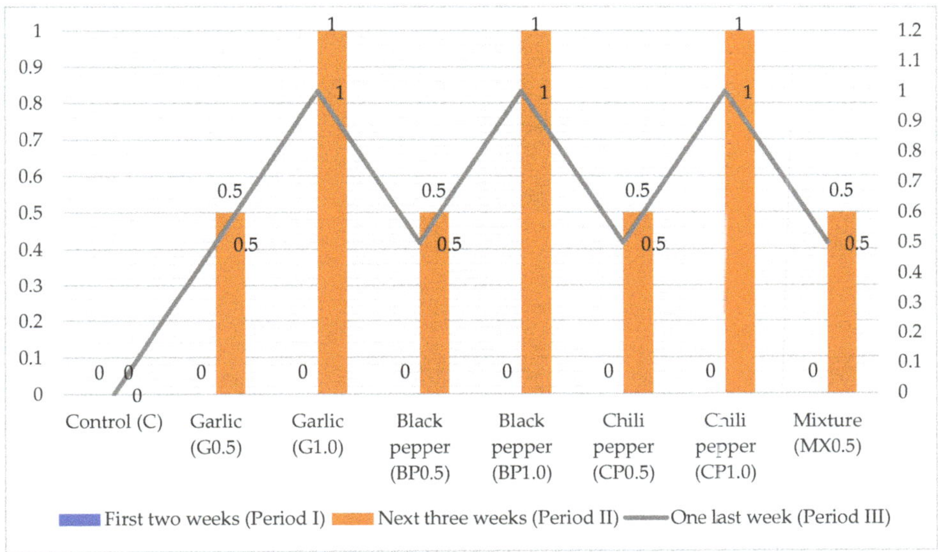

Figure 1. Experimental design with dietary natural or biotic additives, %. Period I—from 1st to 14th day of trial; Period II—from 15th to 35th day of trial; Period III—from 36th to 42nd day of trial; 0.5—concentration of natural feed additive 0.5%; 1.0—concentration of natural feed additive 1%; C—control treatment without additives; G—garlic powder treatment; BP—black pepper powder treatment; CP—chili pepper powder treatment; MX—mixture treatment (G:BP:CP–1:1:1).

Economic and Sustainability Assessment: The economic criteria were focused on the cost of production during the trial. At the end of the 42 days of fattening, the cost of production was determined by the final body mass of the chickens. Cost or divisions of output were calculated based on 1200 broiler chickens, overall. The estimation of economic indicators of production for all experimental treatments by feed periods, as well as for the entire fattening period of chickens, was calculated based on the cost of a kilogram of chicken (CKC), economic efficiency index (EEI), and cost index (CI), and according to the following mathematical Equations (1)–(3):

$$\text{CKC (€/kg)} = \frac{\text{cost of 1kg of feed (€)} \times \text{chicken feed consumption (kg)}}{\text{chicken body weight gain (kg)}} \tag{1}$$

$$\text{EEI (\%)} = \frac{\text{the lowest cost of 1kg of body weight gain (€)}}{\text{cost of 1kg of body weight gain (€)}} \times 100 \tag{2}$$

$$\text{CI (\%)} = \frac{\text{cost of 1kg of body weight gain (€)}}{\text{the lowest cost of 1kg of body weight gain (€)}} \times 100 \tag{3}$$

Statistical Analyses: Data obtained from the trial were analyzed within the statistical software Statistica 13. The data were submitted to analysis of variance (ANOVA) and Fisher's LSD *post hoc* test of significance with Bonferroni correction. The findings were presented as the least square means (LSM) and the standard error (SE_{LSM}). Results were considered statistically significant when $p < 0.05$.

3. Results and Discussion

Based on findings obtained, it was observed that natural additives added to the diet of broiler chickens led to statistically significant ($p < 0.05$) body weight differences (Table 1).

Table 1. Least square means of broiler chickens' productive parameters.

| Treatments | BW, kg | | | FCR, kg/kg | | | | EBI, % | M, % |
| | Period | | | Period | | | | Period | Period |
	I	II	III	I	II	III	Total	Total	Total
C	0.39[a]	1.64[c]	2.08[d]	1.3[a,b]	1.8[ab]	3.0[a]	2.1[a]	220.4[g]	5.1[a]
G0.5	0.39[a]	1.74[b]	2.37[b]	1.4[a,b]	1.7[b]	2.3[b]	1.8[a]	295.1[a,b]	3.2[a,b]
G1.0	0.38[a]	1.74[b]	2.34[b,c]	1.4[a,b]	1.8[b]	2.5[b]	1.9[a]	283.7[c,d]	1.3[b,c]
BP0.5	0.38[a]	1.58[d]	2.08[d]	1.4[a,b]	1.9[a]	2.5[b]	1.9[a]	244.4[f]	1.3[b,c]
BP1.0	0.38[a]	1.50[e]	2.08[d]	1.3[b]	1.9[a,b]	2.3[b]	1.8[a]	260.4[e]	0.6[b,c]
CP0.5	0.38[a]	1.82[a]	2.46[a]	1.4[a]	1.8[ab]	2.4[b]	1.9[a]	298.6[a]	2.6[a,c]
CP1.0	0.38[a]	1.81[a]	2.44[a]	1.4[a,b]	1.8[b]	2.6[b]	1.9[a]	288.6[b,c]	2.6[ac,]
MX0.5	0.38[a]	1.72[b]	2.30[c]	1.4[a,b]	1.8[b]	2.6[b]	1.9[a]	279.6[d]	0.0[c]
SE_{LSM}	3.81	12.02	23.78	0.01	0.05	0.14	0.15	2.77	0.96

[a,b,c,d,e,f] indicated the difference within a row was significant ($p < 0.05$); C—control treatment without additives; G0.5—garlic powder treatment (0.5%); G1.0—garlic powder treatment (1.0%); BP0.5—black pepper powder treatment (0.5%); BP1.0—black pepper powder treatment (1.0%); CP0.5—chili pepper powder treatment (0.5%); CP1.0—chili pepper powder treatment (1.0%); MX—mixture treatment (G:BP:CP–1:1:1); BW–body weight; FCR–feed conversion ratio; EBI–European broiler index; M–mortality; SELSM-standard error of least square means.

In the first two weeks of trial, chickens recorded similar body mass without any statistical significance ($p > 0.05$). At the end of the second trial period, a significant ($p < 0.05$) difference in chicken body mass was noticed. Supplementation with natural feed additives in a powder form of chili pepper in the concentration of 0.5% (CP0.5) and 1% (CP1.0) led to significant differences ($p < 0.05$) in the body mass of chickens compared to control and experimental treatments. At the end of the trial, chickens with the dietary addition of 0.5% of chili pepper recorded the highest body mass of 2.46 kg, followed by treatment with the addition of 1% chili pepper (2.44 kg). Observed differences were statistically significantly ($p < 0.05$) higher when compared with other treatments. The addition of garlic powder in the concentration of 0.5% and 1% led to final body masses (2.37 and 2.34 kg) significantly ($p < 0.05$) higher when compared with final body masses of chickens at treatments C (2.08 kg), BP0.5 (2.08 kg), and BP1.0 (2.08 kg). Similar observations regarding the usage of natural or biotic dietary supplements in broiler chicken nutrition were noticed when used as natural growth promoters [4,38–50]. During the first two weeks of trial, recorded FCR was 1.3 and 1.4 kg/kg. In the first two weeks, chickens achieved uniform FCR, as well the body mass, and entered the second trial period without statistically significant differences. Supplementation of all natural or biotic additives reflected a significant increase ($p < 0.05$) of EBI, when compared to a control treatment C. Control treatment recorded the highest mortality rate and the lowest EBI (Table 1). A mixture of all additives (MX0.5) in the ratio of 1:1:1 in the concentration of 0.5% had the highest rate of survival (100%) during the whole trial period, with recoded EBI of 279.5%, with significant differences ($p < 0.05$) compared to control treatment and both concentrations of black pepper (0.5 and 1%), respectively. The maximum EBI value observed for treatment CP0.5 was 298.6%, and G0.5 was 295.1% without any ($p > 0.05$) discrepancy but significant ($p < 0.05$) differences with other experimental treatments.

A review of used compound mixtures in the trial, costs, feed consumption, and weight gain of chickens are presented in Table 2.

Table 2. Cost of feed mixtures in a trial.

		Feed Cost, €/kg	Feed Consumption, kg	Weight Gain, kg
Starter		0.38	0.44	0.345
Grower	C	0.36	2.25	1.255
Finisher		0.34	1.29	0.432
Starter		0.38	0.48	0.347
Grower	G0.5	0.38	2.30	1.353
Finisher		0.36	1.44	0.628
Starter		0.38	0.48	0.344
Grower	G1.0	0.40	2.43	1.350
Finisher		0.38	1.49	0.598
Starter		0.38	0.47	0.341
Grower	BP0.5	0.41	2.26	1.193
Finisher		0.39	1.24	0.498
Starter		0.38	0.44	0.344
Grower	BP1.0	0.46	2.12	1.117
Finisher		0.44	1.32	0.574
Starter		0.38	0.47	0.342
Grower	CP0.5	0.39	2.57	1.427
Finisher		0.37	1.54	0.630
Starter		0.38	0.47	0.343
Grower	CP1.0	0.42	2.56	1.427
Finisher		0.40	1.63	0.630
Starter		0.38	0.47	0.343
Grower	MX0.5	0.39	2.39	1.332
Finisher		0.37	1.50	0.580

The cost of the feed that was used in the trial with chickens presented in Table 2 was formed based on fattening periods (I, II, and III), respectively. The research of Talpaz et al. [51] shows that the ability to determine the optimum density of poultry compound feed that maximizes feeding margins has a strong economic benefit. The same research highlighted that to assess the optimum feed nutrient content, component costs, meat cost in the market, marketing, and availability of biological efficiency of poultry products must be taken into account. The question can be asked, what is the combined effect of meat and feed cost on optimal dietary energy and protein, and do we need to decrease the concentration of energy and protein in the poultry diet when meat or feed cost are increased? Based on our results, the addition of natural additives has beneficial and stimulating effects in broiler chicken nutrition, so the decrease of certain nutrients in the dietary mixture can be considered. The overall cost of the materials and gain in our experiment is shown in Figures 2–4, and Table 3.

After the calculation of chicken feed consumption during the trial and the costs of feed with the natural or biotic supplementation of each treatment, calculation of EEI and CI was performed as well, respectively. It can be seen that the cost of BWG increased with the addition of natural or biotic supplements to the chicken's daily nutrition.

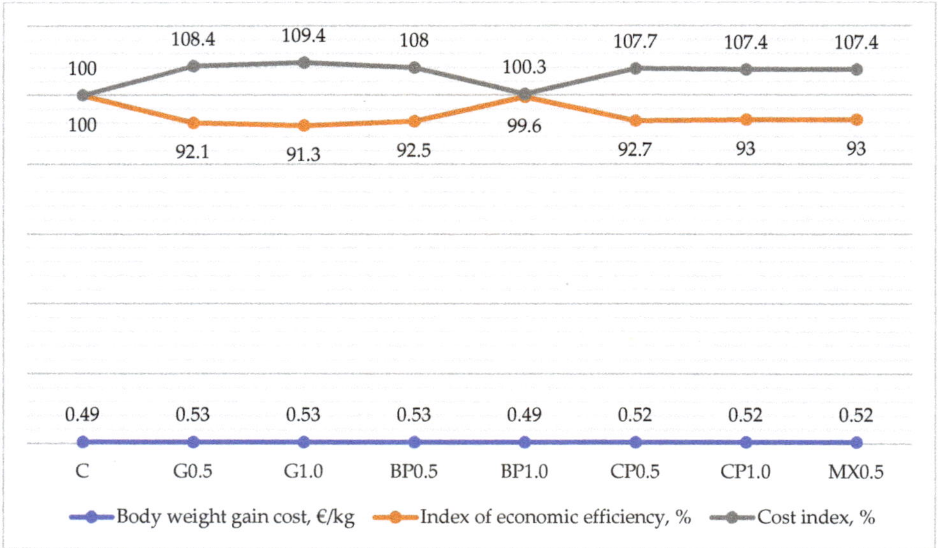

Figure 2. Cost of body weight gain (BWG), economic efficiency index (EEI), and cost index (CI) in period I of the trial.

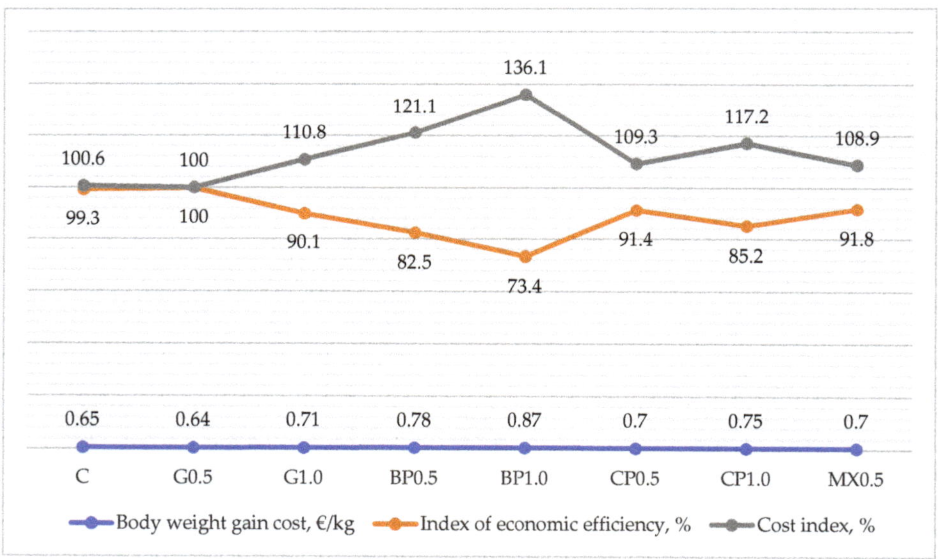

Figure 3. Cost of body weight gain (BWG), economic efficiency index (EEI), and cost index (CI) in period II of the trial.

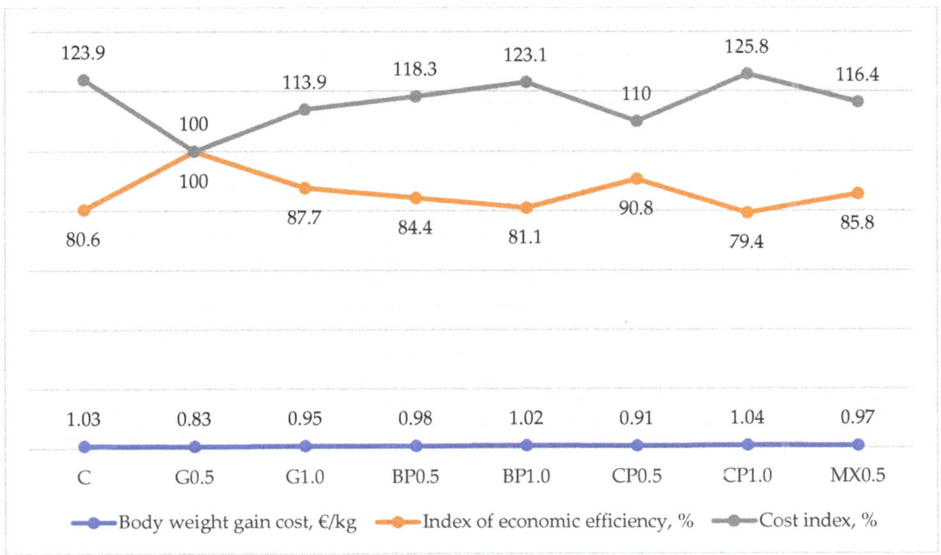

Figure 4. Cost of body weight gain (BWG), economic efficiency index (EEI), and cost index (CI) in period III of the trial.

Table 3. Calculation of cost at the end of the trial.

Production Parameters	Experimental Treatments							
	C	G0.5	G1.0	BP0.5	BP1.0	CP0.5	CP1.0	MX0.5
Average cost of feed, €/kg	0.36	0.37	0.39	0.40	0.43	0.38	0.40	0.38
Average feed consumption, kg	3.98	4.22	4.40	3.97	3.88	4.58	4.66	4.36
Final body weight gain, kg	2.03	2.33	2.30	2.03	2.04	2.40	2.40	2.26
Total cost of body weight gain, €	1.44	1.58	1.70	1.57	1.66	1.75	1.87	1.66
Average cost of body weight gain, €/kg	0.71	0.68	0.74	0.77	0.82	0.73	0.78	0.74
Cost index, %	104.6	100.0	109.2	114.0	120.6	107.5	115.0	108.8

As in the first trial period, the second trial period showed similar results. The cost of BWG was the lowest when garlic powder in the concentration of 0.5% (G0.5) was supplemented to broiler chickens' diet. Supplementation of the black pepper natural additive only at the double concentration of 1% (BP1.0) showed the highest cost of BWG in the trial, respectively. Identical reflections of CI were observed. Some investigations have shown the reduction in feed cost/kg of live BWG of chickens when 1% of mint leaves was supplemented to a basic diet [52]. Also, the results of other authors regarding the calculation of total cost, revenues, and net profit in the economic study indicated that a mixture of garlic and ginger powder in broiler chicken nutrition showed the highest profitability when compared to diets without biotic supplements.

From the results presented in Table 3, it can be seen that the highest cost of feed per total trial period was recorded in treatment with the addition of 1% black pepper (0.43 €/kg), while the lowest cost of feed was recorded in the control treatment (0.36 €/kg). The highest recorded average feed consumption for the entire trial period was recorded in chickens with the addition of chili pepper in the concentration of 1% (4.66 kg), which indicates the highest stimulative effect on feed consumption. At the end of the trial period, the highest final BWG was recorded in chickens on treatments with both concentrations of chili pepper (2.40 kg), but with the highest total costs of BWG (1.75 and 1.87 €), respectively. The lowest average cost of BWG for the entire trial period was recorded in the treatment with the addition of dietary garlic powder in the concentration of 0.5%, with the lowest CI of 100%, respectively.

4. Conclusions

Supplementation of broiler chickens' daily diet with dietary natural or biotic additives has numerous beneficial effects. Most of these effects are reflected in final products with added value which are functional, healthy, and without any residual antibiotics. Nevertheless, the question can be asked whether that kind of production provides economic benefits to producers, and whether it sustainable.

The results of our research have shown and confirmed that the production of safe and healthy meat can satisfy the demands of the market, which increasingly seeks sustainable and organically produced products, with economic benefit to producers of such types of products.

The addition of garlic powder in a lower concentration for four weeks of chickens fattening is the economical solution and the answer to the previously asked question. The increase in production cost with the addition of garlic powder to the chicken's diet compared to chickens without natural additive supplementation is negligible, keeping in mind the product has improved functional properties intended for human consumption. This kind of obtained product is sustainable and part of organic productions.

Limitations of this kind of study lie in not knowing the exact mechanism and mode of actions of natural and biotic additives, so further research in the field of their mode of action and influence on animals, as well on the quality and safety of obtained products for human consumption, is more than necessary soon.

Author Contributions: Conceptualization, N.P. and I.B.; methodology, N.P.; software, R.P.; validation, M.J., S.R.N., and G.R.; formal analysis, D.B.; investigation, N.P.; resources, D.I. (Dragan Ilić); data curation, D.I. (Dragan Ivanišević); writing—original draft preparation, N.P.; writing—review and editing, I.B.; visualization, M.Đ.; supervision, M.J.; project administration, N.P. and S.B.; funding acquisition, N.P. All authors have read and agreed to the published version of the manuscript.

Funding: This research was funded by the Ministry for Education, Science and Technological Development of the Republic of Serbia, grant number III46012.

Acknowledgments: This research was funded by the Ministry for Education, Science and Technological Development of the Republic of Serbia.

Conflicts of Interest: The authors declare no conflict of interest.

References

1. Mottet, A.; de Haan, C.; Falcucci, A.; Tempio, G.; Opio, C.; Gerber, P. Livestock: On our plates or eating at our table? A new analysis of the feed/food debate. *Glob. Food Secur.* **2017**, *14*, 1–8. [CrossRef]
2. Tullo, E.; Finzi, A.; Guarino, M. Review: Environmental impact of livestock farming and Precision Livestock Farming as a mitigation strategy. *Sci. Total Environ.* **2019**, *650*, 2751–2760. [CrossRef]
3. Puvača, N.; Kostadinović, L.; Popović, S.; Lević, J.; Ljubojević, D.; Tufarelli, V.; Jovanović, R.; Tasić, T.; Ikonić, P.; Lukač, D. Proximate composition, cholesterol concentration and lipid oxidation of meat from chickens fed dietary spice addition (*Allium sativum, Piper nigrum, Capsicum annuum*). *Anim. Prod. Sci.* **2016**, *56*, 1920–1927. [CrossRef]
4. Popović, S.; Kostadinović, L.; Đuragić, O.; Aćimović, M.; Čabarkapa, I.; Puvača, N.; Pelić Ljubojević, D. Influence of medicinal plants mixtures (*Artemisia absinthium, Thymus vulgaris, Menthae piperitae* and *Thymus serpyllum*) in broilers nutrition on biochemical blood status. *J. Agron. Technol. Eng. Manag.* **2018**, *1*, 91–98.
5. Puvača, N.; Lika, E.; Tufarelli, V.; Bursić, V.; Ljubojević Pelić, D.; Nikolova, N.; Petrović, A.; Prodanović, R.; Vuković, G.; Lević, J.; et al. Influence of different tetracycline antimicrobial therapy of Mycoplasma (*Mycoplasma synoviae*) in laying hens compared to tea tree essential oil on table egg quality and antibiotics residues. *Foods* **2020**, *9*, 612. [CrossRef] [PubMed]
6. Puvača, N.; Stanaćev, V.; Glamočić, D.; Lević, J.; Perić, L.; Stanaćev, V.; Milić, D. Beneficial effects of phytoadditives in broiler nutrition. *Worlds Poult. Sci. J.* **2013**, *69*, 27–34. [CrossRef]

7. Alabi, O.; Malik, A.; Ng'ambi, J.; Obaje, P.; Ojo, B. Effect of aqueous moringa oleifera (Lam) leaf extracts on growth performance and carcass characteristics of hubbard broiler chicken. *Rev. Bras. Cienc. Avic.* **2017**, *19*, 273–280. [CrossRef]

8. Anadón, A.; Martínez-Larrañaga, M.R.; Ares, I.; Martínez, M.A. Regulatory aspects for the drugs and chemicals used in food-producing animals in the European union. In *Veterinary Toxicology*; Elsevier: New York, NY, USA, 2018; pp. 103–131.

9. Kostadinović, L.; Lević, J. Effects of phytoadditives in poultry and pigs diseases. *J. Agron. Technol. Eng. Manag.* **2018**, *1*, 1–7.

10. Markowiak, P.; Śliżewska, K. The role of probiotics, prebiotics and synbiotics in animal nutrition. *Gut. Pathog.* **2018**, *10*, 21. [CrossRef]

11. Puvača, N.; Lika, E.; Cocoli, S.; Shtylla Kika, T.; Bursić, V.; Vuković, G.; Tomaš Simin, M.; Petrović, A.; Cara, M. Use of tea tree essential oil (*Melaleuca alternifolia*) in laying hen's nutrition on performance and egg fatty acid profile as a promising sustainable organic agricultural tool. *Sustainability* **2020**, *12*, 3420. [CrossRef]

12. Čabarkapa, I.; Puvača, N.; Popović, S.; Čolović, D.; Kostadinović, L.; Tatham, E.K.; Lević, J. Aromatic plants and their extracts pharmacokinetics and *in vitro/in vivo* mechanisms of action. In *Feed Additives*; Elsevier: London, UK, 2020; pp. 75–88.

13. Popović, S.; Puvača, N.; Kostadinović, L.; Džinić, N.; Bošnjak, J.; Vasiljević, M.; Djuragic, O. Effects of dietary essential oils on productive performance, blood lipid profile, enzyme activity and immunological response of broiler chickens. *Eur. Poult. Sci.* **2016**, *80*. [CrossRef]

14. Giannenas, I.; Koidis, A.; Botsoglou, E.; Dotas, V.; Mitsopoulos, I.; Florou-Paneri, P.; Nikolakakis, I. Hen performance and egg quality as affected by dietary oregano essential oil and alpha-tocopheryl acetate supplementation. *Int. J. Poult. Sci.* **2005**, *4*, 449–454. [CrossRef]

15. Giannenas, I.; Bonos, E.; Skoufos, I.; Tzora, A.; Stylianaki, I.; Lazari, D.; Tsinas, A.; Christaki, E.; Florou-Paneri, P. Effect of herbal feed additives on performance parameters, intestinal microbiota, intestinal morphology and meat lipid oxidation of broiler chickens. *Br. Poult. Sci.* **2018**, *59*, 545–553. [CrossRef] [PubMed]

16. Giannenas, I.; Doukas, D.; Karamoutsios, A.; Tzora, A.; Bonos, E.; Skoufos, I.; Tsinas, A.; Christaki, E.; Tontis, D.; Florou-Paneri, P. Effects of *Enterococcus faecium*, mannan oligosaccharide, benzoic acid and their mixture on growth performance, intestinal microbiota, intestinal morphology and blood lymphocyte subpopulations of fattening pigs. *Anim. Feed Sci. Technol.* **2016**, *220*, 159–167. [CrossRef]

17. Hardeland, R.; Pandi-Perumal, S. Melatonin, a potent agent in antioxidative defense: Actions as a natural food constituent, gastrointestinal factor, drug and prodrug. *Nutr. Metab.* **2005**, *2*, 22. [CrossRef]

18. Zeng, Z.; Zhang, S.; Wang, H.; Piao, X. Essential oil and aromatic plants as feed additives in non-ruminant nutrition: A review. *J. Anim. Sci. Biotechnol.* **2015**, *6*, 7. [CrossRef]

19. Dahiya, J.P.; Wilkie, D.C.; Van Kessel, A.G.; Drew, M.D. Potential strategies for controlling necrotic enteritis in broiler chickens in post-antibiotic era. *Anim. Feed Sci. Technol.* **2006**, *129*, 60–88. [CrossRef]

20. Giannenas, I.; Bonos, E.; Christaki, E.; Florou-Paneri, P. Oregano: A Feed Additive with Functional Properties. In *Therapeutic Foods*; Elsevier: London, UK, 2018; pp. 179–208.

21. Yesilbag, D.; Eren, M.; Agel, H.; Kovanlikaya, A.; Balci, F. Effects of dietary rosemary, rosemary volatile oil and vitamin E on broiler performance, meat quality and serum SOD activity. *Br. Poult. Sci.* **2011**, *52*, 472–482. [CrossRef]

22. Saki, A.A.; Aliarabi, H.; Hosseini Siyar, S.A.; Salari, J.; Hashemi, M. Effect of a phytogenic feed additive on performance, ovarian morphology, serum lipid parameters and egg sensory quality in laying hen. *Vet. Res. Forum.* **2014**, *5*, 287–293.

23. Khan, S.A.; McLean, M.K.; Slater, M.R. Concentrated tea tree oil toxicosis in dogs and cats: 443 cases (2002–2012). *J. Am. Vet. Med. Assoc.* **2014**, *244*, 95–99. [CrossRef]

24. Puvača, N.; Ljubojević, D.; Kostadinović, L.; Lukač, D.; Lević, J.; Popović, S.; Đuragić, O. Spices and herbs in broilers nutrition: Effects of garlic (*Allium sativum* L.) on broiler chicken production. *Worlds Poult. Sci. J.* **2015**, *71*, 533–538. [CrossRef]

25. Puvača, N.; Ljubojević, D.; Kostadinović, L.; Lević, J.; Nikolova, N.; Miščević, B.; Könyves, T.; Lukač, D.; Popović, S. Spices and herbs in broilers nutrition: Hot red pepper (*Capsicum annuum* L.) and its mode of action. *Worlds Poult. Sci. J.* **2015**, *71*, 683–688. [CrossRef]

26. Tripathi, D.; Kumar, A.; Mondal, B.C.; Joyti, P. Effect of dietary supplementation of Ajwain, hot red pepper and black pepper on the performance, haemato-biochemical and carcass characteristics of Japanese quail. *Indian J. Poult. Sci.* **2017**, *52*, 288. [CrossRef]

27. Brenes, A.; Roura, E. Essential oils in poultry nutrition: Main effects and modes of action. *Anim. Feed Sci. Technol.* **2010**, *158*, 1–14. [CrossRef]

28. Franz, C.; Baser, K.; Windisch, W. Essential oils and aromatic plants in animal feeding - a European perspective. A review. *Flavour Fragr. J.* **2010**, *25*, 327–340. [CrossRef]

29. Kleer, R.; Piller, F.T. Local manufacturing and structural shifts in competition: Market dynamics of additive manufacturing. *Int. J. Prod. Econ.* **2019**, *216*, 23–34. [CrossRef]

30. Pelletier, N.; Klinger, D.H.; Sims, N.A.; Yoshioka, J.-R.; Kittinger, J.N. Nutritional Attributes, Substitutability, Scalability, and Environmental Intensity of an Illustrative Subset of Current and Future Protein Sources for Aquaculture Feeds: Joint Consideration of Potential Synergies and Trade-offs. *Environ. Sci. Technol.* **2018**, *52*, 5532–5544. [CrossRef]

31. Spasevski, N.; Puvača, N.; Pezo, L.; Tasić, T.; Vukmirović, Đ.; Banjac, V.; Čolović, R.; Rakita, S.; Kokić, B.; Džinić, N. Optimisation of egg yolk colour using natural colourants. *Eur. Poult. Sci.* **2018**, *82*. [CrossRef]

32. Spasevski, N.; Dragojlović, D.; Čolović, D.; Vidosavljević, S.; Peulić, T.; Rakita, S.; Kokić, B. Influence of dietary carrot and paprika on egg physical characteristics and yolk color. *Food Feed Res.* **2018**, *45*, 59–66. [CrossRef]

33. Ashour, E.A.; El-Kholy, M.S.; Alagawany, M.; Abd El-Hack, M.E.; Mohamed, L.A.; Taha, A.E.; El Sheikh, A.I.; Laudadio, V.; Tufarelli, V. Effect of Dietary Supplementation with Moringa oleifera Leaves and/or Seeds Powder on Production, Egg Characteristics, Hatchability and Blood Chemistry of Laying Japanese Quails. *Sustainability* **2020**, *12*, 2463. [CrossRef]

34. Tufarelli, V.; Casalino, E.; D'Alessandro, A.G.; Laudadio, V. Dietary phenolic compounds: Biochemistry, metabolism and significance in animal and human health. *Curr. Drug Metab.* **2018**, *18*. [CrossRef] [PubMed]

35. Kiel, D.; Müller, J.M.; Arnold, C.; Voigt, K.I. Sustainable industrial value creation: Benefits and challenges of industry 4.0. *Int. J. Innov. Manag.* **2017**, *21*, 1740015. [CrossRef]

36. Rauch, E.; Dallasega, P.; Matt, D.T. Sustainable production in emerging markets through Distributed Manufacturing Systems (DMS). *J. Clean. Prod.* **2016**, *135*, 127–138. [CrossRef]

37. Ghisellini, P.; Cialani, C.; Ulgiati, S. A review on circular economy: The expected transition to a balanced interplay of environmental and economic systems. *J. Clean. Prod.* **2016**, *114*, 11–32. [CrossRef]

38. Marcinčák, S.; Klempová, T.; Bartkovský, M.; Marcinčáková, D.; Zdolec, N.; Popelka, P.; Mačanga, J.; Čertík, M. Effect of fungal solid-state fermented product in broiler chicken nutrition on quality and safety of produced breast meat. *BioMed Res. Int.* **2018**, 1–8. [CrossRef]

39. Farahat, M.; Abdallah, F.; Abdel-Hamid, T.; Hernandez-Santana, A. Effect of supplementing broiler chicken diets with green tea extract on the growth performance, lipid profile, antioxidant status and immune response. *Br. Poult. Sci.* **2016**, 1–9. [CrossRef]

40. Christaki, E.; Giannenas, I.; Bonos, E.; Florou-Paneri, P. Innovative uses of aromatic plants as natural supplements in nutrition. In *Feed Additives*; Elsevier: London, UK, 2020; pp. 19–34.

41. Lipiński, K.; Mazur, M.; Antoszkiewicz, Z.; Purwin, C. Polyphenols in Monogastric Nutrition–A Review. *Ann. Anim. Sci.* **2017**, *17*, 41–58. [CrossRef]

42. Bonos, E.; Kasapidou, E.; Kargopoulos, A.; Karampampas, A.; Christaki, E.; Florou-Paneri, P.; Nikolakakis, I. Spirulina as a functional ingredient in broiler chicken diets. *S. Afr. J. Anim. Sci.* **2016**, *46*, 94. [CrossRef]

43. Adaszyńska-Skwirzyńska, M.; Szczerbińska, D. Use of essential oils in broiler chicken production—A review. *Ann. Anim. Sci.* **2017**, *17*, 317–335. [CrossRef]

44. Zhai, H.; Liu, H.; Wang, S.; Wu, J.; Kluenter, A.-M. Potential of essential oils for poultry and pigs. *Anim. Nutr.* **2018**, *4*, 179–186. [CrossRef]

45. Ogbuewu, I.P.; Okoro, V.M.; Mbajiorgu, E.F.; Mbajiorgu, C.A. Beneficial Effects of Garlic in Livestock and Poultry Nutrition: A Review. *Agric. Res.* **2019**, *8*, 411–426. [CrossRef]

46. Ogbuewu, I.P.; Okoro, V.M.; Mbajiorgu, C.A. Meta-analysis of the influence of phytobiotic (pepper) supplementation in broiler chicken performance. *Trop. Anim. Health Prod.* **2020**, *52*, 17–30. [CrossRef] [PubMed]

47. Tashla, T.; Puvača, N.; Pelić, D.L.; Prodanović, R.; Bošković, J.; Ivanišević, D.; Jahić, M.; Mahmoud, O.; Giannenas, I.; Lević, J. Dietary medicinal plants enhance the chemical composition and quality of broiler chicken meat. *J. Hellenic Vet. Med. Soc.* **2019**, *70*, 1823–1832. [CrossRef]

48. Puvaca, N.; Kostadinovic, L.; Djuragic, O.; Ljubojevic, D.; Miscevic, B.; Könyves, T.; Popovic, S.; Levic, J.; Nikolova, N. Influence of herbal drugs in broiler chicken nutrition on primal carcass cuts quality assessments. *Food Feed Res.* **2016**, *43*, 43–49. [CrossRef]

49. Sidhu, N.S.; Singh, U.; Sethi, A.P.S. Growth Performance and Nutrient Utilisation as Affected by Black Pepper or Jaggery Supplementation and Feed Restriction in Broiler Chicken. *Indian J. Anim. Nutr.* **2018**, *35*, 191. [CrossRef]

50. Devi, P.C.; Samanta, A.K.; Das, B.; Kalita, G.; Behera, P.S.; Barman, S. Effect of Plant Extracts and Essential Oil Blend as Alternatives to Antibiotic Growth Promoters on Growth Performance, Nutrient Utilization and Carcass Characteristics of Broiler Chicken. *Indian J. Anim. Nutr.* **2018**, *35*, 421. [CrossRef]

51. Talpaz, H.; Cohen, M.; Fancher, B.; Halley, J. Applying complex models to poultry production in the future—Economics and biology. *Poult. Sci.* **2013**, *92*, 2541–2549. [CrossRef]

52. Khurshid, A.; Banday, M.T.; Adil, S.; Untoo, M.; Afzal, I. Mint Leaves (Mentha piperita) as Herbal Dietary Supplement: Effect on Performance and Economics of Broiler Chicken Production. *Pakistan J. Nutr.* **2016**, *15*, 810–815. [CrossRef]

 *sustainability*

Article

Assessment of Water Buffalo Milk and Traditional Milk Products in a Sustainable Production System

Zsolt Becskei[1,*], Mila Savić[1], Dragan Ćirković[2], Mladen Rašeta[3], Nikola Puvača[4], Marija Pajić[5], Sonja Đorđević[6] and Snežana Paskaš[7]

[1] Department of Animal Breeding and Genetics, Faculty of Veterinary Medicine, University of Belgrade, 18 Bulevar Oslobođenja, 11000 Belgrade, Serbia; milas@vet.bg.ac.rs
[2] Department of Chemical and Technological Sciences, State University of Novi Pazar, bb Vuka Karadžića, 36300 Novi Pazar, Serbia; dr.draganpcela@gmail.com
[3] Institute of Meat Hygiene and Technology, 13 Kaćanskog, 11000 Belgrade, Serbia; mladen.raseta@inmes.rs
[4] Department of Engineering Management in Biotechnology, Faculty of Economics and Engineering Management in Novi Sad, University Business Academy in Novi Sad, 2 Cvećarska, 21000 Novi Sad, Serbia; nikola.puvaca@fimek.edu.rs
[5] Department of Veterinary Medicine, Faculty of Agriculture, University of Novi Sad, 8 Trg Dositeja Obradovića, 21000 Novi Sad, Serbia; marija_vet@polj.uns.ac.rs
[6] Department of Animal Science, Faculty of Agriculture, University of Priština, bb Kopaonička, 38219 Lešak, Serbia; sonja.samardzic@pr.ac.rs
[7] Department of Animal Science, Faculty of Agriculture, University of Novi Sad, 8 Trg Dositeja Obradovića, 21000 Novi Sad, Serbia; snpaskas@gmail.com
* Correspondence: bzolt@vet.bg.ac.rs; Tel.: +381-65-991-1101

Received: 22 July 2020; Accepted: 14 August 2020; Published: 15 August 2020

Abstract: Water buffalo (*Bubalus bubalis*) conservation in Serbia is under an in situ program, but additional efforts are needed to ensure the development of this animal's genetic resources biodiversity. This research aims to describe challenges and possible strategies for sustainable water buffalo milk production. In this study, the physicochemical characteristics of buffalo milk and buffalo dairy products (cheese, butter, and kajmak) were determined. Furthermore, amino and fatty acids composition and the related health lipid indices (atherogenic and thrombogenic) were assessed. The findings support the fact that buffalo milk is a reliable source of high-quality nutrients (dry matter: 16.10%, fat: 6.02%, protein: 4.61%). Leucine, lysine, and valine content were found to be high in buffalo milk and cheese. A substantial quantity of non-essential glutamic and aspartic amino acids was observed in milk, as well as glutamic acid and tyrosine in cheese. It was established that milk protein of buffalo cheese had a favorable proportion of essential and non-essential amino acids (61.76%/38.24%). The results revealed significant differences ($p < 0.05$) in fatty acid profiles among the three dairy products for saturated short-chain, n-3, and n-6 fatty acids. Conversely, no significant difference ($p < 0.05$) was observed in monounsaturated fatty acids content. Kajmak showed the most favorable anti-atherogenic and anti-thrombogenic properties due to lower saturated and higher polyunsaturated fatty acid content. These results confirmed that buffalo milk could be successfully used in producing high-quality traditional dairy products with added value and beneficial characteristics from the aspect of a healthy diet. Furthermore, it could actively contribute to the promotion of sustainable production of buffaloes and strengthen the agricultural production of rural areas and their heritage.

Keywords: sustainable livestock production; added value; buffalo milk; dairy products; chemical composition; nutritional properties

---

1. Introduction

Water buffaloes *(Bubalus bubalis)* are distributed worldwide, with a total population of about 180 million. The majority of populations are located in Asia (around 97%). It is considered that the water buffalo is an efficient converter of poor quality forages into high-quality products [1]. Buffaloes play an essential role in the heritage of the local rural populations and their economy in South-Western Serbia, where these animals are raised under extensive husbandry and sustainable principles. In previous decades, control of productivity and selection for milk production were insufficient. As a result, milk production is lower in comparison with selected populations of buffalo in some Asian countries [1]. The mainly conventional livestock production in Serbia and the popularization of imported, highly productive breeds of livestock contributed to a significant reduction in the number of autochthonous breed populations, and the genetic diversity is eroded. Buffalo population, as well, showed a tendency to decrease. Since the 2000s, a program for the in situ conservation of buffalo population started in Serbia [2], but still, according to the Domestic Animal Diversity Information System of the Food and Agriculture Organization of the United Nations (FAODAD-IS), the estimated population of buffalo is low, and their local risk status is "endangered", with a decreasing trend.

As reported by the same database, there are about 30 bulls and 724 cows of buffalo evidenced in the breeding program in Serbia [3]. Furthermore, the main challenges facing the buffalo breeding sector is low profitability. On the other hand, in many countries, an increasing demand for buffalo dairy products has been noticed, and accordingly, most farms are undertaking a progressive intensification of rearing techniques [4,5]. Besides, non-cattle dairy systems, whether intensive or not, could be a sustainable alternative and potentially a significant added value for producers and the dairy sector [6]. Sheep and goats are essential species in non-cattle milk production. Thus, the polymorphism of selected genes affecting milk composition in small ruminants is well documented [7–9], but efforts are needed in the genetic assessment of buffalo populations.

Buffaloes of the Mediterranean region are phenotypically different from country to country because of the different environment and management practices. Still, the fundamental objective of breeding buffaloes in Europe and the Near East is milk production [5]. Dairy farms have valuable input to a resilient and sustainable food system. They possess the ability to provide not just basic nutrition, but also healthy nutrition. For developed and developing countries, dairy has the potential to reinvigorate rural economies, providing sustainable livelihoods for smallholder farmers and a resilient source of economic growth [10]. According to Borghese and Moioli [5], there is a need to more deeply study dairy products made from buffalo milk in the Mediterranean area as well the variability of their technologies because they represent an important part of global biodiversity.

Buffalo milk, like cow milk, can be used for the manufacture of a wide variety of dairy products [11]. It has long been valued for its fortunate chemical composition which determines its nutritive properties and suitability for the production of traditional as well as industrial dairy products [12]. Compared with cow milk, it contains less water and more fat and is especially useful for the production of fat-based dairy products such as butter and kajmak. Kajmak is an artisanal Serbian dairy product mainly produced from cow's milk, and based on its physicochemical characteristics may be placed between cheese and butter [13]. The specific fatty acid composition of buffalo milk contributes to the unique features of its products. Butter, for example, made from buffalo milk due to its higher level of saturated fatty acids is much harder than that made from bovine milk [11].

Moreover, the high variability in triglyceride and fatty acid composition of buffalo milk makes it possible to separate milk fat into various fractions based on its melting characteristics [14]. Higher proportions of high-melting triglycerides in buffalo milk contribute to its higher density [15] and make it suitable for cheese making. Buffalo cheese is highly prized for its pure white appearance and smooth texture [11]. In Italy, fresh and Pasta Fialata cheeses, especially Mozzarella and Borelli cheeses, are traditionally prepared from buffalo milk. In Balkan countries, several types of white brined cheese and pickled cheese are made from buffalo milk [16]. Because of differences between buffalo and other ruminant milk in compositional and physicochemical properties, some modifications of processing technology and equipment designed for different kinds of milk are necessary for buffalo milk processing [12].

The advantages of buffalo milk compared to cow milk are not only in terms of physicochemical, compositional, and sensory attributes, but also in its nutritional and health aspects [17]. Buffalo milk proteins are complete proteins of high biological value, and they contain all the essential amino acids in the proportions required by the human body [15]. Furthermore, buffalo milk and its products could represent a good source of favorable conjugated linoleic acid (CLA) in human nutrition [18]. In recent decades, many studies have been devoted to improving milk fatty acid (FA) composition by increasing the amount of FA with beneficial effects on human health and with more appropriate technological properties [19]. Findings of FA and triglyceride profiles of low melting fractions of milk fat have suggested that the therapeutic value of low melting fractions of buffalo milk fat was higher than native milk fat [14].

There are insufficient research data on the production potential and dairy product quality of water buffalo in sustainable farming systems in Serbia. In particular, the chemical characteristics of buffalo milk and its dairy products in Serbia have been studied in a limited fashion [2]. In this paper, the composition and physicochemical properties of major constituents of buffalo milk are presented, with particular emphasis on amino and fatty acids. Furthermore, buffalo dairy products are analysed, compared, and discussed in terms of their nutritional, technological, and health aspects to find the added value of the traditional products from sustainable production.

2. Materials and Methods

2.1. Sample Collection

Samples of bulk, raw buffalo milk and dairy products (cheese, butter, and kajmak) were taken in early spring 2017, from small households of the Rashka district (42.989400, 20.332684) in South-Western Serbia. Ten households participated in the study, with an average water buffaloes herd size of 5 milking buffaloes. All the herds were reared in a sustainable, extensive management system, grazing on pastures and meadows. The milking animals each received 0.5 kg oats daily as supplementary feed during milking.

2.2. Milk Chemical Analysis

The standard chemical composition and milk urea (MU) analysis of bulk, raw buffalo milk samples were conducted using a MilkoScan FT + analyzer for routine compositional raw milk analysis employing Fourier Transform Infrared "FTIR". The MilkoScanTM FT + techniques comply with: ISO 9622/IDF 141:2013 [20] and the AOAC official method 972.16 [21]. The energy value of milk was estimated, according to Popović-Vranješ [22].

2.3. Buffalo Milk Dairy Products Physicochemical Analysis

Samples of buffalo dairy products (cheese, butter, and kajmak) were analyzed for chemical composition (dry matter, fat, protein, and moisture) following standard methods of the Association of Official Analytical Chemists [23]. The dry matter was determined with a standard method of measuring weight loss after drying (AOAC 926.08-1927). Protein content was measured using the Kjeldahl-Van Slyke method for the determination of total N (AOAC 2001.14) with a Kjeltec Auto Analyzer (Model 2400, Tecator, Hoganas, Sweden) and converted by a multiplication factor of 6.38. The fat content was quantified according to the Van Gulik method [24]. Fat in dry matter (FDM, %) was calculated as fat/(100 − moisture)×100, moisture in nonfat substance (MFFB, %) was calculated as moisture/(100 − fat) × 100. The active acidity of the dairy product was measured applying a pH meter (WTW, tip inoLab pH 720), and titratable acidity was expressed in terms of Soxhlet-Henkel/°SH value.

2.4. Amino Acid Profile

Amino acid profile of buffalo milk and dairy products was performed following the protocol of Henderson [25]. The prepared samples were analyzed for the content of amino acids after acid hydrolysis using the HPLC method (High-Performance Liquid Chromatograph Chromaster, Zorbax Eclipse-AAA column (4.6 × 250, 5 µm) with DAD-3000 (diode array detector).

2.5. Fatty Acid Profile

To determine the milk fatty acids, a gas chromatograph (GC) with flame ioniyation detector (FID) was used. Determination of the free fatty acids (FA) in the buffalo milk and dairy products was done after methylation with boron trifluoride in methanol, using Shimadzu's gas chromatograph with FID on InterCap WAX (length 30 m, inner diameter 0.25 mm, film thickness 0.25 µm) column (AOAC 996.06). Atherogenic (AI) and thrombogenic indices (TI) were calculated according to Ulbricht and Southgate [26].

2.6. Statistical Analysis

Statistical analysis of the data was carried out using GraphPad Prism v6 (GraphPad, San Diego, CA, USA) software. All data was normally distributed (Shapiro-Wilk normality test $p > 0.05$). The homogeneity of variances was analyzed using Levene's test and then the parametric model of analysis of variances (ANOVA) followed by Tukey's multiple comparison test, and t test. Significance was established at $p < 0.05$.

3. Results

3.1. Milk Chemical Composition

The chemical composition of the buffalo milk samples is presented in Table 1. Results indicated that fat and ash were the most inconsistent components, whereas lactose and MU content showed minimum variation. As a result of a higher degree of fat variation, wide range of fat/protein ratio (CV: 28.36%) was also found. The analysis confirmed that buffalo milk is rich in fat, protein, and lactose. Therefore, the estimated energy value of buffalo milk varied from 81.81 to 106.62 (kcal/100 mL).

Table 1. Chemical composition of buffalo milk.

Parameters	$\bar{x} \pm$ SD	CV (%)	Min–Max
Dry matter (%)	16.60 ± 1.12	6.75	14.91–17.80
Solid nonfat (%)	10.48 ± 0.52	4.96	9.97–11.51
Fat (%)	6.02 ± 1.27	21.10	4.26–7.16
Protein (%)	4.61 ± 0.52	11.28	4.12–5.62
Lactose (%)	5.36 ± 0.16	2.98	5.10–5.63
Ash (%)	0.60 ± 0.14	23.33	0.37–0.80
MU (mg/dL)	35.80 ± 0.88	2.46	33.90–36.70
F/P	1.34 ± 0.38	28.36	0.77–1.72
Energy value (kcal/100 mL)	96.90 ± 10.52	10.86	81.81–106.62
Milk density (g/cm^3)	1.037 ± 0.002	0.19	1.036–1.042

$\bar{x}$—arithmetic mean; SD—standard deviation; CV (%)—coefficient of variation; minimal X (min) and maximal X (max) values of variables; F/P—fat/protein ratio; MU—milk urea.

3.2. Milk Amino Acid Composition

Table 2 summarizes the composition of amino acids in buffalo milk. The results of the amino acid analysis revealed that leucine (0.41%) was the major amino acid, while lysine (0.33%) was second among all essential amino acids. A substantial quantity of other branched amino acids (valine and isoleucine) was also observed. Of the non-essential amino acids (NEAA), glutamic and aspartic acids were found at a high level. Methionine is the first limiting amino acid. Higher variations of NEAA, especially of tyrosine content (CV: 36.93%), were also observed.

Table 2. Amino acids composition of buffalo milk (%).

Essential Amino Acid (EAA)	$\bar{x} \pm SD$	CV (%)	Min–Max	Non-Essential Amino Acid (NEAA)	$\bar{x} \pm SD$	CV (%)	Min–Max
Histidine	0.15 ± 0.02	13.33	0.13–0.17	Asparatic acid	0.32 ± 0.02	5.71	0.29–0.33
Threonine	0.14 ± 0.01	7.14	0.12–0.15	Glutamic acid	1.01 ± 0.01	0.99	0.99–1.02
Valine	0.26 ± 0.01	3.85	0.25–0.27	Serine	0.09 ± 0.02	22.22	0.07–0.11
Methionine	0.05 ± 0.004	8.00	0.05–0.06	Glycine	0.08 ± 0.003	3.75	0.08–0.09
Phenylalanine	0.21 ± 0.01	4.76	0.19–0.23	Arginine	0.11 ± 0.01	9.09	0.10–0.12
Isoleucine	0.24 ± 0.01	4.17	0.23–0.25	Alanine	0.14 ± 0.01	7.41	0.13–0.16
Leucine	0.41 ± 0.01	2.44	0.39–0.42	Tyrosine	0.11 ± 0.04	36.36	0.08–0.16
Lysine	0.33 ± 0.02	6.06	0.30–0.35				
Total	1.79			Total	1.86		

$\bar{x}$—arithmetic mean; SD-standard deviation; CV (%)—coefficient of variation; minimal X (min) and maximal X (max) values of variables.

3.3. Milk Fatty Acid Composition

Saturated (SFA), monounsaturated (MUFA), and polyunsaturated fatty acid (PUFA) content of buffalo milk was found to be 69.04%, 28.46%, and 2.50%, respectively (Table 3). The values found for fatty acid composition also showed that investigated samples contained a considerable amount of saturated long chain fatty acid (SLCFA) content (C16:0, C18:0, C20:0; in total: 45.99 g/100 g). The AI and TI amounted to 2.72 and 3.02, respectively, whereas the n-6/n-3 ratio was 2.42. Results also showed that SLCFAs were the most stable in buffalo milk and varied least (CV: 9.41%), while SSCFA and SMCFA possessed much higher coefficients of variation (CV: 33.05% and 27.48%, respectively).

Table 3. Fatty acid composition of buffalo milk (g/100 g).

Parameters	$\bar{x} \pm SD$	CV (%)	Min–Max
MUFA	25.43 ± 4.38	17.22	19.01–32.09
SSCFA	2.36 ± 0.78	33.05	1.21–3.31
SMCFA	13.32 ± 3.66	27.48	8.04–17.52
SLCFA	45.99 ± 4.33	9.41	40.98–53.58
SFA	61.68 ± 5.81	9.42	52.23–70.35
PUFA	2.23 ± 0.36	16.14	1.7–2.67
PUFA/SFA	0.04 ± 0.01	25.00	0.02–0.05
Total omega-6 fatty acids (n-6)	1.78 ± 0.22	12.36	1.37–2.06
Total omega-3 fatty acids (n-3)	0.77 ± 0.20	25.97	0.45–1.00
n-6/n-3	2.42 ± 0.59	24.38	1.85–3.69
Atherogenicity index (AI)	2.72 ± 1.05	38.60	1.56–4.71
Thrombogenicity index (TI)	3.02 ± 0.76	25.24	2.03–4.47

$\bar{x}$—arithmetic mean; SD-standard deviation; CV (%)—coefficient of variation; minimal X (min) and maximal X (max) values of variables; MUFA—monounsaturated fatty acids (C18:1); SSCFA—saturated short-chain fatty acids (C 4:0, C8:0); SMCFA—saturated medium-chain fatty acids (C10:0, C12:0, C14:0); SLCFA—saturated long-chain fatty acids (C16:0, C18:0, C20:0); SFA—total saturated fatty acids; PUFA—polyunsaturated fatty acids (C18:2, C18:3).

3.4. Chemical Composition of Buffalo Dairy Products

Mean composition values of cheese, butter, and kajmak produced from buffalo milk are shown in Table 4. According to Serbian regulations [27], buffalo cheese is classified as a semi-fat and semi-hard cheese, due to FDM and MFFB values of 43.5% and 60.1%, respectively. Moreover, kajmak samples met the Serbian requirements for mature kajmak, and contained 77.1% of FDM, on average, and appropriate values for pH and moisture (6.6 and 34.0%, respectively). On the other hand, butter did not meet Serbian regulations for solids non-fat (SNF) content (>2%), but fat and moisture contents were within the acceptable range. The results of chemical composition also showed that kajmak represents a special and unique dairy product. It contains less fat compared with butter, as well as less protein than cheese. Cheese is the most varied regarding fat content (CV: 14.51%), while SNF content was the most inconsistent component in butter and kajmak (CV: 63.79% and 25.39%, respectively).

Table 4. Chemical composition of buffalo dairy products.

Parameters	Cheese		Butter		Kajmak	
	$\bar{x} \pm$ SD	CV (%)	$\bar{x} \pm$ SD	CV (%)	$\bar{x} \pm$ SD	CV (%)
Moisture (%)	46.0 ± 1.20	2.61	14.2 ± 0.28	1.97	34.0 ± 1.30	3.82
Dry matter (%)	54.0 ± 1.20	2.22	85.8 ± 0.28	0.33	66.0 ± 1.30	1.97
Protein (%)	27.0 ± 1.62	6.0	0.9 ± 0.08	8.89	5.7 ± 0.34	5.96
Fat (%)	23.5 ± 3.41	14.51	83.0 ± 1.67	2.01	50.8 ± 2.56	5.04
FDM (%)	43.5 ± 5.22	12.0	96.7± 3.93	4.06	77.1 ± 5.39	6.99
MFFB (%)	60.1 ± 1.11	1.85	84.0 ± 8.12	9.67	69.3 ± 6.26	9.03
SNF	30.5 ± 2.21	7.24	2.9 ± 1.85	63.79	15.2 ± 3.86	25.39
pH	5.4 ± 0.12	2.22	6.7 ± 0.08	1.19	6.6 ± 0.12	1.82
°SH	60.0 ± 2.78	4.63	14.0 ± 0.44	3.14	14.0 ± 1.20	8.57

$\bar{x}$—arithmetic mean; SD—standard deviation; CV (%)—coefficient of variation; FDM—Fat on a dry matter basis; MFFB-Moisture on a fat-free basis; SNF—solids non-fat.

3.5. Amino Acid Composition of Buffalo Dairy Products

Buffalo cheese represents a good source of amino acids, and, as expected, it is significantly different ($p < 0.05$) compared with kajmak and butter (Table 5.). In particular, buffalo cheese is abundant in essential amino acids: lysine, leucine, and valine (4.42%, 3.90%, and 2.45%, respectively). The primary non-essential amino acids were glutamic acid (5.09%) and tyrosine (1.97%). Results also revealed that EAA/NEAA ratio in buffalo cheese is very favorable (61.76%/38.24%) and together with a high amount of branched amino acids (leucine, valine, and isoleucine) confirmed the tremendous nutritional value of buffalo cheese. Kajmak and butter, on the other hand, as high-fat dairy products contain a low amount of protein and, consequently, amino acids. Compared with butter, kajmak contained a significantly higher ($p < 0.05$) amount of threonine, leucine, and arginine. However, minor differences were noticed in the isoleucine and glutamic acid content of butter and kajmak, whereas histidine, phenylalanine, valine, and threonine were not detected.

Table 5. Amino acid composition of buffalo dairy products.

Parameters	Cheese	Butter	Kajmak
	$\bar{x} \pm$ SD	$\bar{x} \pm$ SD	$\bar{x} \pm$ SD
Essential Amino Acid (EAA)			
Histidine	0.75 ± 0.05	ND	ND
Threonine	1.13 ± 0.01 [a]	0.02 ± 0.004 [c]	0.13 ± 0.02 [b]
Valine	2.45 ± 0.17 [a]	ND	0.16 ± 0.04 [b]
Methionine	1.68 ± 0.08 [a]	ND	0.10 ± 0.01 [b]
Phenylalanine	1.81 ± 0.04 [a]	ND	0.12 ± 001 [b]
Isoleucine	2.03 ± 0.03 [a]	0.04 ± 0.004 [c]	0.13 ± 0.02 [b]
Leucine	3.90 ± 0.09 [a]	0.06 ± 0.007 [c]	0.26 ± 0.03 [b]
Lysine	4.42 ± 0.18 [a]	0.09 ± 0.07 [b]	0.26 ± 0.03 [b]
Total	18.17	0.21	1.16
Non-Essential Amino Acid (NEAA)			
Aspartic acid	1.46 ± 0.14 [a]	ND	0.23 ± 0.05 [b]
Glutamic acid	5.09 ± 0.20 [a]	0.19 ± 0.01 [c]	0.74 ± 0.05 [b]
Serine	0.99 ± 0.16 [a]	0.04 ± 0.01 [b]	0.17 ± 0.01 [b]
Glycine	0.42 ± 0.01 [a]	0.01 ± 0.002 [c]	0.07 ± 0.01 [b]
Arginine	0.90 ± 0.01 [a]	0.01 ± 0.001 [c]	0.20 ± 0.01 [b]
Alanine	0.42 ± 0.01	ND	ND
Tyrosine	1.97 ± 0.20 [a]	ND	0.11 ± 0.01 [b]
Total	11.25	0.25	1.52

$\bar{x}$—arithmetic mean; SD—standard deviation; ND—Non-detectable; [a,b,c]—different letters indicate statistical differences ($p < 0.05$), while same letters indicate no statistical difference.

3.6. Fatty Acid Composition of Buffalo Dairy Products

The fatty acid content of buffalo dairy products is presented in Table 6. Significant differences ($p < 0.05$) were established between butter and kajmak regarding all groups of saturated fatty acids. In particular, kajmak contained more SSCFA and SMCFA, whereas butter was more abundant in SLCFA. MUFA content did not show a significant difference ($p < 0.05$) between all investigated dairy products. The data showed that the nutritionally most beneficial fatty acids are found in kajmak. Thus, the fatty acid composition grouped as, SFA, MUFA and PUFA were: in cheese: 66.85%, 29.06%, 4.09%; in butter: 70.82%, 26.28%, 2.09%; and in kajmak: 65.30%, 30.52%, 4.18%, respectively. Concerning health-related factors (AI), significant differences ($p < 0.05$) were established between butter and kajmak. Furthermore, it was found that kajmak had the most favourable AI and TI (1.84 and 1.82, respectively). Furthermore, kajmak possessed the highest content of n-3 fatty acids, while cheese contained the highest amount of n-6 fatty acids and had the highest n-6/n-3 ratio.

Table 6. Fatty acid composition of buffalo dairy products (g/100 g).

Parameters	Cheese	Butter	Kajmak
	$\bar{x} \pm$ **SD**	$\bar{x} \pm$ **SD**	$\bar{x} \pm$ **SD**
MUFA	26.85 ± 2.82 [a]	24.06 ± 0.17 [a]	26.89 ± 1.56 [a]
SSCFA	2.53 ± 0.02 [c]	3.5 ± 0.22 [b]	5.41 ± 0.23 [a]
SMCFA	13.35 ± 0.51 [b]	14.88 ± 1.08 [b]	21.48 ± 0.65 [a]
SLCFA	45.90 ± 2.62 [b]	46.46 ± 2.48 [b]	30.09 ± 0.73 [a]
SFA	61.78 ± 3.15 [a]	64.84 ± 3.78 [a]	57.52 ± 1.51 [a]
PUFA	3.78 ± 0.21 [a]	2.66 ± 0.10 [b]	3.68 ± 0.31 [a]
PUFA/SFA	0.06 ± 0.01 [a]	0.04 ± 0.001 [b]	0.06 ± 0.005 [a]
Total omega-6 fatty acids (n-6)	3.77 ± 0.21 [a]	1.92 ± 0.05 [c]	2.65 ± 0.26 [b]
Total omega-3 fatty acids (n-3)	0.31 ± 0.03 [c]	0.74 ± 0.05 [b]	1.30 ± 0.08 [a]
n-6/n-3	12.16 ± 0.54 [a]	2.60 ± 0.11 [b]	2.04 ± 0.07 [b]
Atherogenicity index (AI)	2.31± 0.34 [b]	3.01 ± 0.16 [a]	1.84 ± 0.06 [b]
Thrombogenicity index (TI)	2.99 ± 0.37 [a]	3.12 ± 0.11 [a]	1.82 ± 0.08 [b]

$\bar{x}$—arithmetic mean; SD-standard deviation. MUFA—monounsaturated fatty acids (C18:1); SSCFA—saturated short-chain fatty acids (C 4:0, C8:0); SMCFA—saturated medium-chain fatty acids (C10:0, C12:0, C14:0); SLCFA—saturated long-chain fatty acids (C16:0, C18:0, C20:0); SFA—total saturated fatty acids; PUFA—polyunsaturated fatty acids (C18:2, C18:3); [a,b,c]—different letters indicate statistical differences ($p < 0.05$), while same letters indicate no statistical difference.

4. Discussion

Dairy possesses an important role in the livestock sector and global healthy diets. It makes a significant contribution towards meeting the challenges of nutritional security, sustainability, and reduction in diseases related to poor quality diet [10]. On the other hand, it is very important to consider the economic sustainability of primary production in such a chain, as profitability is a precondition for the food chain to be relevant and sustainable [28]. To achieve economic sustainability dairy farms and producers should meet consumers demand. In particular, for the realization of sustainable non-bovine breeding in Serbia, it is necessary to take into account the habits and preferences of consumers to ensure market supply chains [29]. The study conducted by Paskaš et al. [30] examined the behaviour of consumers in Serbia and confirmed that healthiness and nutritional benefits are amongst the most important factors for consuming non-bovine milk and dairy products.

Even within the same species, milk composition can vary, due to genetics, physiology, nutrition, and environment [31]. According to Borghese and Moioli [5], buffalo milk possesses a higher fat (6–9.5%) and protein (4–5%) content than cow milk. The present study has shown the mean content of fat (6.02%) to fall within this range, but it was considerably less compared with the results of Tiezzi et al. [32] and Liotta et al. [33] (7.56% and 8.78%, respectively). On the other hand, the obtained mean protein content (4.61%) was in agreement with those researches (4.69% and 4.61%, respectively).

The energy value of milk is closely related to the concentration of certain compounds in dry matter, especially the amount of fat [34]. Among the common dairy species, the energy density of buffalo milk is remarkably high [11]. In the research of Mane and Chatli [16], the energetic value of buffalo milk amounted to 117 kcal/100 g while presented results showed a lower value (96.90 kcal/100 mL). Furthermore, our findings have shown that lactose is the second major constituent of buffalo milk, with a minimum and maximum of 5.10% and 5.63%. On the contrary, Gantner et al. [31] reported that the content of lactose in buffalo milk is ≤5%. In general, buffalo milk is a richer source of lactose than cow, goat, sheep, and camel milk. From the health aspect, lactose could be a good source of energy for body functions, particularly for the brain and hormonal regulation [12]. Many factors influence milk quality, and for traditional cheese production milk fat, protein, and lactose content, as well as their ratio, are essential [35]. The optimal fat/protein ratio in buffalo milk is 2:1 [17], while the presented results showed an approximate value of 1.34.

Milk urea measured at the group level can be used to monitor the efficiency of nitrogen utilization in commercial buffalo herds and as an indicator of the protein feeding situation in buffaloes [36]. Observed MU levels in this study ranged between 33.90–36.70 mg/dL and were lower compared with the findings of Di Francia et al. [36] (40.8 mg/dL). Furthermore, Liotta et al. [33] recorded higher MU value in intensive (40.68 mg/dL) than in semi-intensive buffalo herds (37.50 mg/dL). In contrast, Santillo et al. [37] concluded that levels of MU were not different among buffalo groups that were fed low protein diets and diets with flaxseed supplementation.

Buffalo milk can be considered a good source of essential amino acids, and lysine content was found to be the highest, followed by valine and isoleucine. These findings correspond to Ren et al. [38]. Studies also have shown that buffalo milk contained methionine in traces, which is comparable with the findings of Barlowska et al. [34].

The majority of fatty acids in ruminant milk are saturated [19]. Obtained results especially showed a considerable amount of SLCFA (45.99 g/100 g). Buffalo milk contains almost three times more C14:0 (myristic) acid and two times less C16:0 (palmitic) acid than cow, sheep, or goat milk [34]. Regarding fatty acids content, the previous research works are very diverse. Bustamante et al. [39] reported a higher amount of MUFA (33.1% vs. 28.46%) and, at the same time, lower content of n-6 acids than the current study (0.96% vs. 1.78%). In contrast, Pegolo et al. [40] compared with our results, found in Mediterranean buffalo milk less amount of n-3 (0.46%) but approximately the same content of n-6 fatty acids (1.78%), whereas findings of Gantner et al. [31] were in line with our data. Comparative studies indicate that dairy breeds with a high milk fat content often have a less desirable milk fat composition, have higher levels of saturated and hypercholesterolaemic fatty acids, and a lower proportion of PUFA than breeds with a lower milk yield or fat content [41]. The importance of FA profile is reflected in terms of the technological quality of raw milk. It possesses the potential to contribute to the production of dairy products with added value. From a usability point of view, higher proportions of unsaturated fatty acids are preferred (as they increase the spreading ability of butter), but their milk fat content could also cause lower stability, oxidation, and possible sensory changes [19]. Nutritional effects of milk and dairy products are highly influenced by fatty acid profile [42], as well, and FAs possess diverse implications on human health [19]. Considering the two health-related indices, the ratio of essential FA (n-6/n-3), and AI estimated values were higher than those reported by Varricchio et al. [43] (2.42 and 2.71 vs. 2.15 and 2.61, respectively). According to Claeys et al. [44], the n-6/n-3 ratio varies from 1.0 to 4.0 in ruminant milk. Consuming large amounts of n-3 fatty acids is beneficial and contributes to a lower risk of coronary diseases and some types of cancer [34].

Considerable diversity traits of buffalo milk result in various directions of milk utilization. Each buffalo dairy product varies in composition, and the present study discussed their different production, composition, and potential health properties. Thanks to its specific characteristics, buffalo milk needs to be treated differently in the cheese-making process. It is less suitable for the manufacture of hard varieties of cheese, such as Cheddar cheese. Some problems that can occur are slow development of acidity, higher curd tension, shorter renneting period, hard, dry, crumbly, corky body and texture, and slower proteolysis [16]. The buffering capacity, pH, and viscosity of buffalo milk are higher than those of cow milk, while the fermentation and ripening process of buffalo milk is generally slower [11]. Some procedural improvements are necessary for hard cheese production from buffalo milk, such as higher heat treatment, Mucor rennet, and a greater amount of *Streptococcus thermophilus* and *Lactobacterium bulgaricus* cultures [16]. Results of the present study showed the good quality of semi-hard buffalo cheese. Still, certain varieties including Mozzarella and white pickled Domiati cheese possess superior quality when made from buffalo milk [15]. One of the main problems in the production of buffalo semi-hard and hard cheese is the difficulty of converting milk into naturally ripened cheese [45].

Buffalo milk is also very desirable for the manufacture of fat-rich dairy products due to its higher fat content, the bigger size of the globule, and the higher proportion of solid fat [17]. Industrial butter is produced by the churning of cream, often after pasteurization. The homemade product is obtained simply by churning acidified milk. A peculiarity of buffalo butter is the colour, which is much whiter than cows' milk butter, due to the lack of carotenoids [5]. Furthermore, buffalo milk produces butter with a significantly higher yield, and displays more stability than that from cow cream, due to the more solid fat and a slower rate of fat hydrolysis in the former cream [17]. The results of this study confirm the results of [46,47] for butter chemical composition (pH, fat, and protein). Kajmak is a homemade product, it is produced based on the traditional manufacturing procedure [13]. Kajmak is a specific product, characterized by high-fat content, the presence of proteins, and its peculiar ripening process [48]. During the ripening process, which lasts 3–4 weeks, kajmak partly loses the continuity of its moisture phase, while limited fat phase continuity appears, and a specific flavor is developed [13]. Therefore, the ripened kajmak made from cow milk published by Pudja et al. [13] had values of chemical composition in the following ranges: moisture: 15–35%, fat: 50–70%, FDM: 75–90% and proteins: 2–7%, and do not differ from values in the present study.

The present investigation revealed that in cheese, lysine was the major amino acid but also a substantial quantity of branched-amino acids (valine, isoleucine, and leucine) was observed. Branched-amino acids promote protein synthesis in muscle cells, and they are metabolized to generate energy in muscles rather than in the liver [49]. Glutamic acid content was found to be the highest of non-essential amino acids in cheese. Similarly, glutamic acid was the predominant non-essential amino acid in butter made from cow milk, but in a much smaller amount. At the same time, leucine and isoleucine were the main essential amino acids [50]. On the contrary, leucine was not detected in the present study in buffalo butter.

Dairy products, such as butter, ghee, and cream, have been considered as basic nutrient-dense foods that can deliver many energy-rich nutrients [46]. It was reported by Popović et al. [47] that traditional Serbian dairy products, cheese, and kajmak, made from bovine milk, possessed a high content of SFAs (70%), mainly palmitic acid. In contrast, the present study showed lower SFA contents in buffalo milk cheese and kajmak and a more desirable fatty acid pattern. However, saturated fatty acids were the predominant fraction in buffalo butterfat (70.49%). Kwak et al. [50] reported that butter has double the concentration of saturated fat in comparison with that in cream and other dairy products. Therefore, dairy products such as butter very often have been criticized for their unfavorable FA profile [19]. However, stearic acid could be beneficial to health and contribute to the level of low-density lipoprotein (LDL) in the blood [50]. Compared with butter, cheese possesses a more favourable FA profile: in particular, higher proportions of MUFA and PUFA.

The manufacturing process could also affect the nutritional and health characteristics of the cheeses [51]. Even though ripened cheeses contain more fat on a wet basis, their fatty acid profile is more desirable. In particular, the Blu cheese shows a healthier fatty acid profile than Mozzarella and the cheese-making process, and ripening contribute to reducing atherogenic (C12:0 and C14:0) and increasing some beneficial fatty acids (C18:3 n-3, cis-9, trans-11 conjugated linoleic acid) [52]. Furthermore, it is supposed that milk fat with high AI and TI values may be more likely to contribute to the development of atherosclerosis or coronary thrombosis in humans [53]. The value of AI in milk and dairy products is around 2, whereas AI = 1.5 is considered as low, and 2.5 is high [54]. Accordingly, our dairy products, kajmak, and cheese appeared to exhibit stronger anti-atherogenic activities compared with milk and butter. The thrombogenic indices take into account the relationship between the pro-thrombogenic (saturated) and anti-thrombogenic fatty acids (unsaturated) [26]. Despite being a high-fat product, kajmak also showed a low TI value (1.82).

5. Conclusions

The buffalo production under sustainable principles and extensive husbandry plays an essential role in the heritage of local rural populations and their economy in South-West Serbia. Buffalo milk is used for manufacturing of traditional milk products, such as cheese, butter, and kajmak. The results of the chemical composition of buffalo milk showed that it could be utilized for making a variety of good quality traditional dairy products with added value. Consequently, they are important for the dairy industry and especially useful for traditional artisanal production. The analysed parameters of milk and dairy products showed considerable diversity, and nutritional value varies from one product to the other. In particular, the nutritional content of ripened kajmak and cheese was the most favourable. Primarily, kajmak represents a valuable source of essential fatty acids, while cheese was abundant in important branched amino acids (leucine, valine, and isoleucine). These results can contribute to the promotion of the value-added buffalo dairy products from sustainable production systems. Although buffaloes are successfully bred with traditional methods on pastures, to achieve more consistent and sustainable milk production, some husbandry improvements should be required. In particular, more attention should be paid to the uniform quality of the raw milk, selective breeding, and productivity of animals. Thus, these would improve the farm economy and the quality of derived products, which could actively contribute to rural development of the region and more effective conservation of buffalo genetic resources. Furthermore, to enhance buffalo production in more sustainably and holistically, producers should be encouraged to access appropriate market information and take into account more sustainable initiatives in the dairy industry, such as the nutritional and health benefits of dairy products.

Author Contributions: Conceptualization, Z.B., S.P. and M.S.; methodology, S.P.; software, S.P.; validation, S.P.; formal analysis, Z.B., M.R., N.P. and M.P.; investigation and resources, Z.B., M.S., D.Ć., S.Đ. and M.P.; data curation, S.P.; writing—original draft preparation, S.P., Z.B.; writing—review and editing, S.P., Z.B., N.P.; visualization, M.S.; supervision, Z.B.; project administration, M.S. All authors have read and agreed to the published version of the manuscript.

Funding: This research was funded by the Ministry of Education, Science and Technological Development of the Republic of Serbia, grant number TR31085.

Acknowledgments: This study was supported by the Ministry of Education, Science and Technological Development of the Republic of Serbia.

Conflicts of Interest: The authors declare no conflict of interest. The funders had no role in the design of the study; in the collection, analyses, or interpretation of data; in the writing of the manuscript, or in the decision to publish the results.

References

1. Deb, G.; Nahar, T.; Duran, P.; Pressice, G. Safe and sustainable traditional production: The water buffalo in Asia. *Front. Environ. Sci.* **2016**, *38*, 1–7. [CrossRef]
2. Perišić, P.; Bogdanović, V.; Mekić, C.; Ružić-Muslić, D.; Stanojević, D.; Popovac, M.; Stepić, S. The importance of buffalo in milk production and buffalo population in Serbia. *Biotechnol. Anim. Husb.* **2016**, *32*, 255–263.
3. FAODAD-IS. Domestic Animal Diversity-Information System. Percentage of Data Fields Completed by Country. Available online: http://www.fao.org/dad-is/browse-by-country-and-species/en/ (accessed on 10 June 2020).
4. Sabia, E.; Napolitano, F.; Claps, S.; Braghieri, A.; Piazzolla, N.; Pacelli, C. Feeding, Nutrition and Sustainability in Dairy Enterprises: The Case of Mediterranean Buffaloes (*Bubalusbubalis*). In *The Sustainability of Agro-Food and Natural Resources Systems in the Mediterranean Basin*; Vastola, A., Ed.; Springer: Cham, Switzerland, 2015; pp. 57–64.

5.	Borghese, A.; Moioli, B. Buffalo: Mediterranean Region. In *Elsevier Public HealthEmergency Collection, 2016, Update of Buffalo Mediterranean Region. Encyclopedia of Dairy Sciences*, 2nd ed.; Borghese, A., Moioli, B., Eds.; Elsevier Ltd.: San Diego, CA, USA, 2011; pp. 780–784.

6.	Faye, B.; Konuspayeva, G. The sustainability challenge to the dairy sector-The growing importance of non-cattle milk production worldwide. *Int. Dairy J.* **2012**, *24*, 50–56. [CrossRef]

7.	Kusza, S.; Sziszkosz, N.; Nagy, K.; Masala, A.; Kukovics, S.; Jávor, A. Preliminary result of genetic polymorphism of ß-lactoglobulin gene and phylogenetic study of ten Balkan and Central European indigenous sheep breeds. *Acta Biochim. Pol.* **2015**, *62*, 109–112. [CrossRef] [PubMed]

8.	Kusza, S.Z.; Cziszter, L.T.; Ilie, D.E.; Sauer, M.; Padeanu, I.; Gavojdian, D. Kompetitive Allele Specific PCR (KASP™) genotyping of 48 polymorphisms at different caprine loci in the Saanen and French Alpine goat breeds and their association with milk composition. *PeerJ* **2018**, *6*, e4416. [CrossRef] [PubMed]

9.	Kusza, S.; Ilie, D.E.; Sauer, M.; Nagy, K.; Atanasiu, T.S.; Gavojdian, D. Study of LGB gene polymorphisms of small ruminants reared in Eastern Europe. *Czech J. Anim. Sci.* **2018**, *63*, 152–159. [CrossRef]

10.	Food Agriculture Organization (FAO). *8 Sustainable Dairy Goals Achieving the Sustainable Development Goals-the Role of the Dairy Sector Committee on World Food Security Making a Difference in Food Security and Nutrition*; FAO: Rome, Italy, October 2016.

11.	Guo, M.; Hendricks, G. Improving buffalo milk. Improving the Safety and Quality of Milk. Improving Quality in Milk Products. *Woodhead Publ. Ser. Food Sci. Technol. Nutr.* **2010**, *2*, 402–416.

12.	Ahmad, S.; Anjum, F.M.; Huma, N.; Sameen, A.; Zahoor, T. Composition and physicochemical characteristic of buffalo milk with particular emphasis on lipids, proteins, minerals, enzymes and vitamins. *J. Anim. Plant Sci.* **2013**, *23*, 62–74.

13.	Pudja, P.; Djerovski, J.; Radovanović, M. An autochthonous Serbian product-Kajmak Characteristics and production procedures. *Dairy Sci. Technol.* **2008**, *88*, 163–172. [CrossRef]

14.	Khan, T.I.; Nadeem, M.; Imran, M.; Asif, M.; Khan, K.M.; Din, A.; Ullah, R. Triglyceride, fatty acid profile and antioxidant characteristics of low melting point fractions of buffalo milkfat. *Lipids Health Dis.* **2019**, *18*, 1–11. [CrossRef]

15.	Khedkar, C.D.; Kalyankar, S.D.; Deosarkar, S.S. Buffalo Milk. In *The Encyclopedia of Food and Health*; Caballero, B., Finglas, P., Toldrá, F., Eds.; Academic Press: Oxford, UK, 2016; pp. 522–528.

16.	Park, W.Y.; Haenlein, W.F.G. Buffalo milk: Utilization for Dairy Products. Handbook of Milk of Non-Bovine Mammals. In *Technology & Engineering*; Wiley & Sons: New York, NY, USA, 2008; pp. 195–274.

17.	Mane, B.G.; Chatli, M.K. Buffalo Milk: Saviour of Farmers and Consumers for Livelihood and Providing Nutrition. *Agric. Rural Dev.* **2015**, *2*, 5–11.

18.	Khanal, R.C.; Olson, K.C. Factors Affecting Conjugated Linoleic Acid (CLA) Content in Milk, Meat, and Egg: A Review. *Pak. J. Nutr.* **2004**, *2*, 82–98.

19.	Hanuš, O.; Samková, E.; Křížová, L.; Hasonová, L.; Kala, R. Role of Fatty Acids in Milk Fat and the Influence of Selected Factors on Their Variability-A Review. *Molecules* **2018**, *23*, 1636.

20.	ISO (International Organization for Standardization) 9622:2013 (IDF 141:2013). Available online: https://www.iso.org/standard/56874.html (accessed on 14 August 2020).

21.	AOAC (Association of Official Agricultural Chemists) 972.16. Fat, Lactose, Protein and Solids in Milk, 1972, (app. 1996). Available online: http://www.aoacofficialmethod.org/index.php?main_page=product_info&products_id=38 (accessed on 14 August 2020).

22.	Popović Vranješ, A. *Special Cheesemaking*; The University of NoviSad, Faculty of Agriculture, Department of Animal Science, Komazec Press: Inđija, Serbia, 2015; pp. 641–660.

23.	AOAC. *Official Methods of Analysis of the AOAC*, 17th ed.; 33: Methods 926.08, 2000.18, 933.05, 2001.14, 935.42, 920.124; AOAC International: Gaithersburg, MD, USA, 2005; Volume II.

24.	IDF. *International IDF Standard 222:2008*; Cheese and processed cheese products; Determination of Fat Content-Van Gulik method; IDF: Brussels, Belgium, 2008.

25.	Henderson, W.J.; Ricker, D.R.; Bidlingmeyer, A.B.; Woodward, C. Rapid, accurate, sensitive, and reproducible HPLC analysis of amino acid analysis using Zorbax Eclipse-AAA Columns and the Agilent 1100 HPLC. *Agil. Technol.* **2000**, *1100*, 1–10.

26.	Ulbricht, T.L.V.; Southgate, D.A.T. Coronary heart disease: Seven dietary factors. *Lancet* **1991**, *338*, 985–992. [CrossRef]

27. Serbian Regulation. *Ordinance on the Quality of Dairy Products and Starter Cultures*; No. 34/2014; Official Gazetteof the Republic of Serbia: Belgrade, Serbia, 2014.

28. Hessle, A.; Bertilsson, J.; Stenberg, B.; Kumm, I.K.; Sonesson, U. Combining environmentally and economically sustainable dairy and beef production in Sweden. *Agric. Syst.* **2017**, *156*, 105–114. [CrossRef]

29. Petrović, M.P.; Petrovic, V.C.; Muslic, D.R.; Maksimovic, N.; Cekic, B.; Ilic, Z.; Kurcubic, V. Strategy for Sustainable Development and Utilization of Sheep and Goat Resources in Serbia. *KnE Life Sci.* **2017**, *2*, 11–21. [CrossRef]

30. Paskaš, S.; Miočinović, J.; Lopičić-Vasić, T.; Mugoša, I.; Pajić, M.; Becskei, Z. Consumer attitudes towards goat milk and milk products in Vojvodina. *Mjekarstvo* **2020**, *70*, 171–183. [CrossRef]

31. Gantner, V.; Mijić, P.; Baban, M.; Škrtić, Z.; Turalija, A. The overall and fat composition of milk of various species. *Mljekarstvo* **2015**, *65*, 223–231. [CrossRef]

32. Tiezzi, F.; Cecchinato, A.; De Marchi, M.; Gallo, L.; Bittante, G. Characterization of buffalo productionof the northeast of Italy. *Ital. J. Anim. Sci.* **2009**, *8*, 160–162. [CrossRef]

33. Liotta, L.; Chiofalo, V.; Lo Presti, V.; Vassallo, A.; Dalfino, G.; Zumbo, A. The Influence of Two Different Breeding Systems on Quality and Clotting Properties of Milk from Dairy Buffaloes Reared in Sicily (Italy). *Ital. J. Anim. Sci.* **2015**, *14*, 3669. [CrossRef]

34. Barłowska, J.; Szwajkowska, M.; Litwi´nczuk, Z.; Krol, J. Nutritional Value and Technological Suitability of Milk from Various Animal Species Used for Dairy Production. *Compr. Rev. Food Sci. Food Saf.* **2011**, *10*, 291–302. [CrossRef]

35. Eenenneem, A.V.; Medrano, J.F. Milk protein polymorphisms in California dairy cattle. *J. Dairy Sci.* **1991**, *74*, 1730–1742. [CrossRef]

36. Di Francia, A.; Masucci, F.; DiSerracapriola, M.T.; Gioffré, F.; Proto, V. Nutritional factors influencing milk urea in buffaloes. *Ital. J. Anim. Sci.* **2003**, *2*, 225–227.

37. Santillo, A.; Caroprese, M.; Marino, R.; Sevi, A.; Albenzio, M. Quality of buffalo milk as affected by dietary protein level and flaxseed supplementation. *J. Dairy Sci.* **2016**, *99*, 7725–7732. [CrossRef] [PubMed]

38. Ren, D.; Zou, C.; Lin, B.; Chen, Y.; Liang, X.; Liu, J. A Comparison of Milk Protein, Amino Acid and Fatty Acid Profiles of River Buffalo and Their F1 and F2 Hybrids with Swamp Buffalo in China. *Pak. J. Zool.* **2015**, *47*, 1459–1465.

39. Bustamante, C.; Campos, R.; Sanchez, H. Production and composition of buffalo milk supplemented with agro-industrial byproducts of the African palm. *Rev. Fac. Nac. de Agron.* **2017**, *70*, 8077–8082. [CrossRef]

40. Pegolo, S.; Stocco, G.; Mele, M.; Schiavon, S.; Bittante, G.; Cecchinato, A. Factors affecting variations in the detailed fatty acid profile of Mediterranean buffalo milk determined by 2-dimensional gas chromatography. *J. Dairy Sci.* **2017**, *100*, 2564–2576. [CrossRef]

41. Samkova, E.; Spicka, J.; Pesek, M.; Pelikanova, T.; Hanus, O. Animal factors affecting the fatty acid composition of cow milk fat: A review. *S. Afr. J. Anim. Sci.* **2012**, *42*, 83–100.

42. Gordon, H.M. Milk Lipids. In *Milk and Dairy Products in Human Nutrition: Production, Composition and Health*, 1st ed.; Park, Y.W., Haenlein, G.F.W., Eds.; John Wiley & Sons, Ltd.: New York, NY, USA, 2013; pp. 65–79.

43. Varricchio, M.L.; Di Francia, A.; Masucci, F.; Romano, R.; Proto, V. Fatty acid composition of Mediterranean buffalo milk fat. *Ital. J. Anim. Sci.* **2007**, *6*, 509–511. [CrossRef]

44. Claeys, W.L.; Verraes, C.; Cardoen, S.; De Block, J.; Huyghebaert, A.; Raes, K.; Dewettinck, K.; Herman, L. Consumption of raw or heated milk from different species: An evaluation of the nutritional and potential health benefits. *Food Control* **2014**, *42*, 188–201. [CrossRef]

45. Addeo, F.; Alloisio, V.; Chianese, L.; Alloisio, V. Tradition and innovation in the water buffalo dairy products. *Ital. J. Anim. Sci.* **2007**, *6*, 51–57. [CrossRef]

46. Enb, A.; Abou-Donia, A.M.; Abd-Rabou, S.N.; Abou-Arab, K.A.A.; El-Senaity, H.M. Chemical Composition of Raw Milk and Heavy Metals Behavior During Processing of Milk Products. *Global Vet.* **2009**, *3*, 268–275.

47. Abdeldaiem, M.A.; Jin, Q.; Liu, R.; Wang, X. Effects of pH values on the properties of buffalo and cow butter-based low-fat spreads. *Grasasy Aceites* **2014**, *65*, 038. [CrossRef]

48. Dozet, N.; Maćej, O.; Jovanović, S. Autohonous milk products basis for specific, original milk products development in modern conditions. *Biotechnol. Anim. Husb.* **2004**, *20*, 31–48. [CrossRef]

49. Poltronieri, P.; Cappello, M.S.; D'urso, F.D. Bioactive peptides with health benefit and their differential content in whey of different origin. In *Whey Types, Composition and Health Implications*; Benitez, R.M., Ortero, G.M., Eds.; Nova Publisher: Hauppauge, NY, USA, 2012; pp. 153–168.

50. Kwak, H.S.; Ganesan, P.; Mijan, A.M. Butter, Ghee, and Cream Products. In *Milk and Dairy Products in Human Nutrition: Production, Composition and Health*, 1st ed.; Park, Y.W., Haenlein, G.F.W., Eds.; John Wiley & Sons, Ltd.: New York, NY, USA, 2013; pp. 390–411.
51. Popović, B.T.; Arsić, Č.A.; Debeljak-Martačić, D.J.; Petrović, P.G.; Gurinović, A.M.; Vučić, M.V.; Glibetić, D.M. Traditional food in Serbia: Sources, recipes and fatty acids profiles. *Food Feed Res.* **2014**, *41*, 153–157.
52. Martini, M.; Altomonte, I.; Silva Sant'Ana, M.A.; Salari, F. Nutritional composition of four commercial cheeses made with buffalo milk. *J. Food Nutr. Res.* **2016**, *55*, 256–262.
53. Rafiee-Yarandi, H.; Ghorbani, G.R.; Alikhani, M.; Sadeghi-Sefidmazgi, A.; Drackley. J.K. A comparison of the effect of soybeans roasted at different temperatures versus calcium salts of fatty acids on performance and milk fatty acid composition of mid-lactation Holstein cows. *J. Dairy Sci.* **2016**, *99*, 5422–5435. [CrossRef]
54. Bobe, G.; Hammond, E.G.; Freeman, A.E.; Lindberg, G.L.; Beitz, D.C. Texture of Butter from Cows with Different Milk Fatty Acid Compositions. *J. Dairy Sci.* **2003**, *86*, 3122–3127. [CrossRef]

 *sustainability*

Article

Use of Tea Tree Essential Oil (*Melaleuca alternifolia*) in Laying Hen's Nutrition on Performance and Egg Fatty Acid Profile as a Promising Sustainable Organic Agricultural Tool

Nikola Puvača[1,*], Erinda Lika [2], Sonila Cocoli [2], Tana Shtylla Kika [2], Vojislava Bursić [3,*], Gorica Vuković [4], Mirela Tomaš Simin [3], Aleksandra Petrović [3] and Magdalena Cara [5]

[1] Department of Engineering Management in Biotechnology, Faculty of Economics and Engineering Management in Novi Sad, University Business Academy in Novi Sad, Cvećarska 2, 21000 Novi Sad, Serbia
[2] Faculty of Veterinary Medicine, Agricultural University of Tirana, Koder Kamez, 1029 Tirana, Albania; eklarisalika@yahoo.com (E.L.); scocoli@ubt.edu.al (S.C.); tana_kika@hotmail.com (T.S.K.)
[3] Faculty of Agriculture, University of Novi Sad, Trg Dositeja Obradovića 8, 21000 Novi Sad, Serbia; mirela.tomas@polj.edu.rs (M.T.S.); petra@polj.uns.ac.rs (A.P.)
[4] Institute of Public Health of Belgrade, Bulevar despota Stefana 54a, 11000 Belgrade, Serbia; gorica.vukovic@zdravlje.org.rs
[5] Faculty of Agriculture and Environment, Agricultural University of Tirana, Koder Kamez, 1029 Tirana, Albania; mcara@ubt.edu.al
* Correspondence: nikola.puvaca@fimek.edu.rs (N.P.); bursicv@polj.uns.ac.rs (V.B.); Tel.: +381-65-219-1284 (N.P.); +381-63-886-8456 (V.B.)

Received: 3 April 2020; Accepted: 19 April 2020; Published: 22 April 2020

Abstract: The level of production in a variety of organic production systems is often lower than in other traditional production systems. In poultry production, there is also a direct negative effect of the small scale regarding sustainable organic poultry production. Regardless of differences between organic and conventional production systems, this experiment aimed to investigate the usage of tea tree *Melaleuca alternifolia* (Maiden and Betche) Cheel essential oils as a natural alternative to antibiotics in hen nutrition on productive parameters, table egg quality and eggs fatty acid profile as a promising sustainable organic agricultural tool. A total of 360 Lohmann Brown hens, aged 54 weeks, divided into three different treatment diets, were supplemented with 0 (T1), 40 (T2) and 80 mg/kg (T3) of *M. alternifolia* essential oil, respectively. Experimental treatments were replicated four times within 30 birds each. The experiment lasted for a total of 56 days (55 to 62 weeks of hens age). A 56-day experimental had two timetable periods of 28 days each: period 1 (55 to 58 weeks of hen age) and period 2 (59 to 62 weeks of hen age). For compound feed supplemented with *M. alternifolia* essential oil, daily egg production and the efficiency of nutrient utilization (FCR) was improved significantly ($p < 0.05$) until the end of week 58, with a significant ($p < 0.05$) increase in the thickness of eggshell, as well as egg production ($p < 0.05$). However, egg mass, feed consumption, FCR and albumen height, Haugh unit, and eggshell strength did not show any significant ($p > 0.05$) differences influenced by essential oil feed supplementation. Lower concentrations of saturated fatty acid (SFA) and monounsaturated fatty acid (MUFA), and higher concentrations of polyunsaturated fatty acid (PUFA), were recorded with *M. alternifolia* essential oil supplementation, but without significant ($p > 0.05$) differences. At the end of the experiment, the obtained results showed that the addition of *M. alternifolia* essential oil to hen nutrition had a positive effect on production parameters and eggs fatty acid profile, with increased eggshell thickness ($p < 0.05$).

Keywords: medicinal plants; sustainable; poultry; organic; eggs; fatty acids; agriculture

1. Introduction

Nowadays, modern organic agriculture has been described as production in a variety of organic production systems, which is often lower than in other traditional production systems. In poultry production, there is also a direct negative effect of the small scale regarding sustainable organic poultry production [1]. Organic agriculture helps to make farming sustainable, safe, and environmentally friendly, and essential oils and botanicals as a natural remedy serve that purpose.

For more than a hundred years, *M. alternifolia* has been used as a natural remedy all over the world, and primarily in Australia. Nowadays, bioactive ingredients in *M. alternifolia* are used in the manufacture of homeopathic drugs, and thus in pharmaceuticals and agriculture, with a growing interest in Europe, Japan and the United States [1]. Myrtle family *Myrtaceae*, represents the most important and known species for the industrialization of *M. alternifolia* oil because of its beneficial properties [2]. For the distillation of essential oils, flowers, herbs, leaves, roots, and another plant parts could be successfully used. *M. alternifolia* essential oil is a mixture of terpenes, aldehydes, esters, alcohols, and other chemical molecules, and therefore has been used in poultry nutrition for its antimicrobial, antibacterial, antioxidant and digestive stimulant properties. In the last twenty years, essential oils have been ignored as a possible natural remedy or natural antibiotic alternative for poultry [3]. However, they are recently gaining spotlight in the scientific community [4,5]. *M. alternifolia* oil, because of its antibiotic properties, is widely used as an antiseptic tool, often is used to minimize inflammation and can be very useful in fungal infection treatments [2]. *M. alternifolia* oil, due to its toxicity, should not be used in animal nutrition in high concentrations. Deaths regarding the higher dose of consumed *M. alternifolia* oil were not recorded in the known literature. The aforementioned statement was confirmed in the research of Hammer et al. [2]. However, in the right concentrations, *M. alternifolia* oil addition in broiler chickens diet increases body weight gain by 7% and improve nutrient utilization by 6% when compared with results of chickens that were not fed rations with the supplementation of *M. alternifolia* [6]. Olgun [7] research has shown that eggshell thickness was positively affected by essential oil addition to hens' diet, as well as other quality parameters of eggs, while Nikolova and Kocevski [8] in their research have shown its influence on the technological aspects of eggshell quality. In the research of Nadia et al. [9], usage of essential oil increased egg production and improved feed utilization, when the results of experimental treatments were compared with control treatment. Investigations with hens have shown that the dietary addition of thyme, sage, and rosemary essential oil [10] and essential oil mixture [6] increase hens' production, and increases immunity and eggshell quality, in line with the results obtained by Spasevski et al. [11], with the usage of powder form of marigold, paprika, and carrot in hens' diet. Due to the link between dietary lipids and developing coronary heart disease, the lipid composition of table eggs is the primary consumer concern, which is the same regarding dairy products [12]. It has been proven that adequate and well-balanced compound feeds for hens can improve the fatty acid profile of the eggs. Although the essential oils have a positive effect on the body metabolism of lipids [13], there have been no studies reported, or very few, on the effect of *M. alternifolia* dietary essential oil addition on egg fatty acids profile in hens.

Therefore, this research aimed to evaluate the effects of *M. alternifolia* essential oil in hen nutrition on productive performance results, table egg quality, and egg fatty acid profile.

2. Materials and Methods

Ethical Approval: Biological experiment with laying hens was approved by the University EC board and performed following the EU legislation and principle of the Three Rs within Directive 2010/63/EU.

Animals and Experimental Design: The experiment with laying hens was conducted under the principles of the European Union Strategy for the Protection and Welfare of Animals. A total of 360 Lohmann Brown hens aged 54 weeks were divided into three different treatment diets supplemented with 0, 40 and 80 mg/kg of *M. alternifolia* essential oil (Planet Fresh d.o.o., Montenegro), respectively. *M. alternifolia* was provided, including Terpinen-4-ol 40.0%, γ-Terpinene 23.0% and α-Terpinene

10.4% as the active components. Each treatment was replicated four times with 30 hens in each. Hens were in an environmentally controlled experimental facility with a constart temperature of 22 °C. Environmental conditions in the facility were in line with hybrid specifications. Hens were provided with a diet of 110 g of feed/hen/day, while water of an average temperature of 20 °C was provided ad libitum. Compound feed in mash form was used for hen nutrition during the experimental period. The basal feed was balanced according to the hybrid recommendation with and without the *on top* addition of *M. alternifolia* essential oils in the mixtures. The nutritive value of the used diets is given in Table 1.

Table 1. Nutritive value of the basal diet.

Nutrients*	Content,%
Crude protein	15.5
Crude fat	4.6
Calcium	3.8
Phosphorus (available)	0.37
L-Lysine	0.79
Methionine + Cystine	0.54
Threonine	0.60
Tryptophan	0.18
Metabolizable Energy, MJ/kg	11.0

*The value of crude protein, crude fat and calcium was analyzed and the value of metabolizable energy was calculated.

Sample Preparation and Eggs Quality Assessment: Eggs were gathered every day and the egg production was recorded for both period 1 (55 to 58 weeks of hens age), and period 2 (59 to 62 weeks of hens age), and for the complete periods of research. During the experiment, egg mass was registered daily. Consumption of feed was registered daily, and the feed conversion ratio was calculated for different production periods. Every 28 days in a row, 12 eggs per treatment were randomly gathered: three eggs per each replicate to measure albumen height, calculate Haugh units, and measure eggshell thickness and eggshell strength. Albumen height measurement was performed with a digital caliper, while Haugh units were calculated (1):

$$HU = 100 \, Log \, (h - 1.7 \, w^{0.37} + 7.6)$$
$$HU = Haugh \, unit; \, h = albumen \, height \, (mm) \, and \, w = egg \, weight \, (g)$$

(1)

Eggshell thickness was measured at the three points of measurements, top end, bottom end and center, with a micrometer screw gauge. An average of three thickness values from each egg has been used to define the thickness of the eggshell. A test machine, Instron, was used to measure the break strength of uncracked eggs. An egg was laid down on a continually raised load until it broke down. The point of egg break under load is considered as the measured strength of the egg.

Table Egg Yolk Fatty Acids Analyses: Egg yolk fatty acids profile was analyzed by the gas chromatography technique. First, lipids were extracted from egg yolk. The extraction of lipids from the yolk was performed according to the Folch extraction method. Extracted lipids with 14% (w/w) boron trifluoride–methanol solution were used for the preparation of fatty acid methyl esters. Upon obtaining the samples, analyzes by GC were conducted on an Agilent 7890A chromatograph (Agilent Technologies, USA) with an FID, auto-injection module for liquid, equipped with fused silica capillary column (Supelco SP-2560 Capillary GC Column 100 m × 0.25 mm, d = 0.20 µm). As a gas carrier, helium of 99.9997 vol % purity, with a flow rate of 1.5 mL/min and with 1.092 bar pressure was used. The fatty acid profile of egg yolk was determined based on retention times and standard comparison. The fatty acid profile was shown as the % of the total fatty acids.

Statistical Analyses: The data obtained in the experiment were analyzed by one-way analysis of variance within statistical software Statistica 13. When the analysis of variance showed statistical significance, Duncan's multiple range test was used. The significant difference was registered at $p < 0.05$.

3. Results and Discussion

The effect of *M. alternifolia* essential oil on hen productive performance results is given in Table 2. During the 56 days of the experimental trial, only egg production showed significant ($p < 0.05$) results, while egg weight, feed intake and feed conversion ratio did not record any significant ($p > 0.05$) influence of diet supplementation with *M. alternifolia* essential oils. Feed conversion ratio and hen-day egg production were significantly improved ($p < 0.05$) in weeks 59 to 62 of the experimental period.

Table 2. Influence of *M. alternifolia* essential oil on productive performance results of hens.

Parameter	Dietary Treatments with *M. alternifolia* Essential Oil, mg/kg			SEM	*p*-Value
	0 (T1)	40 (T2)	80 (T3)		
Feed intake, g/(hen/day)					
55 to 58 wk	123.0	123.2	124.1	1.354	0.489
59 to 62 wk	120.5	120.5	121.2	1.198	0.483
55 to 62 wk	122.0	121.9	122.7	1.076	0.382
Feed conversion ratio					
55 to 58 wk	2.51	2.45	2.43	0.042	0.672
59 to 62 wk	2.49[a]	2.36[b]	2.40[b]	0.025	0.004
55 to 62 wk	2.50	2.41	2.41	0.027	0.081
Hen-day egg production, %					
55 to 58 wk	78.5	80.2	81.6	1.353	0.622
59 to 62 wk	77.0[b]	79.5[a]	80.5[a]	1.076	0.034
55 to 62 wk	77.8[b]	79.9[b]	81.1[a]	1.098	0.016
Egg weight, g					
55 to 58 wk	62.6	63.0	63.0	0.319	0.558
59 to 62 wk	62.8	62.9	62.7	0.322	0.631
55 to 62 wk	62.7	63.0	62.9	0.297	0.529

[a,b] indicated the difference within a row was significant ($p < 0.05$).

Generally, it is assumed that essential oils of herbs and spices may have a positive influence, and might improve the feed consumption, due to their aromatic properties [14,15]. Bozkurt et al. [16] compared four herbs (thyme, oregano, rosemary, turmeric) at an inclusion level of 5 or 10 g/kg diet in a study with 28-week old laying hens for 12 weeks. They found no differences in feed intake among the treated groups and did not observe any differences when comparing the treatments to the untreated control group, which is in agreement with our research. Similar results were obtained when garlic was included in diets fed to laying hens in the concentrations of 20, 40, 60, 80 or 100 g/kg diet for hens of nearly the same age (27–28 weeks), but for only 5–6 weeks [17]. However, Abdo et al. [18] found that the average feed intake was significantly reduced for hens which were fed either 10 to 50 g/kg of green tea leaves or 5 to 25 mg/kg of green tea essential oil when compared to an unsupplemented control treatment. In our research, feed conversion ratio and hen–day egg production were increased ($p < 0.05$) in weeks 59 to 62, however, during a total of 56 days of feeding, trial egg production record significant ($p < 0.05$) results, while dietary addition of essential oils did not show significant ($p > 0.05$) effects regarding egg weight, feed consumption and feed conversion ratio, in agreement with other research [9,11,16,19]. The essential oils used as "production stimulators" for hens were primarily introduced in their daily diet to increase the utilization of the limit-fed diet and, in turn, improve the production of eggs. Some studies showed positive effects on production results, such as egg production rate, as was the case in our research, and the egg mass, whereas some other studies demonstrated no effect of essential oil addition [20].

The effect of *M. alternifolia* essential oil on egg quality is shown in Table 3. Supplementation of *M. alternifolia* essential oil significantly increased the eggshell thickness ($p < 0.05$) at the end of the experimental period of 56 days, in line with the research of Bozkurt et al. [21], who concluded that dietary addition of essential oil mixture increases the weight of eggshell, eggshell thickness, and strength of shell breaking, while dietary supplementations of *M. alternifolia* essential oil did not affect eggshell thickness ($p > 0.05$) at the beginning of the hen laying period in our study. No significant differences in albumen height, Haugh unit, and eggshell strength between the different dietary treatments T1 and T2 ($p > 0.05$), compared to control treatment T1, were recorded.

Table 3. Influence of *M. alternifolia* essential oil on egg quality parameters.

Parameter	Dietary Treatments with *M. alternifolia* Essential Oil, mg/kg			SEM	*p*-Value
	0 (T1)	40 (T2)	80 (T3)		
Albumen height					
58 wk	7.30	7.35	7.20	0.145	0.755
62 wk	7.20	7.30	7.20	0.321	0.562
Haugh unit					
58 wk	83.60	83.00	82.50	1.235	0.890
62 wk	82.30	82.30	82.30	1.899	0.869
Eggshell strength, kg/cm^2					
58 wk	4.70	4.30	4.20	0.257	0.693
62 wk	4.50	4.20	4.71	0.178	0.813
Eggshell thickness, mm					
58 wk	0.26	0.24	0.25	0.007	0.158
62 wk	0.24[b]	0.28[a]	0.27[a]	0.007	0.031

[a,b] indicated the difference within a row was significant ($p < 0.05$).

Bölükbaşi et al. [22] showed that the supplementation of bergamot essential oil at three inclusion levels (0.25, 0.50, 0.75 mL/kg) had no impact on the proportion of albumen and eggshell thickness of the eggs of hens at the age of 67 weeks. However, at any addition level, bergamot essential oil decreased eggshell percentage by 15% or more when compared with an untreated control group. The dietary addition of black cumin essential oil did not show a significant ($p > 0.05$) influence on shell weight when added in concentrations of 1; 2 or 3 mL/kg, to hens' diet, respectively. Similarly, essential oil of oregano in the amount of 50 or 100 mg/kg, did not show either positive or negative effects on yolk color score, Haugh unit or shell thickness when added to the feed at the age of 32 weeks [19], but when used in broiler chickens diet, oregano oil and powder form mixtures of medicinal plants, and have shown significant effects [4,23]. The Haugh unit value as the main measure of internal quality of eggs was not compromised by the dietary supplementation of bergamot essential oil [22], oregano essential oil [19], a mixture of essential oils [21], green tea leaves or essential oil [18], thyme, oregano, rosemary or turmeric powder [9], or a garlic–thyme combination [24], in agreement with our study with a dietary addition of *M. alternifolia* essential oil to hen nutrition in concentrations of 40 and 80 mg/kg, respectively.

The influence of *M. alternifolia* essential oil on the fatty acid profile of egg yolk of control and experimental treatments is given in Table 4. The addition of *M. alternifolia* essential oil in treatments T2 and T3 compared with control treatment, T1, did not show a significant ($p > 0.05$) influence on the fatty acid profile of the yolk. Lower concentrations of saturated fatty acid (SFA) and monounsaturated fatty acid (MUFA), while higher concentrations of polyunsaturated fatty acid (PUFA), were recorded with *M. alternifolia* essential oil supplementation, but without significant ($p > 0.05$) differences.

Table 4. Influence of *M. alternifolia* essential oil on fatty acid profile of egg yolk, %.

Fatty Acids	Dietary Treatments with *M. alternifolia* Essential Oil, mg/kg			SEM	*p*-Value
	0 (T1)	40 (T2)	80 (T3)		
C14:0	3.3	3.1	3.6	0.322	0.1931
C16:0	27.0	26.9	26.9	0.425	0.3887
C16:1	3.2	3.0	3.1	0.220	0.6228
C18:0	9.6	9.7	9.5	0.351	0.2190
C18:1	41.0	42.0	41.2	0.834	0.1558
C18:2 n-6	11.6	11.4	11.6	0.381	0.8647
C18:3 n-6	0.06	0.09	0.06	0.045	0.8110
C18:3 n-3	0.09	0.03	0.08	0.062	0.1473
C20:3 n-3	0.03	0.05	0.23	0.098	0.9472
C22:6 n-3	0.3	0.5	0.42	0.187	0.2219
SFA	40.6	40.0	39.9	0.511	0.8732
MUFA	46.0	45.2	46.3	0.700	0.2965
PUFA	13.4	14.8	13.8	0.479	0.1632

In our study, different dietary treatments did not show significant ($p > 0.05$) influence on the egg yolk fatty acid profile. The research of Galobart et al. [10] showed that the dietary addition of rosemary essential oil in hen nutrition had no effects on the fatty acid composition of egg yolk. Besides that, the research of Bölükbaşi et al. [22] showed that the ratio of docosahexaenoic fatty acid (DHA) and the ratio of n-3 fatty acids of egg yolk were elevated by the dietary addition of bergamot essential oils. Other studies have shown changes in the lipid metabolism of poultry fed with the addition of essential oils [22,25]. To our knowledge, there have been no reports on the effect of *M. alternifolia* essential oil on the fatty acid composition of egg yolk, which makes this research a very valuable asset to the scientific community. The adequate ratio between n-6 and n-3 polyunsaturated fatty acids is a very important standard in healthy human nutrition. Being aware of n-3 polyunsaturated fatty acids' benefits and health-promoting effects, the nutritionists recommend a diet rich in n-3 fatty acids, as well as a lower ratio between n-6 and n-3 fatty acids, from the currently common 15–20:1 to 1–4:1. According to Simopoulos [26] and Kralik et al. [27], the ratio of n-6 and n-3 polyunsaturated fatty acids in egg yolk lipids less than 4:1 is considered to be beneficial to human health. This desirable ratio of n-6 and n-3 polyunsaturated fatty acids was highlighted by other authors as well [28,29].

4. Conclusions

Obtained results have shown that the addition of *M. alternifolia* essential oil in the hen diet has a significant positive effect on egg production, but did not show a significant effect on the yolk fatty acid profile, except increased eggshell thickness. Overall, *M. alternifolia* essential oil in a lower concentration of 40 mg/kg in may be beneficial and recommended for hen nutrition, but further investigation about *M. alternifolia* essential oil's influence on laying hens' nutrition and its effects and mode of action is more than necessary.

Author Contributions: Conceptualization, N.P. and M.T.S.; methodology, N.P.; software, V.B.; validation, M.C., S.C. and T.S.K.; formal analysis, G.V.; investigation, N.P.; resources, A.P.; data curation, A.P.; writing—original draft preparation, N.P.; writing—review and editing, E.L.; visualization, S.C.; supervision, V.B.; project administration, N.P.; funding acquisition, N.P. All authors have read and agreed to the published version of the manuscript.

Funding: This research was funded by the Ministry for Education, Science and Technological Development of the Republic of Serbia.

Conflicts of Interest: The authors declare no conflict of interest.

References

1. Puvača, N.; Čabarkapa, I.; Bursić, V.; Petrović, A.; Aćimović, M. Antimicrobial, antioxidant and acaricidal properties of tea tree (*Melaleuca alternifolia*). *J. Agron. Technol. Eng. Manag.* **2018**, *1*, 29–38.

2. Hammer, K.A.; Carson, C.F.; Riley, T.V.; Nielsen, J.B. A review of the toxicity of *Melaleuca alternifolia* (tea tree) oil. *Food Chem. Toxicol.* **2006**, *44*, 616–625. [CrossRef] [PubMed]

3. Puvača, N.; Stanaćev, V.; Glamočić, D.; Lević, J.; Perić, L.; Stanaćev, V.; Milić, D. Beneficial effects of phytoadditives in broiler nutrition. *Worlds Poult. Sci. J.* **2013**, *69*, 27–34. [CrossRef]

4. Popović, S.; Puvača, N.; Kostadinović, L.; Džinić, N.; Bošnjak, J.; Vasiljević, M.; Djuragic, O. Effects of dietary essential oils on productive performance, blood lipid profile, enzyme activity and immunological response of broiler chickens. *Eur. Poult. Sci.* **2016**, *80*. [CrossRef]

5. Puvača, N. Bioactive compounds in selected hot spices and medicinal plants. *J. Agron. Technol. Eng. Manag.* **2018**, *1*, 8–17.

6. Khattak, F.; Ronchi, A.; Castelli, P.; Sparks, N. Effects of natural blend of essential oil on growth performance, blood biochemistry, cecal morphology, and carcass quality of broiler chickens. *Poult. Sci.* **2014**, *93*, 132–137. [CrossRef]

7. Olgun, O. The effect of dietary essential oil mixture supplementation on performance, egg quality and bone characteristics in laying hens. *Ann. Anim. Sci.* **2016**, *16*, 1115–1125. [CrossRef]

8. Nikolova, N.; Kocevski, D. Quality of hen's eggshell in conditions of high temperatures. *J. Agron. Technol. Eng. Manag.* **2018**, *1*, 84–90.

9. Radwan Nadia, L.; Hassan, R.A.; Qota, E.M.; Fayek, H.M. Effect of natural antioxidant on oxidative stability of eggs and productive and reproductive performance of laying hens. *Int. J. Poult. Sci.* **2008**, *7*, 134–150. [CrossRef]

10. Galobart, J.; Barroeta, A.C.; Baucells, M.D.; Codony, R.; Ternes, W. Effect of dietary supplementation with rosemary extract and α-tocopheryl acetate on lipid oxidation in eggs enriched with ω3-fatty acids. *Poult. Sci.* **2001**, *80*, 460–467. [CrossRef]

11. Spasevski, N.; Puvača, N.; Pezo, L.; Tasić, T.; Vukmirović, Đ.; Banjac, V.; Čolović, R.; Rakita, S.; Kokić, B.; Džinić, N. Optimisation of egg yolk colour using natural colourants. *Eur. Poult. Sci.* **2018**, *82*. [CrossRef]

12. Puvača, N.; Ljubojević Pelić, D.; Tomić, V.; Radišić, R.; Milanović, S.; Soleša, D.; Budakov, D.; Cara, M.; Bursić, V.; Petrović, A.; et al. Antimicrobial efficiency of medicinal plants and their influence on cheeses quality. *Mljekarstvo* **2020**, *70*, 3–12. [CrossRef]

13. Acamovic, T.; Brooker, J.D. Biochemistry of plant secondary metabolites and their effects in animals. *Proc. Nutr. Soc.* **2005**, *64*, 403–412. [CrossRef] [PubMed]

14. Puvača, N.; Kostadinović, L.; Popović, S.; Lević, J.; Ljubojević, D.; Tufarelli, V.; Jovanović, R.; Tasić, T.; Ikonić, P.; Lukač, D. Proximate composition, cholesterol concentration and lipid oxidation of meat from chickens fed dietary spice addition (*Allium sativum, Piper nigrum, Capsicum annuum*). *Anim. Prod. Sci.* **2016**, *56*, 1920–1927. [CrossRef]

15. Tashla, T.; Puvača, N.; Pelić Ljubojević, D.; Prodanović, R.; Bošković, J.; Ivanišević, D.; Jahić, M.; Mahmoud, O.; Giannenas, I.; Lević, J. Dietary medicinal plants enhance the chemical composition and quality of broiler chicken meat. *J. Hell. Vet. Med. Soc.* **2019**, *70*, 1823–1832. [CrossRef]

16. Bozkurt, M.; Hippenstiel, F.; Abdel-Wareth, A.A.A.; Kehraus, S.; Küçükyilmaz, K.; Südekum, K.-H. Effects of selected herbs and essential oils on performance, egg quality and some metabolic activities in laying hens—A review. *Eur. Poult. Sci.* **2010**, *78*, 15.

17. Chowdhury, S.R.; Chowdhury, S.D.; Smith, T.K. Effects of dietary garlic on cholesterol metabolism in laying hens. *Poult. Sci.* **2002**, *81*, 1856–1862. [CrossRef]

18. Abdo, Z.M.A.; Hassan, R.A.; El-Salam, A.A.; Helmy, S.A. Effect of adding green tea and its aqueous extract as natural antioxidants to laying hen diet on productive, reproductive performance and egg quality during storage and its content of cholesterol. *Egypt. Poult. Sci.* **2010**, *30*, 1121–1149.

19. Giannenas, I.; Koidis, A.; Botsoglou, E.; Dotas, V.; Mitsopoulos, I.; Florou-Paneri, P.; Nikolakakis, I. Hen performance and egg quality as affected by dietary oregano essential oil and alpha-tocopheryl acetate supplementation. *Int. J. Poult. Sci.* **2005**, *4*, 449–454.

20. Saki, A.A.; Aliarabi, H.; Hosseini Siyar, S.A.; Salari, J.; Hashemi, M. Effect of a phytogenic feed additive on performance, ovarian morphology, serum lipid parameters and egg sensory quality in laying hen. *Vet. Res. Forum* **2014**, *5*, 287–293.

21. Bozkurt, M.; Küçükyilmaz, K.; Catli, A.U.; Cinar, M.; Bintas, E.; Cöven, F. Performance, egg quality, and immune response of laying hens fed diets supplemented with mannan-oligosaccharide or an essential oil

mixture under moderate and hot environmental conditions. *Poult. Sci.* **2012**, *91*, 1379–1386. [CrossRef] [PubMed]

22. Bölükbasi, S.C.; Erhan, M.K.; Ürüsan, H. The effects of supplementation of bergamot oil (*Citrus bergamia*) on egg production, egg quality, fatty acid composition of egg yolk in laying hens. *J. Poult. Sci.* **2010**, *47*, 163–169. [CrossRef]

23. Popović, S.; Kostadinović, L.; Đuragić, O.; Aćimović, M.; Čabarkapa, I.; Puvača, N.; Pelić Ljubojević, D. Influence of medicinal plants mixtures (*Artemisia absinthium, Thymus vulgaris, Menthae piperitae* and *Thymus serpyllum*) in broilers nutrition on biochemical blood status. *J. Agron. Technol. Eng. Manag.* **2018**, *1*, 91–98.

24. Ghasemi, R.; Zarei, M.; Torki, M. Adding medicinal herbs including garlic (*Allium sativum*) and thyme (*Thymus vulgaris*) to diet of laying hens and evaluating productive performance and egg quality characteristics. *Am. J. Anim. Vet. Sci.* **2010**, *5*, 151–154. [CrossRef]

25. Abdel-Wareth, A.A.A. Effect of dietary supplementation of thymol, synbiotic and their combination on performance, egg quality and serum metabolic profile of Hy-Line Brown hens. *Br. Poult. Sci.* **2016**, *57*, 114–122. [CrossRef]

26. Simopoulos, A.P. Human requirement for N-3 polyunsaturated fatty acids. *Poult. Sci.* **2000**, *79*, 961–970. [CrossRef]

27. Kralik, G.; Škrtić, Z.; Suchý, P.; Straková, E.; Gajčević, Z. Feeding fish oil and linseed oil to laying hens to increase the n-3 PUFA in egg yolk. *Acta Vet. Brno* **2008**, *77*, 561–568. [CrossRef]

28. Ebeid, T.A. The impact of incorporation of n-3 fatty acids into eggs on ovarian follicular development, immune response, antioxidative status and tibial bone characteristics in aged laying hens. *Animal* **2011**, *5*, 1554–1562. [CrossRef]

29. Puvača, N.; Lukač, D.; Ljubojević, D.; Stanaćev, V.; Beuković, M.; Kostadinović, L.; Plavša, N. Fatty acid composition and regression prediction of fatty acid concentration in edible chicken tissues. *Worlds Poult. Sci. J.* **2014**, *70*, 585–592. [CrossRef]

 *sustainability*

Article

Effect of Dietary Supplementation with *Moringa oleifera* Leaves and/or Seeds Powder on Production, Egg Characteristics, Hatchability and Blood Chemistry of Laying Japanese Quails

Elwy A. Ashour [1], Mohamed S. El-Kholy [1], Mahmoud Alagawany [1], Mohamed E. Abd El-Hack [1,*], Laila A. Mohamed [1], Ayman E. Taha [2], Ahmed I. El Sheikh [3], Vito Laudadio [4] and Vincenzo Tufarelli [4,*]

[1] Poultry Department, Faculty of Agriculture, Zagazig University, Zagazig 44511, Egypt; dr.elwy.ashour@gmail.com (E.A.A.); elkolymohamed@yahoo.com (M.S.E.-K.); dr.mahmoud.alagwany@gmail.com (M.A.); taha_agronomy_1978@yahoo.com (L.A.M.)
[2] Department of Animal Husbandry and Animal Wealth Development, Faculty of Veterinary Medicine, Alexandria University, Edfina 22578, Egypt; ayman.taha@alexu.edu.eg
[3] Department of Veterinary Public Health, Faculty of Veterinary Medicine, King Faisal University, P. O. Box 400 Al-Hasa 31982, Saudi Arabia; aelsheikh@kfu.edu.sa
[4] Department of DETO, Section of Veterinary Science and Animal Production, University of Bari "Aldo Moro", Valenzano, 70010 Bari, Italy; vito.laudadio@uniba.it
* Correspondence: dr.mohamed.e.abdalhaq@gmail.com (M.E.A.E.-H.); vincenzo.tufarelli@uniba.it (V.T.)

Received: 27 January 2020; Accepted: 18 March 2020; Published: 20 March 2020

Abstract: The present study aimed to evaluate the effect of dietary *Moringa oleifera* (*M. oleifera*) leaves and/or seed powder on laying Japanese quail performance in terms of egg production, egg quality, blood serum characteristics, and reproduction. In total, 168 Japanese quails (120 hens and 48 males) at eight weeks of age in laying period were randomly distributed to four treatment groups, with six replicates per group and seven birds (five hens and two males) per replicate. The first group (G1) served as a control group, while G2, G3 and G4 groups were supplemented with *M. oleifera* leaves (ML) and *M. oleifera* seeds (MS) and their combination ((1 g/kg ML; 1 g/kg MS; and 1 ML g/kg + 1 MS g/kg (MSL), respectively). From the results, feed consumption, feed conversion ratio, egg weight, fertility and hatchability from fertile eggs, egg and yolk index, and Haugh unit were not affected by dietary treatments. However, egg production, egg mass, eggshell thickness, and hatchability were significantly increased and blood aspartate transaminase (AST) and urea decreased in the MS treatment. Both triglycerides and total cholesterol were reduced ($p < 0.05$) in all treatments with ML, MS, and MSL, with no significant differences in alanine aminotransferase (ALT), albumin, total protein, globulin, and A/G ratio among dietary treatment. Our results clearly indicated that the inclusion of *M. oleifera* seeds in Japanese quail diet significantly increased egg production and improved hatchability, along with some egg quality parameters, and also lowered some blood biochemical components.

Keywords: *Moringa oleifera*; Japanese quail; fertility; egg production; livestock

---

1. Introduction

Traditional synthetic feed additives such as antibiotics, growth stimulants, antioxidants, antiparasite, and antifungal agents have been used for decades in poultry feed. However, they pose many issues such as residues in animal products and resistance to antibiotics in the consumer, which is a matter of public health [1]. Therefore, the use of antibiotics as a growth stimulant in animal feed was banned in

Europe. Revolution in animal feed production has resulted in the development of feed additives in the forms of phytogenics [2,3]. Herbs and their metabolites (known as bioactive substances) play a good role as feed additives. These bioactive compounds such as carotenoids, flavonoids, and herbal oils help enhance animal health and productivity to yield safe and healthy products [4]. The essential role of these active compounds is to dampen microbes and toxins in the gut and promote effectiveness of the pancreas, resulting in good metabolism of nutrients [3,5]. Medicinal plants contain several phytochemicals and bioactive compounds such as trace metal ions, alkaloids, vitamins, carotenoids, fats, polyphenols, carbohydrates, and proteins, which are useful for long-term health [6]. Compared to antibiotics, the utilization of plants elevates trust of usage. Herbs have been suggested to enhance metabolic processes and the health conditions of livestock [7]. Several plants may improve the effect of digestive enzymes, feed consumption, feed utilization, and carcass traits [8]. However, Halle et al. [9] did not observe significant effects for some additives such as oregano and its essential oils, savory, *Nigella sativa* L. and cacao husks on live weight, and carcass parameters of broilers.

M. oleifera plays a useful role against inflammatory and oxidant effects [10]. The administration of *M. oleifera* leaf extracts hinders the development of pathogenic gram-positive and gram-negative bacteria and antioxidant activity. There was an improvement in Hubbard broiler chicks' performances, immune response, and carcass quality parameters with increased benefits with the usage of *M. oleifera* [11]. The growing popularity of using *M. oleifera* as a feed additive in poultry nutrition necessitates thorough investigation into its nutritional value, as well as its effects on hematological characteristics as a measure of both the nutritional and medicinal importance of its leaves in broiler chicks. It was indicated that many vitamins (A, E, B2, B5, B6, folic acid) and minerals (Ca, Fe) are present in moringa [12], having also a powerful fungicidal and antimicrobial activity. It also has an inhibitory effect on cholesterol levels in blood [13]. Yang et al. [10] indicated that *M. oleifera* improved immunity, lowered *E. coli*, and enhanced Lactobacilli in the gastrointestinal tract of chickens. *M. oleifera* improves feed conversion ratio and increases the immune reception of birds [13]. It also has natural antioxidant components and dissolvable proteins in its leaves [14]. Elkloub et al. [15] found that abdominal fat and plasma cholesterol, especially low-density lipoprotein (LDL), decreased with improved performance of immune organs and blood constituents using *M. oleifera* leaves meal in Japanese quail diets. There are only a few studies on the bioactive constituents of *M. oleifera* leaves and their effect on meat antioxidant status [15,16]. Therefore, the objective of this study was to verify the usefulness of *M. oleifera* leaves and/or *M. oleifera* seed meal as natural feed supplements and as a source of antioxidants on the productive and physiological characteristics of laying Japanese quails.

2. Materials and Methods

The experiment was performed at the Research Farm, Poultry Department, Faculty of Agriculture, Zagazig University in Egypt. All research protocols were conducted with the approval of the Local Experimental Animal Committee and were confirmed by the organized council.

2.1. Analysis of M. oleifera

Moringa oleifera plants have high amounts of crude protein (CP) in the leaves (251 g/kg) dry matter (DM) and an abundant proportion of tannins and some anti-nutrient components and offers an abundant source of proteins for ruminants and non-ruminants [17]. The nutrient structure of *M. oleifera* seeds and leaves were determined according to the Association of Official Analytical Chemists (AOAC) [18], as shown in Table 1. The content of total phenolic compounds and total flavonoids in *M. oleifera* seed and leaf mixture were estimated according to Gurnani et al. [19] and Meda et al. [20], respectively as shown in Table 1.

Table 1. Determined analysis *of M. oleifera* seeds and leaves.

Items (g/kg)	Dry Matter	Ash	Crude Protein	Ether Extract	Crude Fiber
Seeds	965	33.9	395.8	394	46
Leaves	934	137.5	270.1	62	215
Content of total phenolics and total flavonoids in seed and leaf mixture					
Total phenolics (mg GAE/g)			65.24		
Total flavonoids (mg QE/g)			17.58		

GAE: gallic acid equivalent; QE: quercetin equivalent.

2.2. Birds, Experimental Design, and Diet

In total, 168 Japanese quails (120 hens and 48 males) at eight weeks of age in laying phase were randomly divided into four treatment groups in a complete randomized design experiment with four treatments having six replicates of seven birds (five hens and two males) each. The experiment included three dietary treatments including *M. oleifera* leaves (ML), *M. oleifera* seeds (MS) and their combination with a level 1.0 g/kg diet of ML; MS; and ML + MS (MSL), and a basal control-diet without moringa. *Moringa oleifera* leaves (ML) and *M. oleifera* seeds (MS) were purchased from the local market, Egypt. The experimental period spanned from 8 to 20 weeks of age. The birds were housed in 24 cages, and each cage measured 90 × 40 × 40 cm. Each metallic cage had a drinker in the form of nipples and feeders. The light program was 14 h of light daily at the start of the experiments and was increased by 15 min weekly to 16 h of light. The birds were allowed to eat and drink ad libitum at every period of the experiment. The corn-soybean diets were in a mash form and were calculated according to the National Research Council (NRC) [21] (Table 2). Birds were vaccinated with distilled water by a veterinarian at the appropriate age. The highest and lowest ambient temperatures were noted every day at noon (12.00 PM) and ranged from 14 to 23 °C, whereas the relative humidity was approximately 60%–70%. All quails were reared in wire batteries under the same managerial, hygienic, and environmental conditions.

Table 2. Composition and calculated analysis of layer quail diet.

Items	Laying Period 8–20 Weeks
Ingredients %	
Yellow corn	55.00
Soybean meal 44% CP	29.50
Corn gluten 60% CP	3.65
Cotton seed oil	4.30
Dicalcium phosphate	1.70
Limestone	5.00
Salt	0.30
Premix *	0.30
L-lysine	0.08
DL-methionine	0.17
**Calculated analysis ** **	
Crude protein %	19.96
Metabolizable energy MJ/kg	12.61
Calcium %	2.51
Available Phosphorous %	0.37
Lysine %	1.02
Methionine %	0.45
Methionine + Cysteine %	0.77

* Layer vitamin and mineral premix. Each 2.5 kg consisted of vit. A. 12 Miu, E. 15 IU., vit. D_3 4 Miu; vit. B_1 1 g, vit. B_2 8 g, pantothenic acid 10.87 g, nicotinic acid 30 g, vit. B_6 2 g, vit. B_{12} 10 mg, folic acid 1 g, biotin 150 mg, copper 5g, iron 5g, manganese 70 g iodine 0.5 g, selenium 0.15 g, zinc 60 g, antioxidant 10 g. ** Calculated according to National Research Council (NRC) [21].

2.3. Collection of Data

Feed utilization was recorded and measured by grams of feed consumed over 28 days and divided by the number of birds/day, and mortality rates were checked. The feed conversion ratio (g feed/g egg) was determined according to the egg mass value divided by the quantity of feed consumption. Eggs were collected every day and egg production was calculated on a hen-day basis. Egg number and egg weight were recorded daily, and the egg mass (egg number × egg weight) was calculated.

2.4. Egg Characteristics

The exterior and interior egg quality parameters were examined. Three eggs from every replicate were collected, and egg components were measured during different time intervals. The shape index of eggs was calculated according the proportion of egg width to length [22]. The yolk index was calculated as yolk height/yolk diameter (mm) after separating the yolk and albumen according to Keener et al. [23]. The shell thickness of eggs was examined (with shell membrane) using a micrometer. The thickness of the shell was measured at three different places on the eggs (air cell, equator, and narrow end). The Haugh unit was calculated as

$$\text{Haugh unit score} = 100 \times \log (H + 7.57 - 1.7\, W\, 0.37)$$

where H is the height of albumen and W is the weight of egg, as per the formula proposed by Card and Nesheim [24].

2.5. Fertility and Hatchability Percentages

In total, 45 eggs from each treatment were collected after three weeks of the experimental period. Eggs were then sprayed with TH4® solution (2 ml/liter of water) for disinfection and set in an incubator. The incubated eggs were subjected to 37.5 °C and 65% RH for the first 14 days. The eggs were transferred to a hatchery machine at the end of the 14th day of incubation and received 37.4 °C and 70% RH until hatching. After hatching, the chicks were counted and the eggs that were not hatched were counted to calculate the fertility rate (number of hatched chicks + number of fertile non hatched eggs/total number of eggs set in incubator) × 100 and hatchability percent. The hatchability was expressed according to the chicks that hatched from fertile eggs.

At the end of the trial, blood samples were collected from six hens, which had been randomly chosen from each group for slaughter, with a clean antiseptic pipe. Samples were allowed to clot and were centrifuged at 3500 rpm (G-force value = 2328.24) for 15 min to obtain serum. Serum samples were preserved in Eppendorf tubes at −20 °C until further examination. Blood biochemical characteristics were specified as total protein (TP), albumin (ALB), globulin (GLB), aspartate aminotransferase (AST), alanine aminotransferase (ALT), bilirubin, creatinine, urea levels, total cholesterol (TC) and high density lipoprotein (HDL), cholesterol, triglyceride (TG) and measured spectrophotometrically using commercial kits provided by Biodiagnostic Co. (Giza, Egypt). Low-density lipoprotein (LDL) cholesterol was calculated as described by Friedewald et al. [25].

2.6. Statistical Analysis

All the statistical analyses of the obtained results were achieved using the SPSS software program [26]. The average values and standard error of the mean (SEM) are described. All data were evaluated with a one-way analysis of variance (with the diet as the fixed factor) using the post-hoc Newman-Keuls test, and $p < 0.05$ was considered to be statistically significant.

3. Results and Discussion

3.1. Effect of M. oleifera on Productive Performance

Effects of *M. oleifera* on production performance are presented in Table 3. Feed consumption, feed conversion ratio, and egg weight were not affected by dietary *moringa* leaves, seeds, and their combination. However, egg production and egg mass were increased significantly using MS compared to the other groups. Kwariet et al. [27] found no significant effects for *M. oleifera* leaf meal at a level of 1%–2% of the basal diet on feed conversion and egg weight of Vanaraja laying hens. This is not in agreement with Olugbemi et al. [28] who found that using Moringa leaf meal (20%) as a replacement for sunflower seed meal in chicken layer diets led to significant decrease in egg production and whole egg weight. Riry et al. [29] found that feeding Japanese quails on a diet with 5% *M. oleifera* seed meal led to a decrease in feed intake in contrast to the control birds. Authors of previous studies [28,30] postulated that the use of ML up to 10% had no negative effects on the egg production of laying birds, but levels greater than 10% led to adverse effects possibly due to increasing the level of anti-nutritional factors and dustiness of ML and low digestibility of energy and protein.

Table 3. Production performance of laying Japanese quails as affected by dietary treatments.

Items	Control	MS	ML	MSL	SEM	*p* Value
Feed intake (g/d/bird)	33.54	33.23	33.09	33.67	0.10	0.14
Feed conversion ratio (g/g)	3.18	2.83	3.20	3.02	0.20	0.07
Egg production (%)	78.95 [b,c]	83.41 [a]	76.93 [c]	81.73 [a,b]	0.88	0.01
Egg weight (g)	13.39	14.07	13.50	13.63	0.16	0.49
Egg mass (g/d/bird)	10.57 [b]	11.74 [a]	10.38 [b]	11.14 [a,b]	0.06	0.03

Control, the basal diet; MS, 1 g *Moringa* seeds/kg diet; ML, 1 g *Moringa* leaves/kg diet; MSL, 1 g seeds + 1 g leaves/kg. SEM: standard error means. a–d: Means in the same row with no superscript letters after them or with a common superscript letter following them are not significantly different ($p < 0.05$).

3.2. Effect of M. oleifera on Fertility and Hatchability

Data of fertility and hatchability of the eggs are presented in Table 4. The results showed that fertility and hatchability from fertile eggs were not affected by dietary ML, MS, and their combination. But hatchability was significantly increased by using only MS compared with the other groups. Etalem et al. [31] found that the hatchability percentage in the 5% ML group was significantly higher than that of the control. Mahmood and Al-Daraji [32] and Moyo et al. [33] observed that *M. oleifera* leaves have higher levels of zinc and vitamin E, which can be useful to the hatchability of eggs.

Table 4. Fertility and hatchability of eggs from laying Japanese quails as affected by dietary treatments.

Items	Control	MS	ML	MSL	SEM	*p* Value
Fertility (%)	90.35	92.78	89.18	89.95	0.77	0.42
Hatchability (%)	64.62 [b]	71.87 [a]	66.24 [b]	68.03 [a,b]	1.02	0.04
Hatchability (fertile eggs, %)	71.59	77.54	74.29	75.66	1.10	0.30

Control, the basal diet; MS, 1 g Moringa seeds/kg diet; ML, 1 g Moringa leaves/kg diet; MSL, 1 g seeds + 1 g leaves/kg. SEM: standard error means. a, b: Means in the same row with no superscript letters after them or with a common superscript letter following them are not significantly different ($p < 0.05$).

3.3. Effect of M. oleifera on Egg Quality

The results in Table 5 show the effect of *M. oleifera* on egg characteristics. Results showed that shell thickness was significantly increased by adding MS meal compared to the control diet. Whereas, egg index, shell percent, yolk percent, albumen percent, Haugh unit, and yolk index were not affected by the dietary supplementation. There was no difference in Haugh units, eggshell strength, or egg shape index among the groups ($p > 0.05$) in response to dietary *M. oleifera* leaves [34–36]. Ebenebe et al [37] reported that adding MOL had no effect on egg shape index that's correlated with the strength of an eggshell and the grade of eggs. Mabusels et al. [38] found that the addition of *M. oleifera* seed meal to layer diets increased the shell thickness ($p \leq 0.05$), compared to a diet with 10% *Moringa oleifera* seed meal (MOSM) and the control diet. However, both 5% and 7.5% MOSM supplementation was equally effective for eggshell thickness. Generally, the seed contains antioxidants, essential oils, minerals such as Ca, Mg, K, Se, P, and Zn, and vitamins such as A, C, D, K, and E, so it can improve egg quality and improve most of the egg quality parameters. The egg shape index percentage was significantly reduced ($p < 0.05$) in birds fed diets containing 10% MS [39].

Table 5. Egg quality criteria for laying Japanese quails as affected by dietary treatments.

Items	Control	MS	ML	MSL	SEM	*p* Value
Egg index	75.74	79.46	79.46	80.27	0.84	0.24
Shell, %	13.38	13.22	13.45	13.61	0.12	0.75
Yolk, %	32.40	31.64	32.62	31.78	0.24	0.44
Albumen, %	54.23	55.14	53.93	54.61	0.24	0.32
Shell thickness, mm	0.23 [b]	0.26 [a]	0.25 [a,b]	0.25 [a,b]	0.004	0.03
Haugh unit	92.57	95.05	93.87	94.39	0.81	0.79
Yolk index	48.66	49.33	49.06	48.83	0.46	0.97

Control, the basal diet; MS, 1 g Moringa seeds/kg diet; ML, 1 g Moringa leaves/kg diet; MSL, 1 g seeds + 1 g leaves/kg. SEM: standard error means. a, b: Means in the same row with no superscript letters after them or with a common superscript letter following them are not significantly different ($p < 0.05$).

3.4. Effect of M. oleifera on Blood Biochemical Indices

Data of blood biochemical indices for laying Japanese quails are listed in Table 6. The results showed that AST (U/L) was significantly decreased due to ML and MS in comparison to the control group and MSL. The urea level was lowered by only *Moringa* seeds (MS) having a powerful fungicidal and antimicrobial activity. Further, both triglycerides and cholesterol were significantly reduced by all dietary treatments including ML, MS, and MSL compared with the control group. ALT, albumin, total protein, globulin, and the albumin/ globulin ratio (A/G) were not affected due to any dietary treatment. Laying hens in the ML group had a lower concentration of albumen (ALB) and urea (UA) than those in the control group ($p < 0.05$) [36]. AST decreased with *M. oleifera* supplementation. The birds fed *M. oleifera* (ML and MS) recorded significantly ($p \leq 0.05$) reduced cholesterol levels. The reduction in cholesterol levels may be because *M. oleifera* contains hypocholesterolemic agents such the phytoconstituents and β-sitosterol [39]. Yuangsoi et al. [40] found that the levels of ALT and AST were similar in all diets, indicating normal organ function upon feeding with *Moringa* seed meal. Elkloub et al. [15] reported that plasma AST and ALT decreased with all levels of ML. As the liver produces enzymes like ALT and AST and releases them into the blood upon liver damage [37], thus, the absence of significant differences in serum AST values may indicate normal liver function of the birds on diets containing MSL. Lowered AST activity was observed in hens on 0.4% and 0.6% ML, which could suggest that ML can elevate liver health as well. Makanjuola et al. [38] observed that 0.2%, 0.4%, and 0.6% ML did not affect serum total protein, albumin, globulin, and AST levels. However, AST showed significant reduction in the birds fed a diet of (0.4%) *M. oleifera* leaves, with a good effect on the immune responses and development of the intestinal health of birds [28].

Table 6. Blood biochemical indices of laying Japanese quails as affected by dietary treatments.

Items	Control	MS	ML	MSL	SEM	*p* Value
AST (U/L)	193.00 [a]	134.00 [b]	144.00 [b]	178.67 [a]	8.50	0.01
ALT (U/L)	57.00	54.67	50.67	48.00	2.38	0.61
Total protein (g/dL)	5.80	6.30	6.07	6.23	0.08	0.13
Albumin (g/dL)	3.47	3.33	3.67	3.30	0.07	0.22
Globulin (g/dL)	2.33	2.97	2.40	2.93	0.12	0.09
Albumin / Globulin ratio	1.49	1.13	1.56	1.16	0.08	0.12
Triglycerides (mg/dL)	65.33 [a]	27.00 [b]	36.00 [b]	35.33 [b]	4.59	0.0002
Cholesterol (mg/dL)	115.33 [a]	47.67 [c]	85.00 [b]	74.67 [b]	7.76	0.0004
Urea (mg/dL)	59.67 [a]	41.67 [b]	58.33 [a]	46.00 [a,b]	2.92	0.03
Creatinine (mg/dL)	0.70	0.53	0.70	0.60	0.04	0.35
Bilirubin (mmol/L)	0.37	0.60	0.43	0.37	0.04	0.14

Control, the basal diet; MS, 1 g Moringa seeds/kg diet; ML, 1 g Moringa leaves/kg diet; MSL, 1 g seeds + 1 g leaves/kg. SEM: standard error means. a,b: Means in the same row with no superscript letters after them or with a common superscript letter following them are not significantly different ($p < 0.05$).

4. Conclusions

In comparison with the control group, egg production, egg mass, hatchability (%), and shell thickness were significantly higher and blood urea, AST, and lipid profile (cholesterol, and triglycerides) concentrations were significantly lower in birds fed a *M. oleifera* seed supplemented diet (treatment MS). From our results, we recommend the use of Moringa seeds at 1 g/kg in the diet of laying Japanese quails.

Author Contributions: M.A. and M.E.A.E.-H. conceived and designed the experiments. E.A.A., L.A.M. and M.S.E.-K. performed the experiments. M.A., A.E.T. and M.E.A.E.-H. analyzed data. E.A.A. and M.S.E.-K. contributed reagents/materials/analysis tools. M.A., M.E.A.E.-H. and V.T. wrote the paper. M.A., M.E.A.E.-H., V.L., V.T. and A.I.E.S. revised the paper. All authors have read and agreed to the published version of the manuscript.

Funding: The authors acknowledge the Deanship of Scientific Research at King Faisal University for the financial support under Nasher Track (Grant No. 186160).

Conflicts of Interest: All authors declare that they do not have any conflicts of interests that could inappropriately influence this manuscript.

References

1. Abd El-Hack, M.E.; Alagawany, M.; Shaheen, H.; Samak, D.; Othman, S.I.; Allam, A.A.; Taha, A.E.; Khafaga, A.F.; Arif, M.; Osman, A.; et al. Ginger and Its Derivatives as Promising Alternatives to Antibiotics in Poultry Feed. *Animals* **2019**, *10*, 452. [CrossRef]
2. Alagawany, M.; Elnesr, S.S.; Farag, M.R.; Abd El-Hack, M.E.; Khafaga, A.F.; Taha, A.E.; Tiwari, R.; Yatoo, M.; Bhatt, P.; Marappan, G.; et al. Use of licorice (Glycyrrhiza glabra) herb as a feed additive in poultry: Current knowledge and prospects. *Animals* **2019**, *9*, 536. [CrossRef] [PubMed]
3. Khoobani, M.; Hasheminezhad, S.H.; Javandel, F.; Nosrati, M.; Seidavi, A.; Kadim, I.T.; Laudadio, V.; Tufarelli, V. Effects of Dietary Chicory (*Chicorium intybus* L.) and Probiotic Blend as Natural Feed Additives on Performance Traits, Blood Biochemistry, and Gut Microbiota of Broiler Chickens. *Antibiotics* **2020**, *9*, 5. [CrossRef] [PubMed]
4. Adline, J.; Devi, J. A study on phytochemical screening and antibacterial activity of *Moringa oleifera*. *Int. J. Res. Appl.* **2014**, *2*, 169–176.
5. Shewita, R.S.; Taha, A.E. Influence of dietary supplementation of ginger powder at different levels on growth performance, haematological profiles, slaughter traits and gut morphometry of broiler chickens. *S. Afr. J. Anim. Sci.* **2018**, *48*, 997–1008. [CrossRef]
6. Gopalakrishnan, L.; Doriya, K.; Kumar, D.S. *Moringa oleifera*: A review on nutritive importance and its medicinal application. *Food Sci. Hum. Wellness* **2016**, *5*, 49–56. [CrossRef]
7. Makkar, H.P.S.; Becker, K. Nutritional value and antinutritional components of whole and ethanol extracted of *Moringa oleifera* leaves. *Anim. Feed Sci. Technol.* **1996**, *63*, 211–228. [CrossRef]

8. Pietrzak, D.; Mroczek, J.; Antolik, A.; Michalczuk, M.; Niemiec, J. Influence of growth stimulators added to feed on the quality of meat and fat in broiler chickens. *Med. Wet.* **2005**, *61*, 553–557.

9. Halle, I.; Thomann, R.; Bauermann, U.; Henning, M.; Kohler, P. Effects of a graded supplementation of herbs and essential oils in broiler feed on growth and carcass traits. *Landbauforshung Volkenrode* **2004**, *54*, 219–229.

10. Yang, R.Y.; Chang, L.C.; Hsu, J.C.; Weng, B.B.; Palada, M.C.; Chadha, M.L.; Levasseur, V. Nutritional and functional properties of *Moringa* leaves—From germplasm, to plant, to food, to health. In *Moringa Leaves: Strategies, Standards and Markets for a Better Impact on Nutrition in Africa*; Moringa news; CDE, CTA, GFU: Paris, France, 2006.

11. AbouSekken, M.S.M. Performance, Immune Response and Carcass Quality of Broilers Fed Low Protein Diets contained either *Moringa* Oleifera Leaves meal or its Extract. *J. Am. Sci.* **2015**, *11*, 153–164.

12. Biel, W.; Jaroszewska, A.; Lyson, E. Nutritional quality and safety of moringa (*Moringa oleifera* Lam., 1785) leaves as an alternative source of protein and minerals. *J. Elementol.* **2017**, *22*, 569–579.

13. Ghasi, S.; Nwobodo, E.; Ofilis, J.O. Hypocholesterolemic effects of crude extract of the leafs of *Moringa oleifera* Lam. in high-fat diet fed Wistar rats. *J. Ethnopharmacol.* **2000**, *69*, 21–25. [CrossRef]

14. Kakengi, A.M.V.; Kaijage, J.T.; Sarwatt, S.V.; Mutayoba, S.K.; Shem, M.N.; Fujihara, T. Effect of *Moringa oleifera* leaf meal as a substitute for sunflower seed meal on performance of laying hens in Tanzania. *Livest. Res. Rural Dev.* **2007**, *19*, 436–446.

15. Elkloub, K.; Moustafa, M.E.L.; Riry, F.H.; Mousa, M.A.M.; Hanan, A.H. Effect of using *Moringa oleifera* leaf meal on performance of Japanese quail. *Egypt. Poult. Sci.* **2015**, *35*, 1095–1108.

16. Abd El-Hack, M.; Alagawany, M.; Elrys, A.; Desoky, E.S.; Tolba, H.; Elnahal, A.; Swelum, A. Effect of Forage *Moringa oleifera* L.(moringa) on animal health and nutrition and its beneficial applications in soil, plants and water purification. *Agriculture* **2018**, *8*, 145. [CrossRef]

17. Ghebreselassie, D.; Mekonnen, Y.; Gebru, G.; Ergete, W.; Huruy, K. The effects of *Moringa* stenopetala on blood parameters and histopathology of liver and kidney in mice. *Ethiop. J. Health Dev.* **2011**, *25*, 51–57. [CrossRef]

18. AOAC. *Official Method of Analysis of the Association of Official Analytical Chemists*, 15th ed.; AOAC: Washington, DC, USA, 2004.

19. Gurnani, N.; Gupta, M.; Mehta, D.; Mehta, B.K. Chemical composition, total phenolic and flavonoid contents, and in vitro antimicrobial and antioxidant activities of crude extracts from red chilli seeds (*Capsicum frutescens* L.). *J. Taibah Univ. Sci.* **2016**, *10*, 462–470. [CrossRef]

20. Meda, A.; Lamien, C.E.; Romito, M.; Millogo, J.; Nacoulma, O.G. Determination of the total phenolic, flavonoid and proline contents in burkina fasan honey, as well as their radical scavenging activity. *Food Chem.* **2005**, *91*, 571–577. [CrossRef]

21. NRC. *National Research Council. Nutrient Requirements of Poultry*, 9th ed.; National Academy Press: Washington, DC, USA, 1994.

22. Saeed, M.; Abd El-Hack, M.E.; Arif, M.; El-Hindawy, M.M.; Attia, A.I.; Mahrose, K.M.; Hayat, K. Impacts of distiller's dried grains with solubles as replacement of soybean meal plus vitamin E supplementation on production, egg quality and blood chemistry of laying hens. *Ann. Anim. Sci.* **2017**, *17*, 849–862. [CrossRef]

23. Keener, K.M.; McAvoy, K.C.; Foegeding, J.B.; Curtis, P.A.; Anderson, K.E.; Osborne, J.A. Effect of testing temperature on internal egg quality measurements. *Poult. Sci.* **2006**, *85*, 550–555. [CrossRef]

24. Card, L.E.; Nesheim, M.C. Marketing eggs. In *Poultry Production*, 11th ed.; Lea and Febiger: Philadelphia, PA, USA, 1972; pp. 291–295.

25. Friedewald, W.T.; Levy, R.I.; Fredrickson, D.S. Estimation of the concentration of low-density lipoprotein cholesterol in plasma, without use of the preparative ultracentrifuge. *Clin. Chem.* **1972**, *18*, 499–502. [CrossRef] [PubMed]

26. Nie, N.H.; Bent, D.H.; Hull, C.H. *SPSS: Statistical Package for the Social Sciences*; McGraw-Hill: New York, NY, USA, 1970.

27. Kwari, I.; Diarra, S.; Raji, A.; Adamu, S. Egg production and egg quality of laying hens fed raw or processed sorrel (*Hibiscus sabdariffa*) seed meal. *Agric. Biol. J. N. Am.* **2011**, *2*, 616–621. [CrossRef]

28. Olugbemi, T.S.; Mutayoba, S.K.; Lekule, F.P. *Moringa oleifera* leaf meal as a hypocholesterolemic agent in laying hen diets. *Livest. Res. Rural Dev.* **2010**, *84*, 27–37.

29. Riry, F.H.; Elkloub, K.; Moustafa, M.E.L.; Mousa, M.A.M.; Youssef, S.F.; Hanan, A.H. Effect of using *Moringa oleifera* seed meal on Japanese quail performance during growing period. In Proceedings of the 9th International Poultry Conference, Hurghada, Egypt, 7–10 November 2016; pp. 322–337.

30. Abdelnour, S.A.; Abd El-Hack, M.E.; Ragni, M. The Efficacy of High-Protein Tropical Forages as Alternative Protein Sources for Chickens: A Review. *Agriculture* **2018**, *8*, 86. [CrossRef]

31. Etalem, T.; Getachew, A.; Mengistu, U.; Tadelle, D. Cassava root chips and *Moringa oleifera* leaf meal as alternative feed ingredients in the layer ration. *J. Appl. Poult. Res.* **2014**, *23*, 614–624.

32. Mahmood, H.; Al-Daraji, H. Effect of dietary supplementation with different level of zinc on sperm egg penetration and fertility traits of broiler breeder chicken. *Pak. J. Nutr.* **2011**, *10*, 1083–1088.

33. Moyo, B.; Masika, P.; Hugo, A.; Muchenje, V. Nutritional characterization of *Moringa* (*Moringa oleifera* Lam.) leaves. *Afr. J. Biotechnol.* **2011**, *10*, 1292–1293.

34. Durmus, I.; Atasoglu, C.; Mizrak, C.; Ertas, S.; Kaya, M. Effect of increasing zinc concentration in the diets of Brown parent stock layers on various production and hatchability traits. *Arch. Fur Tierz.* **2004**, *5*, 483–489. [CrossRef]

35. Brown, L.; Pentland, S. *Health Infertility Organization: Male Infertility-Improving Sperm Quality*; Acubalance Wellness Centre Ltd.: Vancouver, BC, Canada, 2007.

36. Lu, W.; Wang, J.; Zhang, H.J.; Wu, S.G.; Qi, G.H. Evaluation of *Moringa oleifera* leaf in laying hens: Effects on laying performance, egg quality, plasma biochemistry and organ histopathological indices. *Ital. J. Anim. Sci.* **2016**, *15*, 658–665. [CrossRef]

37. Ebenebe, C.I.; Anigbogu, C.C.; Anizoba, M.A.; Ufele, A.N. Effect of various levels of *Moringa* leaf meal on the egg quality of ISA Brown breed of layers. *J. Adv. Life Sci. Technol.* **2013**, *14*, 45–49.

38. Makanjuola, B.A.; Obi, O.O.; Olorungbohunmi, T.O.; Morakinyo, O.A.; Oladele-Bukola, M.O.; Boladuro, B.A. Effect of *Moringa oleifera* leaf meal as a substitute for antibiotics on the performance and blood parameters of broiler chickens. *Livest. Res. Rural Dev.* **2014**, *26*, 144.

39. Riry, F.H.; Elkloub, K.; Moustafa, M.E.L.; Mousa, M.A.M.; Hanan, A.H.; Youssef, S.F. Effect of partial replacement of soybean meal by *Moringa oleifera* seed meal on Japanese quail performance during laying period. *Egypt. Poult. Sci. J.* **2018**, *38*, 255–267.

40. Yuangsoi, B.; Klahan, R.; Charoenwattanasak, S. Partial replacement of protein in soybean meal by *Moringa* seed cake (*Moringa oleifera*) in bocourti's catfish (*Pangasius bocourti*). *Songklanakarin J. Sci. Technol.* **2014**, *36*, 125–135.

 *sustainability*

Article

Composition and Efficacy of a Natural Phytotherapeutic Blend against Nosemosis in Honey Bees

Romeo Teodor Cristina [1], Zorana Kovačević [2,*], Marko Cincović [2], Eugenia Dumitrescu [1], Florin Muselin [1], Kalman Imre [1], Dumitru Militaru [3], Narcisa Mederle [1], Isidora Radulov [1], Nicoleta Hădărugă [1] and Nikola Puvača [4]

[1] Department of Pharmacology and Pharmacy, Toxicology, Parasitology and Biochemistry, Faculty of Veterinary Medicine, Banat's University of Agricultural Sciences and Veterinary Medicine "King Michael I of Romania" from Timișoara, 300645 Timisoara, Romania; romeocristina@usab-tm.ro (R.T.C.); cris_tinab@yahoo.com (E.D.); florin.muselin@gmail.com (F.M.); kalman_imre27@yahoo.com (K.I.); narcisamederle@usab-tm.ro (N.M.); isidoraradulov@yahoo.com (I.R.); nico_hadaruga@yahoo.com (N.H.)

[2] Department of Veterinary Medicine, Faculty of Agriculture, University of Novi Sad, 21000 Novi Sad, Serbia; mcincovic@gmail.com

[3] Department of Parasitology, Faculty of Biotechnology, University of Agronomic Sciences and Veterinary Medicine of Bucharest, 011464 Bucharest, Romania; pasteurmili@yahoo.com

[4] Department of Engineering Management in Biotechnology, Faculty of Economics and Engineering Management in Novi Sad, University Business Academy in Novi Sad, 21000 Novi Sad, Serbia; nikola.puvaca@fimek.edu.rs

* Correspondence: zorana.kovacevic@polj.edu.rs; Tel.: +381-63-846-8365

Received: 3 July 2020; Accepted: 18 July 2020; Published: 21 July 2020

Abstract: Honey bees are essential to sustaining ecosystems, contributing to the stability of biodiversity through pollination. Today, it is known that the failure of pollination leads irremediably to the loss of plant cultures and, as a consequence, inducing food security issues. Bees can be affected by various factors, one of these being *Nosema* spp. which are protozoans specifically affecting adult honey bees and a threat to bee populations around the world. The composition of the phytotherapeutic product (*Protofil®*) for treating nosemosis was analyzed from a biochemical point of view. The most concentrated soluble parts in the phytotherapeutic association were the flavonoids, most frequently rutin, but quercetin was also detected. Additionally, the main volatile compounds identified were eucalyptol (1.8-cineol) and chavicol-methyl-ether. To evaluate the samples' similarity–dissimilarity, the PCA multivariate statistical analysis, of the gas-chromatographic data (centered relative percentages of the volatile compounds), was applied. Statistical analysis revealed a significant similarity of *Protofil®* with the *Achillea millefolium* (Yarrow) samples and more limited with *Thymus vulgaris* (Thyme) and *Ocimum basilicum* (Basil), and, respectively, a meaningful dissimilarity with *Taraxacum officinale* (Dandelion). The results have shown a high and beneficial active compounds concentration in the analyzed herbs. High similarity with investigated product recommending the *Protofil®*, as the treatment compatible with producing organic honey.

Keywords: *Apis mellifera*; *Nosema* spp.; *Protofil®*; biochemical analysis

1. Introduction

Bees are necessary for maintaining ecosystems, contributing to biodiversity through pollination, a vital factor for a wide range of crops and wild plants. Today, it is known that the failure of pollination will lead irremediably to the loss of plant cultures and, as a consequence, food security concerns [1].

Worldwide, 75% of the crops are pollinated by insects with 57 species (mostly bees) as crucial pollinators for approximately 107 plants [1,2].

Honey bees (*Apis mellifera L.*) are affected by many diseases, the most important being of fungal and viral origin. The main factors affecting disease are small colony population size, extended winter, reduction of cleaning flights, feed supplements, and the hive's excessive humidity [3–5].

Under these circumstances, nosemosis caused by *Nosema apis* Zander and *Nosema ceranae* Fries protozoa became the principal threat and the most commonly found in honey bee populations [6–12].

During the last decade, allopathic drugs against nosemosis were restricted to a few active substances such as Fumagillin (fumidil), an antibiotic obtained from *Aspergillus fumigatus.* Unfortunately, although an efficient product, due to the risk of residues, EMA has excluded this product from use in Europe in February 2016 [6,13–18].

In the given circumstances in the treatment of nosemosis, a reliable backup could be ecologic phytotherapy, the usage of whole herbs or parts, with recognized antiprotozoal activity (like flowers of *Matricaria chamomilla*, *Hypericum perforatum* or *Achillea millefolium*, leaves of *Mentha piperita*, or leaves and flowers of *Ocimum basilicum*) currently being viewed as a great opportunity [19–22].

The Research and Development Institute for Beekeeping has developed an herbal product that presents the blend of essential oils highly efficient against *Nosema* spp [23]. Essential oils used in this product are derived from herbs found in spontaneous flora, which include different cyclic and aliphatic hydrocarbons, triterpenes and sesquiterpenes, phenolic structures, oleanolic acid, flavones, microelements, and the vitamins of B group [24].

This study aimed to analyze the composition of *Protofil®* as commercial product suggested was the usage in honey bees' production, as well as to analyze basil, thyme, yarrow, and dandelion, and to compare them to the aforementioned product, respectively.

2. Materials and Methods

The product *Protofil®* plant association is a brownish solution, with a characteristic aromatic odor and taste, designed to combat *Nosema* spp., and unique advantage is that it has no contraindications (no intoxication or any side effects) to honey bees [25,26].

The sample of the product *Protofil®* was chemically investigated, directly from the producer, the ICDA (Research and Development Institute for Beekeeping, Bucharest, Romania).

Besides *Protofil®*, samples of *Achillea millefolium*, *Thymus vulgaris*, *Ocimum basilicum*, and *Taraxacum officinale*, were chemically investigated as well.

The physicochemical methods used to investigate *Protofil®* and plants were: Reversed-Phase High-Performance Liquid Chromatography (RP-HPLC) of the filtered undiluted or diluted hydro-alcoholic extracts and Mass-Spectrometry (GC-MS) coupled with Gas Chromatography of volatile compounds separated by hydro-distillation-extraction in an organic solvent (SDE).

2.1. RP-HPLC Investigation

RP-HPLC investigation of the flavonoid standards and hydro-alcoholic extract samples was performed on a Jasco apparatus (*Abbl&e-Jasco, Bucharest, Romania*) equipped with: quaternary pump (PU-2080 Plus); mixing unit (LG-2080-04 Quaternary Gradient); degasser (DG-2080-54 4); spectrophotometric detector (UV-2070 Plus Intelligent UV/VIS Detector); acquire and process computer data (JASCO ChromPass Chromatography Data System, Version 1.7.403.1), through an LC-Net II/ADC interface.

The conditions of analysis were:

- Column: Nucleosil 100 C18, 250 × 4.6 mm × mm, 5 μm particle diameter;
- UV wavelength: 254 nm; Mobile phase: Acetonitrile: Water = 50:50; Temperature: 25 °C; Flow rate: 1.0 mL/min; Injected volume: 20 μL.

For flavonoids, evaluation of their concentration in hydro-alcoholic extracts were performed using the obtained HPLC calibration curves. The flavonoids' identification correlated the detection of retention times with the standards matching. Therefore, before analysis, the samples were filtered, and, in most cases, they were diluted (1:100).

The samples' bioactive compounds concentration was measured using the calibration curves for the available flavonoids, results being expressed as mg of flavonoid compound, separated at the retention times corresponding to the standard/mL of sample.

For the *RP-HPLC*, the following standards were used:

- Rutin (≥94%) (Sigma-Aldrich, Taufkirchen, Germany),
- Quercetin (≥95%) (Sigma-Aldrich, Taufkirchen, Germany),
- Chrysene (>98%) (Sigma-Aldrich, Taufkirchen, Germany),
- Flavone (≥99%) (Sigma-Aldrich, Taufkirchen, Germany).

Standard solutions were obtained by dilution in 96% ethanol (*Chimopar, Bucharest, Romania*) also HPLC purity solvents being used for the chromatographic analysis: acetonitrile (HPLC grade) (*Fluka Chemie, München, Germany*) and bidistilled water HPLC (*Fluka Chemie, München, Germany*).

2.2. GC-MS Analysis

The *GC-MS analysis* of SDE-separated volatile compounds implied the use of hexane (GC grade) (*Fluka*) for the extraction of volatile compounds separated, and anhydrous sodium sulfate (>99%) (*Merck*) to dry the hexane extract. The Kovats retention indices were calculated based on GC-MS assays performed under the same conditions for a mixture of linear C8–C20 alkanes (*Fluka Chemie*).

2.3. Separation of Volatile Compounds by Hydrodistillation-Extraction (SDE)

The GC-MS analysis of the separated volatile compounds from hydro-alcoholic extracts by hydrodynamic extraction in hexane (SDE) allowed the relative percentage concentrations of the components to be evaluated using the area method (Equation (1)):

$$\text{Relative concentration } (\%) = \frac{\text{Area (compound)}}{\sum \text{Area}} \times 100 \tag{1}$$

For the analysis of the separated volatile compounds, an HP 6890 Series GC (*Hewlett Packard*), coupled with an HP 5973 Mass Selective Detector mass spectrometer was used.

The GC assay conditions were: Column: HP-5MS, L = 30 m, inner diameter 0.25 mm, film thickness 0.25 µm; Temperature program: 50 to 250 °C at a speed of 6 °C/min; Injector temperature: 280 °C; Detector temperature: 280 °C; Injection volume: 2 µL; Carrier gas: He.

For the MS detector, an EE energy of 70 eV was used, at a source temperature of 150 °C, scanning range of 50–300 amu, with the speed of 1 s^{-1} for mass spectrometry, and the obtained spectra, compared with a NIST/EPA/NIH Mass Spectral Library 2.0 database (2002). For data acquisition, version B.01.00/98, of HP Enhanced Chem Station G1701BA software was used, the data processing, being completed utilizing the HP Enhanced Data Analysis program. Hydro-alcoholic samples (~800 mL) were prepared and the condensed volatile compounds, extracted in an SDE system, in 20 mL hexane. The method lasted four hours, and the separated hexane extract was dried. Dry hexane extracts were then GC-MS analyzed, determining the relative percentage concentration of the volatile compounds.

2.4. Statistical Multivariate Principal Component Analysis (PCA) of GC Data

Multivariate analysis of gas chromatography data for hexane extracts of volatile compounds, allowed a classification of samples based on volatile compounds and their relative concentrations, identifying the similarity of these samples. To assess the investigated samples similarity–dissimilarity, the multivariate statistical data analysis—Principal component analysis (PCA), of gas-chromatographic data, was used, the GC data being used for analysis, and validated by *cross-validation* method.

3. Results

3.1. HPLC Curve Calibration for Standard Compounds

To evaluate the concentration of the flavonoid compounds in hydro-alcoholic extracts, calibration curves for the available flavonoids, rutin, quercetin, chrysene, and flavone, were determined. In the case of *rutin*, the HPLC analysis of the standard solutions indicated the chromatographic peak presence in the retention time range of 2–3 min (most probably, a mixture of isomers due to the presence of two chromatographic peaks that were analyzed together). The quercetin chromatographic peak was detected to 4.2 min, the HPLC examination of *chrysene* and *flavone,* assigning peaks, after 9.8 and, respectively, 15.8 min. The HPLC results for standards are presented in Figure 1.

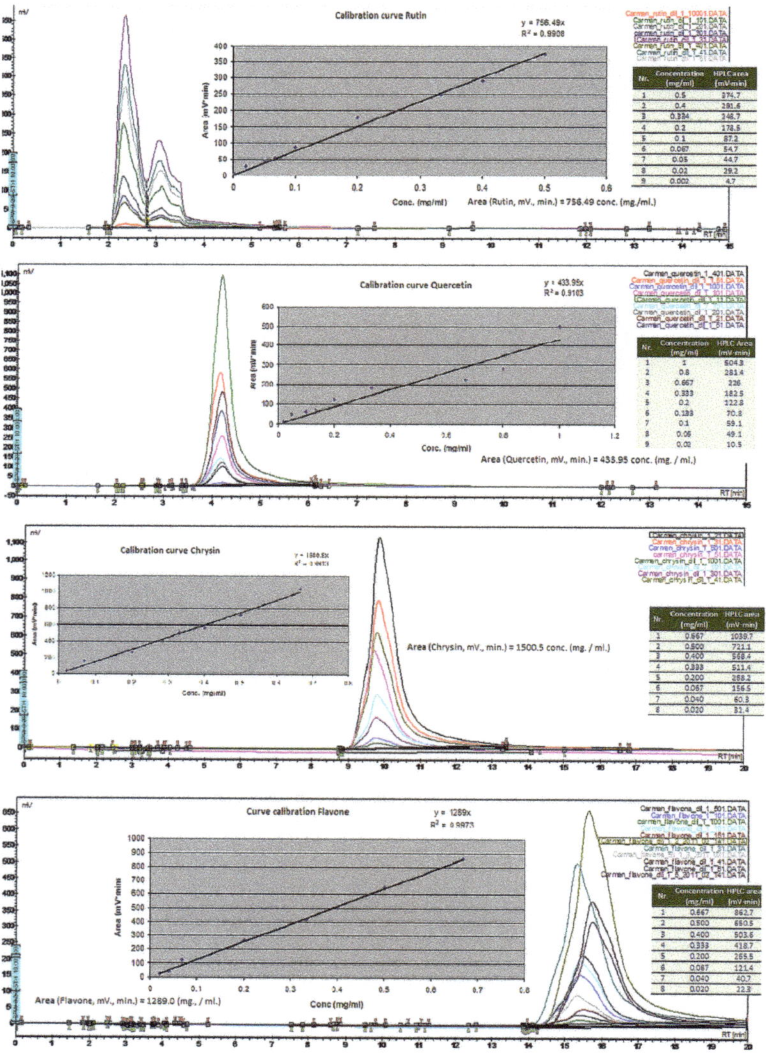

Figure 1. Calibration curves for rutin, quercetin, chrysene, and flavone.

Evaluation of the Flavonoids' Concentration

Concentrations of the studied samples (mean of four replicates, expressed as mg flavonoid available/mL sample) are shown in Table 1.

Table 1. Concentrations of compounds (mean of four replicates), expressed as mg flavonoids/mL, determined from HPLC analyzes.

Nr.	Compound	RT (min)	Conc. (Ba) (mg/mL)	Conc. (Th) (mg/mL)	Conc. (Ya) (mg/mL)	Conc. (Da) (mg/mL)	Conc. (PF) (mg/mL)
1	*Rutin*	2–3.6	1.843	3.437	2.543	1.049	1.540
2	*Quercetin*	4.2	0.017	9.379	0.232	0.029	0.061
3	*Chrysene*	9.8	0.027	0.012	0.004	0.000	0.007
4	*Flavone*	15.8	0.000	0.000	0.010	0.016	0.002

RT-retention time; Conc.-concentration; Ba-basil; Th-thyme; Ya-yarrow; Da-dandelion; PF-*Protofil®*.

HPLC chromatograms of undiluted samples and etalons, for *Ocimum basilicum*, *Thymus vulgaris*, *Achillea millefolium*, and *Taraxacum officinale* are presented in Figure 2, and the chromatogram for the associated conditioning *Protofil®*, in Figure 3.

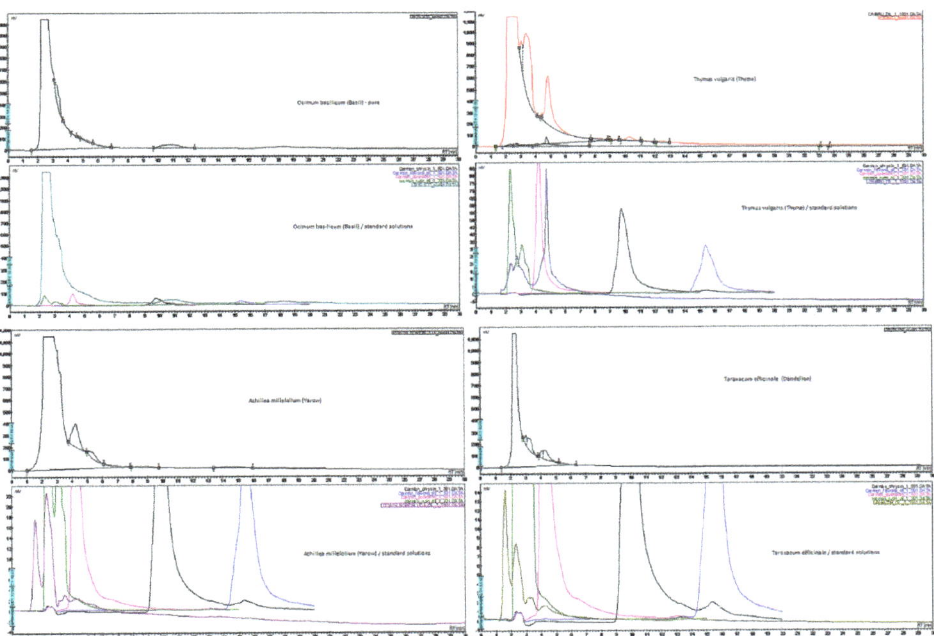

Figure 2. The HPLC chromatograms obtained for the pure samples undiluted (up) and ethalons overlayed (down) for Basil (*Ocimum basilicum*), Thyme (*Thymus vulgaris*), Yarrow (*Achillea millefolium*) and Dandelion (*Taraxacum officinale*).

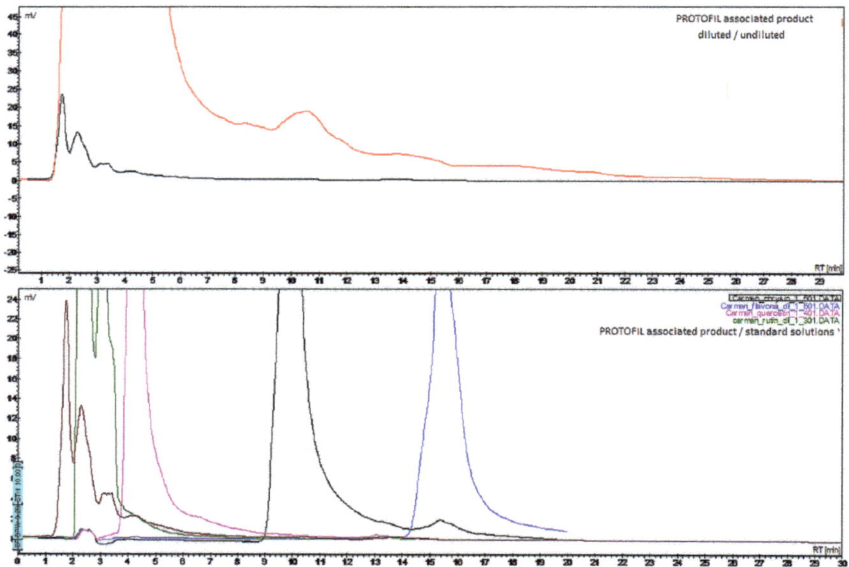

Figure 3. Overlap chromatograms from HPLC analysis for samples/standards (undiluted and diluted) for *Protofil®*.

The most concentrated were the flavonoids (expressed as *rutin*) separated at the beginning of the chromatogram due to the higher hydrophilicity of these compounds, containing saccharide residues, followed by polyphenolic flavonoids of the quercetin type.

Chrysene, a *bis*-phenolic compound, and similar structures separated at high retention times were detected in medium–low concentrations, while flavone a non-phenolic compound were detect in extremely low concentrations. Analyzing the data for the four herb samples leads to results close to the *Protofil's* obtained data, except in the case of quercetin (probably due to inappropriate *rutin* separation).

The HPLC separation of the flavonoid compounds studied, on the C18 nonpolar column, correlates well with their hydrophobicity, with retention times increasing with hydrophobicity, expressed as the logarithm of the octanol/water partition coefficient, calculated with the QSAR Properties program in the *HyperChem* 5.1 package (log P_{rutin} = 1.61, log $P_{quercetin}$ = 0.28, log $P_{chrysin}$ = 1.75 and log $P_{flavone}$ = 2.32). The best correlation is polynomial of order 2 (r^2 = 0.98).

3.2. GC-MS Analysis of Volatile Compounds' Relative Concentration

For basil extract, (the most significant from the set of analyses), 56 components (expressed as *abundance* of 10,000) were separated, the most concentrated compound identified being *chavicol-methyl-ether* (55%) (Table 2).

In the case of volatile compounds in the GC-MS of *Thymus vulgaris* (Thyme), 43 chromatographic peaks were identified, the most concentrated being *eucalyptol* and *γ-terpinene* (Table 3).

Table 2. Results of GC-MS analysis for *Ocimum basilicum* (Basil) samples.

No.	A Compound Identified by GC-MS	RT (min)	Kovats Index	Relative Concentration (%)
1	*Column*	5.198	948	0.00
2	*alpha-Pinene*	5.686	974	1.29
3	*Column*	5.944	987	0.00
4	*Column*	6.173	999	0.00
5	*Camphene*	6.250	1002	0.15
6	*beta-Phellandrene*	6.796	1028	0.50
7	*beta-Pinene*	6.955	1035	1.42
8	*Sabinen/beta-pinene*	7.037	1039	0.64
9	*Bicyclo [3.1.0]hexane, 4-methyl-1-(1-methylethyl)-, didehydro*	7.584	1063	0.05
10	*Terpinolen*	7.830	1074	0.06
11	*Limonene*	8.101	1086	1.10
12	*Dihydrocarveol*	8.195	1090	0.08
13	*3-Carene/alpha-pinene*	8.318	1095	0.27
14	*Eucalyptol*	8.582	1106	7.31
15	*gamma-Terpinen*	9.005	1124	0.14
16	*Terpinolen*	9.740	1155	0.13
17	*2-Cyclohexen-1-ol, 1-methyl-4-(1-methylethyl)-, cis-*	10.116	1171	0.09
18	*Linalool*	10.192	1174	0.31
19	*Fenchone*	10.739	1198	0.08
20	*Tetrahydroactinidiolide*	11.015	1210	0.05
21	*Camphor*	12.607	1280	0.08
22	*Caprylyl acetate*	12.683	1283	0.02
23	*Fenchyl acetate*	13.136	1303	0.05
24	*Chavicol methyl ether*	13.870	1337	54.91
25	*Bornyl acetate*	15.034	1391	0.04
26	*alpha-Cubebene*	15.239	1400	0.04
27	*Copaene*	16.015	1437	0.30
28	*Di-epi-alpha-cedrene*	16.138	1443	0.28
29	*beta-Bourbonene*	16.362	1454	0.24
30	*beta-Elemene*	16.479	1459	0.58
31	*alpha-Bergamotene*	16.984	1484	0.15
32	*alpha-Bergamotene*	17.290	1498	5.79
33	*Caryophyllene*	17.401	1504	6.75
34	*trans-Caryophyllene/Isocaryophyllene*	17.684	1518	0.25
35	*beta-Farnesene*	17.778	1522	1.74
36	*Humulene*	18.300	1548	1.88
37	*gamma-Muurolene*	18.436	1555	0.22
38	*(Z)-beta-Farnesene*	18.706	1568	0.48
39	*Germacrene D*	18.906	1578	3.41
40	*alpha-Himachalene*	19.088	1587	0.54
41	*Eremophilene*	19.293	1598	0.11
42	*Elixene*	19.358	1601	0.23
43	*beta-Cedrene*	19.599	1613	0.10
44	*gamma-Cadinene*	19.658	1616	0.75
45	*beta-Cadinene*	19.758	1621	0.19
46	*1,4,7,-Cycloundecatriene, 1,5,9,9-tetramethyl-, Z,Z,Z-*	19.928	1630	6.33
47	*Calamenene*	20.275	1648	0.09
48	*Caryophyllene oxide*	21.790	1728	0.02
49	*Palmitic acid*	27.519	2068	0.15
50	*Ethyl palmitate*	27.584	2072	0.41
51	*Column*	30.163	2262	0.00
52	*Arachidic acid*	30.598	2299	0.10
53	*Ethyl linolenate*	30.980	2333	0.08
54	*Column*	31.755	2407	0.00
55	*Column*	32.214	2456	0.00
56	*Column*	33.230	2577	0.00

RT-retention time.

Table 3. Results of GC-MS analysis for *Thymus vulgaris* (Thyme) samples.

No.	A Compound Identified by GC-MS	RT (min)	Kovats Index	Relative Concentration (%)
1	*solvent*	5.254	951	0.00
2	*alpha-Thujene*	5.454	962	2.06
3	*alpha-Pinene*	5.683	974	2.94
4	*Camphene*	6.247	1002	1.39
5	*beta-Terpinen*	6.8	1028	0.05
6	*beta-Pinene*	6.952	1035	0.58
7	*beta-Pinene*	7.035	1039	2.51
8	*alpha-Thujene*	7.581	1063	0.43
9	*alpha-Terpinen*	7.828	1074	2.14
10	*Limonene*	8.104	1086	0.99
11	*beta-Phellandrene*	8.321	1095	0.45
12	*Eucalyptol*	8.556	1105	57.63
13	*gamma-Terpinen*	9.015	1124	15.03
14	*Terpinolen*	9.743	1155	0.19
15	*2-Cyclohexen-1-ol, 1-methyl-4-(1-methylethyl)-, cis-*	10.113	1171	0.11
16	*Dehydro-p-cymene*	10.583	1191	0.11
17	*2-Cyclohexen-1-ol, 1-methyl-4-(1-methylethyl)-, cis-*	11.018	1210	0.06
18	*2-Cyclohexen-1-ol, 1-methyl-4-(1-methylethyl)-, cis-*	12.616	1280	0.18
19	*Methyl chavicol*	13.874	1337	0.14
20	*Thymol methyl ether*	13.991	1342	1.68
21	*2-Isopropyl-1-methoxy-4-methylbenzene*	14.15	1350	1.61
22	*Bornyl acetate*	15.031	1391	0.06
23	*Ylangene*	15.883	1431	0.06
24	*Copaene*	16.012	1437	0.20
25	*beta-Bourbonene*	16.359	1453	0.14
26	*Ylangene*	17.293	1499	0.07
27	*Caryophyllene*	17.387	1503	5.10
28	*Alloaromadendren*	17.716	1519	0.15
29	*Humulene*	18.292	1548	0.20
30	*gamma-Muurolene*	18.686	1567	0.48
31	*alpha-Muurolene*	18.803	1573	0.07
32	*Germacrene D*	18.903	1578	0.07
33	*alpha-Muurolene*	19.22	1594	0.57
34	*gamma-Cadinene*	19.661	1617	0.61
35	*beta-Cadinene*	19.761	1622	0.76
36	*alpha-Muurolene*	20.107	1639	0.07
37	*Calamenene*	20.272	1648	0.30
38	*Caryophyllene oxide*	21.788	1728	0.15
39	*5,9,9-Trimethyl-spiro[3.5]non-5-en-1-one*	27.029	2035	0.06
40	*Palmitic acid, ethyl ester*	27.605	2073	0.49
41	*Cholesterol, trifluoroacetate*	30.677	2306	0.00
42	*Linolenic acid, methyl ester*	31.006	2335	0.11
43	*2,4,4,6,6,8,8-Heptamethyl-1-nonene*	32.699	2511	0.00

RT-retention time.

The most concentrated volatile components in the *Achillea millefolium* (Yarrow) specimens were: *camphor* (relative concentration of 37.5%) and *eucalyptol* (25%), the total GC-separated compounds, in this case, being 44, some of which derived from the column (especially at the high separation temperatures cases) (Table 4).

Table 4. Results of GC-MS analysis for *Achillea millefolium* (Yarrow) samples.

No.	A Compound Identified by GC-MS	RT (min)	Kovats Index	Relative Concentration (%)
1	*alpha-Pinene*	5.681	974	0.22
2	*Camphene*	6.251	1002	0.78
3	*Yomogi alcohol*	7.42	1056	0.26
4	*3-Thujene*	7.59	1063	0.17
5	*Terpinolen*	7.837	1074	0.46
6	*4-Oxo-beta-isodamascol*	7.913	1077	0.62
7	*beta-Terpinen*	8.319	1095	0.17
8	*Eucalyptol*	8.577	1106	25.14
9	*2-Carene*	9.012	1124	1.12
10	*2-Norpinanol, 3,6,6-trimethyl-*	9.652	1151	0.43
11	*3-Thujanone*	10.91	1205	3.17
12	*Isopulegol*	11.021	1210	0.69
13	*alpha-Thujone*	11.133	1215	0.79
14	*4-Oxo-beta-isodamascol*	11.679	1239	0.85
15	*cis-Sabinol*	11.779	1243	0.37
16	*Verbenyl ethyl ether*	11.867	1247	0.79
17	*Lavandulol*	12.061	1255	0.57
18	*Lavandulol*	12.208	1262	0.79
19	*4-Oxo-beta-isodamascol*	12.42	1271	0.58
20	*Camphor*	12.596	1279	37.52
21	*E-3,5-Dimethylhex-2-en-1,2-dicarboxylic acid*	13.031	1299	1.25
22	*Isobornyl formate*	13.142	1304	1.01
23	*Chavicol methyl ether*	13.877	1337	3.27
24	*trans-Chrysanthenyl Acetate*	14.288	1356	0.21
25	*Isobornyl acetate*	15.028	1390	0.35
26	*cis-Carvyl Acetate*	16.345	1453	0.36
27	*Capric acid, ethyl ester*	16.82	1476	1.13
28	*9-Cedranone*	19.811	1624	0.34
29	*Spathulenol*	21.656	1721	0.45
30	*Caryophyllene oxide*	21.785	1728	1.95
31	*2-Cyclohexene-1-carboxaldehyde,2,6-dimethyl-6-(4-methyl-3-pentenyl)*	22.437	1763	0.16
32	*4(equatorial)-n-Propyl-trans-3-oxabicyclo[4.4.0]decane*	22.496	1766	0.21
33	*gamma-Eudesmol*	22.614	1773	1.94
34	*beta-Guaiene*	22.772	1782	0.58
35	*alpha-Eudesmol*	23.148	1803	2.47
36	*Humulane-1,6-dien-3-ol*	23.425	1818	0.78
37	*Aristolone*	23.742	1836	0.24
38	*Ethyl myristate*	24.347	1871	0.16
39	*Palmitic acid, ethyl ester*	27.602	2073	3.37
40	*Ethyl Oleate*	30.504	2291	0.28
41	*Linoleic acid ethyl ester*	30.663	2305	2.19
42	*Ethyl linolenate*	30.992	2334	1.79
43	*Column*	31.962	2429	0.00
44	*Column*	32.584	2498	0.00

RT-retention time.

The total concentration of active compounds in the hexane extracts of dandelion was identified. Upon identification of total active compounds, relative concentration was determined. The highest concentration of 5.7% *eucalyptol*, and 62.5% *ethyl palmitate* (Table 5) was recorded, respectively.

Table 5. Results of GC-MS analysis for *Taraxacum officinalis* (Dandelion) samples.

No.	A Compound Identified by GC-MS	RT (min)	Kovats Index	Relative Concentration (%)
1	*Bicyclo[2.1.1]hexan-2-ol, 2-ethenyl-*	5.146	945	0.59
2	*p-Xylene*	5.252	951	2.42
3	*Octane, 1-chloro-*	5.458	962	0.43
4	*alpha-Thujene*	5.693	975	0.53
5	*Isovaleraldehyde, diethyl acetal*	5.945	987	1.72
6	*Linalyl propionate*	6.239	1002	0.64
7	*Pentane, 1,1-diethoxy-*	7.003	1037	1.41
8	*Eucalyptol*	8.572	1106	5.72
9	*trans-Verbenol*	9.03	1125	0.40
10	*Chavicol methyl ether*	14.012	1343	0.46
11	*Bicyclo[2.2.1]heptane, 2-cyclopropylidene-1,7,7-trimethyl-*	17.385	1503	0.66
12	*2,3-Dehydro-4-oxo-beta-ionone*	19.688	1618	0.50
13	*Ethyl laurate*	20.863	1679	1.15
14	*Ethyl myristate*	24.371	1872	2.17
15	*Oxirane, 2-methyl-2-(1-methylethyl)-*	25.117	1916	0.93
16	*Methyl 2-methylhexanoate*	26.016	1971	0.71
17	*Ethyl palmitate*	27.614	2074	62.49
18	*Eicosane*	28.084	2106	1.31
19	*2,6-Pyrazinediamine*	30.522	2292	0.55
20	*Ethyl stearate*	30.622	2301	1.02
21	*Ethyl linolate*	30.687	2307	8.13
22	*Methyl linolenate*	31.016	2336	6.06

RT-retention time.

The volatile compounds GC-MS analysis is presented in Figures 4 and 5.

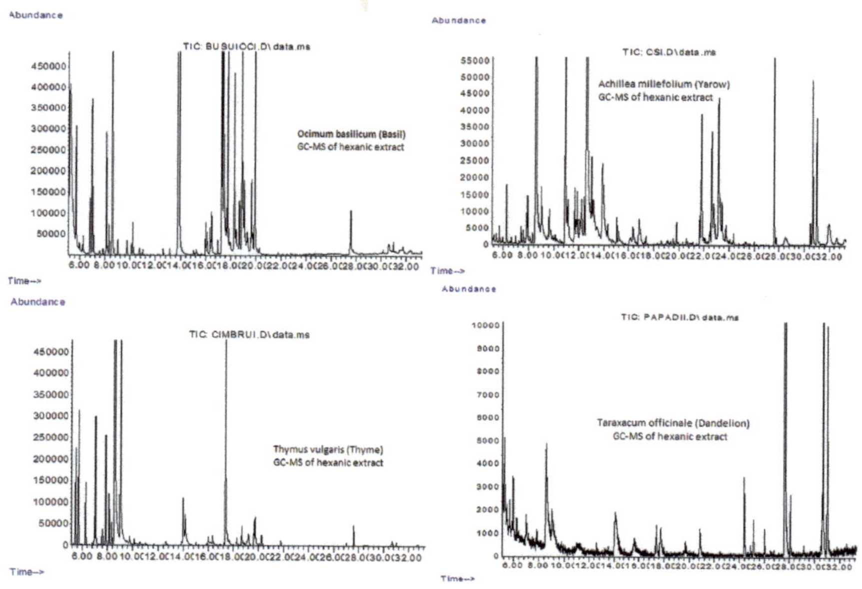

Figure 4. Chromatograms of GC-MS analysis of hexane extract obtained from basil, thyme, yarrow, and dandelion.

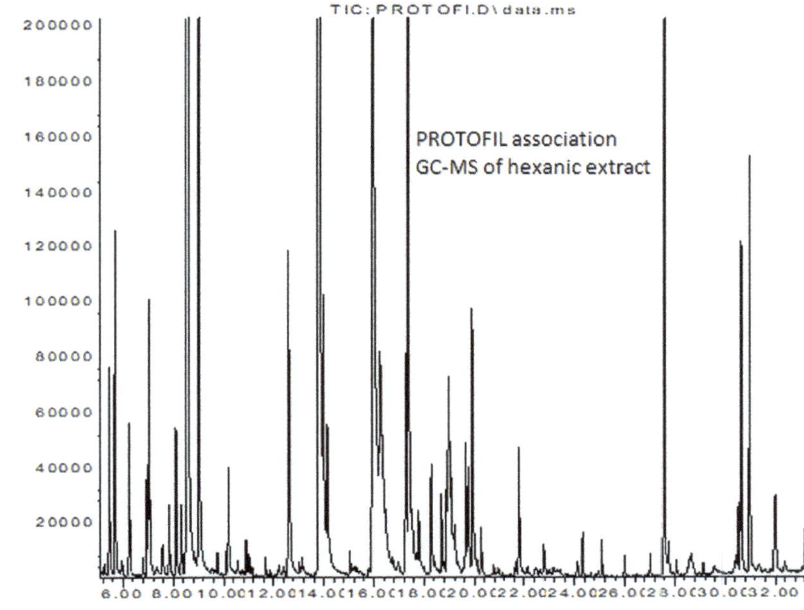

Figure 5. Chromatograms of GC-MS analysis of hexane extract obtained from *Protofil®*.

The *Protofil®* analysis, described the most relevant absolute concentration, totaling 74 different components, as well as their absolute abundance in the hexane extract. The highest recorded concentration of active ingredients was *eucalyptol* (28.6%), followed by *chavicol-methyl-ether* (28.1%), while the lover concentration of *thymol* (7.19%), and *gamma-Terpinen* (5.86%), was also present in the investigated sample (Table 6).

Table 6. Results of GC-MS analysis for *Protofil®* samples.

No.	A Compound Identified by GC-MS	RT (min)	Kovats Index	Relative Concentration (%)
1	*p-Xylene*	5.253	951	0.05
2	*1-Isopropyl-4-methylbicyclo[3.1.0]hex-2-ene/alpha-Phellandrene*	5.458	962	0.72
3	*alpha-Pinene*	5.687	974	1.20
4	*Valeraldehyde, diethyl acetal*	5.946	988	0.07
5	*Camphene*	6.252	1003	0.56
6	*beta-Terpinen*	6.798	1028	0.07
7	*beta-Pinene*	6.957	1035	0.39
8	*beta-Pinene*	7.039	1039	0.95
9	*3-Carene*	7.585	1063	0.15
10	*alpha-Terpinen*	7.832	1074	0.24
11	*Limonene*	8.102	1086	0.55
12	*beta-Phellandrene*	8.32	1095	0.24
13	*Eucalyptol*	8.531	1104	28.61
14	*gamma-Terpinen*	9.013	1124	5.86
15	*Terpinolene*	9.747	1155	0.10
16	*4-Isopropyl-1-methyl-2-cyclohexene-1-ol*	10.118	1171	0.08
17	*Linalool*	10.2	1175	0.44
18	*3,4-Dimethylstyrene*	10.57	1190	0.08

Table 6. *Cont.*

No.	A Compound Identified by GC-MS	RT (min)	Kovats Index	Relative Concentration (%)
19	*Thujone*	10.911	1205	0.12
20	*4-Isopropyl-1-methyl-2-cyclohexene-1-ol*	11.022	1210	0.07
21	*4-Oxo-beta-isodamascol*	11.68	1239	0.06
22	*1,3-Dioxolane, 2,2-dimethyl-4,5-bis(1-methyl phenyl)-*	12.209	1262	0.05
23	*Tetrahydroactinidiolide*	12.415	1271	0.04
24	*Camphor*	12.597	1279	1.73
25	*E-3,5-Dimethylhex-2-en-1,2-dicarboxylic acid*	13.026	1298	0.04
26	*alpha-Terpineol*	13.138	1304	0.07
27	*Chavicol methyl ether*	13.784	1333	28.09
28	*Thymol methyl ether*	13.978	1342	1.43
29	*2-Isopropyl-1-methoxy-4-methylbenzene*	14.136	1349	0.80
30	*Bornyl acetate*	15.035	1391	0.09
31	*Thymol*	15.799	1427	0.05
32	*Thymol*	15.958	1434	7.19
33	*Carvacrol*	16.252	1448	0.77
34	*Carvacrol*	16.739	1472	0.06
35	*1-Cyclopropene-1-pentanol, à,î,î,2-tetramethyl-3-(1-methyl phenyl)-*	16.992	1484	0.07
36	*alpha-Bergamotene*	17.274	1498	0.89
37	*Caryophyllene*	17.386	1503	3.29
38	*beta-Farnesene*	17.785	1523	0.27
39	*Humulene*	18.296	1548	0.49
40	*gamma-Muurolene*	18.696	1568	0.34
41	*Germacrene D*	18.901	1578	0.35
42	*Eugenol methyl ether*	18.978	1582	0.83
43	*gamma-Cadinene*	19.659	1616	0.48
44	*beta-Cadinene*	19.759	1622	0.40
45	*cis-alpha-Bisabolene*	19.924	1630	1.21
46	*Calamenene*	20.27	1648	0.22
47	*Ethyl laurate*	20.764	1674	0.05
48	*Cadala-1(10),3,8-triene*	20.952	1683	0.04
49	*Spathulenol*	21.657	1721	0.07
50	*Caryophyllene oxide*	21.792	1728	0.55
51	*delta-Cadinol*	22.127	1746	0.06
52	*12-Oxabicyclo[9.1.0]dodeca-3,7-diene, 1,5,5,8-tetramethyl-, [1R-(1R*	22.444	1764	0.05
53	*Cubenol*	22.626	1774	0.04
54	*tau-Cadinol*	22.779	1782	0.16
55	*Cadalene*	24.113	1858	0.08
56	*Ethyl myristate*	24.324	1870	0.15
57	*Hexahydrofarnesyl acetone*	25.082	1914	0.12
58	*Ethyl pentadecanoate*	25.987	1969	0.08
59	*Naphthalene, 1,2,3,4,4a,5,6,7-octahydro-4a-methyl-*	27.027	2035	0.08
60	*Ethyl palmitate*	27.591	2072	4.45
61	*Ethyl (9E)-9-hexadecenoate*	27.762	2084	0.13
62	*Nonadecane*	28.061	2104	0.05
63	*cyclopentane carboxylic acid, 4-hexadecyl ester*	28.672	2147	0.08
64	*Ethyl heptadecanoate*	29.125	2181	0.04
65	*2-Piperidinone, N-[4-bromo-n-butyl]-*	29.565	2214	0.06
66	*4-Butoxy-2,4-dimethyl-2-pentene*	30.411	2283	0.08
67	*Ethyl Oleate*	30.494	2290	0.20
68	*Ethyl stearate*	30.594	2298	0.20
69	*9,12-Octadecadienoic acid, ethyl ester*	30.652	2304	1.06
70	*Linolenic acid, ethyl ester*	30.981	2333	1.37
71	*Linolenic acid, ethyl ester*	31.357	2368	0.05
72	*2,4,4,6,6,8,8-Heptamethyl-2-nonene*	31.998	2433	0.57
73	*(Z)-7-Hexadecenal*	32.403	2477	0.04
74	*2,4,4,6,6,8,8-Heptamethyl-1-nonene*	33.226	2577	0.26

RT-retention time.

3.3. Statistical Multivariate Principal Component Analysis (PCA) of GC Data

PCA reveal that *Protofil®* had a significant similarity with yarrow, more limited similarity with thyme and basil, and little similarity with dandelion. Data variation described 53% for PC1, and 34% for PC2 and, for this classification, the *chavicol-methyl-ether*, and *α-muuroien* were essential, as a first main component. Eucalyptol concentration had significance, as a following main component (Figure 6).

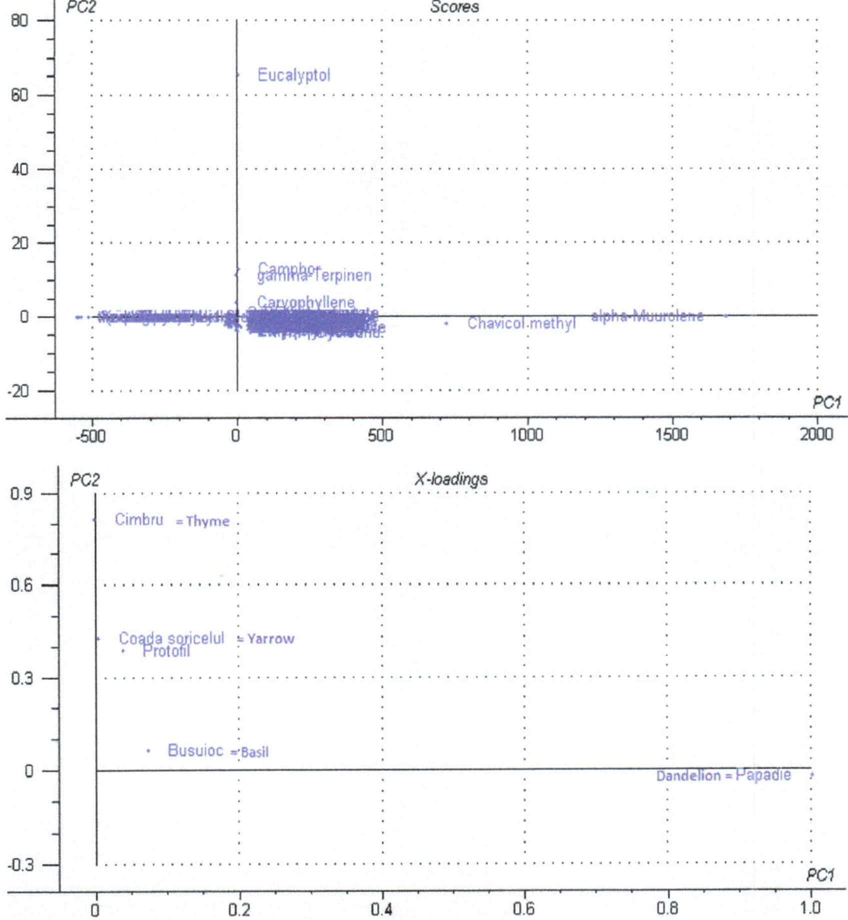

Figure 6. PCA analysis—Graph of records for volatile compounds (**up**) and chart of PCA analysis scores of GC data (**down**), for volatile compounds in the studied samples.

Analyzing the outcome of the chromatographic analyses used in this study, HPLC and GC-MS, an approximately equal proportion for the four distinct studied components in the *Protofil®* association it was ascertained.

4. Discussion

Considering treatment with antibiotics is now forbidden in European countries, control of nosemosis has to be completed mainly by employing defensive and alternative measures. Additionally, if a beehive is critically impaired by nosemosis, the strategy, from an economic point of view and in many countries, is to destroy those colonies, although losses could become sizeable. Nevertheless, in the literature, there are presented efforts to combat nosemosis, original phytotherapeutic conditionings being proposed, a present study trying to be part of this cause by proposing this phytotherapeutic approach.

Research shows that food supplements are common in beekeeping [24]. Research was conducted to evaluate the brood development from colonies, which were fed with different naturals supplement added in supplementary food compared to product *Protofil®*. According to result of this research, after the winter period and during the period of preparation for principal honey harvest, the best results were obtained for *Protofil®* and *Echinacea* [24].

For instance, thymol was among the first natural substances studied in the beehive infections [27–29] as well as various thymol links [30]. In our results, the volatile compounds analyzed in *T. vulgaris* were *eucalyptol* and *γ-terpinene*.

For example, Maistrello et al. [19], had evaluated the effectiveness of different phyto compounds, like resveratrol, thymol, vetiver essential oil, and lysozyme, to control nosema in honeybees. The results revealed that bees, fed especially with thymol, which is also identified in our study, and resveratrol considerably reduced infection rates and extending longevity. Thymol and resveratrol have therefore been shown to be effects for control of nosemosis [19].

Mărghitaş et al. [23] investigated the influence of nettle, thyme and *Echinacea*, fresh juice of onion and garlic, and *Protofil®* as supplementary feed in artificially weakened bee colonies. The most effective results in this field experiment were recorded in bees supplemented with nettle [23].

In another study, *N. ceranae* infection was stopped with the use of oxalic acid syrup, in laboratory and field studies, being proposed by authors, as an alternative control strategy [16].

Yucel and Dogaroglu [31] studied comparatively, for three years, the activity of Fumagillin, and thymol in *N. apis* infection, in 208 honey bee colonies. The results confirm the present investigation with the aim of phytotherapy efficiency and underlining the importance of alternative treatments in honey bees [31].

The observed low mortality, as well as the honey production, which also brings the organic honey's benefits, does validate the *Protofil®* use judiciousness, as a reliable phytotherapeutic choice. This observation is significant from the organic product consumers and the beekeepers' economic point of view because research has shown consumers' higher willingness to pay for organic honey [32]. The efficacy of *Protofil®* for treating nosemosis was demonstrated on 15 colonies. The mortality values compared to the honey production/categories/total quantity, confirmed the judiciousness of treatments with *Protofil®* [33].

Cola [34] tested to caraway, *Protofil®*, fresh juice of onions, garlic, stinging nettle, thyme, *Echinacea*, and selenium on the bee families artificially weakened by removing the existing population of 3/4 from initial. It was found that the most significant influence in this research had a stinging nettle, which was in agreement with earlier findings [24].

5. Conclusions

The chromatographic analyzes completed on plant extracts from different botanical families revealed that the most concentrated soluble components in the alcohol–water mixture were flavonoids, most often rutin, identified in high concentrations in most of the studied samples (except the thyme), but also its corresponding aglycons. The most significant volatile compounds identified were *eucalyptol* (1,8-cineol) and *chavicol-methyl ether,* for *Lamiaceae* (basil and thyme) samples and *camphor* for *Asteraceae* (yarrow) family. Representatives of the *Compositae* family were less concentrated in the volatile compounds (except thyme, significant from this point of view).

The results of our study revealed a considerable similarity of *Protofil®* with with *A. millefolium*, less so with *T. vulgaris* and *O. basilicum*, while they were significantly different from *T. officinale*. The results revealed a high concentration of beneficial active components of herbs in *Protofil®*, and the promised benefits of organic honey, with no residues, plus the lack of undesirable effects, but the further research are still necessary.

Author Contributions: Conceptualization, R.T.C. and E.D.; methodology, F.M.; software, N.P.; validation, R.T.C., K.I., and N.M.; formal analysis, N.H.; investigation, R.T.C.; resources, I.R.; data curation, M.C.; writing—original draft preparation, R.T.C.; writing—review and editing, N.P.; visualization, E.D.; supervision, Z.K.; project administration, R.T.C. and Z.K.; funding acquisition, D.M. All authors have read and agreed to the published version of the manuscript.

Funding: This work was supported by USAMVBT Institutional development projects—Projects for financing excellence in CDI under the Grant 35PFE.

Acknowledgments: One part of this research is supported by COST Action CA18217 European Network for Optimization of Veterinary Antimicrobial Treatment.

Conflicts of Interest: The authors declare no conflict of interest.

References

1. Kevan, P.G.; Phillips, T.P. The Economic Impacts of Pollinator Declines: An Approach to Assessing the Consequences. *CE* **2001**, *5*, 8. [CrossRef]

2. Steffan-Dewenter, I.; Münzenberg, U.; Bürger, C.; Thies, C.; Tscharntke, T. Scale-dependent effects of landscape context on three pollinator guilds. *Ecology* **2002**, *83*, 1421–1432. [CrossRef]

3. Bromenshenk, J.J.; Henderson, C.B.; Wick, C.H.; Stanford, M.F.; Zulich, A.W.; Jabbour, R.E.; Deshpande, S.V.; McCubbin, P.E.; Seccomb, R.A.; Welch, P.M.; et al. Iridovirus and Microsporidian Linked to Honey Bee Colony Decline. *PLoS ONE* **2010**, *5*, e13181. [CrossRef]

4. Botías, C.; Martín-Hernández, R.; Barrios, L.; Meana, A.; Higes, M. Nosema spp. infection and its negative effects on honey bees (Apis mellifera iberiensis) at the colony level. *Vet. Res.* **2013**, *44*, 25. [CrossRef]

5. Ignjatijević, S.; Prodanović, R.; Bošković, J.; Puvača, N.; Tomaš-Simin, M.; Peulić, T.; Đuragić, O. Comparative analysis of honey consumption in Romania, Italy and Serbia. *Food Feed Res.* **2019**, *46*, 125–136. [CrossRef]

6. Goblirsch, M. Nosema ceranae disease of the honey bee (Apis mellifera). *Apidologie* **2018**, *49*, 131–150. [CrossRef]

7. Mederle, N.; Kaya, A.; Balint, A.; Morariu, S.; Oprescu, I.; Ilie, M.; Imre, M.; Ciobanu, G.H.; Dărăbuș, G.H. Epidemiological investigations on honey bees nosemosis in Timis County. *Lucr. Stiintifice Univ. De Stiinte Agric. A Banat. Timis. Med. Vet.* **2017**, *50*, 142–146.

8. Ptaszyńska, A. Monitoring of nosemosis in the Lublin region and preliminary morphometric studies of Nosema spp. spores. *Med. Weter.* **2012**, *68*, 622–625.

9. Stanimirovic, Z.; Stevanovic, J.; Bajic, V.; Radovic, I. Evaluation of genotoxic effects of Fumagillin by cytogenetic tests in vivo. *Mutat. Res. Genet. Toxicol. Environ. Mutagenesis* **2007**, *628*, 1–10. [CrossRef] [PubMed]

10. Stevanovic, J.; Stanimirovic, Z.; Genersch, E.; Kovacevic, S.R.; Ljubenkovic, J.; Radakovic, M.; Aleksic, N. Dominance of Nosema ceranae in honey bees in the Balkan countries in the absence of symptoms of colony collapse disorder. *Apidologie* **2011**, *42*, 49–58. [CrossRef]

11. Webster, T.C.; Pomper, K.W.; Hunt, G.; Thacker, E.M.; Jones, S.C. Nosema apis infection in worker and queen *Apis mellifera*. *Apidologie* **2004**, *35*, 49–54. [CrossRef]

12. Plavša, N.; Stojanović, D.; Stojanov, I.; Puvača, N.; Stanaćev, V.; Đuričić, B. Evaluation of oxytetracycline in the prevention of American foulbrood in bee colonies. *AJAR* **2011**, *6*, 1621–1626. [CrossRef]

13. van den Heever, J.P.; Thompson, T.S.; Curtis, J.M.; Ibrahim, A.; Pernal, S.F. Fumagillin: An Overview of Recent Scientific Advances and Their Significance for Apiculture. *J. Agric. Food Chem.* **2014**, *62*, 2728–2737. [CrossRef] [PubMed]

14. Mendoza, Y.; Diaz-Cetti, S.; Ramallo, G.; Santos, E.; Porrini, M.; Invernizzi, C. *Nosema ceranae* Winter Control: Study of the Effectiveness of Different Fumagillin Treatments and Consequences on the Strength of Honey Bee (Hymenoptera: Apidae) Colonies. *J. Econ. Entomol.* **2016**. [CrossRef]

15. Michalczyk, M.; Sokół, R.; Koziatek, S. Evaluation of the Effectiveness of Selected Treatments of Nosema Spp. Infection by the Hemocytometric Method and Duplex Pcr. *Acta Vet.* **2016**, *66*, 115–124. [CrossRef]

16. Nanetti, A.; Rodriguez-García, C.; Meana, A.; Martín-Hernández, R.; Higes, M. Effect of oxalic acid on Nosema ceranae infection. *Res. Vet. Sci.* **2015**, *102*, 167–172. [CrossRef]

17. Puvača, N.; Stanaćev, V.; Glamočić, D.; Lević, J.; Perić, L.; Stanaćev, V.; Milić, D. Beneficial effects of phytoadditives in broiler nutrition. *Worlds Poult. Sci. J.* **2013**, *69*, 27–34. [CrossRef]

18. Puvača, N.; Lika, E.; Cocoli, S.; Shtylla Kika, T.; Bursić, V.; Vuković, G.; Tomaš Simin, M.; Petrović, A.; Cara, M. Use of Tea Tree Essential Oil (Melaleuca alternifolia) in Laying Hen's Nutrition on Performance and Egg Fatty Acid Profile as a Promising Sustainable Organic Agricultural Tool. *Sustainability* **2020**, *12*, 3420. [CrossRef]

19. Maistrello, L.; Lodesani, M.; Costa, C.; Leonardi, F.; Marani, G.; Caldon, M.; Mutinelli, F.; Granato, A. Screening of natural compounds for the control of nosema disease in honeybees (*Apis mellifera*). *Apidologie* **2008**, *39*, 436–445. [CrossRef]

20. Prodanović, R.; Ignjatijević, S.; Bošković, J. Innovative potential of beekeeping production in AP Vojvodina. *J. Agron. Technol. Eng. Manag.* **2019**, *2*, 268–277.

21. Čabarkapa, I.; Puvača, N.; Popović, S.; Čolović, D.; Kostadinović, L.; Tatham, E.K.; Lević, J. Aromatic plants and their extracts pharmacokinetics and in vitro/in vivo mechanisms of action. In *Feed Additives*; Elsevier: Amsterdam, The Netherlands, 2020; pp. 75–88, ISBN 978-0-12-814700-9.

22. Popović, S.; Kostadinović, L.; Đuragić, O.; Aćimović, M.; Čabarkapa, I.; Puvača, N.; Pelić, D.L. Influence of medicinal plants mixtures (Artemisia absinthium, Thymus vulgaris, Menthae piperitae and Thymus serpyllum) in broilers nutrition on biochemical blood status. *J. Agron. Technol. Eng. Manag.* **2018**, *1*, 91–98.

23. Mărghitaş, L.; Bobiş, O.; Tofalvi, M. The Effect of Plant Supplements on the Development of Artificially Weaken Bee Families. *Sci. Pap. Anim. Sci. Biotechnol.* **2010**, *43*, 402–406.

24. Mărghitaş, L.; Tofalvi, M.; Dezmirean, D.; Moise, A. The Effect of Nourishing Supplements on Bees Colonies Growing after Winter Period. *Bull. UASVM Anim. Sci. Biotechnol.* **2009**, *66*, 242–246.

25. Nica, D.; Bianu, E.; Chioveanu, G. A case of acute intoxication with deltamethrin in bee colonies in romania. *Apiacta* **2004**, *39*, 71–77.

26. Chioveanu, G.; Ionescu, D.; Mardare, A. Control of nosemosis-treatment with protofil. *Apiacta* **2004**, *39*, 31–38.

27. Bogdanov, S.; Kilchenmann, V.; Imdorf, A.; Fluri, P. Residues in honey after application of thymol against varroa using the Frakno Thymol Frame. *Am. Bee J.* **1998**, *138*, 610–611.

28. Chiesa, F.; D'Agaro, M. Effective control of varroatosis using powdered thymol. *Apidologie* **1991**, *22*, 135–145. [CrossRef]

29. Costa, C.; Lodesani, M.; Maistrello, L. Effect of thymol and resveratrol administered with candy or syrup on the development of *Nosema ceranae* and on the longevity of honeybees (*Apis mellifera* L.) in laboratory conditions. *Apidologie* **2010**, *41*, 141–150. [CrossRef]

30. Imdorf, A.; Kilchenmann, V.; Bogdanov, S.; Bachofen, B.; Beretta, C. Toxizität von Thymol, Campher, Menthol und Eucalyptol auf Varroa jacobsoni Oud und Apis mellifera L im Labortest. *Apidologie* **1995**, *26*, 27–31. [CrossRef]

31. Yücel, B.; Doarolu, M. The Impact of Nosema apis Z. Infestation of Honey Bee (Apis mellifera L.) Colonies after Using Different Treatment Methods and their Effects on the Population Levels of Workers and Honey Production on Consecutive Years. *Pak. J. Biol. Sci.* **2005**, *8*, 1142–1145. [CrossRef]

32. Vapa-Tankosić, J.; Ignjatijević, S.; Kiurski, J.; Milenković, J.; Milojević, I. Analysis of Consumers' Willingness to Pay for Organic and Local Honey in Serbia. *Sustainability* **2020**, *12*, 4686. [CrossRef]

33. Cristina, R.T.; Mederle, N.; Soreanu, M.; Moruzi, R.F.; Dumitrescu, E.; Muselin, F.; Dar, A.P.; Imre, K.; Militaru, D. The efficiency of a phytotherapeutic association to combat nosemosis in honey bees in romania. *Rev. Rom. Med. Vet.* **2020**, *30*, 13–18.

34. Cola, M. Effect of bio-stimulator of teas on the development of bee's family. *Ann. Univ. Craiova Agric. Mont. Cadastre Ser.* **2015**, *45*, 99–102.

Communication

Sea Buckthorn and Grape Extract Might Be Helpful and Sustainable Phyto-Resources as Associated Hypolipidemic Agents—Preliminary Study

Erieg A. Mohamed [1], Camelia Tulcan [2], Ersilia Alexa [2], Doru Morar [3], Eugenia Dumitrescu [1], Florin Muselin [4], Isidora Radulov [3], Nikola Puvača [5,*] and Romeo T. Cristina [1,*]

[1] Department of Pharmacology & Pharmacy, Faculty of Veterinary Medicine, Banat's University of Agricultural Sciences and Veterinary Medicine "King Michael I of Romania" from Timișoara, 300645 Timisoara, Romania; erieg_1980@yahoo.com (E.A.M.); eugeniadumitrescu@usab-tm.ro (E.D.)

[2] Department of Biochemistry, Faculty of Veterinary Medicine, Banat's University of Agricultural Sciences and Veterinary Medicine "King Michael I of Romania" from Timișoara, 300645 Timisoara, Romania; cameliatulcan@usab-tm.ro (C.T.); ersiliaalexa@usab-tm.ro (E.A.)

[3] Department of Internal medicine, Faculty of Veterinary Medicine, Banat's University of Agricultural Sciences and Veterinary Medicine "King Michael I of Romania" from Timișoara, 300645 Timisoara, Romania; dorumorar@usab-tm.ro (D.M.); isidoraradulov@usab-tm.ro (I.R.)

[4] Department of Toxicology, Faculty of Veterinary Medicine, Banat's University of Agricultural Sciences and Veterinary Medicine "King Michael I of Romania" from Timișoara, 300645 Timisoara, Romania; florinmuselin@usab-tm.ro

[5] Department of Engineering Management in Biotechnology, Faculty of Economics and Engineering Management in Novi Sad, University Business Academy in Novi Sad, 21000 Novi Sad, Serbia

* Correspondence: nikola.puvaca@fimek.edu.rs (N.P.); romeocristina@usab-tm.ro (R.T.C.)

Received: 16 September 2020; Accepted: 4 November 2020; Published: 9 November 2020

Abstract: Phytotherapy can enhance the beneficial health outcomes in the prevention of obesity and is able to improve the function of the metabolic organs, like the liver and kidneys. Since sea buckthorn (SBT) and grape extracts are known as abundant sources of polyphenol, we assumed that the extracts of these two plants might have a hypolipidemic effect and an improved metabolic function in obese rats treated with atorvastatin. One hundred and twelve white Wistar rats were divided equally into seven groups (G.I–VII) and orally treated as follows: G.I, atorvastatin 20 mg × kg·bw^{-1}; G.II, atorvastatin 20 mg × kg·bw^{-1} + SBT 100 mg × kg·bw^{-1}; G.III, atorvastatin 20 mg × kg·bw^{-1} + grape extract 100 mg × kg·bw^{-1}; G.IV, grape extract 100 mg × kg·bw^{-1}; G.V, SBT 100 mg × kg·bw^{-1}; G.VI, high-fat diet (HFD); group VII was considered the control group. After two and six months of administration, the rats were sacrificed, and blood samples were taken for biochemical analyses. The statistical results (analysis of variance (ANOVA)) showed that a combination of SBT and grape extracts with atorvastatin significantly reduced ($p < 0.001$) the lipid parameters. After six months, the liver and kidneys improved their functioning, showing a statistically significant change ($p < 0.001$) in the grape and sea buckthorn groups compared to the other groups. In addition, grape extract and SBT combined with atorvastatin proved to be potent hypolipidemic agents, so associations with phytodietary supplements can be considered as a valuable means of combating hypolipidemia and decreasing risk factors.

Keywords: dyslipidemia; sea buckthorn; grape; high-fat diet; rats; phytotherapy; atorvastatin

---

1. Introduction

Dyslipidemia is characterized by irregular levels of lipid parameters, such as low-density lipoprotein cholesterol (LDL-c), triglycerides, high-density lipoprotein cholesterol (HDL-c), and atherogenic fractions in the bloodstream. It is one of the current abnormalities and is one of the main components of the metabolic syndrome, as well as a frequent challenge that leads to morbidity and mortality [1]. Lipid and lipoprotein irregularities introduce risk factors for cardiovascular diseases, and the predominance of these in the worldwide community has developed considerably in the last decades. Studies have exposed that dyslipidemia is very widespread among males aged 30–40 years, justifying the raised prevalence of infarctions due to coronary artery disease in this age group [2]. Data has shown that lipid-lowering therapies can diminish the growth of coronary atherosclerosis [3]. With this aim, a major goal of clinical treatment of cardiovascular disease (CVD) is risk minimization by completing healing target levels for any lipid parameter [4]; for regulating these, "statins" are involved, since they can reduce the LDL-c levels by up to 55% [5,6].

Statins are structures with anti-atherosclerotic characteristics; atorvastatin is considered the first choice of drug for the management of heart disease, as it alters the reversible inhibition of hydroxymethylglutaryl coenzyme A (HMG-CoA), but, like all medications, is also has potential adverse effects (the most known is liver toxicity through elevation of transaminases enzymes) [7].

Along with developments in the occurrence of chronic metabolic diseases, there is a growing interest in new natural resources. Phytomedicine with respect to CVDs has acknowledged many phytochemicals, including carotenoids, isorhamnetin, anthocyanins, resveratrol, etc. These were proven to have potential biological activities both in vivo and in vitro, such as eliminating free radicals, anti-tumor potential, and anti-inflammation potential; In addition, they include some amounts of trace elements, vitamins, and essential fatty acids [8].

Recent lifestyle changes with high-fat diets (HFDs) have resulted in increases in the prevalence of obesity, which, unfortunately, is associated with chronic disease, like fatty liver or type II diabetes. HFDs can cause impairment of metabolic status, including altering transaminases and the lipogenesis pathway, which leads to excessive accumulation of triglyceride (TG) in hepatic tissues, fibrosis, hepatic cellular ballooning, steatosis, or inflammation. Therefore, liver protection is linked to the inhibition of hepatic fat accumulation by restraint of lipolysis and lipogenesis pathways [4].

Recently, there has been a lot of attention on HFD-related kidney damage, and it is now increasingly comprehended as an independent risk factor for chronic kidney disease. HFDs, with their associated cardiovascular disease risk, can also affect the kidney, and show a highly significant increase in the concentration of serum urea, uric acid, and creatinine. HFDs induce alteration of the metabolism of renal lipids due to the inequality between lipogenesis and lipolysis in the kidneys, in addition to the systemic metabolic irregularities; following renal lipid accumulation, it leads to renal injury [9].

Sea buckthorn (SBT) (*Hippophae rhamnoides L.*) is a shrub that belongs to the family *Elaeagnaceae*. All parts of this plant (but especially fruits and leaves) are considered to be a rich source of bioactive substances with many different pharmacological effects on wellbeing, such as anti-atherogenic effects, anti-oxidant effects, anti-cancer effects, etc. [10–12]. Currently, research has focused on the effect of Sea buckthorn or its extracts on cardiovascular disease. Furthermore, outcomes from experimental studies with SBT or its extracts on blood lipid profiles were inconsistent due to differences in experimental models, study plans, and health states, but supplemental SBT certainly enhanced the blood lipid profiles in patients suffering from heart disease [13].

Grape (*Vitis Vinifera*) extract is an important reservoir of polyphenols, mainly tannins, resveratrol, and flavonoids [13,14], polyphenols being among the most common phytochemicals added in diets [15]; they are considered as a powerful health promoter in hyperlipidemia [16].

This study aimed to investigate the hypolipidemic effect of Sea buckthorn and grape extract in rats fed for a prolonged period with HFD in relation to the atorvastatin administration. Consequently, comparative observations were made to assess the effectiveness of phytotherapy with respect to induced hyperlipidemia and the opportunity to associate these phyto-structures as a component of rats' diet alone or in combination with the allopathic hypolipidemic structures.

2. Materials and Methods

2.1. Experimental Animals and Design

To assess the protective influences of phytotherapy with respect to hyperlipidemia caused by a high-fat diet (HFD), 112 white Wistar rats weighing between 150 and 165 grams and with ages of 3–4 months were included in this study. They were purchased from the National Research and Development Institute for Microbiology and Immunology "Cantacuzino" (NIRDMI) (Bucharest, Romania). The rats were housed in regular cages (l × w × h = 750 × 720 × 360 mm) and fed ad libitum with a usual diet (Diet, Biovetimix, code 140-501, Romania). For bedding, wood shavings were used. All animals were put in specific free-animal rooms in a controlled environment at 22 ± 2 °C with a humidity of 55 ± 10% and with 12 h light/12 h dark.

The investigation was approved by the Ethical Committee of the Faculty of Veterinary Medicine of Banat's University of Agricultural Science and Veterinary Medicine in Timișoara under no. 6/30.01.2019. Before the inception of the study, animals were put in cages for seventeen days to acclimatize, and were managed under Directive 2010/63/EU [16] on the management of animals used for scientific goals (Directive 2010) and the guidelines of the National Research Council (NRC). Animals were divided into seven experimental groups, each group comprising 16 animals fed with a known regular HFD [17], except for the control group [18]. Animals were dosed orally for the analyzed biochemical tests, as presented in Table 1.

Table 1. Experimental plan and dosage for all experimental groups.

Group	Animals No.	2 Months	6 Months	Dosage
I	16	8	8	Atorvastatin (*Sortis, Pfizer Europe*)—20 mg × kg·bw^{-1}
II	16	8	8	Atorvastatin + sea buckthorn: 20 mg + 100 mg × kg·bw^{-1}
III	16	8	8	Atorvastatin + Antioxivita: 20 mg + 100 mg × kg·bw^{-1}
IV	16	8	8	Antioxivita (*Phenalex, Romania*)—received 100 mg × kg·bw^{-1}
V	16	8	8	Sea buckthorn—received 100 mg × kg·bw^{-1}
VI	16	8	8	Control (positive)—received high-fat diet (HFD)
VII	16	8	8	Control (negative—received only normal diet (ND)

Legend: kg·bw^{-1}—kilogram/body weight; HFD—high-fat diet; ND—normal diet.

2.2. Atorvastatin

Atorvastatin *(Sortis, Pfizer Europe)* belongs chemically to the statin group, with a molecular formula $C_{33}H_{35}FN_2O_5$ and molecular weight 558.65 g/mol; this structure belongs to the group named diphenylpyrrols, as it comprises heterocyclic aromatic molecules with a pyrrole ring attached to two phenyl groups. Atorvastatin is a 3-hydroxy-3-methylglutary CoA (HMG-CoA) reductase inhibitor, acting by lowering LDL cholesterol in hyperlipidemia. It was obtained from Help Net pharmacy, Timișoara, Romania. It was used as an oral suspension in distilled water at a dose level of 20 mg × kg·bw^{-1}, which is considered the lowest therapeutic dose level in humans [19].

2.3. Plant Material and Preparation of Sea Buckthorn Extract

Sea buckthorn fruits were purchased from a natural plant store. The method of extraction was arranged to receive 100 mg polyphenol for each administered dose [20]. Ten grams of fresh fruits were crushed using a laboratory grinder, then weighed and added to 100 mL ethanol 70%, put in a shaker for one hour after that filtered, and transferred to the rotary evaporator at 70 °C until a final solution was obtained. The total polyphenol amount was measured according to the Folin–Ciocalteu method. The method was replicated many times to obtain the total amount of extract necessary for the whole research period [21]. Chemical compounds of total polyphenol obtained from Sea buckthorn extract were measured using HPLC (Table 2). The dose of SBT was calculated such that 10 g of SBT included 26 mg of polyphenol, and 30 mL of the concentrated extract included 624 mg of polyphenol. So, using the rule of three, the given dose was 100 mg $\times$ kg·bw^{-1}, which indicates 1.5 mL of SBT extract/rat.

Table 2. Individual determinations for total polyphenol in the sea buckthorn extract used (μg $\times$ mL^{-1}).

No.	Compound Name	Retention Time	Area	Height	Concentration
1	Gallic acid	0.000	0	0	0.000
2	Protocatechuic acid	0.000	0	0	0.000
3	Caffeic acid	0.000	0	0	0.000
4	Epicatechin acid	0.000	0	0	0.000
5	*p*-coumaric acid	0.000	0	0	0.000
6	Ferulic acid	0.000	0	0	0.000
7	Rutin	0.000	0	0	0.000
8	Rosmarinic acid	28.851	4,542,389.0	116,917.0	43.742
9	Resveratrol	30.235	4,511,738.0	86,437.0	28.385
10	Quercetin	32.087	2,426,515.0	77,684.0	40.534
11	Kaempferol	33.636	109,323.0	4451	6.208

2.4. Antioxivita

Antioxivita *(Phenalex, Romania)* was bought from a local herbal pharmacy; it is formed of water, seed extract, skin, and grapeseed. The total polyphenol content in this product included: 300 mg $\times$ mL^{-1} gallic acid equivalent (GAE), phenolic acids, anthocyanins, flavonoids, tannins, catechins, and resveratrol (Table 3).

Table 3. Individual determinations for total polyphenols in the grape extract used (μg $\times$ mL^{-1}).

No.	Compound Name	Retention Time	Area	Height	Concentration
1	Gallic acid	0.000	0	0	0.000
2	Protocatechuic acid	0.000	0	0	0.000
3	Caffeic acid	0.000	0	0	0.000
4	Epicatechin acid	0.000	0	0	0.000
5	*p*-coumaric acid	24.945	86,367.0	3501	0.101
6	Ferulic acid	0.000	0	0	0.000
7	Rutin	26.522	997,509.0	24,804.0	7.525
8	Rosmarinic acid	28.466	2,728,108.0	71,428.0	26.271
9	Resveratrol	29.154	2,958,778.0	68,286.0	18.615
10	Quercetin	30.890	12,200,171.0	87,510.0	203.798
11	Kaepferol	34.636	4,764,393.0	36,177.0	270.556

2.5. Clinical Biochemistry Analyses

The concentrations of triglyceride (TG), serum total cholesterol, high-density lipoprotein (HDL), low-density lipoprotein (LDL), alanine aminotransferase (ALT), aspartate aminotransferase (AST), alkaline phosphatase (AP), urea, uric acid, and creatinine were measured on a Randox Daytona Plus (Randox, UK) automatic biochemistry analyzer using the corresponding commercial kits (Randox, UK). The calibration was performed with Randox Calibration Serum Level 3 (Cat. No. CAL 2351), and the QC charts for reference materials were created using two levels of control serum: Randox Assayed Multisera Level 2 (Cat. No. HN 1530) and Level 3 (Cat. No. HE 1532), while VLDL was measured by divided triglyceride/5.

2.6. Statistical Analysis

The statistical software used was Graph Pad Prism 6.0 for Windows (Graph Pad Software, San Diego, USA). Values were expressed as mean ± SEM (standard error of means). The estimation of the difference between groups was ascertained using two-way analysis of variance (ANOVA) with Tukey's multiple comparison tests. The statistical values were considered as follows: * $0.01 \leq p < 0.05$, significant; ** $0.001 \leq p < 0.01$, highly significant; *** $p < 0.001$, very high significant; [ns]: not significant.

3. Results

In the present study, phytotherapy was shown to possess a good hypolipidemic and hepatoprotective activity. The treatment with $100\ \mathrm{mg} \times \mathrm{kg \cdot bw^{-1}}$ of both Sea buckthorn and grape extract in combination with the atorvastatin showed a statistically significant reduction in the lipid and liver parameters, especially after six months of observation.

The first sign of hyperlipidemia was the registered values for weight gain, as well as, biochemically, abnormalities in triglyceride (TG) and total cholesterol (TC), so they are investigated in this study.

In Table 4, the weight evolutions of the studied groups (divided according to sex) after two and six months of observation are presented.

Table 4. The weight evolutions of the studied groups at two and six months of observation.

Group	Average Weight of the Rats after:			
	Two Months		Six Months	
	Female	Male	Female	Male
I	265.2	473.5	329.2	496.3
II	252.7	458.2	319.0	463.0
III	241.0	414.7	314.3	425.0
IV	266.0	502.7	325.2	553.2
V	286.5	507.0	360.7	563.2
VI	360.0	522.0	378.0	606.0
VII	312.7	499.0	340.7	552.0

As shown in Table 5, the TC and TG levels of rats in the group fed with an HFD was significantly higher than those of the control group ($p < 0.001$), meaning that the TG levels of rats in the control group showed a favorable tendency after being fed with an HFD. Compared with the control, the mean values of the lipid parameters in the experimental groups I, II, III, IV, and V, which were analyzed at the baseline and measured for two months, showed no statistical differences after two months of treatment, while after six months, the results were totally different; the mean values of the TC, TG, and LDL-*c* in all these groups significantly decreased. The HFD group illustrated a significant rise in TC (95.9 ± 5.1), TG (245.6 ± 30.1), and LDL-*c* (16.2 ± 1.6) as well as a clear decline in HDL-*c* (7.6 ± 0.9) after six months of HFD administration.

Significant alterations were observed in all lipid parameters in the groups with atorvastatin + SBT and atorvastatin + grape extract that received these combinations for six months, as shown in Table 5.

In this study, we ascertained that phytotherapy was proficient in the alteration of liver enzymes in the rats' serum. It was observed that the rats that got grape extract and sea buckthorn alone revealed low values of serum enzymes when compared to the atorvastatin and HFD groups ($p < 0.001$). The liver enzyme levels in atorvastatin- and HFD-fed rats were significantly increased (ALT - 65.5 ± 12.4; AST = 286.3 ± 94.0; AP = $524.04 \pm 72.1 = p < 0.001$), while for HFD, the statistics were ALT = 62.3 ± 6.7; AST = 159.5 ± 21.5; AP = $488.5 \pm 56.4 = p < 0.001$, with respect to rats fed with a normal diet (Table 6).

As shown in Table 7, there were no statistical differences after two months of treatment. However, a significant difference was observed in urea after six months in the atorvastatin group ($n = 8$) (55.7 ± 19.6) and HFD group ($n = 8$) (43.6 ± 2.7) compared with other groups.

Table 5. Lipid profile parameters in the experimental groups after two and six months of phytotherapy.

Group	Mean (mg/dL) ± SE									
	Cholesterol		Triglyceride		HDL-*c*		LDL-*c*		VLDL	
	2 Months	6 Months	2 Months	6 Months	2 Months	6 Months	2 Months	6 Months	2 Months	6 Months
I	85.6 ± 5.3 *	77.8 ± 10.4 ns	166.6 ± 22.2 **	91.8 ± 21.4 ns	19.6 ± 0.7 ns	26.7 ± 1.8 ns	9.6 ± 1.8 ns	7.3 ± 0.1 ns	33.3 ± 4.4 **	18.3 ± 4.1 ns
II	83.9 ± 3.5 *	74.8 ± 7.2 ns	158.2 ± 32.8 **	86.9 ± 16.8 ns	20.6 ± 2.2 ns	28.1 ± 1.6 *	8.6 ± 1.2 ns	7.1 ± 0.4 ns	31.6 ± 6.5 **	17.3 ± 3.3 ns
III	73.5 ± 6.4 ns	57.6 ± 7.6 ns	119.8 ± 9.5	84 ± 16.7 ns	21.1 ± 2 ns	32.7 ± 3.4 ***	9.1 ± 1.1 ns	6.9 ± 0.4 ns	23.9 ± 1.8 ns	16.8 ± 3.3 ns
IV	93.5 ± 4.8 **	80 ± 9.4 ns	169.1 ± 26.8 **	107 ± 12.8 ns	18.1 ± 1 ns	26.2 ± 2.3 *	11.1 ± 1.3 *	7.4 ± 0.5 ns	33.8 ± 5.3 **	21.4 ± 2.5 ns
V	93.8 ± 2.8 **	79.4 ± 11 ns	176.2 ± 44.4 **	143.4± 24.1 ns	18.5 ± 0.6 ns	22.8 ± 1.6 ns	11.2 ± 1.4 *	8.5 ± 0.3 ns	35.2 ± 8.8 ***	28.6 ± 4.8 ns
VI	88.2 ± 7.0 **	95.9 ± 5.1 *ns	212.3 ± 27.3 ***	245.6 ± 30.0 ***ns	9.3 ± 0.8 ns	7.6 ± 0.9 ***ns	13.7 ± 2.4 ***	16.2 ± 1.6 ***ns	42.4 ± 5.4 ***	49.1 ± 6 ***ns
VII	46.9 ± 13.9	53.5 ± 5.9	33.5 ± 7.5	52.7 ± 7.5	24.7 ± 1.5 ***	25.7 ± 2.8 ns	4.9 ± 0.7	6.8 ± 0.4 ns	6.7 ± 1.5	10.5 ± 1.5 ns

Comparison with control: ns $p > 0.05$, * $p < 0.05$, ** $p < 0.01$, *** $p < 0.001$; Comparison of two months vs. six months: ns $p > 0.05$.

Table 6. Liver enzyme activity in the experimental groups after two and six months of treatment.

Group	Mean (U/L) ± SE					
	ALT		AST		AP	
	2 Months	6 Months	2 Months	6 Months	2 Months	6 Months
I	41.1 ± 5.7 ns	65.5 ±12.4 ***	171.5 ± 39 ns	286.3 ± 94 ***ns	469.3 ± 84.5 ***ns	524 ± 72.1 ***ns
II	41.8 ± 11.6 ns	38.7 ± 6.5 ns	157.9 ± 15.1 ns	120.8 ± 7.4 ns	333.1 ± 38.7 ns	216.3 ± 23.6 ns
III	37.3 ± 6.2 ns	27.4 ± 3.0 ns	129.8 ± 21 ns	106.2 ± 7.9 ns	197.2 ± 38.2 ns	179.9 ± 39 ns
IV	27.8 ± 1.3 ns	25.4 ± 2.6 ns	100.6 ± 7.7 ns	90.8 ± 7.3 ns	167.9 ± 34.7 ns	167.7 ± 30.6 ns
V	31.1 ± 3.9 ns	27.3 ± 2.6 ns	115 ± 16.5 ns	95.8 ± 4.6 ns	177.3 ± 22.9 ns	174.7 ± 19.5 ns
VI	39.2 ± 4.5 ns	62.3 ± 6.7 ***ns	138.6 ± 27.3 ns	159.5 ± 21.5 ns	395.8 ± 35.7 **ns	488.5 ± 56.4 ***ns
VII	20.3 ± 1.6	24.8 ± 1.6	79.6 ± 4.6	95.2 ± 6.9	161.9 ± 33.6	169.6 ± 41.7

Comparison with control: ns $p > 0.05$, * $p < 0.05$, ** $p < 0.01$, *** $p < 0.001$; Comparison of two months vs. six months: ns $p > 0.05$.

Table 7. Kidney function tests in experimental groups after two and six months of treatment.

Group	Mean (mg/dL) ± SE					
	Urea		Uric Acid		Creatinine	
	2 Months	6 Months	2 Months	6 Months	2 Months	6 Months
I	54.5 ± 15 *	55.7 ± 19.6 *	1.8 ± 0.0 [ns]	2.1 ± 0.636 **	0.6 ± 0.0 ***	0.7 ± 0.0 ***
II	43.3 ± 1.8 [ns]	28.6 ± 0.7 [ns]	1.4 ± 0.0 [ns]	0.9 ± 0.059 [ns]	0.7 ± 0.0 ***	0.6 ± 0.0
III	42.3 ± 2.8 [ns]	29 ± 0.4 [ns]	1.4 ± 0.2 [ns]	0.9 ± 0.1 [ns]	0.7 ± 0.0 ***	0.6 ± 0.0
IV	26.3 ± 1.8 [ns]	23.6 ± 1 [ns]	1 ± 0.1 [ns]	0.9 ± 0.0 [ns]	$0.6 = 0.0$	0.6 ± 0.0
V	35 ± 3.2 [ns]	26.7 ± 1.8 [ns]	1.1 ± 0.0 [ns]	0.7 ± 0.1 [ns]	0.7 ± 0.0 ***	0.6 ± 0.0
VI	38.6 ± 2.1 [ns]	43.6 ± 2.7 [ns]	1.3 ± 0.2 [ns]	1.7 ± 0.3 [ns]	0.6 ± 0.0 *	0.7 ± 0.0 ***
VII	23.3 ± 1.2	24.4 ± 1.2	0.7 ± 0.1	0.9 ± 0.0	$0.5 = 0.0$	0.5 ± 0.0

Comparison with control: [ns] $p > 0.05$, * $p < 0.05$, ** $p < 0.01$, *** $p < 0.001$; Comparison of two months vs. six months: [ns] $p > 0.05$.

4. Discussion

Due to their relationship with atherogenic dyslipidemia, obesity and metabolic syndrome are linked with an enhanced danger of CVD [21,22].

Dyslipidemias comprise a wide-ranging variety of lipid abnormalities, some of which are of great importance in CVD. High levels of TC and LDL-*c* have been determined to be of main importance, mostly because they can be tailored by treatments. Randomized controlled trials (RCTs) have proven that reducing TC and LDL-*c* can prevent CVD; therefore, TC and LDL-*c* are the primary targets of therapy [23].

In addition, some other types of dyslipidemias emerge that cause predisposition to early CVD. Among these, the co-existence of augmented VLDL residues (mirrored by slight TG presence), increased LDL-*c* elements, and low HDL-*c* levels—recognized as the "lipid triad"—is frequently considered as major target of CVD prevention [23].

Our outcomes demonstrated a highly significant reduction in all lipid values in the atorvastatin + SBT and atorvastatin + grape extract groups when compared to the HFD and control groups. These reductions indicate that the added phytotherapy could improve the process of managing hyperlipidemia.

Another notable result was an enhancement in HDL-*c* level, which is considered the good cholesterol, since it has a protective influence against heart disease. Atorvastatin significantly decreased lipid values in the atorvastatin group due to its inhibition of HMG-CoA reductase in the microsomal cells, as well as its alteration of the serum LDL-*c* by improving LDL-receptor-mediated LDL uptake [24].

The present study demonstrated that the addition of $100 \text{ mg} \times \text{kg·bw}^{-1}$ of sea buckthorn in combination with atorvastatin, or only as a dietary supplement in obese rats, could reduce the serum TC and increase HDL-*c*. These findings were in agreement with those in a previous study done on the hamsters, which demonstrated that SBT seed oil could effectively reduce plasma LDL-*c* and exhibited anti-atherogenic activity [25]. This study supports our observations that hyperlipidemia emerges from long-term irregularities in lipid metabolism, and reducing serum lipid levels through dietary treatment alone or with drug treatment appears to be associated with a certain decrease in the risk of vascular disease and the related complications [25].

The hypocholesterolemic and LDL-*c* lowering properties of SBT extract were observed in this study and could be valuable information for the prevention of CVD through the improvement of dyslipidemia [26]. The SBT-supplemented rats demonstrated a remarkable lowering in the lipid parameters, which might be partially mediated by inhibition of cholesterol synthesis. Furthermore, polyphenols and polyunsaturated fatty acids can decrease cholesterol by suppressing hepatic cholesterol synthesis, as other authors also stated [27,28].

In addition to that, from the point of view of nutritional value, SBT contains a large amount of chemical structures that are valuable for health, including kaempferol, phenolics, quercetin, iso-rhamnetin, fat-soluble vitamins (A, K, and E), soluble vitamins (C, B1, B2, and B11), carotenoids, and other helpful nutrients [29], which have certain hypolipidemic properties.

Therapy with grape extract demonstrated a significant decrease in the lipid values induced by HFD in rats due to inhibition of lipogenesis and VLDL in the liver, which were described as results of the HFD. Therefore, consumption of proanthocyanin- and polyphenol-rich foods (such as grapes) can manage atherogenic dyslipidemias associated with obesity [30]. These results are in agreement with those of Del Bas et al. [31].

The literature reported that an HFD is one of the main constituents that causes obesity in human and animal models [32]; in addition to causing obesity and excessive intake of fat, it also raises the state of co-morbidities, such as hypertension, hyperlipidemia, CVD, and diabetes mellitus [33]. Accordingly, the consumption of phyto-compounds supports the physiological balance of the redox status, since these natural sources have many antioxidant compounds capable of fighting oxidative stress [34].

Abnormal hepatic transaminase levels are generally identified as a result of atorvastatin therapy. In our case, it was certain that the transaminases increased within the experimental period. Our trial revealed significantly elevated ALT and AST values—three times the normal limit, which is considered a toxic endpoint. The present study assessed the hepatoprotective potential of the continuous use of grape extract on aminotransferase enzymes in the livers of obese rats. Grapes have been studied extensively because of their antioxidant properties [35]. Now, it is well described in the literature that grape extract can offer a good protection against neurodegenerative and metabolic disease [36]. *Vitis vinifera* ethanol extract was also found to have a significant protective effect by reducing the serum levels of AP and total bilirubin [37].

The current study also proved that rats fed with an HFD for a six-month period produced a change in the aminotransferase activity. In addition, ALT—a marker for hepatic injury—was significantly raised in the HFD group (62.3 ± 6.7) in comparison with the control (24.8 ± 1.6). This elevation was due to the free radical injury that induced the oxidative process, which is considered as a process responsible for the change toward inflammation, fibrosis, and hepatocellular damage [38].

However, both the 100 mg × kg·bw^{-1} grape extract alone (25.4 ± 2.6) and 100 mg × kg·bw^{-1} SBT in SBT (27.3 ± 2.6) supplementation evidently improved the ALT activities. This research also suggested that the presence of polyphenols (and phytotherapy supplementation) may reduce hepatic injury by enhancing the antioxidant defense system and by controlling the aminotransferases' value. The presented data are in line with previous studies explaining that mice fed with an HFD and grape phytochemicals had a significant reduction in hepatic aminotransferase in obese animals [39].

Dani et al. [34] observed that purple grape juice is antitoxic, preventing the damage caused by carbon tetrachloride (CCL4) in the rats' livers. This prevention was attributed to the rich polyphenol content of the grape. Our study confirmed that with HFD-induced hepatotoxicity through elevated aminotransferase enzymes, grape extract combined with atorvastatin therapy reduced the liver damage, proving the grape's effective protection.

Kidneys are dynamic organs that serve as major systems for maintaining the body's homeostasis; they are influenced by various chemicals and drug structures. By evaluating the kidney biochemical parameters, our study also investigated the potential reno-protective outcome of SBT and grape extract against renal damage caused over time by atorvastatin administration in obese rats. Our results showed that there was no statistically significant difference with a dose of 20 mg × kg·bw^{-1} of atorvastatin in uric acid and creatinine values during the whole experimental period. There are authors who found that a four-time atorvastatin dose (80 mg × kg·bw^{-1}) lowered the blood urea nitrogen and creatinine levels in rats [40,41]. In our case, the mixture of 100 mg × kg·bw^{-1} grape extract associated with 20 mg × kg·bw^{-1} atorvastatin had the most positive result on the renal function values after six months of therapy compared to the HFD group.

The obese rats from the HFD group had a large significant increase in urea, uric acid, and creatinine values in comparison with the control group, which is in agreement with the results of Cindik et al. [40] and Amin and Nagy [42].

This change can be attributed to the effect of the HFD on the renal lipid metabolism due to irregularity between lipogenesis and lipolysis in the kidney, as well as systemic metabolism variations, leading to renal lipid mass and to renal injury, as other authors observed [43]. In addition, activation of the rennin–angiotensin system, glomerular hyperfiltration, and kidney structural change may lead to more severe glomerular injury, which is linked with continued obesity, a fact that was observed especially after six months, and that is confirmed by other studies [44]; however, this is in disagreement with another study performed on HFD rats treated with grape juice for three months [45].

Additionally the grape extract has been reported to have a broad variety of therapeutic effects, such as anti-inflammatory effects and decreasing apoptotic cell death. The grape extract is known to have high antioxidant activity and to contain different polyphenols, with real demonstrated effects against the vascular damage. These are also associated with free radical scavenging and with anti-mutagenic activity [46]. Grape seed oil and extract have a variety of biologically active classes used for protection, including a multiplicity of biologically active forms used for protection from oxidative stress induced by free radicals and reactive oxygen species (ROS), which are capable of supporting oxidative stress both in vitro and in vivo [47,48].

Fewer studies have explained the effects of SBT on kidney function. From our point of view, the presence of valuable chemicals, nutrients, and numerous biologically active substances, especially rosmarinic acid, resveratrol, quercetin, and kaempferol, explains the high hepatoprotective effect of this plant.

5. Conclusions

The results proved that plants rich in polyphenolic compounds are able to successfully support the metabolic organs' functioning, and that the constant intake of SBT or grape extract—alone or in combination with atorvastatin therapy—may be considered as an adjuvant in the treatment of patients suffering from metabolic-syndrome-like obesity.

SBT and grape extract were shown to be valuable means to reduce injuries produced by the lipid peroxidation in combination with long-term atorvastatin administration, which was attributed to the flavonoid glycosides, including quercetin, kaempferol, resveratrol, and rosmarinic acid. So, phytotherapy can improve biochemical parameters, and it thus proved to be valuable in the preservation of metabolic organs like the liver and kidneys. In our study, rats fed an HFD gained more weight than those that were provided a normal diet. However, the rats fed an HFD along with phytotherapy did not achieve any significant weight gain (Table 4); this indicated that SBT and grape extracts were effective in controlling weight gain.

The supplementation of phytotherapy significantly improved hyperlipidemia, as evidenced by the reduced level of serum total cholesterol and triglyceride as well as the improved HDL-c after six months of therapy (Table 5). This finding is supported by a previous study by Arai et al [49]. There is an inverse link between quercetin intake and TC and LDL-c concentration in Japanese women due to the reduction in hepatic lipogenesis by quercetin [50].

The excessive consumption of fat leads to a number of conditions of damage, mainly in the liver, thus inducing a loss of metabolic processes in the organ and releasing liver enzymes into the blood [51]. In our study (in Table 6), we observed that the HFD provoked an increase in liver enzymes, which is in agreement with a study that analyzed ten weeks of HFD in mice [52]. This could be related to the accumulation of liver lipids, which could cause damage to cellular homeostasis, leading to cytotoxicity [52].

The grape extract was able to prevent increases in AP, ALT, and AST, which was attributed to the polyphenol content; this is in line with another work that also observed that resveratrol was able to attenuate hepatic steatosis in mice [34].

In summary, our data demonstrated that SBT and grape extracts were beneficial in decreasing biochemical parameters and were effective in producing hypolipidemia effects in rats, including controlling body weight and with improving hepatic enzymes; thus, they may have potential as phytochemicals for managing hyperlipidemia.

Author Contributions: Conceptualization, E.A.M. and R.T.C.; methodology, R.T.C., C.T., and E.A.; software, F.M. and E.D.; validation, C.T., I.R., D.M., and E.A.; formal analysis, E.A.M.; investigation, R.T.C., C.T., and E.A.; resources, I.R.; data curation, R.T.C. and F.M.; writing—original draft preparation, E.A.M.; writing—review and editing, R.T.C. and N.P.; visualization, E.A.M. and R.T.C.; supervision, E.D., D.M., and F.M.; project administration, I.R.; funding acquisition, R.T.C. and I.R. All authors have read and agreed to the published version of the manuscript.

Funding: This work was supported by research grants from USAMVBT code 35PFE projects for financing excellence in CDI. The funders had no role in the design of the study; in the collection, analyses, or interpretation of data; in the writing of the manuscript, or in the decision to publish the results.

Conflicts of Interest: The authors declare no conflict of interest.

References

1. Vashishtha, V.; Barhwal, K.; Kumar, A.; Hota, S.K.; Chaurasia, O.P.; Kumar, B. Effect of seabuckthorn seed oil in reducing cardiovascular risk factors: A longitudinal controlled trial on hypertensive subjects. *Clin. Nutr.* **2017**, *36*, 1231–1238. [CrossRef]
2. Gupta, R.; Rao, R.S.; Misra, A.; Sharma, S.K. Recent trends in epidemiology of dyslipidemias in India. *Indian Heart J.* **2017**, *69*, 382–392. [CrossRef] [PubMed]
3. Nissen, S.; Tuzcu, E.; Al, P.S. Effect of intensive compared with moderate lipid-lowering therapy on progression of coronary atherosclerosis: A randomized controlled trial. *ACC Curr. J. Rev.* **2004**, *13*, 18. [CrossRef]
4. Wang, X.; Hasegawa, J.; Kitamura, Y.; Wang, Z.; Matsuda, A.; Shinoda, W.; Miura, N.; Kimura, K. Effects of Hesperidin on the progression of hypercholesterolemia and fatty liver induced by high-cholesterol diet in rats. *J. Pharmacol. Sci.* **2011**, *117*, 129–138. [CrossRef]
5. Gotto, A.M. Risk factor modification: Rationale for management of dyslipidemia. *Am. J. Med.* **1998**, *104*, 6S–8S. [CrossRef]
6. Stancu, C.; Sima, A. Statins: Mechanism of action and effects. *J. Cell. Mol. Med.* **2001**, *5*, 378–387. [CrossRef] [PubMed]
7. Golomb, B.A.; Evans, M.A. Statin adverse effects. *Am. J. Cardiovasc. Drugs* **2008**, *8*, 373–418. [CrossRef]
8. Forman, H.J.; Davies, K.J.A.; Ursini, F. How do nutritional antioxidants really work: Nucleophilic tone and para-hormesis versus free radical scavenging in vivo. *Free Radic. Biol. Med.* **2014**, *66*, 24–35. [CrossRef]
9. Kume, S.; Uzu, T.; Araki, S.-I.; Sugimoto, T.; Isshiki, K.; Chin-Kanasaki, M.; Sakaguchi, M.; Kubota, N.; Terauchi, Y.; Kadowaki, T.; et al. Role of altered renal lipid metabolism in the development of renal injury induced by a high-fat diet. *J. Am. Soc. Nephrol.* **2007**, *18*, 2715–2723. [CrossRef]
10. Suomela, J.-P.; Ahotupa, M.; Yang, B.; Vasankari, T.; Kallio, H.P. Absorption of flavonols derived from sea buckthorn (*Hippophaë rhamnoides* L.) and their effect on emerging risk factors for cardiovascular disease in humans. *J. Agric. Food Chem.* **2006**, *54*, 7364–7369. [CrossRef]
11. Kostadinović, L.; Lević, J. Effects of phytoadditives in poultry and pigs diseases. *J. Agron. Technol. Eng. Manag.* **2018**, *1*, 1–7.
12. Puvača, N. Bioactive compounds in selected hot spices and medicinal plants. *J. Agron. Technol. Eng. Manag.* **2018**, *1*, 8–17.
13. Larmo, P.S.; Kangas, A.J.; Soininen, P.; Lehtonen, H.-M.; Suomela, J.-P.; Yang, B.; Viikari, J.; Ala-Korpela, M.; Kallio, H.P. Effects of sea buckthorn and bilberry on serum metabolites differ according to baseline metabolic profiles in overweight women: A randomized crossover trial. *Am. J. Clin. Nutr.* **2013**, *98*, 941–951. [CrossRef]
14. Fuleki, T.; Ricardo-Da-Silva, J. Effects of cultivar and processing method on the contents of catechins and procyanidins in grape juice. *J. Agric. Food Chem.* **2003**, *51*, 640–646. [CrossRef]
15. Maante-Kuljus, M.; Rätsep, R.; Moor, U.; Mainla, L.; Põldma, P.; Koort, A.; Karp, K. Effect of vintage and viticultural practices on the phenolic content of hybrid winegrapes in very cool climate. *Agriculture* **2020**, *10*, 169. [CrossRef]

16. Frederiksen, H.; Krogholm, K.S.; Poulsen, L. Dietary proanthocyanidins: Occurrence, dietary intake, bioavailability, and protection against cardiovascular disease. *Mol. Nutr. Food Res.* **2005**, *49*, 159–174. [CrossRef]

17. Directive 2010/63/EU of the European Parliament and of the Council of 22 September 2010 on the Protection of Animals Used for Scientific PurposesText with EEA Relevance. p. 47. Available online: https://eur-lex.europa.eu/legal-content/EN/TXT/?uri=celex%3A32010L0063 (accessed on 9 September 2020).

18. Doucet, C.; Flament, C.; Sautier, C.; Lemonnier, D.; Mathon, G.; N'Diaye, A. Effect of an hypercholesterolemic diet on the level of several serum lipids and apolipoproteins in nine rat strains. *Reprod. Nutr. Dev.* **1987**, *27*, 897–906. [CrossRef]

19. Schmechel, A.; Grimm, M.; El-Armouche, A.; Höppner, G.; Schwoerer, A.P.; Ehmke, H.; Eschenhagen, T. Treatment with atorvastatin partially protects the rat heart from harmful catecholamine effects. *Cardiovasc. Res.* **2009**, *82*, 100–106. [CrossRef]

20. Gao, S.; Hu, G.; Li, D.; Sun, M.; Mou, D. Anti-hyperlipidemia effect of sea buckthorn fruit oil extract through the AMPK and Akt signaling pathway in hamsters. *J. Funct. Foods* **2020**, *66*, 103837. [CrossRef]

21. Council of Europe. *European Pharmacopoeia Commission European Pharmacopoeia*; Council of Europe: Strasbourg, France, 2007; ISBN 9789287160546.

22. Bamba, V.; Rader, D.J. Obesity and Atherogenic Dyslipidemia. *Gastroenterology* **2007**, *132*, 2181–2190. [CrossRef]

23. Reiner, Z.; Catapano, A.L.; De Backer, G.; Graham, I.; Taskinen, M.-R.; Wiklund, O.; Agewall, S.; Alegria, E.; Chapman, M.J.; Durrington, P.N.; et al. ESC/EAS Guidelines for the management of dyslipidaemias: The Task Force for the management of dyslipidaemias of the European Society of Cardiology (ESC) and the European Atherosclerosis Society (EAS). *Eur. Heart J.* **2011**, *32*, 1769–1818. [CrossRef]

24. Maron, D.J.; Fazio, S.; Linton, M.F. Current perspectives on statins. *Circulation* **2000**, *101*, 207–213. [CrossRef]

25. Hao, X.Y.; Yu, S.C.; Mu, C.T.; Wu, X.D.; Zhang, C.X.; Zhao, J.X. Replacing soybean meal with flax seed meal: Effects on nutrient digestibility, rumen microbial protein synthesis and growth performance in sheep. *Animal* **2020**, *14*, 1841–1848. [CrossRef]

26. Kumar, G.; Murugesan, A. Hypolipidaemic activity of *Helicteres isora* L. bark extracts in streptozotocin induced diabetic rats. *J. Ethnopharmacol.* **2008**, *116*, 161–166. [CrossRef]

27. Price, P.T.; Nelson, C.M.; Clarke, S.D. Omega-3 polyunsaturated fatty acid regulation of gene expression. *Curr. Opin. Lipidol.* **2000**, *11*, 3–7. [CrossRef]

28. Puvača, N.; Kostadinović, L.; Popović, S.; Lević, J.; Ljubojević, D.; Tufarelli, V.; Jovanović, R.; Tasić, T.; Ikonić, P.; Lukač, D. Proximate composition, cholesterol concentration and lipid oxidation of meat from chickens fed dietary spice addition (*Allium sativum, Piper nigrum, Capsicum annuum*). *Anim. Prod. Sci.* **2016**, *56*, 1920. [CrossRef]

29. Bal, L.M.; Meda, V.; Naik, S.; Satya, S. Sea buckthorn berries: A potential source of valuable nutrients for nutraceuticals and cosmoceuticals. *Food Res. Int.* **2011**, *44*, 1718–1727. [CrossRef]

30. Quesada, H.; Del Bas, J.M.; Pajuelo, D.; Díaz, S.; Fernandez-Larrea, J.; Pinent, M.; Arola, L.; Salvadó, J.; Bladé, C. Grape seed proanthocyanidins correct dyslipidemia associated with a high-fat diet in rats and repress genes controlling lipogenesis and VLDL assembling in liver. *Int. J. Obes.* **2009**, *33*, 1007–1012. [CrossRef]

31. Del Bas, J.M.; Fernández-Larrea, J.; Blay, M.; Ardèvol, A.; Salvadó, M.J.; Arola, L.; Bladé, C. Grape seed procyanidins improve atherosclerotic risk index and induce liver CYP7A1 and SHP expression in healthy rats. *FASEB J.* **2005**, *19*, 1–24. [CrossRef]

32. World Health Organization. Obesity and Overweight. 2018. Available online: http://www.who.int/news-room/fact-sheets/detail/obesity-and-overweight (accessed on 21 August 2020).

33. Perichart-Perera, O.; Balas-Nakash, M.; Rodriguez-Cano, A.; Muñoz-Manrique, C.; Monge, A.; Vadillo-Ortega, F. Correlates of dietary energy sources with cardiovascular disease risk markers in mexican school-age children. *J. Am. Diet. Assoc.* **2010**, *110*, 253–260. [CrossRef] [PubMed]

34. Dani, C.; Oliboni, L.; Vanderlinde, R.; Bonatto, D.; Salvador, M.; Henriques, J. Phenolic content and antioxidant activities of white and purple juices manufactured with organically-or conventionally-produced grapes. *Food Chem. Toxicol.* **2007**, *45*, 2574–2580. [CrossRef]

35. Park, Y.K.; Park, E.; Kim, J.-S.; Kang, M.-H. Daily grape juice consumption reduces oxidative DNA damage and plasma free radical levels in healthy Koreans. *Mutat. Res. Mol. Mech. Mutagen.* **2003**, *529*, 77–86. [CrossRef]

36. Venturini, C.D.; Merlo, S.; Souto, A.A.; Fernandes, M.C.; Gomez, R.; Rhoden, C.R. Resveratrol and red wine function as antioxidants in the nervous system without cellular proliferative effects during experimental diabetes. *Oxidative Med. Cell. Longev.* **2010**, *3*, 434–441. [CrossRef]

37. Sharma, S.K.; Suman; Vasudeva, N. Hepatoprotective activity of Vitis vinifera root extract against carbon tetrachloride-induced liver damage in rats. *Acta Pol. Pharm. Drug Res.* **2012**, *69*, 933–937.

38. Chitturi, S.; Wong, V.W.-S.; Farrell, G. Nonalcoholic fatty liver in Asia: Firmly entrenched and rapidly gaining ground. *J. Gastroenterol. Hepatol.* **2011**, *26*, 163–172. [CrossRef]

39. Buchner, I.; Medeiros, N.D.S.; Lacerda, D.D.S.; Normann, C.A.B.M.; Gemelli, T.; Rigon, P.; Wannmacher, C.M.D.; Henriques, J.A.P.; Dani, C.; Funchal, C. Hepatoprotective and antioxidant potential of organic and conventional grape juices in rats fed a high-fat diet. *Antioxidants* **2014**, *3*, 323–338. [CrossRef]

40. Jiang, H.; Zheng, H. Efficacy and adverse reaction to different doses of atorvastatin in the treatment of type II diabetes mellitus. *Biosci. Rep.* **2019**, *39*, 20182371. [CrossRef]

41. Cindik, N.; Baskin, E.; Agras, P.I.; Kinik, S.T.; Turan, M.; Saatci, U. Effect of obesity on inflammatory markers and renal functions. *Acta Paediatr.* **2007**, *94*, 1732–1737. [CrossRef]

42. Amin, K.A.; Nagy, M.A. Effect of Carnitine and herbal mixture extract on obesity induced by high fat diet in rats. *Diabetol. Metab. Syndr.* **2009**, *1*, 17. [CrossRef]

43. Puvača, N.; Ljubojević Pelić, D.; Popović, S.; Ikonić, P.; Đuragić, O.; Peulić, T.; Lević, J. Evaluation of broiler chickens lipid profile influenced by dietary chili pepper addition. *J. Agron. Technol. Eng. Manag.* **2019**, *2*, 318–324.

44. Ben Gara, A.; Kolsi, R.B.A.; Chaaben, R.; Hammami, N.; Kammoun, M.; Patti, F.P.; El Feki, A.; Fki, L.; Belghith, H.; Belghith, K. Inhibition of key digestive enzymes related to hyperlipidemia and protection of liver-kidney functions by *Cystoseira crinita* sulphated polysaccharide in high-fat diet-fed rats. *Biomed. Pharmacother.* **2017**, *85*, 517–526. [CrossRef]

45. Lacerda, D.D.S.; De Almeida, M.G.; Teixeira, C.; De Jesus, A.; Édison, D.S.P.J.; Bock, P.M.; Henriques, J.A.P.; Gomez, R.; Dani, C.; Funchal, C. Biochemical and physiological parameters in rats fed with high-fat diet: The protective effect of chronic treatment with purple grape juice (Bordo Variety). *Beverages* **2018**, *4*, 100. [CrossRef]

46. Çetin, A.; Kaynar, L.; Koçyiğit, I.; Hacioğlu, S.K.; Saraymen, R.; Oztürk, A.; Orhan, O.; Sağdiç, O. The effect of grape seed extract on radiation-induced oxidative stress in the rat liver. *Turk. J. Gastroenterol.* **2008**, *19*, 92–98.

47. Baiges, I.; Palmfeldt, J.; Bladé, C.; Gregersen, N.; Arola, L. Lipogenesis is decreased by grape seed Proanthocyanidins according to liver proteomics of rats fed a high fat diet. *Mol. Cell. Proteom.* **2010**, *9*, 1499–1513. [CrossRef]

48. Şehirli, Ö.; Ozel, Y.; Dulundu, E.; Topaloglu, U.; Ercan, F.; Şener, G. Grape seed extract treatment reduces hepatic ischemia-reperfusion injury in rats. *Phytother. Res.* **2007**, *22*, 43–48. [CrossRef]

49. Arai, Y.; Watanabe, S.; Kimira, M.; Shimoi, K.; Mochizuki, R.; Kinae, N. Dietary intakes of flavonols, flavones and isoflavones by Japanese women and the inverse correlation between Quercetin intake and plasma LDL cholesterol concentration. *J. Nutr.* **2000**, *130*, 2243–2250. [CrossRef]

50. Odbayar, T.-O.; Badamhand, D.; Kimura, T.; Takahashi, Y.; Tsushida, T.; Ide, T. Comparative studies of some phenolic compounds (quercetin, rutin, and ferulic acid) affecting hepatic fatty acid synthesis in mice. *J. Agric. Food Chem.* **2006**, *54*, 8261–8265. [CrossRef]

51. Ferramosca, A. Modulation of hepatic steatosis by dietary fatty acids. *World J. Gastroenterol.* **2014**, *20*, 1746–1755. [CrossRef] [PubMed]

52. Ezhumalai, M.; Radhiga, T.; Pugalendi, K.V. Antihyperglycemic effect of carvacrol in combination with rosiglitazone in high-fat diet-induced type 2 diabetic C57BL/6J mice. *Mol. Cell. Biochem.* **2013**, *385*, 23–31. [CrossRef]

Publisher's Note: MDPI stays neutral with regard to jurisdictional claims in published maps and institutional affiliations.

 *sustainability*

Article

Plant Protection Products Residues Assessment in the Organic and Conventional Agricultural Production

Vojislava Bursić [1], Gorica Vuković [2], Magdalena Cara [3], Marija Kostić [4], Tijana Stojanović [1], Aleksandra Petrović [1], Nikola Puvača [5], Dušan Marinković [1,*] and Bojan Konstantinović [1,*]

[1] Department for Phytomedicine and Environmental Protection, Faculty of Agriculture, University of Novi Sad, 21000 Novi Sad, Serbia; bursicv@polj.uns.ac.rs (V.B.); tijana.stojanovic@polj.edu.rs (T.S.); aleksandra.petrovic@polj.uns.ac.rs (A.P.)

[2] Center for Hygiene and Human Ecology, Institute of Public Health of Belgrade, 11000 Belgrade, Serbia; gorica.vukovic@zdravlje.org.rs

[3] Department of Plant Protection, Faculty of Agriculture and Environment, Agricultural University of Tirana, 1029 Tirana, Albania; mcara@ubt.edu.al

[4] Faculty of Hotel Management and Tourism, University of Kragujevac, 36210 Vrnjacka Banja, Serbia; marija.kostic@kg.ac.rs

[5] Department of Engineering Management in Biotechnology, Faculty of Economics and Engineering Management in Novi Sad, University Business Academy in Novi Sad, 21000 Novi Sad, Serbia; nikola.puvaca@fimek.edu.rs

* Correspondence: dusan.marinkovic@polj.uns.ac.rs (D.M.); bojan.konstantinovic@polj.uns.ac.rs (B.K.)

check for **updates**

Citation: Bursić, V.; Vuković, G.; Cara, M.; Kostić, M.; Stojanović, T.; Petrović, A.; Puvača, N.; Marinković, D.; Konstantinović, B. Plant Protection Products Residues Assessment in the Organic and Conventional Agricultural Production. *Sustainability* **2021**, *13*, 1075. https://doi.org/10.3390/su13031075

Received: 26 November 2020
Accepted: 4 January 2021
Published: 21 January 2021

Publisher's Note: MDPI stays neutral with regard to jurisdictional claims in published maps and institutional affiliations.

Abstract: The organic food is progressively enticing purchasers' attention, as it is recognized to be better than the food produced by the conventional agriculture and more sustainable for the natural environment. Pesticides and their metabolites can enter the human body via food and water. In the food production, over 60 thousand chemical agents are applied, while 90% of the harmful substances are consumed. The organic production is based on the qualitative and healthy food using the natural resources in an ecologically sustainable way. The European Regulations set the maximum pesticide levels (MRLs) in the organic products, which are also regulated by The United States Department of Agriculture in their National program supported by The United States Environmental Protection Agency. It is imperative to bear in mind that in the products from the organic production, the multiple detections cannot be tolerated, i.e., that one product cannot contain more than two detected pesticide residues. In this paper, a multi-residue pesticide method has been developed to determine the pesticides in the agricultural products from the organic and conventional production. In this work, 60 pesticides were analyzed using a simple QuEChERS sample preparation procedure, followed by LC-MS/MS. The tomato, potato, apple, and carrot samples from the organic and conventional products were collected from the market and the pesticide residues assessment comparing the organic to the conventional was done.

Keywords: plant protection product residues; organic and conventional agriculture; LC-MS/MS

1. Introduction

The organic food is increasingly attracting the interest of the consumers, as it is perceived to be healthier than the food produced by the conventional agriculture and more sustainable for the environment [1].

It is well-known that the pesticides and their metabolites can be brought into the human body through food and water. There are many efforts from the EU to achieve sustainable use of these compounds to avoid the increase of pesticide levels in the environment and food [2]. Today, in the food production, over 60 thousand chemical agents are used, whereas 90% of the harmful substances are taken with food. However, the increased use of pesticides is of concern to the agricultural workers and food consumers and threatens the environment [3]. That is why, in the last decades, there have been the growing interest and demand for the organic production [4–8]. The organic production is based on the qualitative and healthy food through the use of natural resources in an ecologically sustainable way [9–11]. This way of the agricultural production, different from the conventional, eliminates the application of the pesticides, growth regulators, synthetic minerals, hormones, fertilizers, antibiotics and additives [1,11]. Additionally, the use of genetically modified organisms is forbidden [12,13]. The ban on the synthetic chemical formulations, which are frequently used in the conventional production for the control of weeds, pests and diseases, represents the greatest problem for the organic food producers [14]. The pesticide contamination of organic products can be induced due to the use of water and soil with pesticide residues [15].

The occurrence of pesticide residues in the organic fruit and vegetables is not enough stated in the scientific literature [16–18]. Tobin et al. [19] detected one or more pesticide residues in 15 out of 27 tested organic samples, with one pesticide being above the LOQ (imazalil in organic onion, 0.11 mg/kg). In the case of the conventional samples, the pesticide residues were present in 17 out of 27 samples in total, with 12 of them being above the LOQ with the concentrations between 0.01 and 0.154 mg/kg. Out of 136 tested organic samples, the authorized pesticide residues were detected in 4 samples, while the non-authorized pesticides were discovered in 61 samples, which was in accordance with the study from Ireland. Namely, the authors detected the pesticide residues in 15 out of the 27 tested organic samples [2].

The studies conducted in Belgium (1995–2001) determined the presence of the pesticide residues in 12% of organic food samples and 49% of the conventional food samples. The monitoring of the German market from Baden-Württemberg (2002–2009) showed that 88% of the conventional raw materials and 27% of organic product samples contained the pesticide residues. The contamination of the organic crops in some European countries is determined as follows: the Czech Republic 14%, Ireland 11%, Finland 5%, Denmark 3% and New Zealand 22% [20,21]. This statistic is also shown in the last EFSA report [22], where organic food encompassed 6.5% of the total samples [23].

It is important that no specific MRLs are established for the organic food produced in accordance with Regulation (EC) No 2018/848 [24]. The MRLs set in Regulation (EC) No 396/2005 [25] apply to the conventional foodstuff.

The maximum residue limits (MRL) in the foodstuff, which represent the maximum residue concentration allowed in the food agricultural commodity, are being controlled by the established legislative framework. Being in accordance with the MRLs is now an obligatory norm for the food security. Depending on the country and the particular commodity, the MRLs can vary, which can be noted in the online databases that contain the summary of their regulatory status in the world [26–28].

We are not able to claim that the organic crops do not contain pesticide residues, as well as that they are truly produced according to the good agricultural practice in the organic production, since the products which authenticity of organic origin cannot be confirmed may be found everywhere throughout the market. There is no doubt that the organic products lack the certification, the continuous supply and a proper retail space, while the consumers rightfully expect the certification, quality and product attributes according to their price. Therefore, the aim of this case study was to compare the detected pesticide residues in organic fruit and vegetable samples with those from the conventional production. For this purpose, a monitoring study was conducted based on 92 commercial samples from the conventional (50) and the organic (42) products from 4 different commodity groups (tomato, potato, apple and carrot). The pesticide residues were analyzed using a simple QuEChERS sample preparation procedure, followed by the liquid chromatography coupled with tandem quadrupole mass spectrometry (LC-MS/MS).

2. Materials and Methods

Chemicals and reagents: Acetonitrile and methanol (HPLC grade) were purchased from J.T.Baker (Deventer, Netherlands), acetone was purchased from Merck (Kenilworth,

NJ, USA). The QuEChERS extract tubes (Par No. 5982-5650), as well as the dispersive SPE 15 mL kits for fruits and vegetables, EN (Part No. 5982-5056), were purchased from Agilent Technologies (Santa Clara, CA, USA). The water was purified by Mili-Q plus system from Millipore (18.2 MΩ–cm, A10 FOCN53824k, USA). The pesticides (60 active substances) and internal standards (IS, carbofuran-D3 and acetamipride-D3) were obtained from Dr. Ehrenstorfer (Munich, Germany) and Sigma Aldrich (Schnelldorf, Germany) and were prepared in acetone, methanol, or acetonitrile (depending on the solubility of the compound) at the concentration nearest to 1.0 mg/mL. Stock solutions were used to prepare working standard solutions (the mix of 60 pesticide active substances in acetonitrile at 1 and 10 µg/mL) for the calibration. The calibration curves were prepared in the mobile phase as well as matrix-matched calibration (MMC) used in order to minimize the matrix effects because matrix constituents may increase or decrease the analytical signal. MMC was prepared for each matrix separately, namely for tomato, potato, apple and carrot. For obtaining the analytical curves in the solvent and matrix (recovery calibration) the concentration ranged from 0.005 to 0.10 µg/mL.

Sample collection: Tomato, potato, apple and carrot samples from the organic and conventional production for multi-pesticide residues quantification were collected from the Serbian largest cities open markets (Belgrade, Novi Sad, Subotica, Niš, Kragujevac and Čačak) (Table 1) according to SANTE/12682/2019. Randomly sampled units in the amount of 1 kg were rapidly (within one day) transported in the polypropylene bags in the clean containers to the laboratory for the homogenization. In case of each sample the information considering the market location, purchase date and variety has been recorded. Until the moment of the preparation and the analysis, which were carried out within 3 days from the purchase date, the samples were stored at 4 °C.

Table 1. Number of samples from organic and conventional production.

Commodity Group	Organic Production	Conventional Production
Tomato	10	10
Potato	9	11
Apple	18	21
Carrot	5	8
Total	42	50

Samples extraction and clean-up procedures: The agricultural samples were extracted by the QuEChERS method described by Anastassiades et al. [15] and Bursić et al. [29]. For the extraction, the homogenized samples (10.0 g) were weighed into a polypropylene centrifuge tube (50 mL) and spiked with 100 µL of ISs. Next, 10 mL of acetonitrile were added, and the mixture was shaken vigorously for 1 min using a vortex mixer. A liquid-liquid partitioning step was performed by adding the QuEChERS extraction kit to the tube and the solution was stirred again for 1 min. After that the mixture was centrifuged for 5 min (at 4000 rpm–1900 g). After the centrifugation, the clean-up step was done based on which an aliquot of 6 mL was transferred to a 15 mL polypropylene centrifuge tube containing dispersive SPE kits for fruits and vegetables. The extract was vigorously shaken for 1 min and centrifuged for 5 min at 4000 rpm (1900 g). Finally, an aliquot of supernatant was filtrated through a PTFE 0.45 µm filter and transferred to a vial followed by injecting into the LC-MS/MS.

LC-MS/MS analysis: The detection and quantification were performed by the liquid chromatography tandem mass spectrometry equipped with the electrospray ionization (LC(ESI)-MS/MS), 6410B Agilent Technologies. In terms of chromatographic conditions, a Zorbax Eclipse XDBC18 column (50 mm × 4.6 mm id 1.8 µm) was used and kept at 25 °C. The mobile phase consisted of the gradient using methanol with 0.1% formic acid (solvent A) and 0.1% formic acid in water (solvent B), with the following gradient: 0 min–90% B; 2 min–90% B; 15 min 20% B; 20 min–15% B; 25 min–5% B and then returning to the initial conditions in 5 min. The total run time was 30 min. The flow rate of the mobile phase was 0.4 mL/min and the volume of 5 μL of sample extract was injected into the column. In terms of mass spectrometry, the MS source temperature was set at 350 °C, nitrogen gas flow 10 L/min and nebulizer pressure 40 psi. The data acquisition in the multiple reaction monitoring mode (MRM) was optimized after direct infusion of each pesticide. The instrument uses MassHunter software (vB.06.00, Agilent Technologies, Santa Clara, CA, USA) for the acquisition and quantification [30].

Method validation: All the validation parameters were evaluated following the Document N° SANTE/12682/2019 [31]. The analytical curves linearity was evaluated by injecting the analytical solutions prepared in the solvent and the matrix (tomato, potato, apple and carrot–matrix match calibration-MMC) at 0.005, 0.01, 0.05 and 0.1 μg/mL. The recovery was obtained by spiking the samples with a known amount of the mixture solution in the concentration range at 0.005 and 0.1 mg/kg. For each concentration five replicates were performed. The limit of detection (LOD) was approximated in the MRM mode analysis as the lowest concentration level that yielded a signal-to-noise ratio S/N ratio greater than 5. The limit of quantification (LOQ) of the method was set on 0.005 mg/kg as the most common default LOQ value for pesticide residues, i.e., which is below the MRLs for most pesticides in food [32].

3. Results and Discussion

The fragmentation of the protonated molecular ion obtained by LC-MS/MS in the positive electrospray ionization (ESI+) of the examined pesticides is given in Table 2. The selected reaction monitoring mode (SRM) was carried out to obtain the maximum sensitivity for each pesticide detection, while the confirmation of pesticides, two SRM transitions and a correct ratio between the optimized SRM transitions abundance were used taking into account the matching of the Rt (pesticide retention time).

Table 2. MRM transitions, fragmentation, and collision energies.

Pesticide	Molecular Formula	M g/Mol	Precursor ion m/z	Product ion m/z	Frag (V)	CE (V)	Rt (min)
Acetamiprid	$C_{10}H_{11}ClN_4$	223	223.0 223.0	125.8 55.7	120 120	10 10	11.45
Azoxystrobin	$C_{22}H_{17}N_3O_5$	403	404.1 404.1	372.0 344.1	100 100	9 25	13.17
Aldicarb	$C_7H_{14}N_2O_2S$	190	213 213	116 89	120 120	10 15	13.90
Azinphos-ethyl	$C_{12}H_{16}N_3O_3PS_2$	346	346 346	132 77.1	120 120	16 16	6.20
Bitertanol	$C_{20}H_{23}O_2N_3$	337.4	338 338	145 117	120 120	20 30	7.68
Dimethomorph	$C_{21}H_{22}ClNO_4$	388.1	388.1 388.1	301.1 165	120 120	30 20	17.30
Epoxiconazole	$C_{17}H_{13}ClFN_3O$	329.7	330.1 330.1	121 101	130 130	21 50	18.13
Ethiofencarb	$C_{11}H_{15}NO_2S$	225.3	226.1 226.1	164.1 107	80 80	5 5	14.95
Fenarimol	$C_{17}H_{12}Cl_2N_2O$	331.2	331 331	268 81	80 80	10 25	18.40
Fenoxycarb	$C_{17}H_{19}NO_4$	301.4	302.1 302.1	116.1 88	100 100	5 20	18.30
Fenpropathrin	$C_{22}H_{23}NO_3$	349. 4	350.1 350.1	125 97	135 135	24 34	7.51
Fenpropimorph	$C_{20}H_{33}NO$	303.5	304.2 304.2	147.1 57.2	120 100	30 28	5.53

Table 2. *Cont.*

Pesticide	Molecular Formula	M g/Mol	Precursor ion m/z	Product ion m/z	Frag (V)	CE (V)	Rt (min)
Fluroxypyr-meptyl	$C_{15}H_{21}Cl_2FN_2O_3$	367.2	367 367	254.9 181	80 80	11 32	7.44
Flusilazole	$C_{16}H_{15}F_2N_3Si$	315.4	316.1 316.1	247.1 165	110 110	12 20	18.21
Flutriafol	$C_{16}H_{13}F_2N_3O$	301.2	302.1 302.1	70.2 123.1	100 100	18 29	5.24
Phoxim	$C_{12}H_{15}N_2O_3PS$	298.3	299 299	129 77	80 80	10 20	17.52
Hexaconazole	$C_{14}H_{17}Cl_2N_3O$	314.2	314.1 314.1	159 70.1	100 130	20 17	18.80
Imazalil	$C_{14}H_{14}Cl_2N_2O$	297.1	297.1 297.1	255 159	100 100	15 23	14.80
Imidacloprid	$C_9H_{10}ClN_5O_2$	255.7	256 256	208.7 174.6	100 100	15 20	11.60
Indoxacarb	$C_{22}H_{17}ClF_3N_3O_7$	527.8	528.1 528.1	203 150	120 120	36 16	18.80
Isoproturon	$C_{12}H_{18}N_2O$	206.3	207 207	78 123	135 135	17 17	12.40
Carbaryl	$C_{12}H_{11}NO_2$	201.2	202.1 202.1	145 127	100 100	10 35	15.50
Carbendazim	$C_9H_9N_3O_2$	191.1	192.1 192.1	160.1 132	104 104	18 34	9.35
Carbofuran	$C_{12}H_{15}NO_3$	221.2	222.1 222.1	165.1 123	90 90	20 15	15
Carboxin	$C_{12}H_{13}NO_2S$	235.3	236 236	87 143	120 120	20 20	6.19
Carbosulfan	$C_{20}H_{32}N_2O_3S$	380.5	381.2 381.2	118.1 160.1	31 31	33 22	5.52
Clothianidin	$C_6N_5H_8SO_2Cl$	249.6	250 250	169.1 132.1	90 90	10 15	11.80
Kresoxim-methyl	$C_{18}H_{19}NO_4$	313.3	336.2 336.2	246.2 229.2	120 120	15 15	18.40
Quintozene	$C_6Cl_5NO_2$	295.3	237 237	143 119	30 30	10 10	13.61
Myclobutanil	$C_{15}H_{17}ClN_4$	288.7	289.2 289.2	125.1 70.2	150 150	20 15	17.78
Linuron	$C_9H_{10}Cl_2N_2O_2$	249.0	249 249	182 160	70 70	18 18	9.72
Malathion	$C_{10}H_{19}O_6PS_2$	330.3	331.1 331.1	127 99	90 90	5 21	17.6
Metalaxyl	$C_{15}H_{21}NO_4$	279.3	280.2 280.2	220.1 192.1	120 120	10 15	16.3
Metamitron	$C_{10}H_{10}N_4O$	202.2	203.1 203.1	175 104	115 115	14 22	12.78
Methidathion	$C_6H_{11}N_2O_4PS_3$	302.3	303 303	165 127	120 120	10 20	12.76
Methiocarb	$C_{11}H_{15}NO_2S$	225.3	226.1 226.1	169 121	62 62	6 18	17.36
Metconazole	$C_{17}H_{22}ClN_3O$	319.8	320 320	125 70	100 100	20 20	18.88
Methoxyfenozide	$C_{22}H_{28}N_2O_3$	368.5	369.2 369.2	149.1 133	100 90	20 25	17.2
Methomyl	$C_5H_{10}N_2O_2S$	162.2	163.1 163.1	106 88	80 80	5 5	9.8
Nicosulfuron	$C_{15}H_{18}N_6O_6S$	410.4	411 411	182 106	100 100	32 32	4.57

Table 2. *Cont.*

Pesticide	Molecular Formula	M g/Mol	Precursor ion m/z	Product ion m/z	Frag (V)	CE (V)	Rt (min)
Oxadixyl	$C_{14}H_{18}N_2O_4$	278.3	279.1 279.1	219.1 133.3	80 80	10 15	14.35
Oxamyl	$C_7H_{13}N_3O_3S$	219.2	237.1 237.1	90 72	60 60	5 10	9
Pencycuron	$C_{19}H_{21}ClN_2O$	328.8	329.1 329.1	125.1 99.1	120 130	38 35	17.62
Pymetrozine	$C_{10}H_{11}N_5O$	217.2	218 218	105 78	120 100	30 20	3.61
Pyraclostrobin	$C_{19}H_{18}ClN_3O_4$	387.8	388.1 388.1	194 163	100 100	10 10	18.6
Pyrimethanil	$C_{12}H_{13}N_3$	199.2	200.1 200.1	107.1 82.1	136 136	26 30	16
Pirimiphos-methyl	$C_{11}H_{20}N_3O_3PS$	305.3	306 306	164 108	20 20	20 39	7.49
Pirimicarb	$C_{11}H_{18}N_4O_2$	238.2	239.2 239.2	182.1 72	120 120	15 20	12
Pyriproxyfen	$C_{20}H_{19}NO_3$	321.3	322.1 322.1	227.1 185.1	120 120	10 10	20
Prochloraz	$C_{15}H_{16}C_{13}N_3O_2$	376.6	376 376	308 266	80 80	10 10	18.39
Propamocarb	$C_9H_{20}N_2O_2$	188.2	189.1 189.1	102 144	120 100	20 20	1.82
Propiconazole	$C_{15}H_{17}Cl_2N_3O_2$	342.2	342.1 342.1	159 69	120 120	20 20	18.60
Propyzamide	$C_{12}H_{11}Cl_2NO$	236.3	256.1 256.1	190 173	120 120	23 31	5.98
Propoxur	$C_{11}H_{15}NO_3$	209.2	210.1 210.1	168.1 111	60 60	5 10	15.10
Spiroxamine	$C_{18}H_{35}NO_2$	297.4	298 298	144 100	120 100	32 20	5.44
Tebufenpyrad	$C_{18}H_{24}ClN_3O$	333.8	334.2 334.2	145.1 117	175 175	24 32	19.70
Tebuconazole	$C_{16}H_{22}ClN_3O$	307.8	308.1 308.1	125 70	100 100	25 25	18.58
Tefluthrin	$C_{17}H_{14}ClF_7O_2$	418.7	177 177	137 127	10 10	15 15	14.99
Thiodicarb	$C_{10}H_{18}N_4O_4S_3$	354.4	355.1 355.1	108 88	80 80	10 15	15.50
Thiacloprid	$C_{10}H_9ClN_4S$	252.7	253 253	186 126	110 110	10 20	13.40
Trifloxystrobin	$C_{20}H_{19}F_3N_2O_4$	408.3	409.1 409.1	206.1 186.1	120 120	10 15	18.95

The obtained results indicate a good response linearity in the range of 0.005 to 0.1 µg/mL for all the investigated analytes. Therefore, the method is selective, showing good linearity, expressed by the values of determination coefficient (r^2)>0.99 for all 60 pesticides. The matrix effect (ME) was estimated on matrix and solvent calibration graph slopes and it indicated that tomato, potato, apple and carrot matrix have a strong influence on 60 pesticides. The ME was compensated with MMC.

The LOQ as the lowest concentration that will be detected and quantified by an outstanding analytical method with sufficient precision and accuracy was established on 0.005 mg/kg for every pesticide and was confirmed experimentally. The LODs were calculated by MassHunter software and all the values were in the range of 0.001 to 0.003 mg/kg.

The recovery studies were appraised at two levels, spiking blank tomato, apple, carrot and potato samples at 0.01 and 0.1 mg/kg in five replicates (Figure 1). The 53 out of 60 analyzed pesticides showed the recovery ranging from 67.4 to 118.5%. The obtained results are in accordance with those published by Mao et al. [18], whose values for recovery varied from 61.6 to 119.4%. The repeatability, expressed as a relative standard deviation (%RSD), was between 1.87 and 14.73%. Broadly, the accuracy and precision results were tolerable to all investigated pesticides, according to the Document N° SANTE/12682/2019 [31].

According to the validation parameters, LC-MS/MS is a suitable technique for the qualitative and quantitative analysis of 60 pesticide residues in selected matrices-samples. TIC and MRM chromatograms of the pesticides determined in the apple samples from the organic production are given in Figures 1 and 2.

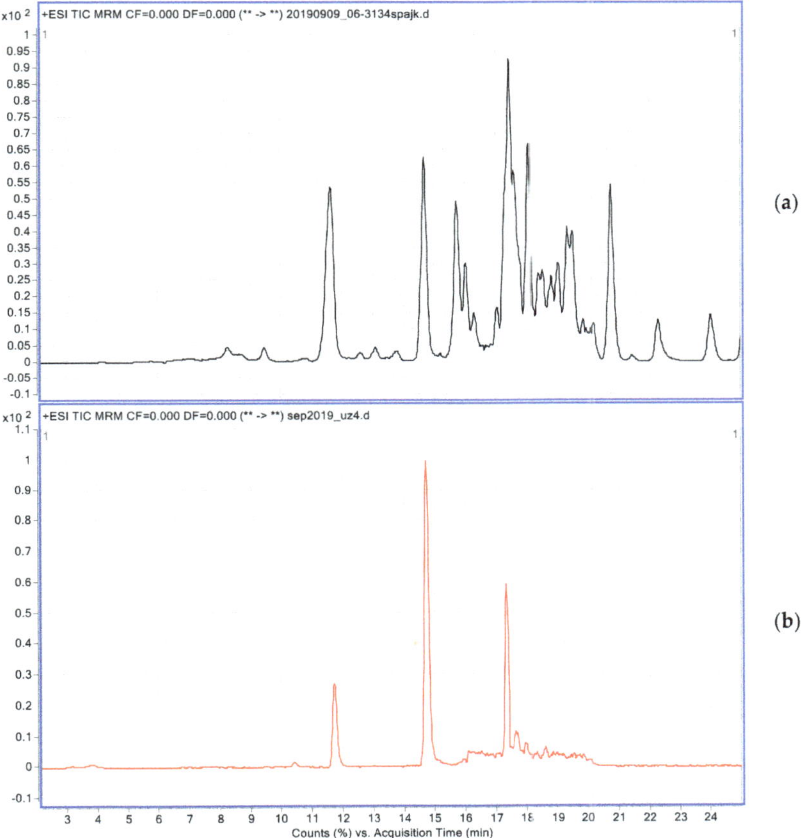

Figure 1. TIC chromatograms of (**a**)−spiked apple sample (0.01 mg/kg) and (**b**)−analyzed apple sample from organic production.

The results presented in Figures 3 and 4 show the pesticide residues in the investigated samples from the organic (Figure 3) and conventional production (Figure 4) with no detections (meaning<LOD), the samples with the determinations below LOQ, the determinations compliant with the MRLs and the determinations exceeding the MRLs.

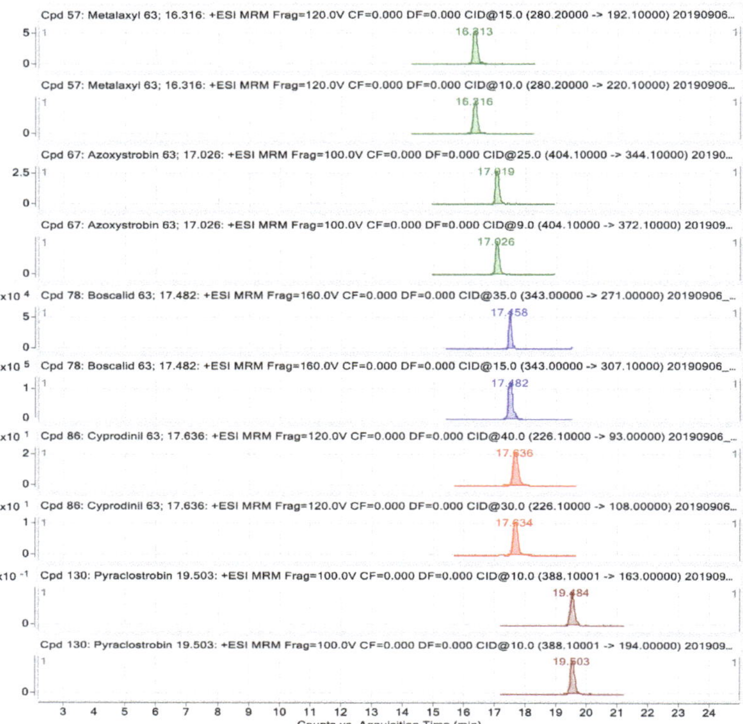

Figure 2. MRM chromatograms of determined pesticides in apple samples from organic production.

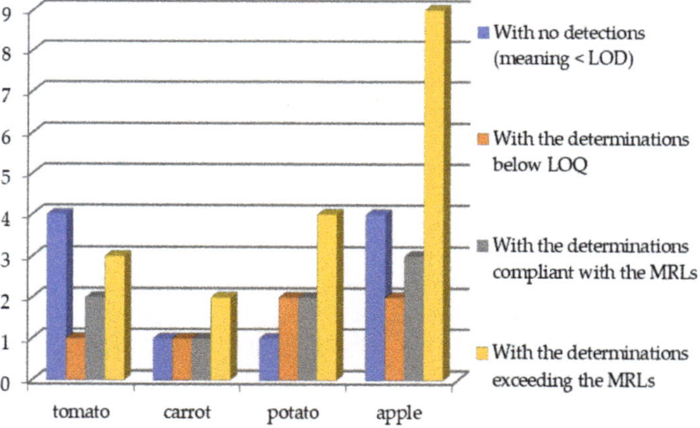

Figure 3. Samples from organic production.

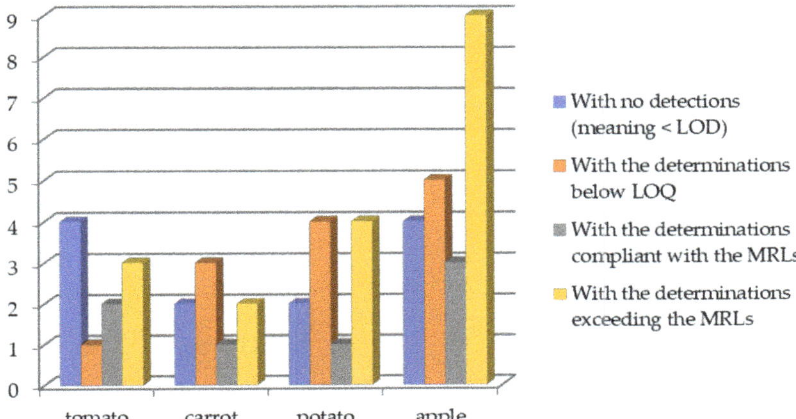

Figure 4. Samples from conventional production.

The EU-harmonized MRLs are set for more than 500 pesticides covering 370 food products/food groups. A default MRL of 0.01 mg/kg is applicable for pesticides not explicitly mentioned in the MRL legislation. The Regulation (EC) No 396/2005 [25] imposes on the Member States the obligation to carry out the controls to ensure that the food placed on the market is compliant with the legal limits. For the organic food items produced following Regulation (EC) No 834/2007 [33] no specific MRLs are established. However, in the Regulation (EC) No 396/2005 [25] in Article 18 it is stated that 0.01 mg/kg is the MRL value for those products for which no specific MRL is set out in Annexes II or III, or for the active substances not listed in Annex IV. The value of 0.01 mg/kg is the usually accepted MRL for organic products. According to Regulation (EC) No 834/2007 [33], the plant protection products should only be used if they are compatible with the objectives and principles of the organic production following the provisions laid down in Article 16(3)(c). Regulation (EC) No 889/2008 [20] lays down the detailed rules for the implementation of Council Regulation (EC) No 834/2007 [33] on organic production and labelling of the organic products. It defines the restricted list of plant protection products that may be used in the organic farming. Most of these substances are exempted from the setting of legal limits under Regulation (EC) No 396/2005 [25], as these substances are listed in Annex IV of the MRL regulation. The EOCC (European Organic Certifiers Council) is an organization of the organic certifiers in Europe. The EOCC has formed a "task force residues", which developed the "EOCC pesticide residues guideline", and presented it to the public in 2012. This guideline also follows the BNN (Bundesverband Naturkost Naturwaren) concept of the orientation value of 0.010 mg/kg, but the value is called 'action level'. This guideline emphasizes the procedural aspects in which certifiers should handle pesticide residues. Together with this guideline, the 'EOCC task force residues' has also published a discussion paper in which the possibilities of applying a maximum pesticide level for the organic products are discussed [34]. This maximum level is called 'critical level'. The task force proposed that the critical level might be set at the value of 10% of the MRL, but does not insist on this particular value. It is extremely important to bear in mind the fact that in the products from the organic production the multiple detections cannot be tolerated, i.e., that one product cannot contain more than two detected pesticide residues concerning the BNN.

The most detected pesticides from the conventional production were fluopyram, difenoconazole, metalaxyl, pyrimethanil, azoxystrobin, boscalid, cyprodinil, pyraclostrobin and delthametrine. The concentrations were in the range from 0.003 to 0.154 mg/kg. In the samples from the organic production the most frequently detected were fluopyram, difenoconazole, azoxystrobin, boscalid and cyprodinil.

According to Montiel-León et al. [35] the pesticides of great concern these days imply carbamates, neonicotinoids, organophosphates and triazines.

The similar results to those obtained in our case study were published by Mao et al. [18], where the conventional vegetable samples contained multiple pesticide residues compared with those in the organic vegetable samples and most of these residues were detected at higher levels in the conventional than in the organically produced vegetables.

According to Mansour et al. [36], the organic potato tubers sampled from the market have had higher pesticide residue levels than those collected from a specific organic farm. Therefore, along with our results, these findings may give an indication that the data obtained from a single supervised farm may not reflect the market quality where the products from the different agricultural producers could be found. Although the pesticide residues uptake from soils depended on plant variety, the preparation of the products for sale on the market could have a significant influence. For example, Zohair et al. [37] emphasized that washing and peeling carrots or potatoes removed 52–100% of the contaminant residues, which also varied with the crop type and the contaminant amount and properties.

Considering the fact that our samples were taken simultaneously during a week in April, the interesting fact that should not be neglected is the seasonal dynamic of pesticide residue levels. According to Mansour et al. [36], the highest pesticide residue peaks in the conventional potato production were noticeably raised in August, December, February and April, and for the organic potatoes in September. The total pesticide contamination level showed different arrangements: winter > summer > fall > spring in the conventional and fall > summer > winter > spring in the organic potato production.

The tomato, carrot and potato samples are considered to be the organic products based on the pesticide residues. However, the analyzed organic apple sample contained six pesticide residues, with the pyrimethanil and pyraclostrobin residues above the MRLs (for the conventional production) of 0.05 and 0.02 mg/kg, respectively. This sample cannot implement the state established in SANTE/11945/20, as well as IFOAM [38], which allows the pesticide residue detection concerning the measurement of the uncertainty of 50% because we have detections of six pesticide residues.

The apple samples from the conventional production contained four pesticide residues, with azoxystrobin concentration over MRL of 0.01 mg/kg [32]. The conventional tomato and potato did not contain pesticide residues, all detections were under the LOQs. The carrot sample contained fluopyram and difenoconazole with residues being below the MRLs.

Montiel-León et al. [39] conducted the research on 37 samples of apples and determined that 57% of the tested samples contained at least one of the studied pesticides. The most common detected pesticide was acetamiprid, with the detection frequency being 41% and the maximum concentration of 24 µg/kg in the case of the Cortland apple, which was sampled from the conventional production. They also detected carbendazim (detection frequency of 19%), carbaryl (3%) and simazine (5%), as well as some other neonicotinoids: clothianidin (detection frequency of 3%), imidacloprid (16%) and thiacloprid (5%). Their research also comprised the analysis of the tomato samples, the results of which showed that 17% of the tested samples contained at least one of the studied pesticides, all of which were classified as neonicotinoids. The acetamiprid was detected in one sample (detection frequency of 3%) at 16 µg/kg, dinotefuran was found in two samples (concentrations of 13 and 20 µg/kg), while the imidacloprid was registered in 10% of the tested tomato samples (concentrations of 7.6, 10 and 11 µg/kg).

The analysis of the pesticide residues in food is subject to constant modification owing to matrix complexity, low concentrations of the compounds of interest and the increasing number of pesticides approved for use [40]. Namely, LC coupled with a QQQ tandem mass spectrometer, working in the multiple reaction monitoring (MRM) mode is the most frequently applied platform used in the analysis of pesticide residues in food. The most important advantages of validated LC-MS/MS in this study include high sensitivity and selectivity, short duration of analysis, which enables the separation and determination of a considerable number of compounds (60 pesticides with internal standard) during a single analytical run. The obtained results indicate good response linearity in the range of 0.005 to 0.1 µg/mL for all 60 pesticides (r^2)>0.99. The MMC reduces the matrix effect on the quantification results, especially taking into account that the amount of pesticide residues is in/on the trace levels. Very low LOQ set on 0.005 mg/kg for every pesticide, with the LODs values in the range of 0.001 to 0.003 mg/kg, potentiate the quantification of pesticide residues in the organic food below the 0.01 mg/kg. Additionally, the recovery studies on two spiking levels (0.01 and 0.1 mg/kg) indicate that 88.3% of the investigated pesticides have the recovery in the interval from 67.4 to 118.5%, with the RSD between 1.87 and 14.73%.

The obtained results of the present study provide an indication regarding the pesticide residues in the organic apples. However, they cannot be responsible for the de-characterization of apples as an organically-produced commodity. The amount of the analyzed samples is not high; still, the results of our results accentuate the need for the constant monitoring of the products from the organic, as well as from the conventional production.

Author Contributions: Conceptualization, V.B. and G.V.; methodology, G.V.; software, N.P. and M.K.; validation, D.M., B.K. and M.C.; formal analysis, G.V. and V.B.; investigation, T.S.; resources, B.K.; data curation, T.S.; writing—original draft preparation, V.B.; writing—review and editing, G.V.; visualization, V.B.; supervision, M.C. and A.P.; project administration, N.P.; funding acquisition, N.P. All authors have read and agreed to the published version of the manuscript.

Funding: This research was funded by the Ministry for Education, Science and Technological Development of the Republic of Serbia.

Informed Consent Statement: Not applicable for studies not involving humans.

Data Availability Statement: Data available in a publicly accessible repository.

Acknowledgments: This research was supported by the Ministry for Education, Science and Technological Development of the Republic of Serbia.

Conflicts of Interest: The authors declare no conflict of interest.

References

1. Gomiero, T. Food quality assessment in organic vs. conventional agricultural produce: Findings and issues. *Appl. Soil. Ecol.* **2018**, *123*, 714–728. [CrossRef]
2. Del Mar Gómez-Ramos, M.; Nannou, C.; Martínez Bueno, M.J.; Goday, A.; Murcia-Morales, M.; Ferrer, C.; Fernández-Alba, A.R. Pesticide residues evaluation of organic crops. A critical appraisal. *Food Chem. X* **2020**, *5*, 100079. [CrossRef] [PubMed]
3. Chinnici, G.; D'Amico, M.; Pecorino, B. A multivariate statistical analysis on the consumers of organic products. *Brit. Food J.* **2002**, *104*, 187–199. [CrossRef]
4. Nguyen, H.; Nguyen, N.; Nguyen, B.; Lobo, A.; Vu, P. Organic Food Purchases in an Emerging Market: The Influence of Consumers' Personal Factors and Green Marketing Practices of Food Stores. *IJERPH* **2019**, *16*, 1037. [CrossRef]
5. Aarset, B.; Beckmann, S.; Bigne, E.; Beveridge, M.; Bjorndal, T.; Bunting, J.; McDonagh, P.; Mariojouls, C.; Muir, J.; Prothero, A.; et al. The European consumers' understanding and perceptions of the "organic" food regime: The case of aquaculture. *Br. Food J.* **2004**, *106*, 93–105. [CrossRef]
6. Cristina, R.T.; Kovačević, Z.; Cincović, M.; Dumitrescu, E.; Muselin, F.; Imre, K.; Militaru, D.; Mederle, N.; Radulov, I.; Hădărugă, N.; et al. Composition and Efficacy of a Natural Phytotherapeutic Blend against Nosemosis in Honey Bees. *Sustainability* **2020**, *12*, 5868. [CrossRef]
7. Vapa-Tankosić, J.; Ignjatijević, S.; Kiurski, J.; Milenković, J.; Milojević, I. Analysis of Consumers' Willingness to Pay for Organic and Local Honey in Serbia. *Sustainability* **2020**, *12*, 4686. [CrossRef]
8. Tomaš-Simin, M.; Glavaš-Trbić, D.; Petrović, M. Organic production in the Republic of Serbia: Economic aspects. *Ekon. Teor. Praksa* **2019**, *12*, 88–101. [CrossRef]
9. Browne, A.W.; Harris, P.J.C.; Hofny-Collins, A.H.; Pasiecznik, N.; Wallace, R.R. Organic production and ethical trade: Definition, practice and links. *Food Policy* **2000**, *25*, 69–89. [CrossRef]
10. Ostapenko, R.; Herasymenko, Y.; Nitsenko, V.; Koliadenko, S.; Balezentis, T.; Streimikiene, D. Analysis of Production and Sales of Organic Products in Ukrainian Agricultural Enterprises. *Sustainability* **2020**, *12*, 3416. [CrossRef]
11. Ferreira, S.; Oliveira, F.; Gomes da Silva, F.; Teixeira, M.; Gonçalves, M.; Eugénio, R.; Damásio, H.; Gonçalves, J.M. Assessment of Factors Constraining Organic Farming Expansion in Lis Valley, Portugal. *AgriEngineering* **2020**, *2*, 111–127. [CrossRef]
12. Hartung, U.; Schaub, S. The Regulation of Genetically Modified Organisms on a Local Level: Exploring the Determinants of Cultivation Bans. *Sustainability* **2018**, *10*, 3392. [CrossRef]

13. Bošković, J. Influence of Genetic Variability of Grapes to Produce High-Quality Wines. *J. Agron. Technol. Eng. Manag.* **2020**, *3*, 483–488.

14. Van Bruggen, A.H.; Gamliel, A.; Finckh, M.R. Plant disease management in organic farming systems. *Pest. Manag. Sci.* **2016**, *72*, 30–44. [CrossRef]

15. Benbrook, C.; Baker, B. Perspective on Dietary Risk Assessment of Pesticide Residues in Organic Food. *Sustainability* **2014**, *6*, 3552–3570. [CrossRef]

16. Baker, B.P.; Benbrook, C.M.; Iii, E.G.; Benbrook, K.L. Pesticide residues in conventional, integrated pest management (IPM)-grown and organic foods: Insights from three US data sets. *Food Addit. Contam.* **2002**, *19*, 427–446. [CrossRef]

17. Winter, C.K.; Davis, S.F. Organic Foods. *J. Food Sci.* **2006**, *71*, R117–R124. [CrossRef]

18. Mao, X.; Wan, Y.; Li, Z.; Chen, L.; Lew, H.; Yang, H. Analysis of organophosphorus and pyrethroid pesticides in organic and conventional vegetables using QuEChERS combined with dispersive liquid-liquid microextraction based on the solidification of floating organic droplet. *Food Chem.* **2020**, *309*, 125755. [CrossRef]

19. Tobin, R. Detection of Pesticide Residues in Organic and Conventional Fruits and Vegetables Available in Ireland Using Gas Chromotography/Tandem Mass Spectrometry (GC-MS/MS) and Liquid Chromotography/Tandem Mass Spectrometry (LC-MS/MS) Detection. *J. Nutr. Health Food Sci.* **2014**, *2*. [CrossRef]

20. *Laying Down Detailed Rules for the Implementation of Council Regulation (EC) No 834/2007 on Organic Production and Labelling of Organic Products with Regard to Organic Production, Labelling and Control, No 889/2008 of 5 September 2008*; European Commission: Brussels, Belgium, 2008; Volume 250.

21. Rembiakowska, E.; Badowski, M. Pesticide Residues in the Organically Produced Food. In *Pesticides in the Modern World-Effects of Pesticides Exposure*; Stoytcheva, M., Ed.; InTech: London, UK, 2011; ISBN 978-953-307-454-2.

22. Authority, European Food Safety (EFSA). The 2017 European Union report on pesticide residues in food. *EFSA J.* **2019**, *17*. [CrossRef]

23. Authority, European Food Safety. Trusted Science for Safe Food. Available online: https://www.efsa.europa.eu/en (accessed on 25 November 2020).

24. *Regulation (EU.) 2018/848 of the European Parliament and of the Council of 30 May 2018 on Organic Production and Labelling of Organic Products and Repealing Council Regulation (EC) No 834/2007*; European Commission: Brussels, Belgium, 2018; Volume 150.

25. *Regulation (EC.) No 396/2005 of the European Parliament and of the Council of 23 February 2005 on Maximum Residue Levels of Pesticides in or on Food and Feed of Plant and Animal Origin and Amending Council Directive 91/414/EECText with EEA Relevance.*; European Commission: Brussels, Belgium, 2005; Volume 070.

26. Botitsi, H.; Tsipi, D.; Economou, A. Current legislation on pesticides. In *Applications in High Resolution Mass Spectrometry*; Elsevier: Amsterdam, The Netherlands, 2017.

27. Handford, C.E.; Elliott, C.T.; Campbell, K. A review of the global pesticide legislation and the scale of challenge in reaching the global harmonization of food safety standards. *Integr. Environ. Assess. Manag.* **2015**, *11*, 525–536. [CrossRef] [PubMed]

28. Anastassiades, M.; Lehotay, S.J.; Stajnbaher, D.; Schenck, F.J. Fast and easy multiresidue method employing acetonitrile extraction/partitioning and "dispersive solid-phase extraction" for the determination of pesticide residues in produce. *J. AOAC Int.* **2003**, *86*, 412–431. [CrossRef] [PubMed]

29. Bursić, V.; Vuković, G.; Đukić, M.; Petrović, A.; Cara, M.; Marinković, D.; Đurović-Pejčev, R. Article Entitled: Determination of Multi-Class Pesticide Residues in Sour Cherries by LC-MS/MS. *Contemp. Agric.* **2018**, *67*, 227–232. [CrossRef]

30. Vuković, G.; Đukić, M.; Bursić, V.; Stojanović, T.; Petrović, A.; Kuzmanović, S.; Starović, M. Development and validation of LC-MS/MS method for the citrinin determination in red rice. *J. Agron. Technol. Eng. Manag.* **2019**, *2*, 192–199.

31. Main Changes Introduced in Document N° SANTE/12682/2019 with Respect to the Previous Version (Document N° SANTE/11813/2017). Available online: https://ec.europa.eu/food/sites/food/files/plant/docs/pesticides_mrl_guidelines_wrkdoc_2019-12682.pdf (accessed on 25 November 2020).

32. EUR-Lex-32005R0396-EN-EUR-Lex. Available online: https://eur-lex.europa.eu/legal-content/EN/ALL/?uri=celex%3A32005R0396 (accessed on 23 April 2020).

33. *Council Regulation (EC) No 834/2007 of 28 June 2007 on Organic Production and Labelling of Organic Products and Repealing Regulation (EEC) No 2092/91*; European Commission: Brussels, Belgium, 2007; Volume 189.

34. Milan, M.A.; Bickel, R.; Speiser, B. *Improving the Handling of Residue Cases in Organic Production–Part 1 "Quick Scan"*; Research Institute of Organic Agriculture FiBL: Frick, Switzerland, 2019.

35. Montiel-León, J.M.; Munoz, G.; Vo Duy, S.; Do, D.T.; Vaudreuil, M.-A.; Goeury, K.; Guillemette, F.; Amyot, M.; Sauvé, S. Widespread occurrence and spatial distribution of glyphosate, atrazine, and neonicotinoids pesticides in the St. Lawrence and tributary rivers. *Environ. Pollut.* **2019**, *250*, 29–39. [CrossRef]

36. Mansour, S.A.; Belal, M.H.; Abou-Arab, A.A.K.; Ashour, H.M.; Gad, M.F. Evaluation of some pollutant levels in conventionally and organically farmed potato tubers and their risks to human health. *Food Chem. Toxicol.* **2009**, *47*, 615–624. [CrossRef]

37. Zohair, A.; Salim, A.-B.; Soyibo, A.A.; Beck, A.J. Residues of polycyclic aromatic hydrocarbons (PAHs), polychlorinated biphenyls (PCBs) and organochlorine pesticides in organically-farmed vegetables. *Chemosphere* **2006**, *63*, 541–553. [CrossRef]

38. IFOAM Organic Europe. Guideline for Pesticide Residue Contamination for International Trade in Organic. Available online: https://www.organicseurope.bio/content/uploads/2020/09/ifoameu_regulation_guideline_for_pesticide_residue_contamination_for_international_trade_in_organic_2012.pdf?dd (accessed on 13 December 2020).

39. Montiel-León, J.M.; Duy, S.V.; Munoz, G.; Verner, M.-A.; Hendawi, M.Y.; Moya, H.; Amyot, M ; Sauvé, S. Occurrence of pesticides in fruits and vegetables from organic and conventional agriculture by QuEChERS extraction liquid chromatography tandem mass spectrometry. *Food Control.* **2019**, *104*, 74–82. [CrossRef]
40. Stachniuk, A.; Fornal, E. Liquid Chromatography-Mass Spectrometry in the Analysis of Pesticide Residues in Food. *Food Anal. Methods* **2016**, *9*, 1654–1665. [CrossRef]

 *sustainability*

Article

Sustainable Organic Corn Production with the Use of Flame Weeding as the Most Sustainable Economical Solution

Miloš Rajković [1], Goran Malidža [1], Mirela Tomaš Simin [2,*], Dragan Milić [2], Danica Glavaš-Trbić [2], Maja Meseldžija [2] and Sava Vrbničanin [3]

[1] Institute of Field and Vegetable Crops, Maksima Gorkog 30, 21000 Novi Sad, Serbia; milos.rajkovic@nsseme.com (M.R.); goran.malidza@ifvcns.ns.ac.rs (G.M.)
[2] Faculty of Agriculture, University of Novi Sad, Trg Dositeja Obradovića 8, 21000 Novi Sad, Serbia; mdragan@polj.uns.ac.rs (D.M.); danicagt@polj.uns.ac.rs (D.G.-T.); maja@polj.uns.ac.rs (M.M.)
[3] Faculty of Agriculture, University of Belgrade, Nemanjina 6, 11080 Belgrade, Serbia; sava@agrif.bg.ac.rs
* Correspondence: mirela.tomas@polj.edu.rs; Tel.: +381-62-240-123

Abstract: Flame weeding is an alternative method of weed control. Essentially, it is a supplement to other physical and mechanical processes used in organic production. Weed control costs have a large share of the total cost of crop production. This study aimed to investigate hand weed hoeing's cost-effectiveness, accompanied by inter-row cultivation and flame weeding applied in organic maize production using two different machines to determine the economically best solution. For this purpose, the prototype flame weeder and commercial flame-weeding machinery were used. Designed primarily for smaller fields, the prototype flame weeder was equipped with a cultivator and a 70 kg propane bottle. Commercial Red Dragon flame weeder, fitted with an 800 kg propane tank and featuring no cultivation implements, is designed for larger areas. The analysis has shown that hand hoeing produced a higher yield (8.3 t/ha in total), but it contributed significantly to the production costs. The costs per hectare decreased when the prototype flame weeder and the commercial Red Dragon flame weeder were used compared to hand hoeing. More beneficial economic impacts were recorded when the prototype flame weeder was used (439.39 €/ha) than in applying the Red Dragon flame weeder (456.47 €/ha). The efficacy of flame weeding is somewhat limited and could be enhanced by additional hand hoeing, if the effect of the machine in terms of weeding is observed. However, the analysis has shown that, in this case, investments in additional hand hoeing are not economically justified because the operating costs incurred therein (168 €/ha) were not met by a yield increase of 500 kg/ha, i.e., a surplus revenue of 100 €/ha. Moreover, the economic impacts of flame weeding would be considerably more significant in larger fields.

Keywords: cost-effectiveness; flame weeding; maize; organic agricultural production

 check for updates

Citation: Rajković, M.; Malidža, G.; Tomaš Simin, M.; Milić, D.; Glavaš-Trbić, D.; Meseldžija, M.; Vrbničanin, S. Sustainable Organic Corn Production with the Use of Flame Weeding as the Most Sustainable Economical Solution. *Sustainability* **2021**, *13*, 572. https://doi.org/10.3390/su13020572

Received: 30 October 2020
Accepted: 6 January 2021
Published: 9 January 2021

Publisher's Note: MDPI stays neutral with regard to jurisdictional claims in published maps and institutional affiliations.

1. Introduction

Increased demand for food has altered the methods of agricultural production. "Nowadays, it is obvious that conventional (industrial) methods of agricultural production, in addition to procuring sufficient sustenance and an array of products, precipitate a series of negative not only ecological but also social and economic repercussions" [1]. Hodge [2] summarized specific negative trends in contemporary agriculture, which prompted such production systems' long-term sustainability. He further argued that agriculture is inclined to utilize inputs originating from distant spatial and sectoral sources. Additionally, it consumes increasing energy supplies generated from non-renewable sources. Agriculture depends on a deteriorating genetic base, and it exerts detrimental effects on the environment. All the considerations mentioned above are mainly manifested in (1) a strengthening link between agriculture and the chemical industry (under artificial fertilizers and pesticides), (2) dependence of agriculture on subsidies and price control, and (3) an increasing number of grave repercussions affecting habitats, animal and plant species, the environment, and human health and well-being. As part of the socio-economic sub-systems, modern agricultural production has proven adverse effects on the environment [3–5] due to increasing dependence on the industry (in terms of fertilizers and pesticides) and the introduction of monoculture for the sake of profit. The adverse effects of contemporary agricultural production have emphasized the importance of alternative production systems. A different approach to the environment inherently characterizes such systems. Organic agriculture is an alternative production system considered more beneficial to the environment than conventional production systems [1]. Lampkin and Padel [6] define organic agriculture as a philosophy and a production method to create integrated, humane, economically sustainable, and environment-friendly agriculture. Organic agriculture tends to maximize the use of renewable on-farm resources. It represents a system of managing ecological and biological processes to achieve acceptable crop yields, animal gains, and essential nutrient production. Notwithstanding the conceptual differences in organic farming, such a system's primary objective is sustainable agricultural production.

The notion of sustainability is used in a broad sense, encompassing economic, social, and environmental sustainability [7]. Considering that sustainability comprises an economic aspect (among others) [8,9], it is crucial to develop cost reduction methods in organic agriculture, which would exert positive effects on the total financial results. Although organic production requires higher investments, it ultimately results in higher quality products [10]. Farmers are often prompted to convert and conform to an organic farming system by economic sustainability [11–13].

According to some estimates, in conventional agriculture, weed control costs are higher than the costs of disease and pest control put together, and weeds account for a maize yield decrease of 13.2% worldwide [14]. Weed control is one of the most expensive steps in crop production [15] and is considered one of the main challenges responsible for significantly reducing yields in agricultural farming, especially in organic systems, where weed control is cited as the most critical production problem [15].

Consequently, weed control is of paramount importance to cost-effective maize production. Organic farmers cite weeds as the most severe production problem they encounter, and total crop losses from weeds can occur under the organic system [16].

Farmers often perceive the fear of ineffective weed control as one of the significant obstacles to conversion from conventional to organic farming [17]. Pannacci [18] stated that weed control has a significant effect on maize growth because the competition ability of maize is relatively low at early crop growth stages [18]. Weed control is a challenge in all systems of agricultural production, but it is especially pronounced in organic crop production. The use of herbicides is unallowed [19,20], so weed control is often limited to physical methods where hand weeding and mechanical cultivation are the most popular [21]. Organic producers often utilize hand weeding, but it is expensive, time-consuming, and challenging to organize [22]. Mechanical cultivation is one of the most commonly used weed control practices in row crops. Additionally, it has been reported that cultivation leaves a strip of uncontrolled weeds on either side of the crop row, influencing crop yield [23],and repeated cultivation increases the chance of soil erosion, destroys soil quality, and promotes the emergence of new weed flushes [24,25]. So, there is a need to review existing methods and evaluate alternative approaches that could be utilized for weed control in organic crop systems [22]. Flame weeding using propane has been proposed as an alternative to these traditional techniques both in organic and conventional agricultural systems [26].

In 1852, flame weeding was first introduced in sugar cane production in the USA. In the mid-1940s, selective flaming was used in many crop productions, including corn, soybeans, alfalfa, cotton, and different fruit and vegetable productions [27].

The greatest challenge in organic row crop production is intra-row weed control [28]. Flaming with cross burner orientation weeds can be controlled in a row of the crop with minimal damage to cultivated crops [15]. The main reason is the distribution of temperatures that are the highest at lower canopy (<6 cm tall) that gradually decreased with increase in canopy height [29].

Bearing in mind the fact that practices of thermal methods, e.g., propane flaming, can significantly contribute to weed control on certain surfaces, which, at the same time, affects the results of production in a positive way, made them popular among organic producers since the 1980s. According to previous research, flaming can reduce the dependence on synthetic herbicides, and it can essentially complement mechanical methods of weed control in organic and conventional production [15,30]. In 1990, flame weeding was applied to more than 4000 ha in the USA [31]. The method was also used in maize fields in Germany (totaling 75,000 ha), in the areas where weeds developed resistance to atrazine [32]. Maize is highly tolerant to flaming and can endure two treatments without a significant decrease in yields [33,34].

Propane flaming exposes weeds to heat stress, causing the denaturation of membrane proteins, resulting in a loss of cell function and dehydration, leading to their death or reducing their competition [21,26]. Flaming is a transfer of heat from the flame to the plant tissues resulting in boiling water molecules inside the cell [35]. This method leaves no chemical residues in soil, plants, water, or air; does not disturb the soil surface, an does not bring more weed seeds to the soil surface [19,20]. Temperatures in the range of 95–100 degrees Celsius can be lethal to weed leaves and stems when applied for at least 0.1 s, which further results in cell desiccation and ultimately the loss of cell function [23].Plant responses to flaming depend on their heat tolerance, protective layers of hair and wax, lignification, and water conditions [36].Weeds are most sensitive to flame heat when they are in the three to the five-leaf stage. Broad-leaf weeds are more susceptible to control with flaming than grasses that are more tolerant to flaming. Ulloa et al. reported a propane dose of about 60 kg ha^{-1}whichprovided almost 80% control of grass weeds and 90% control of broad-leaf species [37]. Thus, flaming has the potential to be used in crops such as maize (*Zea mays* L.) and sorghum (*Sorghum bicolor* [L.] Moench) because these crops have problems with the mentioned weeds, especially in the organic cultivation system, but it can also be used in soybean at the appropriate growth stage [20].

From an economic point of view, the cost of a single flame application emitted under crop canopy could range from 30 to 40USD/ha, not taking into account equipment and labor costs. In comparison, application in the form of layers (over the crop row) for the flame can cost 12–20 USD/ha due to lower propane use rates (30–40 kg/ha) [23]. To be cost-effective and sustainable, organic production has to be financially rewarding for individual producers, providing them with a decent living standard. This research's importance is high, having in mind the mentioned restrictions in controlling and suppressing weeds in organic corn production and the fact that weed control costs have a large share in the total cost of crop production. This study aimed to examine the cost-effectiveness of three methods of weed control used in organic maize production: hand weed hoeing accompanied by inter-row cultivation, and two types of flame weeding, one with a prototype flame weeder and the other with the commercial Red Dragon flame weeder [38], and to determine which one of the methods contributes the most to cost savings.

As for this paper's contribution to the literature, this study is even more critical since no economic analysis on flame weeding control used in organic maize production has been conducted in Serbia and only just a few in the world. There are studies of the economic aspects of flame weed control applied in conventional agricultural production, but more knowledge is needed in the field of organic production. For the first time, research was done with a new prototype flame weeder constructed in Serbia [29,39]. The study is expected to provide data for future organic maize producers and other relevant participants. This research aims to indicate the cost-effectiveness of investment in the machinery for intra-row flame weeding in organic maize production to achieve economic sustainability for the household.

The structure of the paper is as follows. The first part (Introduction) lists the primary determinants that are the subject of the paper's research (organic agriculture, sustainability, flame weeding). After that, Materials and Methods are presented, explaining the cost-effectiveness of three different methods of weed control in corn production. The Results section shows the obtained values for three different models.The obtained results are compared with other research in the Discussions, and recommendations for possible further research in the paper's Conclusion.

2. Materials and Methods

Two-year field experiments were established on two locations in Serbia–Rimski Šančevi (45°20′31.5″ N 19°51′39.7″ E) and Čurug (45°28′16.9″ N 20°01′11.1″ E), and based on the results of this research, all expenses and revenues are determined. The cost-effectiveness of three different methods of weed control in organic maize production was calculated as a comparison of three treatments: (1) intra-row hand hoeing, (2) intra-row flame weeding with a prototype flame weeder [29,39] and (3) intra-row flame weeding with the commercial Red Dragon flame weeder [38].

For this paper, a method of differential calculation was used (Equation (2)). This method is used to determine whether a change in the business of the farm is economically justified.

Analytical calculations were used more often in agriculture. Still, it was shown that they can be unsuitable for agriculture because individual agricultural productions are firmly connected and dependent. It makes no sense to calculate success of individual lines of production but, rather, to calculate the farm's success as a whole. The analytical calculation determines all revenues and all production costs on the farm, their difference (financial result), and the obtained products' cost price. The general scheme of such a calculation is

$$p - t = d \tag{1}$$

where p represents the product's market value, t is the sum of all costs, and d is the financial result.

The differential calculation determines only the changes in income and changes in costs resulting from some planned or already taken economic measures on the farm. The general equation of such a calculation is

$$\Delta U - \Delta T = \Delta D \tag{2}$$

ΔU is the change in the total revenue or the value of a specific production line's production. This instance represents the change in the value of organic maize production caused by the change in yields. ΔT is the change in the total costs calculated for each method of weed control. Here, it denotes the difference in the costs of intra-row hand hoeing and flame weeding using two flame weeders. ΔD is the change in financial results or economic impacts.

Expenses entail a price valuation of the input costs incurred in producing a good or service [40]. According to Majcen, fees represent the amount of actual labor expressed in monetary terms and the embodiment of work and stipulated contractual and legal obligations required for producing outputs [41].Markovski defines expenses as a valuation of the asset and labor costs incurred, which are constituents of the output cost price [42]. From a broader theoretical perspective, an expense is a valuation of the asset and labor costs incurred in producing a good or service. Such valuation is necessary for finding the common denominator and assessing the financial performance of a business.

The specifics of agriculture, i.e., the process of agricultural production, determine the specificity of costs concerning the costs of other activities. For that reason, when calculating the costs of the agricultural output, the method of standardized costs was applied, which implies a logical verification of the research resultsconcerning the hitherto valid values in the field of research. The starting point for determining these costs was technological operations or agro-technical measures performed in the production of the tested production methods.Thecosts of work operations in various models are calculated based on experiment monitoring. Operating costs of different weed control methods were calculated on the basis of the costs of fuel (the price and consumption), maintenance cost, registration, and other title fees, depreciation, and labor. The economic impacts were calculated (per hectare) based on the labor and materials' employed market price. It is noteworthy that the annual net income computed in this paper was based on the data about yields produced in a dry farming system, i.e., in areas without irrigation.

The depreciation, accounting reserve for the equipment and machine replacement, was calculated using time-dependent depreciation Equation (3).

$$Annual\_depreciation\_expense = Vo/n \qquad (3)$$

where Vo is the depreciation basis, and n is the total number of years in use.

The interest costs on the funds invested were included in the Other Costs category.

The maintenance costs were calculated using a normative method based on the total maintenance costs incurred during the equipment's useful life [43]. The method was developed at the KTBL Institute (The KuratoriumfürTechnik und Bauwesen in der Landwirtschafte.V. (KTBL) is a registered association whereagriculture, science, commercial economy, administration and consulting belong. It is institutionally funded by the Federal Ministry of Food and Agriculture. https://www.ktbl.de/wir) and adjusted to the present research. The schedule of the given costs during the equipment's useful life was designed according to the change model in the maintenance and repair costs, also developed at the KTBL Institute [44].

The machinery fuel costs were calculated on the basis of the cost data obtained during the research. These costs were allocated according to the average market price of fuel. The data validation and normalization were performed in comparison with the fuel consumption norms for the standard operating mode of the engine employed [45]. Admittedly, the specific fuel consumption varies considerably depending on the operation method, reaching a minimum at approximately an engine load of 80%. To determine fuel consumption, it is requisite to establish a structure of fuel efficiency and specific fuel consumption. One of the models suitable for determining fuel consumption is a model, according to Renijus [46].

A standard tractor operating mode for low-intensity cultivation was used in this research. The fuel consumption was determined on the basis of the nominal (driving) engine power and the engine load percentage required for a specific operation [46]. Other machinery costs (which mostly include the costs of insurance, storage, oils, lubricants, taxes, and fees) were calculated on the basis of empirical norms for a specific type of machinery.

The income accounting was based on gross income, using Serbian agriculture's average income as a point of reference. All the other costs were allocated based on empirical norms. The application of modern agricultural machinery and implements in crop production encompasses requisite parameters for the optimal exploitation of tractors or engine units [47].

As previously stated, the cost-effectiveness of three different models (or methods) of intra-row weed control in organic maize were assessed in this research, where the first model entailed intra-row hand hoeing in organic maize production, following inter-row cultivation. This method needs a great deal of human labor because weeds have to be eradicated promptly to stop them from competing with the crops for sunlight, water, and soil nutrients.

The second model involved a prototype intra-row flame weeder (Figure 1a), a modified 4-row inter-row cultivator equipped with two 35 kg propane bottles, gas installations, a pressure regulator, a gas flow meter, and torches. Two torches were mounted on the sides of each cultivator row for intra-row flame weeding, whereas the previously fitted subsoilers and rear sweeps were used for inter-row cultivation [29,48]. The prototype was mounted on a 60 kW tractor. The testing of this prototype in row-crop weed flaming has been conducted since 2010 at the test field RimskiŠančevi (Novi Sad, AP Vojvodina, Serbia) of the Institute of Field and Vegetable Crops Novi Sad.

(a)   (b)

Figure 1. (a) The prototype flame weeder for smaller fields designed in Serbia; (b) The commercial Red Dragon flame weeder manufactured by Flame Engineering, Inc., La Crosse, Kansas, USA.

The third model encompassed the use of a commercial Red Dragon flame weeder (Figure 1b), which was mounted on a 90 kW tractor. The company AD Budućnost has used this machine—Global Seed (Čurug, AP Vojvodina, Serbia) since 2013, accounting for more than 1000 ha of organic maize production [49]. In contrast with the prototype, the Red Dragon flame weeder features an operating width of 8 rows (without inter-row cultivation implements) and an 800 kg propane tank, which contributes significantly to the overall performance [38].

3. Results

In this research, Model 1 entailed hand hoeing, representing the fundamental method of weed control in organic maize production. Under such conditions, the average maize yield was 8.3 t/ha. Daily wages are closely dependent on the extent of weed infestation. For cost accounting, 25 daily wages per single hoeing were taken as an average, i.e., 50 daily wages per two hoeings. The gross wages amounted to 14 € per day. Table 1 displays the accounting basis for Model 1.

Table 1. The basic parameters of the cost accounting of Model 1 (hand hoeing), Model 2 (using the prototype flame weeder without (2a) and with (2b) additional hoeing), and Model 3 (using the Red Dragon flame weeder without (3a) and with (3b) additional hoeing).

Parameter	Unit of Measurement	(1)	(2a)	(2b)	(3a)	(3b)
Maize yield	t/ha	8.3	7.8	8.3	7.8	8.3
Working day/hand hoeing/	days	50	0	12	0	12
Gross wage	€	14	14	14	14	14
Required tractor power	kw	-	60	60	90	90
Tractor purchase price	€		22,500	22,500	36,000	36,000
Propane consumption	kg	-	80	80	80	80
Propane price	€/kg		1.05	1.05	1.05	1.05
Mounted machine purchase price 01	€	-	2400	2400	19,500	19,500
Mounted machine purchase price 02	€	-	0	0	3350	3350
Operating width	m	-	3	3	6	6
Travel speed	km/h	-	4	4	4	4
Diesel fuel price	€/l		0.95	0.95	0.95	0.95
Other costs			5.50%	5.50%	5.50%	5.50%
Organic maize price	€/kg	0.20	0.20	0.20	0.20	0.20

A comparison of the basic parameters used in the cost accounting of Model 1, 2, and 3 are displayed in Table 1.

The cost accounting of Model 2 and 3 involved maize yields produced both with and without additional hand hoeing after intra-row flame weeding with the machines tested.

Model 2 included the cost accounting of intra-row flame weeding in organic maize production using the prototype flame weeder. Under such conditions, the average maize yield was 7.8 t/ha without additional hand hoeing (shown as Model 2a in Table 1). Upon a single passage of the prototype flame weeder, two additional hand hoeings can follow, with a forecast of 12 daily wages required, resulting in an average maize yield of 8.3 t/ha (shown as Model 2b in Table 1). The gross wages also amounted to 14 € per day. The prototype flame weeder simultaneously performs intra-row flame weeding and inter-row cultivation. Thus, the costs of inter-row cultivation were excluded from the costaccounting of Model 2. The prototype was mounted on a 60 kW tractor with a purchase price of 22,500 €, useful life of 10 years, and an annual depreciation rate of 10%. A diesel fuel price of 0.95 € per liter was used in the Model 2 cost accounting. Table 2 shows an overview of the costs of using a 60 kW tractor.

Table 2. The costs of using a 60 kW tractor (€).

COSTS	TOTAL	%	Per Hour of Use
Fixed assets depreciation	2250.00	24.71	2.50
Interest costs	680.63	7.48	0.76
Maintenance costs	918.00	10.08	1.02
Fuel costs	4655.73	51.14	5.17
Other costs	600.00	6.59	0.67
TOTAL	9104.35	100.00	10.12

The propane consumption amounted to 80 kg per two machine passages, which are required for a single production cycle. The propane price was 1.05 €/kg. The mounted machine's purchase price was 2400 €, the travel speed was 4 km/h, and the operating width was 3 m. Table 3 displays an overview of the costs of using the prototype flame weeder.

Table 3. The costs of using the prototype flame weeder (€).

COSTS	TOTAL	%	Per Hour of Use
Fixed assets depreciation	240.00	49.48	0.80
Interest costs	72.60	14.97	0.24
Maintenance costs	122.40	25.24	0.41
Other costs	50.00	10.31	0.17
TOTAL	485.00	100.00	1.62

The cost accounting of Model 3 included the costs of using the commercial Red Dragon flame weeder. The average maize yield was 7.8 t/ha without additional hand hoeing (shown as Model 3a in Table 1). Upon a single passage of the Red Dragon flame weeder, two additional hand hoeings can follow, with a forecast of 12 daily wages required, amounting gross to 14 € per day (shown as Model 3b in Table 1). The use of the Red Dragon flame weeder also required additional inter-row cultivation, the costs of which were included in Model 3a and Model 3b. The Red Dragon flame weeder was mounted on a 90 kW tractor with a purchase price of 36,000 €, useful life of 10 years, and an annual depreciation rate of 10%. A diesel fuel price of 0.95 € per liter and an interest rate of 5.5% were used in the Model 3 cost accounting. Table 4 shows an overview of the costs of using a 90 kW tractor.

Table 4. The costs of using a 90 kW tractor (€).

COSTS	TOTAL	%	Per Hour of Use
Fixed assets depreciation	3600.00	21.26	3.00
Interest costs	1089.00	6.43	0.91
Maintenance costs	1836.00	10.84	1.53
Fuel costs	9311.45	54.98	7.76
Other costs	1100.00	6.49	0.92
TOTAL	16,936.45	100.00	14.11

The propane consumption amounted to 80 kg per two passages of the Red Dragon machine (the same as in the prototype application). The propane price was 1.05 €/kg. The mounted machine's purchase price was 22,850 €, the travel speed was 4 km/h, and the operating width was 6 m. Table 5 displays an overview of the costs of using the Red Dragon flame weeder.

Table 5. The costs of using the Red Dragon flame weeder (€).

COSTS	TOTAL	%	Per Hour of Use
Fixed assets depreciation	2285.00	65.43	7.62
Interest costs	691.21	19.79	2.30
Maintenance costs	466.14	13.35	1.55
Other costs	50.00	1.43	0.17
TOTAL	3492.35	100.00	11.64

Based on previous research and using various methods, most commonly work process recording [40], a working time in crop production was established. The structure used in the cost accounting performed herein is shown in Figure 2.

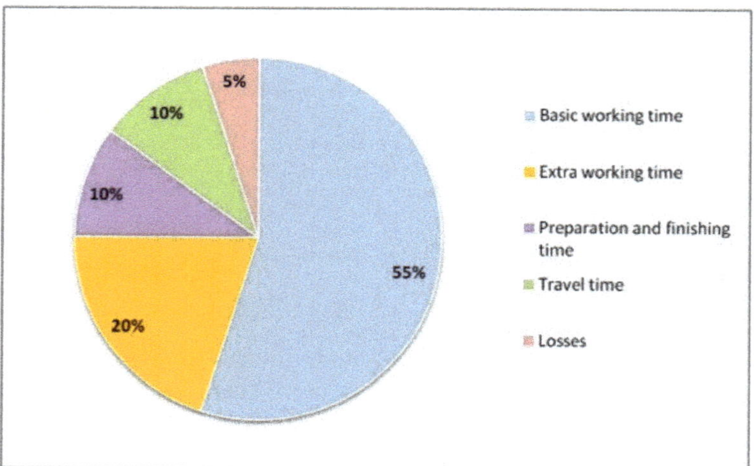

Figure 2. The structure of working time.

The price of organic maize varies according to the demand and supply in the market. On average, it is 30% higher than the cost of conventional corn kernels. A price of 0.20 €/kg, established in the Fruit and Vegetable Market of Bologna [50], was used in the differential calculations of Model 1, 2, and 3 (Table 6).

Table 6. The differential calculations of Model 1 (hand hoeing), Model 2 (using the prototype flame weeder without (2a) and with (2b) additional hoeing) and Model 3 (using the Red Dragon flame weeder without (3a) and with (3b) additional hoeing).

Parameter	Unit of Measurement	(1)	(2a)	(2b)	(3a)	(3b)
Yield	t/ha	8.3	7.8	8.3	7.8	8.3
Selling price	€/kg	0.20	0.20	0.20	0.20	0.20
Revenue	€/ha	1660.00	1560.00	1660.00	1560.00	1660.00
Revenue change	€/ha	0.00	−100.00	0.00	−100.00	0.00
Operating cost	€/ha	729.26	139.87	307.87	172.79	340.79
Expense change	€/ha	0.00	589.39	421.39	556.47	388.47
Economic impact	€/ha	0.00	489.39	421.39	456.47	388.47

To achieve optimum or maximum economic results, the efficiency of a flame weeder application should be considered [51,52]. Should intra-row flame weeding be applied at the V3-V4 and V6-V7 stages of maize growth, using propane (40 kg/ha) in combination with inter-row cultivation, a total of 90–95% of weeds could be eradicated, resulting in a yield of 7.8 t/ha (compared to a yield of 8.3 t/ha in the hand-hoed control field). A decrease in yields is not a consequence of the flaming damage but of the persistence of weeds to endure the treatment.

The cost accounting performed indicated a decrease in the costs per hectare when the prototype flame weeder and the commercial Red Dragon flame weeder were used compared to hand hoeing. However, greater cost savings, were recorded when the prototype flame weeder was used (489.39 €/ha in Model 2a) than in the application of the Red Dragon flame weeder (456.47 €/ha in Model 3a). If we look at Model 2 in particular, the comparison of yields shows that hand hoeing's introduction achieves a difference in yields of 500 kg (8300 kg for 2b, 7800 kg for 2a). At the same time, additional income of 100 €/ha (1660 for 2b, 1650 for 2a) is realized, while additional costs are 168 €/ha (307.87 for 2b, 139.87 for 2a). Given that in this case, the marginal cost exceeds the marginal revenue, investments in additional hand hoeing are not economically justified.

4. Discussion

These results are consistent with Nemming [50], who also proved the cost-effectiveness of flame-weeding machinery in organic agricultural production compared to hand weed hoeing [51]. A flame weeder can be constructed more cost-effectively by modifying an existing cultivator. However, such a machine ought to be applied to an area of 6–20 ha to justify the investments [51].

To be cost-effective and sustainable, organic production has to be financially rewarding for individual producers, providing them with a decent living standard. The study's objective was to investigate the cost-effectiveness of three different weed control methods used in organic maize production and determine the best solution economically. From the individual producer's perspective, weed control is one of the significant issues in organic agriculture. Due to a ban on the use of chemical agents, organic producers are faced with a limited choice of weed control methods, resulting commonly in the use of poorly cost-effective hand hoeing.

5. Conclusions

Following a comparative analysis of hand weed hoeing, the use of Prototype Flame Weeder and Red Dragon flame weeder is economically justified in reducing labor costs of weed control in organic maize production. The prototype flame weeder contributed to a significant cost reduction per unit of area, approximating 490 €/ha per hectare. Moreover, the analysis has also shown that additional hand hoeing investments are not economically justified because the operating costs incurred therein (168 €/ha) were not met by a yield increase of 500 kg/ha, i.e., a surplus revenue of 100 €/ha. The high price and limited efficacy of both the prototype and Red Dragon flame weeder render this method more expensive than the application of herbicides in conventional production. Conversely, flame weeding is economically justified in organic production due to high labor costs. Labor costs are very high in organic production. They are of immense importance to organic farms' financial results, especially in plant production. On average, such prices are from 8 to 25% higher than the labor costs incurred in conventional maize production. Flame weeding can substantially reduce labor costs in organic farming and, therefore, the total production cost.

Additionally, the economic impacts of flame weeding would be considerably more significant in larger fields. However, the cost reduction recorded in smaller fields can affect the overall financial results of organic farms, which contributes to the economic sustainability of organic farming.

Although this study confirmed and justified the use of flame weeding in weed control in organic maize production, it is necessary to extend the research to other crops that are most commonly grown in the organic system and whose weed control costs are high. Moreover, given the small number of studies conducted so far with the prototype, it would be desirable to repeat the research in different natural conditions.Given the importance of organic farming for sustainable development, development policies should pay more attention to this production system. Introducing producers who are already in the organic production system with ways to improve their economic sustainability, the introduction of new production technologies, such as flame weeding, can be achieved through various forms of informal education that can be organized by local governments.

Author Contributions: Conceptualization and methodology, M.R. and G.M.; software, M.T.S.; validation, D.G.-T.; data curation, D.M.; writing—original draft preparation, M.R. and M.T.S.; writing—review and editing, M.M.; supervision, S.V. All authors have read and agree to the published version of the manuscript.

Funding: This research was supported by the Ministry of Education, Science and Technological Development of the Republic of Serbia, grant number 451-03-68/2020-14/200032 and 451-03-68/2020-14/200117.

Institutional Review Board Statement: Not applicable.

Informed Consent Statement: Not applicable.

Data Availability Statement: No new data were created or analyzed in this study. Data sharing is not applicable to this article.

Conflicts of Interest: The authors declare no conflict of interest.

References

1. Kovačević, D.; Lazić, B.; Milić, V. The Effects of Agriculture on Environment. In Proceedings of the International Scientific Symposium of Agriculture "Agrosym Jahorina 2011", Jahorina, Bosnia and Herzegovina, 10–12 November 2011; pp. 34–47. Available online: http://agrosym.ues.rs.ba/agrosym/agrosym_2011/pdf/Plenary_lectures/Kovacevic_D_et_al.pdf (accessed on 21 April 2019).
2. Hodge, I. Sustainability: Putting Principles into Practice. In *An Application to Agricultural Systems, Paper Presented to Rural Economy and Society Study Group*; Royal Holloway College: Egham, UK, 1993.
3. Rodriguez, E.; Sultan, R.; Hilliker, A. Negative Effects of Agriculture on Our Environment. *Eff. Agric. Traprock* **2004**, *3*, 28–32.
4. Praneetvatakul, S.; Schreinemachers, P.; Pananurak, P.; Tipraqsa, P. Pesticides, external costs and policy options for Thai agriculture. *Environ. Sci. Policy* **2013**, *27*, 103–113. [CrossRef]
5. Krajewski, P. Agricultural Biodiversity for Sustainable Development. *Probl. Sustain. Dev.* **2016**, *12*, 135–141.
6. Lampkin, N.; Padel, S. *The Economics of Organic Farming: An International Perspective*; University of Wales: Aberystwyth, UK; CABInternational: Oxford, UK, 1994; p. 467.
7. Tomaš-Simin, M.; Janković, D. Applicability of diffusion of innovation theory in organic agriculture. *Econ. Agric.* **2014**, *61*, 517–529. [CrossRef]
8. Mccann, E.; Sullivan, S.; Erickson, D.; DeYoung, R. Environmental Awareness, Economic Orientation, and Farming Practices: A Comparison of Organic and Conventional farmers. *Environ. Manag.* **1997**, *21*, 747–758. [CrossRef]
9. Patil, S.; Reidsma, P.; Shah, P.; Purushothaman, S.; Wolf, J. Comparing conventional and organic agriculture in Karnataka, India: Where and when can organic farming be sustainable? *Land Use Policy* **2014**, *37*, 40–51. [CrossRef]
10. Pimentel, D. Economics and energetic of organic and conventional farming. *JAGR Environ. Ethic* **1993**, *6*, 53–60. [CrossRef]

11. Darnhofer, I.; Schneeberger, W.; Freyer, B. Converting or not converting to organic farming in Austria: Farmer types and their rationale. *Agric. Hum. Values* **2005**, *22*, 39–52. [CrossRef]
12. Brenes-Muñoz, T.; Lakner, S.; Brümmer, B. What Influences the Growth of Organic Farms? Evidence from a Panel of Organic Farms in Germany. *Ger. J. Agric. Econ.* **2016**, *65*, 1–15. [CrossRef]
13. Läpple, D.; Kelley, H. Understanding the uptake of organic farming: Accounting for heterogeneities among Irish farmers. *Ecol. Econ.* **2013**, *88*, 11–19. [CrossRef]
14. Oerke, E.C.; Dehne, H.W.; Schonbeck, F.; Weber, A. *Crop Production and Crop Protection: Estimated Losses in Major Food and Cash Crops*, 1st ed.; ElsevierScience: Amsterdam, The Netherlands, 1994; p. 808. ISBN 9780444597946.
15. Stepanovic, S.; Datta, A.; Neilson, B.; Bruening, C.; Shapiro, C.A.; Gogos, G.; Knezevic, S.Z. Effectiveness of flame weeding and cultivation for weed control in organi cmaize. *Biol. Agric. Hortic* **2015**, *32*, 47–62. [CrossRef]
16. Bond, W.; Turner, R.; Grundy, A. *A Review of Non-Chemical Weed Management*; Technical Report; HDRA, Ryton Organic Gardens: Coventry, UK; HRI: Wellesbourne, UK, 2003; Available online: https://gardenorganic.org.uk/sites/www.gardenorganic.org.uk/files/updated_review_0.pdf (accessed on 2 May 2019)Technical Report.
17. Barberi, P. Weed management in organic agriculture: Are we addressing the right issues? *Weed Res.* **2002**, *42*, 177–193. [CrossRef]
18. Pannacci, E. Optimization of for am sulfur on doses for post-emergence weed control in maize (ZeamaysL.). *Span. J. Agric. Res.* **2016**, *14*, 1005. [CrossRef]
19. Loni, R.; Loghavi, M.; Jafari, A. Design, Development and Evaluation of Targeted Discrete–Flame Weeding for Inter-row Weed Control Using Machine Vision. *Am. J. Agric. Sci. Technol.* **2014**, *2*, 17–30. [CrossRef]
20. Stepanovic, S.; Datta, A.; Neilson, B.; Bruening, C.; Shapiro, C.; Gogos, G.; Knezevic, S. The effectiveness of flame weeding and cultivation on weed control, yield and yield components of organic soybean as influenced by manure application. *Renew. Agric. Food Syst.* **2015**, *31*, 288–299. [CrossRef]
21. Datta, A.; Knezevic, S.Z. Flaming as an Alternative Weed Control Method for Conventional and Organic Agronomic Crop Production Systems: A Review. *Adv. Agron.* **2013**, *118*, 399–428. [CrossRef]
22. Kruidhof, H.; Bastiaans, l.; Kropff, M. Ecological weed management by cover cropping: Effects on weed growth in autumn and weed establishment in spring. *Weed Res.* **2008**, *48*, 492–502. [CrossRef]
23. Knezevic, S.Z. Flame Weeding in Corn, Soybean and Sunflower. In Proceedings of the 8th International Conference of Information and Communication Technologies in Agriculture, Food and Environment (HAICTA 2017), Chania, Greece, 21–24 September 2017; pp. 390–394.
24. Wszelaki, A.L.; Doohan, D.J.; Alexandrou, A. Weed control and crop quality in cabbage [Brassica oleracea (capitatagroup)] and tomato (Lycopersiconly copersicum) using a propane flamer. *Crop. Prot.* **2007**, *26*, 134–144. [CrossRef]
25. Ulloa, S.; Datta, A.; Malidza, G.; Leskovsek, R.; Knezevic, S. Timing and propane dose of broad cast flaming to control weed population influenced yield of sweet maize (*Zea mays* L.var.*rugosa*). *Field Crop. Res.* **2010**, *118*, 282–288. [CrossRef]
26. Abou Chehade, L.; Fontanelli, M.; Martelloni, L.; Frasconi, C.; Raffaelli, M.; Peruzzi, A. Effects of flame weeding on organic garlic production. *Horttechnology* **2018**, *28*, 502–508. [CrossRef]
27. Lague, C.; Gill, J.; Lehoux, N.; Peloquin, G. Engineering performances of propane flamers used for weed, insect pest, and plant disease control. *Appl. Eng. Agric.* **1997**, *13*, 7–16. [CrossRef]
28. Van Der Weide, R.Y.; Bleeker, P.O.; Achten, V.T.; Lotz, L.A.; Fogelberg, F.; Melander, B. Innovation in mechanical weed control in crop rows. *Weed Res.* **2008**, *48*, 215–224. [CrossRef]
29. Rajković, M.; Malidža, G.; Stepanović, S.; Kostić, M.; Petrović, K.; Urošević, M.; Vrbničanin, S. Influence of Burner Position on Temperature Distribution in Soybean Flaming. *Agronomy* **2020**, *10*, 391. [CrossRef]
30. Bond, W.; Grundy, A.C. Non-chemical weed management in organic farming systems. *Weed Res.* **2001**, *41*, 383–405. [CrossRef]
31. Melander, B.; Rasmussen, I.A.; Bàrberi, P. Integrating physical and cultural methods of weed control – examples from European research. *Weed Sci.* **2005**, *53*, 369–381. [CrossRef]
32. Hoffmann, M. New Experiences in Thermal Treatment of Indian Corn and Potatoes. In Proceedings of the Third International Conference on Non-Chemical Weed Control, Linz, Austria, 5 October 1990; pp. 19–25.
33. Knezevic, S.; Ulloa, S. Potential new tool for weed control in organically grown agronomic crops. *J. Agric. Sci.* **2007**, *52*, 95–104. [CrossRef]
34. Knezevic, S.Z.; Datta, A.; Bruening, C.; Gogos, G.; Stepanovic, S.; Neilson, B.; Nedeljkovic, D. *Propane-Fueled Flame Weeding in Corn, Soybean, and Sunflower*; University of Nebraska: Lincoln, NE, USA, 2012; pp. 1–32.
35. Lague, C.; Gill, J.; Peloquin, G. Thermal Controlin Plant Protection. In *Physical Control Methods in Plant Protection*; Vincent, C., Panneton, B., Fleurat-Lessard, F., Eds.; Springer: Berlin, Germany, 2001; pp. 35–46.
36. Ascard, J. Effects of flame weeding on weed species at different developmental stages. *Weed Res.* **1995**, *35*, 397–411. [CrossRef]
37. Ulloa, S.M.; Datta, A.; Knezevic, S.Z. Growth stage influenced differential response of foxtail and pig weed species to broadcast flaming. *Weed Technol.* **2010**, *24*, 319–325. [CrossRef]
38. Flame Engineering: Row Crop Flamer Red Dragon. Available online: https://flameengineering.com/products/red-dragon-row-crop-flamers (accessed on 1 May 2020).
39. Rajković, M.; Malidža, G.; Kostić, M.; Petrović, K.; Tančić Živanov, S.; Pavlović, D.; Vrbničanin, S. Flame Cultivator, MP-2019/25, Reg.no.1632, Petty Patents, Intellectual Property Gazette 11/2019, p. 56. Available online: http://www.zis.gov.rs/upload/documents/pdf_sr/pdf/glasnik/GIS_2019/Glasnik_11_2019.pdf (accessed on 28 November 2019).

40. Krmpotić, T.; Nikolić, R.; Kiš, T.; Ivančević, S.; Cilger, M.; Jerinkić, E.; Gligorić, R. *Menadžment Poljoprivrednih Mašina*, 1st ed.; Univerzitet u Novom Sadu, Ekonomski fakultet: Subotica, Srbija, 1997; pp. 1–374.
41. Majcen, Ž. *Troškovi u teoriji i praksi*, 4th ed.; Informator: Zagreb, Hrvatska, 1988; pp. 292–293.
42. Markovski, S. *Troškovi u Poslovnom odlučivanju*, 1st ed.; Naučna knjiga: Beograd, Srbija, 1991; pp. 1–342.
43. Funk, M. *KTBL-Taschenbuch Landwirtschaft*; KTBL: Darmstadt, Germany, 1996; pp. 1–275.
44. Schmid, A. *Wirtschaftliche Betriebsführung und Kalkulationim Lohnunternehmen*; KTBL: Darmstadt, Germany, 1995; pp. 1–247.
45. Nikolić, R.; Furman, T.; Gligorić, R.; Popović, Z.; Oparnica, S.; Savin, L. *Potrošnja dizel Goriva u Ratarstvu*, 1st ed.; Univerzitet u Novom Sadu, Poljoprivredni fakultet, Institut za poljoprivrednu tehniku: Novi Sad, Srbija, 1995; pp. 1–113.
46. Renijus, K. *Traktoren–Technik und ihre Anwendung*, 2nd ed.; Verlagsunion–Agrag, Verlagsgesellschaft, BVL: Műnchen, Germany, 1985; pp. 1–230.
47. Ružičić, L.N.; Petrović, P.; Gligorević, K.; Oljača, M.; Ružičić, T. Ekonomsko-tehnološki parametric optimalnog korišćenja traktora (Economic and Technological Parameters for Optimal Use of Tractors). *Agro-Know. J.* **2012**, *13*, 259–266. [CrossRef]
48. Rajković, M.; Malidža, G.; Gvozdenović, Đ.; Vasić, M.; Gvozdanović-Varga, J. Susceptibility of bean and pepper to flame weeding. *Acta Herbol.* **2011**, *19*, 67–76.
49. Global Seed. Available online: http://www.globalseed.info (accessed on 1 May 2020).
50. La Borsa Merci Di Bologna. Available online: http://www.agerborsamerci.it (accessed on 3 May 2020).
51. Nemming, A. Costs of flame cultivation. *Acta Hortic.* **1994**, *372*, 205–212. [CrossRef]
52. Teodinov, D.; Bošnjak, D.; Radojičić, N. Possibilities for improving sugar beet sowing process using the time study analysis. *Field Veg. Crop. Res.* **2012**, *49*, 208–213.

Article

Analysis of Plant-Production-Obtained Biomass in Function of Sustainable Energy

Siniša Škrbić [1], Aleksandar Ašonja [1], Radivoj Prodanović [1,*], Vladica Ristić [2], Goran Stevanović [1], Miroslav Vulić [1], Zoran Janković [1], Adriana Radosavac [3] and Saša Igić [1]

[1] Faculty of Economics and Engineering Management in Novi Sad, University Business Academy in Novi Sad, Cvećarska 2, 21000 Novi Sad, Serbia; sinisa.skrbic@in-tech.rs (S.Š.); asonja.aleksandar@fimek.edu.rs (A.A.); meckags@gmail.com (G.S.); miroslav.vulic@fimek.edu.rs (M.V.); zoran.050164@gmail.com (Z.J.); sasaigic65@gmail.com (S.I.)

[2] Faculty of Applied Ecology—FUTURA, University Metropolitan, Požeška 83a, 11000 Belgrade, Serbia; vladicar011@gmail.com

[3] Faculty of Applied Management, Economics and Finance, University Business Academy in Novi Sad, Cvećarska 2, 21000 Novi Sad, Serbia; adrianaradosavac@gmail.com

* Correspondence: rprodanovic@fimek.edu.rs; Tel.: +381-21-400-484

Received: 19 May 2020; Accepted: 6 July 2020; Published: 7 July 2020

Abstract: This research analyzed the degree of utilization of the agricultural biomass for energy purposes (combustion), in order to indicate the reasons that limit its use. The biomass potential was studied by means of the methodology of the biomass potential, whereas the factors suggesting a low degree of biomass utilization were identified by means of factor analysis. The research results reveal that there is an enormous potential of the unused agricultural biomass. This dissertation research significantly contributes to the establishment of a genuine mathematical model based on multiple linear regression. The solution obtained by this analysis, in both a mathematical and a scientific manner, conveys the primary reasons for an insufficient utilization of the biomass for energy purposes. Moreover, the paper suggests the measures to be applied for a more substantial use of this renewable source of energy and presents the expected benefits to be gained.

Keywords: biomass; crop residue; crop production; energy purposes; sustainability

---

1. Introduction

The Republic of Serbia has been estimated to possess around 5,069,000 ha of the agricultural land: 3,298,000 ha (65%) of the arable land and gardens, 239,000 ha (4.72%) of orchards, 50,000 ha (0.99%) of vineyards, 653,000 ha (12.88%) of meadows and 829,000 ha (16.35%) of pastures [1]. The proportion of the agricultural land and the population in the Republic of Serbia is 0.47 ha/person, which is a rather high figure even for the state members of the EU, when compared, for instance, with these proportions in certain countries: Hungary 0.51 ha/person, Denmark 0.50 ha/person, France 0.33 ha/person, Italy 0.20 ha/person, Germany 0.19 ha/person, Holland 0.06 ha/person [2].

The potential of biomass in the Republic of Serbia is estimated at 2.58 Mtoe and consists of agricultural biomass (about 60%) and forest biomass (about 40%) [3,4]. The Republic of Serbia, and especially its northern region (the Autonomous Province of Vojvodina), has a relatively large potential of agricultural biomass, especially in crop production. That is the sole reason why the region of Autonom Province (AP) Vojvodina was selected for the research on the use of plant-production-obtained biomass for energy purposes. The total potential of the biomass in the AP Vojvodina has been estimated at 6.45 Mt per year, out of which, 2 Mt of the harvesting crop residue and around 0.45 Mt of the fruit growing, grape growing and forestry residue could be used for annual energetic purposes [5].

Today, the chief problem of productivity and competitiveness of agricultural production in the Republic of Serbia lies in fragmented agricultural holdings, the average size of which is 5.77 ha [6]. Among other noted barriers that limit agricultural biomass utilization, the following stand out: the non-existence of incentive measures/feed-in tariffs for wider use (thermal energy) [7], limitations and availability of biomass as a source of energy [8], long waiting time for building permits, extremely low purchase price of electricity for non-privileged suppliers, etc.

The goal of this paper was to identify, by way of an analysis of biomass utilization for energy purposes, the chief causes of low biomass utilization in AP Vojvodina. The starting hypothesis in the research was that less than 20% of the biomass technical potential is used for energy purposes.

The research presented in this paper comprises the following chapters: introduction, the problems related to harvesting crop residue, the justification for the utilization of the agricultural biomass for energy purposes, materials and methods, results, discussion and conclusion.

2. The Problems Related to Harvesting Crop Residue

The process of decomposition of organic matter in the soil releases nutrients, which improves soil fertility and increases crop yields [9]. For example, for an average wheat target yield of 4 t/ha per year, each ton of organic carbon (SOC) added to a 15-cm deep plow layer contributed to the formation of 4.75 kg N/ha [10].

Whether, and to what extent, crop residue will be used for plowing and soil erosion prevention and for the increase in soil carbon content, or for energy purposes, depends on several factors. This relationship must be pre-defined and the biomass amounts used for other purposes must be known in advance. This process of primarily harvesting crop residue must be sustainable, because removing too much residue may cause exposure of the soil to excessive erosion, while too little or no residue removal may lead to residue preventing the soil from drying in the spring, which may affect the planting season [11]. The crop residue removal rate depends on several factors, including maintaining the fertility of soil, the availability of mechanization, plant varieties and yield. [12].

The method and the purpose of handling the remaining plant-production-obtained biomass remains a matter of controversy around the globe. Essentially, it is a highly complex problem depending on several variables, including soil quality (the humus content in the soil), the crop rotation plan, and management methods (fertilization and cultivation). Thus, based on the research [13], it has been pointed out that, for the purpose of maintaining the soil fertility, nothing should be removed from the field, or that only 25 to 50% [14,15] may be removed, or 30% to 60% [16], or 33% [17] etc. Research [16] has indicated the existence of a rate of sustainable crop residue removal from agricultural soil regardless of its location, and this rate of sustainable crop residue removal can go up to 40% for wheat, barley, oats and rye, and about 50% for corn, rapeseed, rice and sunflower.

3. The Justification for the Utilization of Agricultural Biomass for Energy Purposes

The use of fossil fuels for energy purposes leads to a constant increase in the concentration of pollutants (CO_2, CO, SO_x, NO_x, and other harmful oxides in the atmosphere) which cause global warming and have other negative impacts, such as acid rain and photochemical smog [18]. The last century was predominantly marked by the use of fossil fuels in energy production [19], but the beginning of the twenty-first century saw enormous efforts made in order to mitigate the effects of global warming caused by CO_2 emissions in the atmosphere [20]. Those were the exact endeavors in which sustainable development was grounded, a development that implies keeping a balance between the use, saving, and renewal of all resources [21], and not their uncontrolled expenditure to the detriment of future generations [22].

Globally, more than 2 Gt of plant residue are burned unreasonably, generating about 18% of total global CO_2 emissions in the process [23]. The production and use of biomass for energy purposes reduces the emission of harmful gases and contributes to the protection of soil and water. With regard to environmental impact, biomass is a highly acceptable fuel as it contains very little, or even no, toxic substances like sulfur and heavy metals, which are usually found in fossil fuels and are emitted into the air through their combustion, posing a danger to human health and natural resources. The main advantage of biomass over fossil fuels lies in its global availability and renewability [24,25]. Calculations show that atmospheric pollution by biomass combustion is negligible, since the amount of CO_2 emitted during biomass combustion is equal to the amount of CO_2 absorbed during plant growth [26,27].

The significant energy production from agricultural biomass can have a negative impact on food supply and prices, soil erosion and biodiversity [28]. On the other hand, agricultural residue can be an environmentally friendly and renewable source of energy [29]. The thermal energy obtained by burning these fuels can be used for heating households, industrial processes, crop drying, etc., and even for electricity production [30].

The traditional method of biomass utilization is mainly present in developing countries, in which this energy source provides between 34% and 40% of the total energy requirements. Modern biomass processing plants, on the other hand, are primarily used in developed countries [31,32], including European, Asian, and North American countries [33].

Globally, the potential for energy production from biomass is quite large, but also insufficient to replace the current energy production sources [34]. Biomass currently secures the largest share of renewable energy with over 50% of global renewable energy consumption [35] (i.e., over 10% of world primary energy consumption) [36]. In the near future, biomass is expected to be the most beneficial of all renewable energy sources [37]. In this context, there is a possibility that there will be a significant increase in agricultural residue globally if the developing countries continue to intensify agricultural production, in which case it is estimated that about 998 million tons of agricultural residue will be generated annually [38]. For example, it is estimated that, by 2030, about 155 million tons of residue will be used in the production of bioenergy in the United States, while not taking into account the formation of additional agricultural properties [39].

4. Materials and Methods

The first step in the process of defining the research hypothesis, i.e., the calculation of the technical potential of biomass for the purpose of satisfying all agricultural activities, was the adoption of the generally accepted scientific assumption prevailing in the Republic of Serbia, according to which, ¼ solid biomass should be used to increase the soil fertility (it is plowed either immediately or in the form of a mulch, after use), ¼ for the production of animal feed, ¼ for industrial processing (in the production of paper, cardboard, packaging, alcohol, cosmetics, etc.), and ¼ used for energy production [40,41].

The regional testing criteria were the following:

- The investigations were limited to the locality of AP Vojvodina;
- An equal presence of investigations in all three regions (Srem, Banat, and Bačka);
- The smallest number of represented agricultural properties by district was five, so that properties from all districts (South Bačka, North Bačka, West Bačka, South Banat, North Banat, Central Banat, and Srem) could be examined, Figure 1.

Figure 1. AP Vojvodina with Districts [42].

Basic information on the surveyed properties:

- The activity of the surveyed agricultural farms was plant production;
- The number of samples: 75 agricultural farms;
- The sample size: 25 farms with the size of <100 ha, 25 farms with the size of 100–1000 ha, and 25 farms with the size of >1000 ha;
- The farm selection: by employing the random sampling method.

The total area under crop production on the observed properties was 55,880 ha, Table 1. However, the quantities of biomass (area) that were purposely grown for livestock feed production (hay, alfalfa, and clover), those used in biogas production (silage corn, silage sorghum, etc.), and the residue of other crops that could not be utilized in the combustion process were all excluded from the analysis. The total analyzed areas for calculating the total available biomass in the research amounted to 51,382 ha, Table 2. The average size of the surveyed properties was 745.07 ha, the average number of workers per farm was 27.05, and the average number of workers per ha of cultivated area was 0.041 workers/ha, Table 3.

Table 1. Data related to the size of arable land and the number of workers on the observed agricultural properties.

Property Size	Total Arable Land (ha)	Total Number of Workers (Person/ha)
≤100	1073	61
100–1000	8278	249
>1000	46,529	1719
Σ=	55,880	2029

Table 2. Analyzed crops and cultivated area [43].

Plants					
Farming		Vegetables		Fruits and Grapes	
Analyzed Crops	Cultivated Area (ha)	Analyzed Crops	Cultivated Area (ha)	Analyzed Crops	Cultivated Area (ha)
Wheat	11,059.00	Peas	467.00	Apples	927.00
Triticale	290.00	Green beans	250.00	Pears	359.00
Barley	1881.00	Total=	717.00	Apricots	192.00
Corn	13,282.50			Peaches	375.00
Hybrid seed corn	3605.00			Cherries	422.00
Soya	10,267.50			Plums	102.00
Rapeseed	2587.00			Raspberries	1.00
Sunflower	5123.00			Quince	2.00
Tobacco	25,00			Hazelnuts	8.00
Total=	48,120.00			Walnuts	1.00
				Grapevines	156.00
				Total=	2545.00
		Sum total=	51,382		

Table 3. The average data related to the size of arable land and the number of workers on the observed agricultural properties.

Property Size	Average Property Size	Average Number of Workers per Property	Average Number of Workers per Hectare of Arable Land (Person/ha)
≤100	42.92	2.44	0.057
100–1000	331.12	9.96	0.030
>1000	1861.16	68.76	0.037
Σ=	745.07	27.05	0.041

After the results of the utilization of agricultural biomass had been collected, a quantitative assessment of the amount of biomass utilization for energy-related and other purposes was designed. For this assessment, the methodology on biomass potential was used [44] (an Excel application for the development of agricultural biomass balance). The methodology used in this research is identical to other methodologies used in the Republic of Serbia and ountries in the region. It basically calculates the biomass potential based on the determination of the proportion of the total quantity of the crops produced and the residues. Unlike other methodologies that base their estimations on the predicted and expected crop production, this methodology uses the official statistical data on the crops production related to the period in which this research was conducted, i.e., the year of 2018.

The theoretical potential of E_{teo} in this paper was used to present the total amount of biomass in the observed area, i.e., to calculate the technical potential of biomass further, Equation (1)

$$E_{teo} = \sum_{i=1}^{n} P_{polj.(i)} \cdot O_p P_{o(i)} \ (\text{t/year}) \tag{1}$$

where:

- $P_{polj.(i)}$—quantity of crops produced (t/year);
- $O_p P_{o(i)}$—mass ratio of basic products—agricultural residue(t/t).

The technical potential of biomass E_{teh} obtained during the research was calculated on the basis of the calculated theoretical potential E_{teo} and the sustainability factor F_0. The technical potential is basically the part of the theoretical potential that can be used in practice, and thus can be employed in the process of the practical use of energy. The technical potential of the agricultural residue utilization is significantly lower than the theoretical potential. A certain amount of harvest residue must be left in the soil and plowed in order to preserve the soil productivity, or it must be returned to the soil by way of the biomass intended for food and mulch

$$E_{teh} = \sum_{i=1}^{n} E_{teo(i)} \cdot F_{o(i)} \ (\text{t/year}) \tag{2}$$

where:

- $F_{o(i)}$—the sustainability factor (%).

The technical potential of biomass presented in this manner in the research results was compared with the actually used biomass energy potential.

Due to the need to maintain the fertility of agricultural soil in the course of research, it was adopted/calculated that $\frac{1}{4}$ or 25% ($F_o = 0.25$) can be removed from the fields in field and vegetable production. On the other hand, in the calculation of the residue of prune kernels in fruit and vineyard production of seed corn cobs and sunflower husk, the value of $F_o = 1$ was adopted.

The energy potential of biomass E_{pot} is part of the technical potential, in which case the available biomass amount is shown in relation to energy, Equation (3)

$$E_{pot} = \sum_{i=1}^{n} E_{teh(i)} \cdot H_{d(i)} \ (\text{GJ/year}) \tag{3}$$

where:

- $Hd_{(i)}$—the lower thermal power (MJ/kg).

The factor analysis was used to identify the main factors contributing to a low degree of the utilization of biomass for energy purposes in the AP Vojvodina. Primary data related to the identification of the key factors were collected by means of a specially designed initial questionnaire.

The questionnaire was constructed using the rational method and comprised 60 questions concerned with the economic, social, educational and technological aspects of the use of biomass for energy purposes. It is important to emphasize that the respondents who participated in the validation of the questionnaire were not involved in collecting the data related to the influence of the identified factors on the degree of biomass utilization. Thus, partial responses were avoided. A total of 600 respondents participated in the validation of the questionnaire. The *Kaiser–Meyer–Okin* (KMO) test was used for testing the adequacy of the sample size.

The factor analysis was conducted on the basis of the results obtained from the initial questionnaire. It was based on the supposition that the data were interval data, thus satisfying the assumption of normal distribution. For the purpose of identifying the key factors, the *Promax* rotation (with the *Kaiser* normalization) was used, which rotates the orthogonally rotated solution again in order to enable the correlations among the factors. Based on this rotation, 18 key factors were identified which determined the degree of use of biomass for energy purposes.

The obtained results were used in the model of multiple linear regression of Equation (4) to examine the influence and significance of individual factors on the degree of biomass utilization for energy purposes, where the energy purposes are expressed through the percentage of biomass utilization for energy purposes—*y* (%)

$$y = \alpha + \sum_{i=1}^{n} \beta_i x_i \tag{4}$$

where:

- α—the section coefficient within the model;
- β_i—the regression coefficient with the ith independent variable;
- x_i—the independent variable of the i th factor influencing the degree of biomass utilization.

The data related to the degree of biomass utilization were collected from 75 agricultural properties, which were divided into three groups according to the size of the area from which biomass was collected. The least square method (*Ordinal Least Square*) was used for the model evaluation.

5. Results

5.1. The Analysis of Plant-Production-Obtained Biomass Utilization Results in AP Vojvodina

Based on the presented calculation analysis, it was concluded that at the observed location the total theoretical potential of biomass was 481,326.90 t/year, the technical potential of biomass 152,318.49 t/year, and the energy potential 2,201,389.03 GJ/year (Table 4). In addition to the fact that 25% of biomass from field and vegetable production could be used for energy purposes, the analysis also includes the biomass potentials that can be fully utilized, as is the case with crops in fruit and wine production, seed corn cobs, and sunflower, walnut and hazelnut husks.

Table 4. The potentials of biomass [43].

Plant Production	E_{teo} (t/Year)	E_{teh} (t/Year)	E_{pot} (GJ/Year)
For field production	466,259.87	139,205.21	2,000,486.81
For vegetable production	2605.00	651.25	9117.50
For fruit and vineyard production	12,462.03	12,462.03	191,784.72
Total=	481,326.90	152,318.49	2,201,389.03

Research has shown that of the total (theoretical) potential of 481,326.90 t/year, 15,149.42 t/year was used for energy purposes direct combustion), 13,252.90 t/year for livestock, 440,208.68 t/year for mulching and plowing, and 12,715.90 t/year was burned on site, Table 5. The study did not record the amount of biomass that could be used for industrial purposes.

Table 5. The analysis of the utilized potentials of solid biomass [43].

Plant Cultures	Total Biomass Available (t/Year)				
	Total Available Potentials	Energy Utilized	Used to Feed Livestock	Used for Mulching and Plowing	Burned in the Field
For field production	466,259.87	14,813.69	13,252.90	437,603.68	589.60
For vegetable production	2605.00	0.00	0.00	2605.00	0.00
For fruit and vineyard prod.	12,462.03	335.73	0.00	0.00	12,126.30
Total=	481,326.90	15,149.42	13,252.90	440,208.68	12,715.90

Of the total amount of biomass utilization for various purposes, 3.15% was used for energy purposes (direct combustion), 91.46% for mulching and plowing, 2.75% for livestock feeding, and 2.64% was burned in the field, Figure 2. If, in the course of analysis, the amounts of biomass that were combusted/burned on any basis are combined, and if the industrial utilization is included, the following data are obtained: 0.00% of biomass is used for industrial purposes, 2.75% for livestock feed, 91.46% for mulching and plowing, and 5.79% was burned, Figure 3.

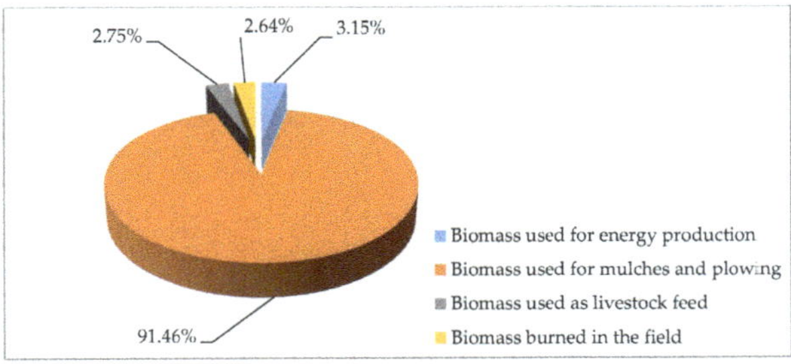

Figure 2. The actual utilization of multipurpose biomass—a separate display of burned biomass (thermal energy production and on-site burning) [43].

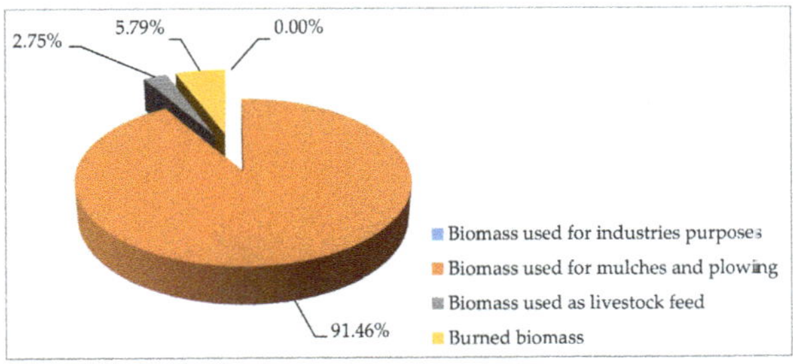

Figure 3. The utilization of multipurpose biomass the summary of burned biomass (thermal energy production and on-site burning).

The problems related to crop residue burning have been noted on smaller farms, because sometimes farmers burn the residue in order to get rid of huge amounts of biomass before the basic tillage. The remains from the fruit and vineyard production are also burned because there is no technological process of pruning collection and utilization.

The ratio of biomass for energy purposes and biomass for other purposes—the investigated state and the assumed state—is shown in Figure 4. The technical potential of biomass intended for energy purposes was 26.94% (mark 1) in relation to the theoretical biomass potential, while 73.06% of biomass could be used for other purposes. The results of the research showed that the usable energy potential was 3.15% (mark 1) in relation to the theoretical biomass potential, while 96.85% of biomass was used for other purposes.

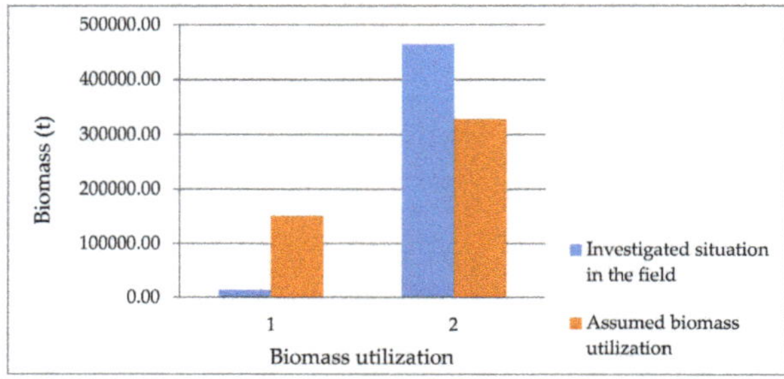

Figure 4. Biomass utilization, the investigated state—the assumed state [43].

Biomass used for energy purposes was 15,149.42 t/year, while an additional 137,169.07 t/year could be used without affecting the sustainability of the agricultural production. These results indicate that the hypothesis is confirmed.

5.2. The Investigation of Key Factors of the Utilization of Biomass for Energy Purposes

The Factor Analysis Identified 18 Factors with their particular values higher than 1, Table 6. The analysis confirmed that these 18 identified factors explained 59% of the variation.

Table 6. Identified factors.

Factor	List of Factors
Q1	Lack of economic justification for the use of biomass for energy purposes;
Q2	Emergence of problems related to technical equipment and technology applied in the production, manipulation and storing of biomass;
Q3	Economic justification for the production of biomass for energy purposes is debatable;
Q4	Lack of precise government measures that might considerably initiate the use of the agricultural biomass for energy purposes;
Q5	Official educational system has not yet recognized the necessity of educating professionals to occupy themselves with the agricultural biomass;
Q6	Investing in agricultural machinery, equipment and energy manufacture is expensive;
Q7	Agricultural biomass is characterized by limiting factors in comparison to other renewable energy sources;
Q8	Biomass power generation (biomass used for combustion) is short-term without the presence of manpower;
Q9	There is no active free market of biomass;
Q10	There is a shortage of either national or international projects that might support the construction of manufacture operating on biomass;
Q11	Shortage of workers t obe employed in the biomass sector;
Q12	The low price of electricity in the Republic of Serbia is a limiting factor that affects the use of biomass;
Q13	The need for the plants that use biomass and produce thermal energy is limited;
Q14	Certain interest groups are not in favor of the use of agricultural biomass for energy purposes;
Q15	Biomass transport is costly, which greatly determines the criteria for defining the economic justification of this type of production;
Q16	The problem of a continuous placement of energy in the course of the year;
Q17	Problems related to the sale of the produced biomass are rather pronounced;
Q18	Stimulating tariffs for the production of energy from the agricultural biomass are hard to get.

In order to identify the main causes of low biomass utilization, Table 7 shows the estimates of the model parameters according to Equation (4), while taking into account the importance of the size of the agricultural property at which biomass for energy purposes is generated.

The results of the model parameter evaluation show that, in the case of agricultural producers who cultivate up to 100 ha, two factors affecting the utilization of biomass for energy purposes are important. Those are factor 6, related to high investment costs for agricultural machinery, equipment, and modern power plants, and factor 14, related to the influence and activities of certain interest groups (decision-makers, lobbyists, etc.) which are not in favor of the utilization of agricultural biomass that could be used for energy production during the combustion process.

Table 7. The results of model evaluation depending on the size of the agricultural property [43].

Variables	Coefficient	Standard Error	T-Test Value	p-Value
		Up to 100 ha		
Constant	0.719	0.151	4.763	0.000
F6	−0.132	0.049	−2.702	0.013
F14	−0.056	0.024	−2.276	0.033
R^2	0.399			
		Over 1000 ha		
Constant	−0.003	0.045	−0.071	0.944
F16	−0.037	0.011	−3.297	0.004
F1	0.019	0.008	2.504	0.021
R^2	0.403			

In the case of farmers cultivating lands of a size between 100 and 1000 ha, the parameter evaluation factors showed that the factors identified using the PCA analysis had no significance for them.

In the case of agricultural producers cultivating lands of over 1000 ha, the analysis indicated that two factors were significant: factor 1, related to a lack of economic justification for the utilization of biomass for energy purposes, and factor 16, related to technical problems, or more precisely to continuous energy placement within a year. A lack of economic justification is related not only to the non-existence of a free market for biomass, guaranteed purchase prices for the sale of biomass, and guaranteed purchase prices for privileged energy producers, but also to the fact that increasing biomass utilization requires additional labor force, which would incur costs higher than the benefits of increasing biomass utilization. In order for such processing plants to be economically justified as soon as possible, it is necessary for them to operate as many days as possible during the year.

6. Discussion

The results of the analysis of the utilization of biomass for various purposes in AP Vojvodina showed that 3.15% was used for energy purposes, 91.46% for mulching and plowing, 2.75% for livestock feed, and 2.64% was burned on site. It is very interesting to compare those results with the countries that are leaders in the utilization of agricultural biomass for energy purposes, for example with the results in Sweden and Denmark. In 2012, Scott Bentsen, N. et al., 2016, analyzed the utilization of straw biomass in Sweden. Their research showed that straw was collected on 40% of the total area planted with crops, i.e., ~1.02 million ha, where 73% was used for mulching/plowing, 13% for livestock feeding, while 9% was used for heating energy requirements (i.e., 3.6% of the total area or 36,000 ha) [45]. Ericsson and Nilsson, 2006, state that in Denmark, 20–40% of crop residue from agricultural production is used for the production of energy [46].

Research has shown that the average used biomass for energy purposes (direct combustion) was 3.15% in AP Vojvodina. Those results are similar to the results of the biomass potential utilization in the Republic of Serbia presented in the *Strategy, 2015*, according to which the utilization of agricultural biomass is merely 2% [47].

Research has shown that 2.64% of crop residue was burned in the field. Problems related to crop residue burning have been noted on smaller farms because farmers sometimes burn the residue in order to get rid of huge amounts of biomass before basic tillage. Burning crop residue in fields is a problem in many countries worldwide. In China, 20.5% of crop residue is discarded or uncontrollably burned on site, 37.5% are used for energy purposes (of which 37% of crop residue are directly burned by farmers, and 0.5% is used for biogas), 23% is used for animal feed, 15% is lost during the process of collection, 4% for industrial substances [48]. In India, the share of burned crop residue ranges from 8 to 80% [49,50], with the crop residues most frequently burned coming from rice (43%), wheat (about 21%), sugar cane (about 19%), and oilseeds (about 5%) [50,51]. In Europe, specifically in Greece, it has been recorded that the remaining amounts of biomass in the fields are burned [52].

While bearing in mind the obtained results, and especially the high values of the determining factor (R^2), it is possible to propose recommendations which could contribute to the utilization of biomass for energy purposes to a significant degree, and those would include the following:

1. Providing farmers with financial incentives for the purchase of machinery and equipment and the construction of energy plants, and for investing in the education of agricultural producers;
2. The introduction of guaranteed quotas for the purchase of produced energy through legislation would eliminate the influence of certain interest groups that disfavor this type of energy;
3. Securing a free biomass market, providing guaranteed biomass purchase prices, guaranteed energy purchase prices for privilege energy producers, etc. The opening of a free biomass market (more regional biomass purchase centers) would allow more people to benefit from biomass utilization (biomass producers, traders, distributors, end users, etc.);
4. Securing heat consumption for the produced energy throughout the year would solve the problems of continuous energy placement and justification of this type of energy production. Using thus produced thermal energy to help heat cities and municipalities could significantly reduce the use of fossil fuels on the one hand, while, on the other, locally available energy sources would be used. That type of thermal energy placement is fully justified, due to the fact that local self-government bodies could provide thermal energy producers with incentive feed-in tariffs for heat production.

With all recommended measures applied, the following benefits are expected to be gained in the forthcoming period: the production of considerable quantities of energy; diversification of the national energy market, revival of rural settlements [53]; creation of reduced dependence on fossil fuels and alleviation of shortages thereof on the market [54]; placement of the produced thermal energy as support in heating cities and municipalities; gaining profit; options for biomass export; considerable number of citizens will benefit from the use of biomass (producers of biomass, businessmen, logistics employees, end users, etc.) and protection of natural resources (water, air, soil, fauna, etc.).

Justification of the conducted research fits into the current needs and goals of the Republic of Serbia in terms of new sources of energy. The needs for energy in the Republic of Serbia are enormous —on the one hand, imported fossil fuels are used, and on the other, locally available renewable resources, such as crop residue, are almost not used at all. The needs for heat energy in the sector of agriculture alone (hothouses, greenhouses, smokehouses etc.) are 420,851.8 toe [55]. The increased usability level of crop residue for the needs of energy production fits into the planned goals of the Republic of Serbia by the end of 2020, with the construction of new power plants that are supposed to achieve energy production in the amount of 209 ktoe (of which, 75 ktoe is from the production of electrical energy and 134 ktoe from the production of energy in the heating and cooling sector) being expected in the biomass sector alone [56]. In addition, the Republic of Serbia, as a European Union candidate, is expected to invest a certain amount of effort in order to close Chapter 27 as referring to "the environment and climate change". Using a considerable amount of the technical potential of agricultural biomass would significantly reduce the level of SO_2 emission (which, at this point, is 4 to 16 times higher than the permitted level in the sector of thermal energy [57]), given that biomass usually contains small amounts of sulfur in comparison with other solid fuels [58,59].

7. Conclusions

The results of the research indicate that there is a huge potential in unused agricultural biomass for energy purposes in AP Vojvodina. Considering the surveyed 75 agricultural properties, the total technical potential of biomass was 152,318.49 t/year, while only 15,149.42 t/year was used for energy purposes. A significant part of biomass is burned in the fields instead of being used for energy purposes. The negative consequences of burning crop residue in the fields certainly include the destruction of flora and fauna, the negative impact on the state of the environment, and the reduction in organic matter in the soil.

The use of this source of energy can be improved in the future by conducting research into the economic advantages of the biomass production, reasons for the field burning of agricultural residues, technology, equipment and plants for biomass production, benefits of the use of biomass in comparison to the use of other fossil fuels, etc.

A greater use of the agricultural biomass resources for energy purposes of the thermal energy production is accorded with the goals established by the Republic of Serbia. Being the official candidate country to become a member state of the European Union, the Republic of Serbia is required to activate the mechanisms for a faster implementation of the projects pertaining to the use of renewable energy sources and environmental protection in the future. Considering that the production of thermal energy in the Republic of Serbia is under the jurisdiction of local self-government bodies, and that the legislation enables them to provide incentive feed-in tariffs for the production of thermal energy, it is realistic to expect that agricultural biomass (harvest residue) in the near future will secure a share in the production of thermal energy for heating certain cities and municipalities in the Republic of Serbia.

Author Contributions: Conceptualization, S.Š. and A.A.; Methodology, R.P.; Software, A.A.; Validation, V.R., G.S. and Z.J.; Formal Analysis, R.P.; Investigation, S.Š.; Resources, Z.J.; Data Curation, M.V.; Writing—Original Draft Preparation, S.Š.; Writing—Review and Editing, A.A.; Visualization, A.R.; Supervision, S.I.; Project Administration, A.A.; Funding Acquisition, S.Š. All authors have read and agreed to the published version of the manuscript.

Funding: This research received no external funding.

Conflicts of Interest: The authors declare no conflict of interest.

References

1. Savin, L.; Nikolić, R.; Tomić, M.; Mileusnić, Z.; Radosavjević, D.; Stelja, Ž. Tractors and Power Machines—Condition and demand 2017–2030. *TrAC Power Mach.* **2016**, *21*, 8–15.
2. Petrović, P.; Petrović, M.; Obradović, D. Serbian Potential in Production of Agricultural Products. *TrAC Power Mach.* **2018**, *23*, 61–72.
3. Pekez, J.; Radovanovic, L.; Desnica, E.; Lambic, M. Increase of exploitability of renewable energy sources. *Energy Sources Part B Econ. Plan. Policy* **2016**, *11*, 51–57. [CrossRef]
4. Tolmac, J.; Josimovic, Lj.; Prvulovic, S.; Cvejic, R.; Radovanovic, Lj.; Blagojevic, Z.; Tolmac, D. Results of research on the energetic and economic efficiency of the use of biomass for heating an agricultural farm. *Energy Sources Part B Econ. Plan. Policy* **2016**, *11*, 96–101. [CrossRef]
5. Nakomčić-Smaragdakis, B.; Cepić, Z.; Dragutinović, N. Analysis of Solid Biomass Energy Potential in Autonomous Province of Vojvodina. *Renew. Sustain. Energy Rev.* **2016**, *57*, 186–191. [CrossRef]
6. Đurić, K.; Cvijanović, D.; Prodanović, R.; Čavlin, M.; Kuzman, B.; Lukač Bulatović, M. Serbian Agriculture Policy: Economic Analysis Using the PSE Approach. *Sustainability* **2019**, *11*, 309. [CrossRef]
7. Ašonja, A.; Vuković, V. The Potentials of Solar Energy in the Republic of Serbia: Current Situation, Possibilities and Barriers. *Energy Sources Part B Econ. Plan. Policy* **2018**, *3*, 90–97. [CrossRef]
8. Pinna-Hernández, M.G.; Martínez-Soler, I.; Díaz Villanueva, M.J.; Acien Fernández, F.G.; López, J.L.C. Selection of biomass supply for a gasification process in a solar thermal hybrid plant for the production of electricity. *Ind. Crop. Prod.* **2019**, *137*, 339–346. [CrossRef]
9. Chauhan, B.S.; Mahajan, G.; Sardana, V.; Timsina, J.; Jat, M.L. Productivity and Sustainability of the Rice–Wheat Cropping System in the Indo-Gangetic Plains of the Indian subcontinent. *Adv. Agron.* **2012**, *117*, 315–369. [CrossRef]
10. Benbi, D.K.; Chand, M. Quantifying the effect of soil organic matter on indigenous soil N supply and wheat productivity in semiarid sub-tropical India. *Nutr. Cycl. Agroecosyst.* **2007**, *79*, 103–112. [CrossRef]

11. Mohammed, N.I.; Kabbashi, N.; Alade, A. Significance of Agricultural residue in Sustainable Biofuel Development. *Agric. Waste Residue* **2018**, 71–88. [CrossRef]

12. Ašonja, A.; Škrbić, S. Analysis of the Biomass Potentials of Hazelnuts and Nuts in the Republic of Srebia. *TrAC Power Mach.* **2019**, *24*, 101–106.

13. Kastori, R.; Tešić, M. Use of harvest residue as alternative fuel - advantages and disadvantages. *Agric. Eng. Rep. Southeast. Eur.* **2005**, *11*, 22–27.

14. Powlson, D. Cereals straw for bioenergy: Environmental and agronomic constraints. In Proceedings of the Expert Consultation: Cereals Straw Resources for Bioenergy in the European Union, Pamplona, Spain, 14–15 October 2006; pp. 45–59.

15. Rosentrater, K.A.; Todey, D.; Persyin, R. Quantifying Total and Sustainable Agricultural Biomass Resources in South Dakota—A Preliminary Assessment. *Agric. Eng. Int.* **2009**, *11*, 1059.

16. Scarlat, N.; Martinov, M.; Dallemand, J.-F. Assessment of the availability of agricultural crop residue in the European Union: Potential and limitations for bioenergy use. *Waste Manag.* **2010**, *30*, 1889–1897. [CrossRef]

17. Martinov, M.; Brkić, M.; Janić, T.; Đatkov, Đ.; Golub, M.; Bojić, S. Biomass in Vojvodina—RES 2020. *Savrem. Poljopr. Teh.* **2011**, *37*, 119–134.

18. Milićević, A.R.; Belošević, S.V.; Tomanović, I.D.; Crnomarković, N.Đ.; Tucaković, D.R. Development of mathematical model for co-firing pulverized coal and biomass in experimental furnace. *Therm. Sci.* **2018**, *22*, 709–719. [CrossRef]

19. Zafar, M.H.; Kazmi, M.; Tabish, A.N.; Ali, C.H.; Gohar, F.; Rafique, M.U. An investigation on the impact of demineralization of lignocellulosic corncob biomass using leaching agents for its utilization in industrial boilers. *Biomass Convers. Biorefinery* **2019**. [CrossRef]

20. Skerlic, J.; Nikolic, D.; Cvetkovic, D.; Miškovic, A. Optimal Position of Solar Collectors: A Review. *Appl. Eng. Lett.* **2018**, *3*, 129–134. [CrossRef]

21. Jokanović, M.; Golubović, D.; Šupić, B.; Koprivica, A. Application of Renewable Energy Sources in Terms of Economic, Environmental and Social Sustainability. *Appl. Eng. Lett.* **2018**, *3*, 34–39. [CrossRef]

22. Fadele, O.; Otieno, M. Sustainable use of supplementary cementitious materials from agricultural wastes—A review. In Proceedings of the 5th International Conference on Sustainable Construction Materials and Technologies (SCMT5), London, UK, 14–17 July 2019; Kingston University London: London, UK; pp. 1–13.

23. Tripathi, N.; Hills, C.D.; Singh, R.S.; Atkinson, C.J. Biomass waste utilisation in low-carbon products: Harnessing a major potential resource. *Npj Clim. Atmos. Sci.* **2019**, *2*, 1–10. [CrossRef]

24. Ašonja, A.; Brkić, M. Ecological Aspects of Using Corncobs as a Biofuel at Seed Centers. In Proceedings of the International Scientific Conference ETIKUM 2014—Metrology and Quality in Production Engineering and Environmental Protection, Novi Sad, Serbia, 19–20 June 2014; Faculty of Technical Science: Novi Sad, Serbia, 2014; pp. 221–224.

25. Ajimotokan, H.A.; Ibitoye, S.E.; Odusote, J.K.; Adesoye, O.A.; Omoniyi, P.O. Physico—Mechanical Characterisation of Fuel Briquettes made from Blends of Corncob and Rice Husk. *J. Phys. Conf. Ser.* **2019**, *1378*, 1–11. [CrossRef]

26. Schwerz, F.; Eloy, E.; Elli, E.F.; Caron, B.O. Reduced planting spacing increase radiation use efficiency and biomass for energy in black wattle plantations: Towards sustainable production systems. *Biomass and Bioenergy* **2019**, *120*, 229–239. [CrossRef]

27. Ašonja, A.; Brkić, M. The Energy Efficiency of the Cobbed Seed Corn Dryer. In *Proceedings—I Scientific-Professional Meeting "Energy Efficiency", 25 October 2013*; Higher Technical School of Professional Studies: Novi Beograd, Serbia, 2013; pp. 34–52.

28. Jiang, Y.; Havrysh, V.; Klymchuk, O.; Nitsenko, V.; Balezentis, T.; Streimikiene, D. Utilization of Crop Residue for Power Generation: The Case of Ukraine. *Sustainability* **2019**, *11*, 7004. [CrossRef]

29. Gustavsson, L.; Holmberg, J.; Dornburg, V.; Sathre, R.; Eggers, T.; Mahapatra, K.; Marland, G. Using biomass for climate change mitigation and oil use reduction. *Energy Policy* **2007**, *35*, 5671–5691. [CrossRef]

30. Sperling, K.; Hvelplund, F.; Mathiesen, B.V. Centralisation and decentralisation in strategic municipal energy planning in Denmark. *Energy Policy* **2011**, *39*, 1338–1351. [CrossRef]

31. Garcia, D.P.; Caraschi, J.C.; Ventorim, G.; Vieira, F.H.A.; de Paula Protásio, T. Assessment of plant biomass for pellet production using multivariate statistics (PCA and HCA). *Renew. Energy* **2019**, *139*, 796–805. [CrossRef]

32. Andrew, A.A.; Hakan, Y. The renewable energy consumption by sectors and household income growth in the United States. *Int. J. Green Energy* **2019**, *16*, 1414–1421. [CrossRef]

33. *Renewables 2018 Global Status Report*; REN21 Secretariat: Paris, France, 2018.

34. Popp, J.; Lakner, Z.; Harangi-Rákos, M.; Fári, M. The effect of bioenergy expansion: Food, energy, and environment. *Renew. Sustain. Energy Rev.* **2014**, *32*, 559–578. [CrossRef]

35. Surendra, K.C.; Ogoshi, R.; Zaleski, H.M.; Hashimoto, A.G.; Khanal, S.K. High yielding tropical energy crops for bioenergy production: Effects of plant components, harvest years and locations on biomass composition. *Bioresour. Technol.* **2018**, *251*, 218–229. [CrossRef]

36. Mukhtar, H.; Feroze, N.; Munir, M.S.; Javed, F.; Kazmi, M. Torrefaction process optimization of agriwaste for energy densification. *Energy Sources Part A: Recovery Util. Environ. Eff.* **2019**, 1–19. [CrossRef]

37. Asonja, A.; Desnica, E.; Radovanovic, Lj. Energy efficiency analysis of corn cob used as a fuel. *Energy Sources Part B Econ. Plan. Policy* **2017**, *12*, 1–7. [CrossRef]

38. Obi, F.; Ugwuishiwu, B.; Nwakaire, J. Agricultural Waste Concept, Generation, Utilization and Management. *Niger. J. Technol.* **2016**, *35*, 957–964. [CrossRef]

39. Crop residue removal for biomass energy production: Effects on soils and recommendations. In *Soil Quality-Agronomy Technical Note*; United States Department of Agriculture USDA: Washington, DC, USA, 2006; pp. 1–7.

40. Carić, M.; Soleša, D. *Biomass as Renewable Energy Source and Waste Treatment Technologies for Biogas Production*; DAI: Leskovac, Serbia, 2014; p. 13.

41. Palková, Z.; Topisirović, G.; Adamovský, F.; Lukáč, O.; Jeremić, M.; Baláži, J.; Zarić. V. *Agriculture Biomass: Its Potential in Slovakia and Serbia*; Slovak University of Agriculture: Nitra, Slovakia, 2016; p. 30.

42. Available online: http://desnamisao.blog.rs/blog/desnamisao/arhiva/2008/12/18 (accessed on 31 December 2019).

43. Škrbić, S. The Level of the Plant Production-Obtained Solid Biomass for Energy Purposes in AP Vojvodina. Ph.D. Thesis, Faculty of Economics and Engineering Management in Novi Sad, University of Business Academy in Novi Sad, Novi Sad, Serbia, 2020. (in press).

44. Janjić, T. *An Excel Application for the Development of Agricultural Biomass Balance, "The Reducing Barriers on the Path to Accelerated Development of the Biomass Market in Serbia" of the United Nations*; Development Program (UNDP) Serbia: Belgrade, Serbia, 2019.

45. Scott Bentsen, N.; Nilsson, D.; Larsen, S.; Stupak, I. Agricultural residue for Energy in Sweden and Denmark—Differences and Commonalities. *IEA Bioenergy* **2016**, *43*, 1–28.

46. Ericsson, K.; Nilsson, L. Assessment of the potential biomass supply in Europe using a resource-focused approach. *Biomass Bioenergy* **2006**, *30*, 1–15. [CrossRef]

47. *The Energy Development Strategy of the Republic of Serbia until 2025 with Projections until 2030*; The Official Gazette of the Republic of Serbia: Belgrade, Serbia, 2015; Volume 101, p. 7.

48. Liu, H.; Jiang, G.; Zhuang, H.; Wang, K. Distribution, utilization structure and potential of biomass resources in rural China: With special references of crop residue. *Renew. Sustain. Energy Rev.* **2008**, *12*, 1402–1418. [CrossRef]

49. Jain, N.; Bhatia, A.; Pathak, H. Emission of Air Pollutants from Crop Residue Burning in India. *Aerosol Qual. Res.* **2014**, *144*, 22–430. [CrossRef]

50. Bhuvaneshwari, S.; Hettiarachchi, H.; Meegoda, J. Crop Residue Burning in India: Policy Challenges and Potential Solutions. *Int. J. Environ. Res. Public Health* **2019**, *16*, 832. [CrossRef]

51. Sahai, S.; Sharma, C.; Singh, S.K.; Gupta, P.K. Assessment of Trace Gases, Carbon and Nitrogen Emissions from Field Burning of Agricultural residue in India. *Nutr. Cycl. Agroecosyst.* **2011**, *89*, 143–157. [CrossRef]

52. Alatzas, S.; Moustakas, K.; Malamis, D.; Vakalis, S. Biomass Potential from Agricultural Waste for Energetic Utilization in Greece. *Energies* **2019**, *12*, 1095. [CrossRef]

53. Rossi, A.M.; Hinrichs, C.C. Hope and skepticism: Farmer and local community views on the socio-economic benefits of agricultural bioenergy. *Biomass Bioenergy* **2011**, *35*, 1418–1428. [CrossRef]

54. Devi, S.; Gupta, C.; Jat, S.L.; Parmar, M.S. Crop residue recycling for economic and environmental sustainability: The case of India. *Open Agric.* **2017**, *2*, 486–494. [CrossRef]

55. Nikolić, R.; Brkić, M.; Savin, L.; Furman; Tomić, M.; Simikić, M. Needed Amounts of Fuel for Agriculture, Waterpower Engineering, and Forestry. *TrAC Power Mach.* **2016**, *17*, 7–14.

56. *Renewable Energy Action Plan of the Republic of Serbia, Republic of Serbia*; Ministry of Energy, Development and Environmental Protection: Belgrade, Serbia, 2013; p. 18.

57. *Chapter 27 in Serbia: Report on (Non) Progress*; Young Researchers of Serbia: Belgrade, Serbia, 2018; p. 24.

58. Horvat, I.; Dović, D. Combustion of agricultural biomass—Issues and solutions. *Trans. FAMENA* **2018**, *42*, 75–86. [CrossRef]

59. Lak, M.; Minaei, S.; Rafiei, A. Temporal and spatial field management using crop growth modeling: A review. *J. Agron. Technol. Eng. Manag.* **2020**, *3*, 375–387.

 *sustainability*

Article

Socioeconomics Determinants to Adopt Agricultural Machinery for Sustainable Organic Farming in Pakistan: A Multinomial Probit Model

Muhammad Waqar Akram [1,2,*], Nida Akram [1], Hongshu Wang [1,*], Shahla Andleeb [3], Khalil Ur Rehman [3], Umair Kashif [1] and Syed Farhaan Hassan [4]

[1] College of Economics & Management, North East Forestry University, Harbin 150040, China; nida.akram88@yahoo.com (N.A.); kashifumair@nefu.edu.cn (U.K.)
[2] Department of Business Administration, Ilma University, Main Ibrahim Hyderi Road, Korangi Creek, Karachi 75190, Pakistan
[3] Department of Environmental Science, Government College Women University, Sialkot 51310, Pakistan; shahla.andleeb@yahoo.com (S.A.); khalil.lastnamerehman@yahoo.com (K.U.R.)
[4] Department Business Administration, Allama Iqbal Open University, Islamabad 44000, Pakistan; farhaan.powerway@gmail.com
*   Correspondence: Waqar.akram13@yahoo.com (M.W.A.); lwanghongshu@163.com (H.W.); Tel.: +86-13654555395 (M.W.A.); +86-15946002021 (H.W.)

Received: 6 November 2020; Accepted: 21 November 2020; Published: 24 November 2020

Abstract: Urbanization and low productivity are real threats to the sustainability of organic farming. The adoption of farm machinery plays a vital role in overcoming these threats to ensure a sustainable and more profitable organic farming model. Farm machinery can also increase farmland yield and reduce the need for labor, although the requirement of significant capital investment often prevents small farmers from buying machinery. There is an increased need to comprehend all relevant elements associated with farming machinery procurement and service delivery. In this article, we provide insight into the impact of different variables of farmers on the adoption of agricultural equipment. A total of 301 organic farmers were surveyed in three districts of Punjab, Pakistan. It was found that the most common machinery concerned herein are tube-well/pumps, tractors, tillage machinery, and thrashers/harvesters. Results from a multinomial probit estimation showed that farm machinery ownership is positively correlated with capital assets, civil infrastructure, alternative sources of power, and credit facility. The findings indicated that policymakers and stakeholders should not concentrate merely on short term planning, such as improving agricultural machinery's adoption rate. Still, they should also strive to upgrade physical infrastructure and facilities and provide credit services to create an enabling environment that can empower the citizen in adopting large scale use of agricultural machinery for long term sustainability of organic farming.

Keywords: sustainability; organic farming; farm development; farm machinery; labor; farm investment; productivity

---

1. Introduction

According to current projections, the global population is expected to reach 9.6 billion people in 2050 [1]. Consequently, the consumption of staple foods such as wheat, rice, maize, meat, and fish is predicted to proliferate, especially in the third world countries and developing nations [2]. These are the same areas where many people also live below the poverty line of USD 1.90 per day; this group of an impoverished population is predicted to increase to 274.5 million persons [3]. On the contrary, the available area of agricultural land per capita is decreasing sharply. For example, in South Asian countries such as Pakistan, India, Bangladesh, and Nepal, the farmland area has reduced from

0.17–0.36 hectors per capita in 1961 to 0.12–0.24 hectors in 2012 [4]. The rapid increase in population and the concurrent decrease in per capita farmland creates serious concern regarding whether organic agriculture can produce a higher yield and sustain the increasing demand. This problem is incredibly real for the South Asia region, which has a high population density. The pressure of food security and the low yield of organic farming may worsen the situation. Furthermore, the simultaneous expansion of other sectors and employment opportunities has resulted in rural-urban migration among the farming community. Consequently, the agriculture sector suffers from a seasonal labor force in rural areas [5,6].

Recently, organic farming (OF) has been facing two main problems. Firstly, the efficiency of OF has often been undermined. Organic farming productivity is 19.3% lower than its counterpart, which translates into an increased threat to food security and organic practice sustainability [7]. The second problem is related to the shortage of rural labor. This is a pertinent issue because OF is considered an incredibly labor-intensive farming method [8]. These arguments strongly support the need to provide specific agricultural machinery for farms that fall short of their limited resource capacity. This will increase their yields and pave the way for more intensified sustainable farming practices [9,10]—usually, organic farmers work on farms of smaller sizes scattered around different areas. For example, the average size of an organic farm in Pakistan is four hectares, much smaller than the Region's average farm size [11]. These small farmers will be interested in small scale machinery that can replace manual work and old traditional tools to save production costs and reduce difficult or labor-intensive work [10,12]. In addition, organic production can be a good alternative for small and medium-sized farms that have no chance of competing with large-scale and modern farms.

However, OF farmers' low purchasing power restricts them from adapting and procuring modern technologies, thus limiting the farm mechanization among Pakistani farmers [11,13]. As Ponisio et al. [7] suggested, the productivity of OF can increase by 9% through better operations and management practice. Specific farm machinery may be able to make OF more sustainable too. As agricultural yield can be improved through farm mechanization, machinery is increasingly designed and made to accommodate farmers with small and scattered lands to promote agricultural resources' sustainable use [14–16]. A good example is the utilization of strip and zero tillage, which requires investment in specialized equipment but, in return, leads to a reduction in fuel expenses, the use of water for irrigation, and time [17–19]. Despite the promised advantages, organic farmers still face problems associated with the adoption of specific machinery. These challenges must be addressed for increased mechanization of OF in Pakistan, which can be achieved via promoting the mechanization program's development.

To overcome problems associated with increased mechanization for OF, it is imperative to comprehend organic farmers' attributes regarding their adoption of or investment into agriculture machinery; in other words, whether the farmers are utilizing such small scale or essential kinds of machinery themselves, as well as renting the machinery to others. This factor is more relevant for organic farmers. Their farms' size is not as big as conventional farmers, leading to a lower preference for purchasing and owning machinery before this. However, in recent years, an increasing number of small farmers have been able to access agricultural mechanization through the enlistment of individual components [20]. Reviewing organic farmers who possess agricultural machinery can give insight into the influential factors that encourage investment in agricultural machinery. With that information, policymakers and development authorities such as commercial and agricultural bankers and administrators can allocate resources or plan investment more efficiently. Furthermore, this will create more significant insights and contribute to the scarce information surrounding farming mechanization and its associated factors.

There are several other reasons as to why this subject is essential and worthy of research. Agriculture is one of Pakistan's most significant economic sectors, contributing one fifth or 19.3% of its GDP [21]. While recent trends showed that OF is practiced increasingly in Pakistan, it contributes to only 0.1% of the total organic agricultural output worldwide [22]. Recently, Pakistan has begun exporting organic products to the European Union. It is a lucrative business as the total market size of organic products in the European Union is 8.96 billion dollars [23]. Apart from that, China is another big potential market for Pakistani organic products through the China Pakistan Economic Corridors (CPEC) project. China's organic market size is nearly 4.19 billion dollars and multiplying [23]. To exploit such opportunities, Pakistan's OF must overcome the various challenges and improve its efficiency. The main challenges of OF included the high labor intensiveness and the shrinking rural labor forces due to rapid urbanization [5].

It must be highlighted that agricultural automation is an integral approach and advancement objective for Pakistan. Compared to other nations in the region, Pakistan's agricultural mechanization is considered under-prioritized, relying only on tractors and specific land-preparing instruments due to the sector's lack of technology [24]. Based on a 2004 survey, Pakistan had a total of 0.41 million tractors and 1.71 million farming implement or machinery, including tillage machines, cultivators, disk harrows/plows, trolleys/trailers, and tube wells/water pumps [25]. These machines' horsepower is only 0.90 HP, much lower than the international standard of 1.7HP, as described by the Food and Agriculture Organization [26]. This lack of resources leads to a lower yield and it affects the productivity of OF more severely than conventional farming as OF is a lower yield farming method to begin with [27]. Besides, Pakistan is also facing a water scarcity problem, especially considering surface water via canal irrigation in the country experienced a 19% decline over the last year. As a result, tube-well/pump irrigation plays a critical role in Pakistan's agricultural sector [28]. Other farm machinery can also help the sustainable use of water. For instance, leveling machinery can save a significant amount of water and add USD 143.5 per hector to a farmer's gross margin. This is especially useful for countries that are heavily dependent on flood irrigation, Pakistan being one of them [29,30] Consequently, research concerning the mechanization of Pakistan OF can provide valuable information regarding factors influencing its uptake.

Despite an extensive search, we found no large-scale studies conducted in recent years that examined organic farms' mechanization. This study aims to fill this gap in the literature. Through contextual analysis, this research will assess the organic farmers in Pakistan and classify the relevant determinants leading to the adoption of necessary and small scale machinery in OF. We will also examine factors leading to the ownership of agricultural farm machinery. In short, this study will analyze available survey data to investigate the adoption of necessary agricultural types of machinery such as tractors, harvesters/threshers, tillers (generally used for land preparation through the utilization of a tractor), and tube wells/water pumps. We will start with an analysis and discussion of the survey data, followed by an econometric model. After presenting the main findings, we will take a more in-depth look at these findings and evaluate farmers' ownership of agricultural machinery's main political implications.

2. Materials and Methods

2.1. Data Description

The data were acquired through a comprehensive survey at the districts of Toba Tek Singh, Khanewal, and Jhang in the Punjab province of Pakistan (Figure 1). These districts were selected based on the suggestion of the Lok Sanjh Foundation, the pioneer in the introduction of OF in Pakistan back in 1996. The community in these districts, such as organic farmers, villagers, and other principal stakeholders, were included in the survey.

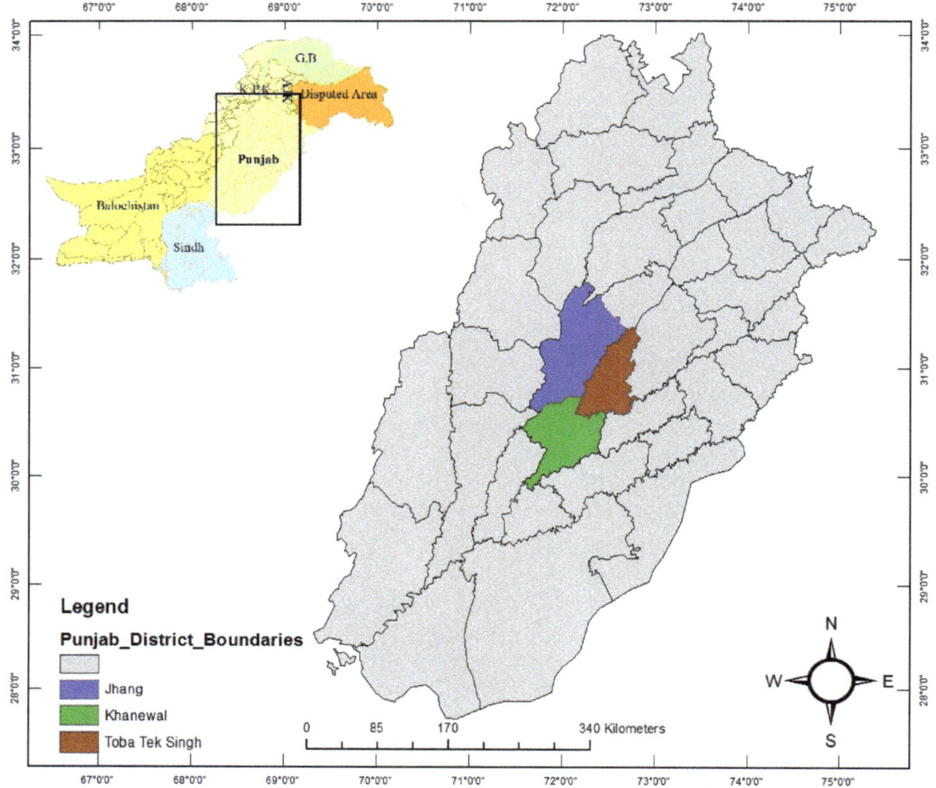

Figure 1. Survey area.

Training sessions that focused on the use of and transition to OF means and techniques were given to the farmers in the aforementioned areas through the help of the FFSs (Farmers Field Schools) and the Lok Sanjh Foundation [31]. All data concerning these farmers who were in the process of assuming OF methods were collected and analyzed to extrapolate a conclusion or result. As the environment and other cultivation conditions of other regions of Punjab Pakistan and the northern part of India are similar, the findings from this specific Region can be extended to other parts of the region.

With the help of the Lok Sanjh Foundation, we identified and selected the 400 farming households that had already adopted OF methods. The farmers were then assigned a number in alphabetical order of their names. A stratified random sampling technique was used to sample the households to ensure that all households had an equal probability of being selected regardless of the household size. This was a fundamental criterion for the identification of an organic approach and machinery use. A questionnaire was distributed to collect the data on personal characteristics, management and structure of the farms, infrastructure of the Region, opinions of the farmers, family information and characteristics, output, and input data. Overall, 400 households received a questionnaire through a cooperative effort with the Lok Sanjh Foundation. Data collection was conducted from September 2017 to February 2018. A total of 344 households returned the questionnaire, which was 89% of the total sample size of households that have adopted OF. Only 301 respondents were included in the final analysis due to missing data.

2.2. Model Specification and Estimation

The following Equation (1) was developed to investigate the determinants of agricultural machinery ownership by organic farmers in Pakistan:

$$Y_i = \alpha_0 + (HHC_i)\varnothing + \alpha_1(Road_{ra}) + \alpha_2(Labor_d) + \sum_{d=1}^{\partial} \beta_j(R_j) - \varepsilon_i. \tag{1}$$

The dependent variable was expressed as Y_i; a zero-based value means that the farmer does not own any farm machinery (including tube wells, pumps, tractors, tillage equipment, thrashers, and harvesters). If a farmer owns a tube well or any other kind of water pump used for irrigation, the dummy value given would be "1", "2" if they owned a tractor, "3" for tillage equipment, and "4" for the ownership of thrashers/harvesters. As Mottaleb et al. [32] explained in his study, this coding method was devised based on the farmer's priority and need, i.e., water is the basic necessity for agriculture, followed by a tractor as the primary source of power for tillage and harvesters. In our study area, farmers who did not have a tractor would not be able to use the other two types of machinery for tillage and harvesting purposes. The explanatory variable of household characteristics (HHC_i) is an independent variable composed of the farmer's age, education, farm size, family members, livestock, access to credit, member of agriculture organization, other jobs, land on lease and ownership, other sources of power, and Region of the farm. Age was further categorized as young (35 years or under), middle-aged (36–56 years), and old (over 56 years). Education was also expressed according to categories, including: "illiterate," "middle school," "secondary school," and "university level." A farmer is defined as a member of any agricultural organization and has livestock assumed with a value of 1 (otherwise zero). The farmer's farmland (Ha) size would be given a dummy value of 1 if the land was leased.

Accordingly, the size of the farm per hectare in our model was determined by the area of land on lease. We also evaluated the impact of ownership types (leased or owned) on the ownership of agricultural machinery. If a household had access to credit from an institution or any personal source, the value was assumed to be 1; otherwise, a zero value was applied. A further dummy value of 1 was given if a household managed to use at least one agricultural machine through an energy source that did not include the machine's engine, such as renewable sources of energy or electricity; otherwise, a zero value was given. In Equation (1), $Road_{ra}$ was the second independent variable that captured the effect of having road access on the farm. If a farm had a gravel or paved road, a value of 1 was given; otherwise, a zero value was assigned. Finally, the impact of the availability of working labor on mechanical adoption was evaluated via the independent variable $Labor_d$. This variable was obtained from an equation that captured two kinds of dummy variables. Firstly, the number of days of labor in which the family spent working on the farm and secondly, the use of hired or outsourced labor (if any). Labor accessibility can have different effects on machine ownership. For instance, if a tube well is utilized for irrigation to increase the crop growth in dry seasons, this will increase labor requirements.

Conversely, tractor, tillage, and threshers may decrease labor needs. Regional infrastructure and facilities influence the adoption of machinery. Hence, to determine whether regional factors affect farming mechanization, the independent variable R_j, which expressed the parameters of the three districts of our selected study area, was included. The scalar parameter was a_0. The vector parameters were a, β, and $\varnothing$; *in which i* represented the household, *ra* represented road access of the farm, *d* represented labor days worked on the farm, and ε stood for random error.

The most widely recognized machine owned by farmers is the tube well or water pump, though it is possible for farmers to own additional farming machinery (Table 1). To explain the utilization of multiple technologies by farmers, a multinomial probit model was used. Two previous studies have adopted a similar model. Mottaleb et al. [33] applied a multinomial probit model to evaluate the determinants for hybrid rice adoption for Bangladesh, whereas Quayum et al. [34] used a single logit estimation to evaluate the adoption of power tillers.

Table 1. Farm Machinery ownership and sample of Household Farm with their Region.

Description	Toba Tek Singh	Jhang	Khanewal	Total
Household Farm Numbers	102	98	101	301
No Farm machinery [a]	10	4	33	47
Tube-well/Water Pump [b]	23	16	24	63
Tractor [c]	10	23	12	45
Tillage [d]	21	8	9	38
Thrasher [e]	38	47	23	108

[a] Farm household has no machinery; [b] Farm household has Tube-well or any other kind of water pump for irrigation purposes; [c] Farm household has a Tractor for agriculture purposes; [d] Farm household has. Tillage or any other kind of implement for land preparation to cultivate the crop; [e] Farm household has a Thrasher or any other kind of harvester.

For this study, we evaluated four models to estimate the effect of these factors on the adoption of agricultural machinery by farmers and to control possible endogenous problems within the data set. First, we included all the aforementioned explanatory variables in the unlimited full model. Three other limited models (L1–L3) were also constructed whereby the nominated variables were excluded to avoid possible endogenous and redundancy issues apart from making sure that other variables of explanatory interests were isolated. For example, the size of the farm was included in the unlimited model and it can be argued that the addition of the dummy variable for owners of farmland and their related multipliers may be superfluous. Hence, in L1, we excluded the variables of farm size and land use (both owned and leased). In L2, we excluded farm size and its related multipliers and also livestock ownership. These factors were excluded because it was possible for a farmer to initially invest in mechanized farming equipment for their farm and purchase a machine before using the same machine to offer or provide services to other farmers in exchange for money. Additional investments in other resources, such as land or livestock, may also occur. Therefore, the credit facility was excluded from the L3 model because such credit could be obtained based on physical capital available through collateral means.

3. Results

3.1. General Survey Results

The characteristics of the study samples are detailed in Table 2. It shows that 78%, 85%, and 70% of surveyed farming households in Toba Tek Singh, Jhang, and Khanewal had livestock. Our study samples comprised of 40% owned land farms and 60% leased land farms, whereby leased land farms included both rent and share-cropper arrangements. It was very surprising to find very low credit access among the households in the areas in which only 22% of households had access to credit. The infrastructure of the Region is quite good, considering that 57% of the farms included in the study had road access. Renewable energy such as alternative sources of energy which can help to reduce the cost of input was used by 35% of farmers in our sample. The usage is the lowest in Khanewal, with only 21%, compared to the other two areas.

Water is a basic need for agriculture. Due to the scarcity of surface water, many farmers either use their own tube-wells or other means to extract groundwater for irrigation. Data from our study showed that 84% of households owned a tube-well and the tractor was the primary source of power. Overall, 76% of farmers in our study owned a tractor, 60% had tillage machinery, and 48% had a thrasher/harvester (Table 2). The demographic features of households are shown in Table A1. The data showed that most household heads were of middle-age and had obtained a high school education. The average household size was found to be 6–7 persons and most families lived in a joint family system.

Table 2. Resource endowments of household's samples by Region.

Description	Region Wise Sample			Total Sample (%)
	Toba Tek Singh (%)	Jhang (%)	Khanewal (%)	
Households that have livestock	78	85	70	78
Households that have farmland ownership	41	36	43	40
Households with access to credit	29	25	12	22
Households that use an alternative source of power	40	45	21	35
Households farm with access to the main road	58	67	45	57
Households with own tube well/water pump	90	96	67	84
Households with own tractor	77	84	68	76
Households with own tillage	67	60	53	60
Households with own thrasher/harvester	46	55	43	48

3.2. Estimation of the Unlimited Model

All household characteristics and other possible explanatory variables were determined for the unlimited estimation model with a 1%, 5%, and 10% significance level followed by the coefficient which explained the ownership of tube-well/pump, tractor, tillage, and thrasher (Table 3). Household characteristics such as education, membership of an agriculture organization, possession of another job outside farming, and livestock ownership were all found to have a positive correlation coefficient. Age, however, was found to have a negative correlation coefficient at 1% confidence interval. Education was revealed to have a positive relationship with tillage and thrasher ownership at the 10% significance level. This relationship was found to be non-significant when it came to tube-wells/pumps and tractors. Membership of an agricultural organization was also found to have a positive but non-significant effect on machinery ownership.

Table 3 shows the variables that had a positive correlation coefficient with the adoption of all farming machinery at a 1% level of significance, other than the possession of a job outside farming and the possession of livestock. In contrast, household members (family size) was found to have a negative correlation coefficient with the adoption of farming machinery.

The variable of farm size (ha) was found to have a positive correlation coefficient with the ownership of all types of farming machines. Overall, farm size was found to have a significant ($p < 0.001$) influence on machinery ownership, while the ownership of farmland was found to have a significant ($p < 0.001$) impact on the ownership of tillage and thrasher/harvester. Additionally, leased farmland was not found to have a significant impact on any kind of machine ownership. As for the influence of credit access, there was no correlation with machinery ownership except for tillage machinery ($p < 0.10$) (Table 4).

The variable concerning accessibility due to infrastructure, such as road access, was revealed to have a significant relationship ($p < 0.001$) with machinery ownership. Furthermore, alternative sources of energy including renewable energy and electricity were also found to have a statistically significant positive relationship ($p < 0.10$) with farmers' adoption of machinery and the use of two kinds of machinery, specifically tillage machinery and tractors (Table 3). Our study also found that hired labor and familial members were negatively correlated with the adoption and use of all farming machinery (Table 3). Family labor days were found to have a significant negative relationship ($p < 0.001$) with machinery adoption. Labor hired or sourced from outside of the family was also found to have a statistically significant ($p < 0.001$) negative relationship with the ownership and use of a thrasher, but this relationship was not detected with other types of machines.

Table 3. Multinomial probit estimation to evaluate farm machinery ownership.

Model Specification	Unlimited Model			
Dependent Variable	**Tube-Well/Pump**	**Tractor**	**Tillage**	**Thrasher**
Age	−0.790 ***	−0.602 **	−0.826 *	−0.580 **
	(−2.81)	(−1.96)	(−2.57)	(−2.09)
Education	0.201	0.342	0.470 *	0.274
	(0.80)	(1.29)	(1.75)	(1.10)
Household Members	0.053	−0.021	−0.01	−0.046
	(0.46)	(−0.17)	(−0.07)	(−0.40)
Member of farming organization	−0.501	0.121	0.184	0.07
	(−1.17)	(0.26)	(0.38)	(0.16)
Other job apart from agriculture	1.227 ***	1.530 ***	1.088 **	1.098 **
	(2.78)	(3.20)	(2.21)	(2.50)
Livestock	0.911 **	1.706 ***	1.342 ***	1.42 ***
	(2.10)	(3.33)	(2.57)	(3.24)
Size of the farm (Hectares)	0.676 ***	0.714 ***	0.477 ***	0.611 ***
	(4.09)	(4.28)	(2.77)	(3.69)
Access to credit	0.718	1.141 *	0.653	0.867
	(1.12)	(1.73)	(0.98)	(1.36)
Alternative sources like electricity	0.484	0.658	0.963 *	1.110 **
	(0.91)	(1.19)	(1.74)	(2.13)
Road Access	0.012 **	1.459 ***	1.998 ***	1.265 ***
	(0.03)	(3.05)	(3.87)	(2.86)
Family Labor days	−0.873 **	−1.049 **	−1.169 ***	−1.267 ***
	(−2.09)	(−2.38)	(−2.62)	(−3.11)
Hired Labor days	0.165	−0.501	−0.11	−0.763 ***
	(0.52)	(−1.51)	(−0.59)	(−2.45)
Farm land ownership	0.773	1.128	2.495 ***	1.676 ***
	(1.19)	(1.62)	(3.28)	(2.56)
Farm land on lease	−0.145	0.49	1.461 **	0.646
	(−0.26)	(0.82)	(2.16)	(1.17)
Region	−0.333	−0.276	−0.808 **	−0.545 **
	(−1.21)	(−0.91)	(−2.60)	(−1.94)
Constant	−0.665	−1.649	−1.633	1.338
	(−0.38)	(−0.89)	(−0.86)	(0.78)
Number of Households	301			
Wald Chi^2 (60)	141.64			
Log-likelihood ratio	−313.51			
Prob > Chi^2	0.00			

Robust standard errors are in parentheses *, **, *** shows the level of significance at 10%, 5%, and 1%, respectively.

Table 4. Multinomial Probit estimation to evaluate farm Machinery ownership.

Model Specification	Limited Model L-1				Limited Model L-2				Limited Model L-3			
Dependent Variable	Tube-Well/Water-Pump	Tractor	Tillage	Thrasher	Tube-Well/Water-Pump	Tractor	Tillage	Thrasher	Tube-Well/Water-Pump	Tractor	Tillage	Thrasher
Age	−0.884 ***	−0.715 ***	−0.853 ***	−0.693 ***	−0.953 ***	−0.828 ***	−0.974 ***	−0.803 ***	−1.021 ***	−0.932 ***	−1.039 ***	−0.860 ***
	(−3.78)	(−2.79)	(−3.19)	(−2.99)	(−4.12)	(−3.28)	(−3.59)	(−3.54)	(−4.63)	(−3.90)	(3.99)	(−3.98)
Education	0.200	0.432 **	0.506 **	0.316	0.267	0.387 *	0.420 **	0.282	0.262	0.394 *	0.419 **	0.280
	(0.96)	(1.95)	(2.27)	(1.52)	(1.27)	(1.75)	(1.85)	(1.35)	(1.32)	(1.87)	(1.93)	(1.41)
Household Members	−0.001	−0.089	−0.045	−0.101	0.088	0.003	0.045	−0.015	0.072	−0.026	0.027	−0.026
	(−0.01)	(−0.90)	(−0.44)	(−1.08)	(0.99)	(0.11)	(0.45)	(−0.17)	(0.84)	(−0.29)	(0.28)	(−0.31)
Member of farming organization	0.177	0.953 ***	0.934 **	0.816 **	0.257	0.829 **	0.767 *	0.723 **	0.245	0.794 **	0.729 **	0.702 **
	(0.53)	(2.55)	(2.40)	(2.39)	(0.77)	(2.26)	(1.95)	(2.17)	(0.76)	(2.24)	(1.91)	(2.18)
Other job apart from agriculture	0.668 *	0.980 ***	0.818 **	0.603 *	0.791 **	1.036 ***	0.830 **	0.648 **	-	-	-	-
	(1.94)	(2.60)	(2.11)	(1.74)	(2.33)	(2.79)	(1.79)	(1.92)				
livestock	0.799 **	1.584 ***	1.393 ***	1.391 ***	-	-	-	-	-	-	-	-
	(2.22)	(3.67)	(3.16)	(3.75)								
Size of the farm (Hectares)	-	-	-	-	-	-	-	-	-	-	-	-
Access to credit	0.952 *	1.172 **	0.933 *	1.084 **	0.942 *	1.204 **	0.973 *	1.097 **	-	-	-	-
	(1.86)	(2.18)	(1.71)	(2.10)	(1.86)	(2.28)	(1.79)	(2.16)				
Alternative sources like electricity	0.785 *	1.049 **	1.185 ***	1.349 ***	0.862 **	1.113 **	1.25 ***	1.418 ***	0.884 **	1.154 ***	1.264 ***	1.425 ***
	(1.85)	(2.36)	(2.65)	(3.21)	(2.02)	(2.51)	(2.76)	(3.38)	(2.15)	(2.24)	(2.89)	(3.54)
Road Access	-	-	-	-	−0.317	1.007 ***	1.510 ***	0.862 ***	−0.293	0.998 ***	1.517 ***	0.863 ***
					(−0.90)	(2.75)	(3.65)	(2.56)	(−0.87)	(2.83)	(3.76)	(2.65)
Family Labor days	−0.493	−0.754 **	−0.739 **	−0.848 ***	−0.481	−0.723 **	−0.627 *	−0.804 ***	−0.367	−0.585 *	−0.508	−0.696 **
	(−1.51)	(−2.18)	(−2.10)	(−2.66)	(−1.49)	(−2.11)	(−1.75)	(−2.57)	(−1.20)	(−1.82)	(−1.48)	(−2.35)
Hired Labor days	−0.057	−0.553 **	−0.373	−0.856 ***	0.094	−0.432 *	−0.210	−0.727 ***	0.126	−0.386	−0.152	−0.654 ***
	(−0.23)	(−2.12)	(−1.37)	(−3.51)	(0.37)	(−1.66)	(−0.76)	(−2.99)	(0.53)	(−1.57)	(−0.59)	(−2.86)
Farm land ownership	-	-	-	-	-	-	-	-	-	-	-	-
Farm land on lease	-	-	-	-	-	-	-	-	-	-	-	-
Region	−0.350 *	−0.185	−0.598 ***	−0.436 **	−0.393 **	−0.222	−0.671 ***	−0.492 ***	−0.543 ***	−0.426 **	−0.831 ***	−0.645 ***
	(−1.71)	(−0.81)	(−2.59)	(−2.09)	(−1.95)	(−0.99)	(−2.85)	(−2.41)	(−2.82)	(−2.02)	(−3.72)	(−3.34)
Constant	1.66	1.808	1.668	4.257 ***	1.504	1.768	1.008	4.121 ***	2.016 *	2.780 **	1.675	4.594 ***
	(1.25)	(1.26)	(1.13)	(3.20)	(1.15)	(1.27)	(0.68)	(3.20)	(1.79)	(2.18)	(1.23)	(3.84)
Number of Household	301				301				301			
Wald Chi^2 (60)	110.64				126.64				123.77			
Log pseudo likelihood ratio	−369.43				−358.89				−367.04			
Prob > Chi^2	0.000				0.000				0.000			
Log Likelihood Ratio Chi2 (16)	111.83 [a]				90.77 [b]				107.06 [c]			
Probability > Chi2	0.000				0.000				0.000			

The robust standard error is in parentheses *, **, *** shows the level of significance at 10%, 5%, and 1%, respectively; Likelihood ratio with the assumptions ([a]) Limited model (L-1) nested in the unlimited model; ([b]) Likelihood ratio with the assumption of the Limited model (L-2) nested in the unlimited model; ([c]) Likelihood ratio with the assumption of the Limited model (L-3) nested in the unlimited model.

3.3. Estimation of the Limited Model

The first limited model, L-1, excluded the explanatory variables such as road access, size of the farm, and land ownership (owned and leased by the farmer). In the second limited model, L-2 excluded livestock, farm size, and related variables seen in the first model. In the third limited model, L-3 excluded farm size and associated variables, external sources of funding variables (such as access to credit), and the non-farming job variable. The variables relevant to financing (credit access and possession of job outside of agriculture), alternative sources of power (renewable energy and electricity), and infrastructure (road access) were all found to be statistically significant at a 0.01, 0.05, and 0.1 level within the first (L-1) and second (L-2) limited models (Table 4).

Therefore, the estimation functions in the L-1 and L-2 models showed that external financing, road access, and alternative power sources have a strong relationship with the adoption of farm machinery, specifically with the ownership of a tractor, tillage machinery, and a thrasher. The findings of the L-2 model supported the argument that credit access and road access do not play a significant role in the ownership of farm machinery. The ownership of farm machinery still depends mainly on farm size and land usage rights.

In the third limited model (L-3), in addition to excluding any possible outstanding endogenous factors, we also excluded other jobs, credit access, ownership of livestock, and the size of the farm and associated variables. Similar to the predicted function in the L-1, L-2, and L-3 models, after statistical regression analysis, the effect of power from alternative sources and road access to the farm on the ownership of farm machinery was found to be statistically significant ($p < 0.001$). The personal attributes such as age, education level, and membership of agriculture organization were also significant in all three limited models (L-1 to L-3), so there was no endogeneity problem. We have proved that the regional variable had a significant ($p < 0.001$) impact on machinery adoption based on the unlimited model (Table 3) and all three of the limited models (Table 4). This means that the infrastructure and the flow of information are critical in farm mechanization.

To verify the predicted model, we performed a post estimation's log-likelihood ratio test by placing all limited models (L-1, L-2, and L-3) within an unlimited model. The likelihood ratio (LR) was 111.83 for the limited models L-1, 90.77 for L-2, and 107.06 for L3. All three ratios were found to be statistically significant at a 1% level. Table 4 shows that the unlimited model estimation is more satisfactory when estimating machinery ownership and adoption.

4. Discussion

Farmers with insufficient capital can be facilitated by arranging agricultural equipment and machinery on rent to promote sustainable agriculture. The availability of various machinery and technical support from agrarian professionals is the main concern for achieving and expanding organic agricultural trends in countries with a poor economic background but rich agricultural potential, like in South Asia. Literature is scarce on this issue and only a few published studies have investigated these questions. Previous empirical studies in the field examined this issue in the general context of the whole agriculture sector but have not emphasized OF specifically.

Moreover, literature also lacked other significant parameters like domestic characteristics, socioeconomic conditions, and infrastructure status related to agricultural machinery ownership. Anyhow, this study included a series of variants observed previously. Considering the brief temporal developments of the appropriate policies in Pakistan's agricultural fields, the conversion of the agricultural sector from a traditional setup to mechanization has always remained a priority. This is why the findings of the present study will be a big step towards bridging the documentation and knowledge gaps in this area. Our research highlights the most critical factors affecting organic farm ownership of necessary agricultural machinery in Pakistan, including well pumps/tubes, tractors, plowing, and harvesting machines.

This study's overall results showed that it is difficult for organic farmers to obtain financial support through proper channels. This is because of a shortness in Government funding for organic farmers. Most of the farmers are dependent on private capital and resources to finance their agricultural activities. Kaleem et al. [35] stated that most farmers obtain loans from intermediaries or agents and are thus compelled to sell their products to the same vendors or middlemen at cheaper rates than market rates. Literature also represented that only 10% of all financial deals are done in cash due to the complex documentation system of public banks and institutions; furthermore, they are charging higher transfer rates.

Many studies highlighted the issues of road infrastructure with agricultural development to access various necessary resources for the agricultural sector like energy and other relevant applications (mainly for tube-well/irrigation pumps), which has a close association with the ownership of all types of agricultural machinery at the domestic level. However, the connection between road access and ownership of the irrigation pump is insignificant. This may be observed due to fixed attributes of the irrigation pump/tube-well. On the contrary, the road network is believed to be an essential factor in tractor ownership. The roads allow farmers to transport their crops to the market and increase income-generating procedures. Besides, specific carpet roads or expressways must transport agricultural machinery such as threshers/harvesters from one city to another.

Furthermore, the transport infrastructure in the present study area is pretty good because the designated area is in the middle of Punjab. This region connects the southern provinces of Punjab and Sindh to the country's main powers; here, agricultural infrastructure and access to roads is significantly better. Besides, a project started by China, the Pakistan Economic Corridor (CPEC), has also helped to improve infrastructure across the country by developing road and communication construction projects [36]. There has not been much usage of substitute energy sources, such as energy resources from natural/renewable resources, in this region. The OF industry remains highly dependent on oil use, which increases input costs and affects the financial status of organic farming development. Moreover, in Pakistan, electricity consumption in the agricultural sector is also much lower than in other developed countries because of poor maintenance in the energy and power sectors to fulfill the requirement sustainably.

Water resources are an essential requirement for the agricultural sector. Pakistan relies heavily on the Indus Watershed Irrigation System (IBIS), which was formulated and designed almost a hundred years ago to increase water availability for community settlements and the farming industry [37]. Over the past few decades, water availability from IBIS has decreased rapidly. Studies concluded that only 10% of total agricultural water is required in Pakistan from underground water resources, but that is not true for the Punjab Province, where almost 90% of all farming land is watered with underground water by using pumping tube wells [38]. These results are consistent with our research, with most farmers relying on groundwater for irrigation and having tube wells to meet their water needs.

Among various tools of agricultural mechanization, tractors play a vital role in the agriculture sector of Pakistan. Tractors are the primary source of power for many farmers, representing the first step toward farming mechanization. It also helps the farmers use different types of agriculture tools like tillage, harvesters, or trolleys. In this study, we found that tractors are the priority for farmers in the study area. Other types of machinery are linked with design and the specific requirement with reference to farmers (Table 2).

The unlimited model results show that the household's specific characteristics play a more significant role in the application of agricultural machinery, especially the age of the farmer. Old age farmers are less likely to use the machines. The same findings were reported in the study conducted about OF by Ullah et al. [39]. Middle-aged farmers are more interested in implementing OF than older farmers in Pakistan. Membership of agricultural organizations has a positive effect on machine ownership. A large number of families in our study were members of agricultural organizations, possibly due to a large number of NGOs working to promote OF in cooperation with the Lok Sanjh Foundation.

Moreover, our research finds that farmers running small businesses or doing some other jobs and agricultural farming can generate more financial resources than farmers only linked with agricultural farming. Likewise, if the farmer's household size is small, the family workforce may not be enough to manage the farm optimally. Therefore, this can motivate farmers to convert traditional farming into mechanized farming. The results of this study on family specifications and applications of agricultural machinery support this argument. This finding is also consistent with small-scale machine adaptation in Bangladesh [32] and the implementation of contemporary techniques in rice cultivation in the Philippines [40].

The farmers' economic condition and farming size have a positive and vital relationship with their modern agricultural equipment possession. Research shows that agriculturalists with more arable land and livestock are more likely to own their own farm machinery. This is not surprising because robust control of underlying assets will best reflect the total means accessible to farmers. Furthermore, the result is much more fascinating. The farmers who own the land are more likely to hold agricultural machinery than those who rent it. Besides, larger farms seem to have more support for machinery; this is why policymakers and stakeholders need to emphasize expanding the use of appropriate mechanisms to broaden the access of small landholders and private investors and people looking to invest in agriculture on a small scale. Such projects aim to foster the advancement and development of the business model to lead the private sector by encouraging the purchase of agricultural machinery via home-grown markets. One example is the M4P or core value chain project, which is about "creating markets for the poor" [41]. The results of this study highlighted that farmers with small scale agricultural land ownership, even though they rent a large area of land for farming, are less interested in investing in agricultural machinery. Still, this does not hinder their ability or rights to use the tools. OF development plans should enhance the administrative arrangements and allocation procedures to ensure agricultural machinery's availability at an affordable price for farmers with smaller agricultural landholders.

Another important finding from this study is an insignificant relationship between credit access of the farmers for buying machinery and machine ownership. The farmers prefer to take loans from local agents/middlemen to sell their commodities to that specific agent at lower rates. A significant factor of this situation is the tough documentation process of the Bank to release money. Another possible explanation for this is the religion of the study population. Most Pakistani farmers are Muslims, so they tend to steer towards traditional banking systems to avoid usury. As stated in the Holy Qur'an, Surat Al-Omran, Verse 130: "O you who believe, indulge in double and double indifference, and fear God, in the hope that you will be blessed with good." The local agent/middleman model often causes financial crises to farmers. So, they avoid purchasing machinery with a credit facility. Furthermore, the study mentioned the importance of easy access to roads. Aside from better corporal connectivity, access to roads also improves the flow of information to farmers and agricultural workers [42].

In organic farming, mainly tractors and irrigation pumps utilize diesel for running engines in the case of energy consumption. OF is very sensitive to water and its availability. Due to severe electricity shortages, sudden and unexpected power cuts, and large initial investments in installing renewables, most farmers still wanted to depend on diesel engines. However, the present study highlighted that other energy sources' influence on farmers' adoption of agricultural tools and equipment could not be ignored. It can be concluded that saving operating costs by using renewable energy resources could motivate advanced farmers to own agricultural machinery. It is also discussed in scientific literature [43] that electricity or renewable energy sources have lower operating costs than current energy sources.

Due to the rapid decline in the countryside workforce, agriculturists are also in a difficult situation if they were to choose agricultural machinery instead of labor. Current results show that farmer families with more family members who wish to work in the fields are less likely to purchase and own farm machinery. A farmer who can employ low-cost family labor without having to outsource may incur additional costs. Therefore, some farmers do not feel the need to invest in agricultural machinery. On the contrary, if the farmer does not receive help from his family and needs to hire outside help from people outside the home, he will feel more entitled to use the available capital for investment and machines to save time and money [32].

5. Conclusions

Mechanization in organic farming can play an essential role in promoting and maintaining practices widely known as labor-intensive agricultural technologies. A comprehensive view of the economic and social variables affecting farmers' purchasing and use of agricultural machinery for sustainable OFs will be invaluable for decisionmakers to make useful policies and will help budget distribution, as well as resource planning and management. The infrastructure improvements necessary in rural areas of Pakistan are basics to enhancing the possession of farm machines and equipment. One way is to rent agricultural machinery to small, organic farmers. Easy access to credit as an essential share of mechanization efforts can enable sustainable development and reduce production risks for farmers. In short, policymakers/decisionmakers and developers wishing to improve farmers' farming mechanization in the country are required to consider increasing their needs and conditions as needed. Besides, it is desirable to use small scale and cost-effective agricultural machinery in stages, facilitating organic agricultural mechanization and increasing the total production of organic productivity.

Author Contributions: M.W.A. and H.W. conceptualized the study design idea, performed the statistical analysis, and wrote the manuscript. N.A., S.F.H., U.K., S.A., and K.U.R. helped in the data collection and provided their intellectual insights. All authors have read and agreed to the published version of the manuscript.

Funding: This research funded by Northeast Forestry University Harbin.

Acknowledgments: We would like to acknowledge the Lok Sanjh Foundation Pakistan for providing technical data related to this study.

Conflicts of Interest: The authors of this research declare no conflict of interest.

Appendix A

Table A1. Descriptive Statistics.

Variables	Description	Mean	Std. Deviation
Age	1 = Young, up to 35 years; 2 = Middle age, "36 To 55" Years; 3 = Old, more than 55 years	1.88	0.72
Education	0 = Illiterate, 2 = Middle school, 3 = middle to secondary school, 4 = University degree	1.52	0.86
Household	Number of household members	6.88	1.84
Member of farming organization	Member of any agriculture organization 0 = No, 1 = yes	0.66	0.48
Another job apart from agriculture	Household head has another job with agriculture farming 0 = No, 1 = Yes	0.60	0.49
Livestock	Household have livestock 0 = No, 1 = Yes	0.77	0.42
Size of the farm (hectares)	Size of agriculture farm in hectares	4.50	3.72
Access to credit	Household have access to Institutions or personal sources of credit; 0 = No, 1 = Yes	0.22	0.41
Alternative sources like electricity	Household have alternative source of power; 0 = No, 1 = Yes	0.35	0.48
Road Access	Farm situated on main road; 0 = No, 1 = Yes	0.57	0.50
Family Labor days	Family members spend days for labor work on farm; 0 = 1 to 10 days, 1 = 11 To 15 days, 2 = 16 to 20 days, 3 = More than 20 days	1.51	0.53
Hired Labor days	Hire labor spend days for labor work on farm; 0 = 1 to 10 days, 1 = 11 to 15 days, 2 = 16 to 20 days, 3 = More than 20 days	2.72	0.73
Farm land ownership	Household owned the farmland; 0 = No, 1 = Yes	0.40	0.49
Farm land on lease	Household cultivate land under lease contract; 0 = No, 1 = Yes	0.41	0.49
Household has a water pump/tube well	Household have a Tubewell/waterpump; 0 = No, 1 = Yes	0.84	0.36
Household has a tractor	Household own a tractor; 0 = No, 1 = Yes	0.76	0.43
Household has a tillage	Household own a Tillage; 0 = No, 1 = Yes	0.60	0.49
Household has a thrasher/harvester	Household own a Thrasher/Harvester; 0 = No, 1 = Yes	0.48	0.50

References

1. Gerland, P.; Raftery, A.E.; Ševčíková, H.; Li, N.; Gu, D.; Spoorenberg, T.; Alkema, L.; Fosdick, B.K.; Chunn, J.; Lalic, N. World population stabilization unlikely this century. *Science* **2014**, *346*, 234–237. [CrossRef]
2. Godfray, H.C.J.; Garnett, T. Food security and sustainable intensification. *Phil. Trans. R. Soc. B* **2014**, *369*, 20120273. [CrossRef] [PubMed]
3. WorldBank. The World Bank Poverty and Equity. Available online: http://databank.worldbank.org/data/reports.aspx?source=poverty-and-equity-database (accessed on 23 October 2018).
4. WorldBank. World Development Indicators 2018: Employment in agriculture (% of total employment). Available online: http://databank.worldbank.org/data/reports.aspx?Code=PAK&id=556d8fa6&report_name=Popular_countries&populartype=country&ispopular=y# (accessed on 23 October 2018).
5. Zhang, X.; Rashid, S.; Ahmad, K.; Ahmed, A. Escalation of real wages in Bangladesh: Is it the beginning of structural transformation? *World. Dev.* **2014**, *64*, 273–285. [CrossRef]
6. Akram, N.; Akram, M.W.; Wang, H.; Mehmood, A. Does land tenure systems affect sustainable agricultural development? *Sustainability* **2019**, *11*, 3925. [CrossRef]
7. Ponisio, L.C.; M'Gonigle, L.K.; Mace, K.C.; Palomino, J.; de Valpine, P.; Kremen, C. Diversification practices reduce organic to conventional yield gap. *Proc. R. Soc. B* **2015**, *282*, 20141396. [CrossRef] [PubMed]
8. Demiryurek, K.; Ceyhan, V. Economics of organic and conventional hazelnut production in the Terme district of Samsun, Turkey. *Renew. Agric. Food. Syst.* **2008**, *23*, 217–227. [CrossRef]
9. FAO (Food and Agriculture Organization of the United Nations). Agricultural Mechanization in Africa, Time for Action. Planning Investment for Enhanced Agricultural Productivity Report of an Expert Group Meeting. Vienna and Rome, Food and Agriculture Organization of the United Nations and (UNFAO) United Nations Industrial Development Organization (UNIDO). 2008. Available online: http://www.fao.org/docrep/017/k2584e/k2584e.pdf (accessed on 23 November 2020).
10. Kienzle, J.; Ashburner, J.E.; Sims, B.G. *Mechanization for Rural Development*; FAO: Rome, Italy, 2013.
11. Akram, M.W.; Akram, N.; Hongshu, W.; Mehmood, A. An assessment of economic viability of organic farming in Pakistan. *Custos e Agronegocio Online* **2019**, *15*, 141–169.
12. WorldBank. *Agriculture for Development, World Development Report 2008*; The International Bank for Reconstruction and Development/The World Bank: Washington, DC, USA, 2014.
13. Ur Rehman, T.; Khan, M.U.; Tayyab, M.; Akram, M.W.; Faheem, M. Current status and overview of farm mechanization in Pakistan–A review. *Agric. Eng. Int.* **2016**, *18*, 83–93.
14. Krupnik, T.J.; Valle, S.S.; Islam, S.; Hossain, A.; Gathala, M.K.; Qureshi, A.S. Energetic, hydraulic and economic efficiency of axial flow and centrifugal pumps for surface water irrigation in Bangladesh. *Irrig. Drain.* **2015**, *64*, 683–693. [CrossRef]
15. Krupnik, T.J.; Santos Valle, S.; McDonald, A.; Justice, S.; Hossain, I.; Gathala, M. *Made in Bangladesh: Scale-Appropriate Machinery for Agricultural Resource Conservation*; CIMMYT: Mexico City, Mexico, 2013.
16. Baudron, F.; Sims, B.; Justice, S.; Kahan, D.G.; Rose, R.; Mkomwa, S.; Kaumbutho, P.; Sariah, J.; Nazare, R.; Moges, G. Re-examining appropriate mechanization in Eastern and Southern Africa: Two-wheel tractors, conservation agriculture, and private sector involvement. *Food Secur.* **2015**, *7*, 889–904. [CrossRef]
17. Erenstein, O.; Laxmi, V. Zero tillage impacts in India's rice–wheat systems: A review. *Soil Tillage Res.* **2008**, *100*, 1–14. [CrossRef]
18. Fileccia, T. *Importance of Zero-Tillage with High Stubble to Trap Snow and Increase Wheat Yields in Northern Kazakhstan*; Food and Agriculture Organization of the United Nations (FAO), Investment Centre Division: Rome, Italy, 2009; Available online: ftp://ftp.fao.org/docrep/fao/012/ak166e/ak166e00.pdf (accessed on 14 May 2015).
19. Krupnik, T.; Yasmin, S.; Shahjahan, M.; McDonald, A.; Hossain, K.; Baksh, E.; Hossain, F.; Kurishi, A.; Miah, A.; Mamun, M.A. Productivity and farmers' perceptions of rice-maize system performance under conservation agriculture, mixed and full tillage, and farmers' practices in rainfed and water-limited environments of southern Bangladesh. In Proceedings of the 6th World Congress on Conservation Agriculture, Winnipeg, MB, Canada, 22–25 June 2014.
20. Biggs, S.; Justice, S. Rural and Agricultural Mechanization: A History of the Spread of Small Engines in Selected Asian Countries. 2015. Available online: https://papers.ssrn.com/sol3/papers.cfm?abstract_id=2623612 (accessed on 23 November 2020).

21. MOF. *Pakistan Economic Survey 2016–2017*; Government of Pakistan, Finance Division, Economic Affairs Wing: Islambad, Pakistan, 2016–2017.
22. IFOAM. *The World of Organic Agriculture, Statistics & Emerging Trend 2018*; IFOAM: Bonn, Germany, 2018.
23. Willer, H.; Lernoud, J. *The World of Organic Agriculture. Statistics and Emerging Trends 2018*; 303736307X; Research Institute of Organic Agriculture FiBL and IFOAM Organics International: Nuremberg, Germany, 2019.
24. FAO. *Food and Agriculture Organization Statistical Year Book 2014*; Regional Office for Asia and the Pacific: Bangkok, Thailand, 2014.
25. GOP. *Pakistan Agricultural Machinery Census 2004*; Agriculture Census Organization, Pakistan Bureau of Statistics, Eds.; Government of Pakistan: Islambad, Pakistan, 2004.
26. MOF. *Pakistan Economic Survey 2010–2011*; MOF, Ed.; Government of Pakistan, Finance Division, Economic Affairs Wing: Islambad, Pakistan, 2010–2011.
27. Reganold, J.P.; Wachter, J.M. Organic agriculture in the twenty-first century. *Nat. Plants* **2016**, *2*, 15221. [CrossRef] [PubMed]
28. MOF. *Pakistan Economic Survey 2018–2019*; MOF, Ed.; Government of Pakistan, Finance Division, Economic Affairs Wing: Islambad, Pakistan, 2018–2019.
29. Aryal, J.P.; Mehrotra, M.B.; Jat, M.; Sidhu, H.S. Impacts of laser land leveling in rice–wheat systems of the north–western indo-gangetic plains of India. *Food Secur.* **2015**, *7*, 725–738. [CrossRef]
30. Magnan, N.; Spielman, D.J.; Lybbert, T.J.; Gulati, K. Leveling with friends: Social networks and Indian farmers' demand for a technology with heterogeneous benefits. *J. Dev. Econ.* **2015**, *116*, 223–251. [CrossRef]
31. Akram, M.W.; Akram, N.; Hongshu, W.; Andleeb, S.; ur Rehman, K.; Kashif, U.; Mehmood, A. Impact of land use rights on the investment and efficiency of organic farming. *Sustainability* **2019**, *11*, 7148. [CrossRef]
32. Mottaleb, K.A.; Krupnik, T.J.; Erenstein, O. Factors associated with small-scale agricultural machinery adoption in Bangladesh: Census findings. *J. Rural Stud.* **2016**, *46*, 155–168. [CrossRef]
33. Mottaleb, K.A.; Mohanty, S.; Nelson, A. Factors influencing hybrid rice adoption: A Bangladesh case. *Aust. J. Agric. Resour. Econ.* **2015**, *59*, 258–274. [CrossRef]
34. Quayum, M.A.; Ali, A.M. Adoption and diffusion of power tillers in Bangladesh. *Bangladesh J. Agric. Res.* **2012**, *37*, 307–325. [CrossRef]
35. Kaleem, A.; Abdul Wajid, R. Application of Islamic banking instrument (Bai Salam) for agriculture financing in Pakistan. *Br. Food J.* **2009**, *111*, 275–292. [CrossRef]
36. MOP. *Long Term Plan for China-Pakistan Economic Corridor (2017–2030)*; Ministry of Planning, D.R.P., Ed.; China-Pakistpan Economic Corridor: Islamabad, Pakistan, 2017.
37. Jurriëns, M.; Mollinga, P.P.; Wester, P. *Scarcity by Design: Protective Irrigation in India and Pakistan*; ILRI: Nairobi, Kenya, 1996.
38. Qureshi, A.S.; McCornick, P.G.; Sarwar, A.; Sharma, B.R. Challenges and prospects of sustainable groundwater management in the Indus Basin, Pakistan. *Water Resour. Manag.* **2010**, *24*, 1551–1569. [CrossRef]
39. Ullah, A.; Shah, S.N.M.; Ali, A.; Naz, R.; Mahar, A.; Kalhoro, S.A. Factors affecting the adoption of organic farming in Peshawar-Pakistan. *Agric. Sci.* **2015**, *6*, 587. [CrossRef]
40. Mariano, M.J.; Villano, R.; Fleming, E. Factors influencing farmers' adoption of modern rice technologies and good management practices in the Philippines. *Agric. Syst.* **2012**, *110*, 41–53. [CrossRef]
41. Akram, N.; Akram, M.W.; Hongshu, W. Study on the Socioeconomic Factors Affecting Adoption of Agricultural Machinery. *J. Econ. Sustain. Dev.* **2020**, *11*. [CrossRef]
42. Mottaleb, K.A.; Mohanty, S.; Nelson, A. Strengthening market linkages of farm households in developing countries. *Appl. Econ. Perspect. Policy* **2014**, *37*, 226–242. [CrossRef]
43. Qureshi, A.S.; Akhtar, M.; Sarwar, A. Effect of electricity pricing policies on groundwater management in Pakistan. *Pak. J. Water Resour.* **2003**, *7*, 1–9.

Article

Analysis of the Quality of the Employee–Bank Relationship in Urban and Rural Areas

Snežana Lekić [1], Jelena Vapa-Tankosić [2,*], Slavica Mandić [1], Jasmina Rajaković-Mijailović [3], Nemanja Lekić [1] and Jelena Mijailović [2]

1   Belgrade Business and Arts Academy of Applied Studies, Kraljice Marije 73, 11050 Belgrade, Serbia; snezana.lekic@bpa.edu.rs (S.L.); slavica.mandic@bpa.edu.rs (S.M.); nemanja.lekic@bpa.edu.rs (N.L.)
2   Faculty of Economics and Engineering Management, University Business Academy, Cvećarska 2, 21000 Novi Sad, Serbia; j.mijailovic@fimek.edu.rs
3   Faculty of Law for Commerce and Judiciary, University Business Academy, Geri Karolja 1, 21102 Novi Sad, Serbia; jasminarajakovic@yahoo.com
*   Correspondence: jvapa@libero.it; Tel.: +381-64-8821282

Received: 10 June 2020; Accepted: 30 June 2020; Published: 6 July 2020

Abstract: Banking sector performance is directly related to the economic performance of the country. This research is an effort to establish the parameters of job satisfaction among bank employees and to ascertain whether there were differences in job satisfaction between employees in urban and rural branches. A randomly selected sample was made of bank employees in the Republic of Serbia. To date, the relative job satisfaction of bank employees in urban and rural areas has not been investigated, and for this reason, it is important to analyze the different facets of job satisfaction such as salaries, cooperation with closest associates, promotion, remuneration policy, cooperation and relationship with superiors, and the nature of the job. The bank employees' satisfaction with their salaries has a major influence on total job satisfaction. Perceptions of teamwork effectiveness and its relationship to overall job satisfaction were analyzed. Team quality has the greatest influence on the bank employees' job satisfaction. These insights can offer guidance for future action on building the quality of the employee–organization relationship.

Keywords: bank; job satisfaction; team; rural; urban; the Republic of Serbia

---

1. Introduction

Human resource management plays an increasingly important role in promoting the success and business sustainability of contemporary organizations. The most challenged seem to be employees in complex organizational systems. Bank employees have to reach set targets and produce great results while adhering to strict business procedures and tight deadlines. If the banks have not made the necessary investments in human resources, but have focused on cutting labor costs [1], it is necessary to determine the factors which impact most significantly upon job satisfaction, then identify activities to enhance it, thus contributing to the accomplishment of the goals of the organization as a whole. The banking sector has a dominant role in the overall economy of the Republic of Serbia, since it has developed substantially in the last ten years. Today there are twenty-one banks on the market, the majority of which are under foreign ownership, and the remainder owned by domestic companies (or by the Republic of Serbia). Of the foreign-owned banks, the largest share in total assets is owned by shareholders from Italy and Austria, followed by Greece and France, while their organizational network consists of 1627 business units and 23,055 employees [2]. The majority of these financial organizations are located in urban areas, but their branch network also extends to rural areas. Some 85% of the territory of Serbia can be considered rural, 55% of the population live in rural areas, and these areas generate 40% of GDP even though they continue to be burdened with high unemployment rates, depopulation, low economic activity, and a decline in natural resources [3].

In the rural areas, we find that the financing of agriculture is an important segment of everyday banking activities. However, agricultural producers mostly live in rural areas where banks have smaller branch offices. Direct financing for organic production is particularly significant in poorer countries and 'new' EU Member States because it is seen as a way of boosting producers' earnings, which are still relatively modest in these countries [4]. The state is striving to maximize its support through various incentives provided to encourage organic agriculture. In the last three years, the types of incentives the Republic of Serbia is providing for this kind of production include payment of plantation incentives per hectare, provision of fertilizers and support for cultivation, incentives for organic livestock production, and incentives for organic product certification. The commercial banking sector also organic farming through provision of agricultural loans, but no lines of credit specifically for organic production have yet been identified.

Bearing in mind the scarcity studies centered on determining the factors of employee commitment in Serbian banking institutions, the authors have focused on a deeper understanding of this particular phenomena, concentrating particularly on predictors of employee job satisfaction. To the best of our knowledge, there are no previous studies comparing the job satisfaction of banking staff in urban and rural areas in the Republic of Serbia. The survey included a sample of bank employees at entry level positions (junior and senior bank officers) in banks' in urban and rural areas in four banking institutions in the Republic of Serbia, which have branch offices in the capital city of Belgrade and in the rural areas. The research has three objectives; first, to investigate whether there were differences in job satisfaction of bank employees in banks' in urban and rural areas; second, to determine whether the bank employees' satisfaction with salary, cooperation with closest associates, possibility of promotion, remuneration policy, cooperation and good relationships with superiors, and the nature of the job had a statistically significant effect on the overall level of job satisfaction of bank employees in urban and rural areas; and third, to examine teamwork factors and its influence on total job satisfaction.

The paper is organized as follows; in the literature review, the authors review the recent findings on the various aspects of job satisfaction and teamwork. In the next sections, we present the methodology and the research results. Finally, the concluding observations summarize the research findings. In particular, we emphasize that our findings can serve as a starting point for the creation of further human resource strategies in the banking sector with a special focus on the motivation and retention of employees in urban and rural areas.

2. Literature Overview

The profitability of banks is influenced by the organizational commitment of employees. It is the responsibility of the bank's management to attract and retain dedicated staff that can contribute to the overall performance of the institution. Organizational commitment is the degree to which an employee identifies with a particular organization and its goals and wants to remain in the organization [5], as it represents the positive attitudes that an employee feels about the organization as a whole or certain of its members. The organizational commitment shows that employees feel a deep attachment to the organization. This attachment is much broader than ordinary satisfaction and includes the willingness of employees to subordinate their own interests to those of their organization [6], and it is directly related to the bank's profitability and its competitive position on the market [7]. The basic dimensions of organizational commitment are commitment objects, commitment basics, and commitment effects. Commitment objects can be different, because people can relate to different entities in their organizations—to the organization itself, but also to specific individuals or groups within it. In an attempt to categorize some of these objects, individual authors have distinguished between those whose commitment is concentrated at lower organizational levels—the task force, and those focused at higher levels—leadership and the organization as a whole [8]. The basis of commitment refers to the sources or causes of commitment. It answers the question why someone is dedicated to their organization. Research has confirmed that there are three groups of causes of commitment: Affective, continuous, and normative commitment [9]. Affective commitment is defined as emotional attachment to an organization, identification with the organization, and involvement in the organization. An employee can be dedicated to an organization because they share its goals and interests and see the best way to achieve individual goals in achieving organizational goals. Continuous commitment is related to the perceived cost of leaving the organization. Normative commitment is linked to an individual's sense of obligation to the organization, and may reflect his or her inner values and family upbringing. Influential individuals in an employee environment can exert strong pressure on an employee to feel moral responsibility to the organization. A high degree of organizational commitment can be linked to a low rate of absenteeism and voluntary leave. Organizational commitment is linked to a high degree of readiness for community and sacrifice, which can be very valuable for an organization in crisis. In addition, commitment has been found to produce certain positive consequences for the individual themselves. On the one hand, it leads to a better career, and thus greater rewards and higher quality jobs, and on the other, dedicated employees are more satisfied with their jobs, which also has a positive impact on their private lives. In the realization of a business strategy, job satisfaction plays an important role. If the employee is satisfied with his/her work, he/she will be motivated to give the maximum in achieving organizational goals, which will directly affect the performance of the organization [10]. Research related particularly to job satisfaction has tackled the positive influence of family-friendly policies of organizations [11] and the employees' self-determination (autonomy, competence, and relatedness) [12] on employee job satisfaction.

Edmondson contends that in conditions of increased uncertainty and deteriorating job security "teams are in a position to provide an important source of psychological safety for individuals at work" [13] (p. 380). The team is defined as a work group formed for a specific task, the members of which have complementary skills and are committed to common goals and tasks [6]. The work team is a set of employees with various character traits, skills, and experience, put together to implement defined goals, project decisions, and specific organizational and business problems in the company. An effective and efficient work team will successfully realize the defined goals with the minimum expenditure in time and other resources. The effectiveness of a team is influenced by numerous human, organizational, and work-related factors [14]. Human factors include: Job satisfaction; mutual trust and team spirit; good communication; minimization of unresolved conflicts; and competition for dominance of the team. Effectively solving threats to the team and within the team will create the impression that the workplace team members are secure. Organizational factors include: Building a stable organization and job security; support of leadership in teamwork; appropriate awards and recognitions for completed assignments; determining stable goals and priorities. Factors related to work tasks include: Setting clear objectives, giving precise instructions and design plans; appropriate professional guidance; job independence and demanding work assignments; appointment of experienced and qualified team members; encouragement of teamwork; and ensuring that the team's work is recognized within the organization. Various types of teams are commonly mentioned in the literature: Quality teams, work teams, problem solving teams, management teams, product development teams, and virtual teams [15]. Each team goes through certain stages of development: Forming; storming; norming; and performing and adjourning [14]. Forming is the initial phase in which team members meet, evaluate each other, begin to define team norms, and compare their expectations with what they are potentially waiting for. Storming is a phase in the development of a team characterized by conflicts and disagreements. During norming, the team members start to get used to their roles in the team, group cohesion grows, and positive team norms develop. Performing is a team development phase in which performance improves as the team matures and becomes an effective and functional unit. During this phase, team members become extremely loyal to each other and feel mutual responsibility for the success or failure of the team. The adjourning phase comes after the completion of the organizational assignment, but only in teams that are formed for special assignments and a limited time period, and need to be integrated in other teams after their work is completed. Adams et al. [16] identify clearly defined goals, common purpose, role clarity, psychological safety, mature communication, productive conflict resolution and accountable interdependence as the features of an effective team. The five most significant associations with team performance were professionally stimulating and challenging work environments, the opportunity for accomplishment and recognition, the ability to resolve conflict and problems, clearly defined organizational objectives, and job skills and expertise of the team members [17]. Ross, Jones, and Adams [18] developed a mathematical model as an empirical function of performance, behavior, attitude, team member style, and corporate culture. In order to have a high-performing team, the following minimum requirements need to be put in place: "The existence of interdependence among team members, the ability of members to resolve conflict, an environment that allows members to take risks and the development of a clear team purpose set by the team" [16] (p. 11).

One of the first theoretical approaches to job satisfaction is found in the Locke value theory [19,20]. According to this theory, job satisfaction is "a pleasurable or positive emotional state resulting from the appraisal of one's job or job experiences" [19] (p. 1304). Locke defined job satisfaction as a positive emotional feeling, a result of the individual's evaluation of his job or his experience of work comparing between what he expects from his job and what he actually gets from it. Locke [21] identifies three factors in action in any job appraisal process: Perception of a facet of the job, the value system, and the evaluation of the relationship between perception and value system. One employee may be strongly influenced by the physical aspects of the job, whilst another may be influenced by the challenge and variation inherent in the job [19]. Baron and Greenberg [20] point out that the degree of emphasis placed on values suggests that job satisfaction may arise from such factors. The Job Descriptive Index that has been designed and copyrighted by Smith, Kendall, and Hulin [22] encompasses a 72-item adjective checklist questionnaire for measuring job satisfaction, as they considered job satisfaction to be comprised of five dimensions: Pay, promotions, co-workers, supervision, and work. The index has been widely used in scientific research [23]. The factor structure of job satisfaction 30-question multiple-choice format, based on the copyrighted Job Descriptive Index [22], was utilized to assess the job satisfaction of public accountants [24]. His findings indicated that male certified public accountants were more satisfied with the promotion dimension of job satisfaction than female, that the pay dimension of job satisfaction was positively associated with tenure with firm, and that the measure of job satisfaction was highly negatively associated with intent to turnover [25].

The research findings show that scholars have chosen to research various factors of job satisfaction prevalent in the particular business environment of their country. The findings on job satisfaction of bank employees in North Cyprus have shown that policies, practices, compensation, and advancement as extrinsic aspects of the job are greater sources of dissatisfaction [26]. The study on personal and organizational determinants of job satisfaction shows that salary has the strongest impact on the level of job satisfaction, while relationships with co-workers have the weakest impact on it. The opportunities for promotion and recognition (and rewards) have proved to be major sources of dissatisfaction [27]. Non-financial rewards were found to influence job satisfaction in commercial bank branches [28]. Research on job satisfaction in the Balearic Islands has shown that the status of the worker within the hierarchic structure determines their job satisfaction, while the bank managers are satisfied with the tasks accomplished and salaries, but are least satisfied in the area of human relationships [29]. On the other hand, an analysis of the job satisfaction of Lebanese bank employees found no significant relationship between job satisfaction and the employees' performance. The results also showed that males are more satisfied with the work itself, promotion, and co-workers, while female employees were found to gain more satisfaction from pay, those with lower educational qualifications were least satisfied, and performance at work increased with tenure [30]. High–low job satisfaction may not be associated with the employees' performance [31]. The job satisfaction of public sector bank employees in India was significantly higher than that in the private sector, although the satisfaction regarding salary, compensation, benefits, and promotion were significantly higher among the private sector bank employees [32]. Employees with a high level of satisfaction are less frequently absent from work, they are less likely to go to another organization, are more productive, more loyal to their organization, and more satisfied with their work [33]. The findings on the banking sector employees of Pakistan have shown a significant positive correlation between job satisfaction and organizational commitment [34]. The banking sector employees of Bangladesh rated social status, supportive colleagues, and feeling secure about the job as the top three best reasons for working in banks [35]. The findings on the organizational commitment dimension in Islamic banks show that male employees' commitment level is higher and that the older employees demonstrate a higher level of commitment [36]. In the sample of Spanish banks, the findings confirm that the greater the proportion of women managers in the organization, the higher employee productivity, while female representation at the board contributes to women being promoted [37]. The job satisfaction of Kuwaiti women bank employees is affected by educational background and educational level, and only women bank employees with degrees in finance are satisfied with their jobs in terms of pay and security [38]. British employees are less satisfied with their pay, but more satisfied with their job achievements, while married employees have lower job satisfaction levels than the unmarried [39]. The findings on worker's sense of job satisfaction in US commercial banks show that satisfaction can be attributed to workplace learning opportunities, as informal learning, incidental learning, and formal learning predict overall job satisfaction [40]. Pay and promotion is the most influential factor on the job satisfaction of banking sector employees in India [41]. The findings on government-linked bank institution employees in Malaysia show that job satisfaction with extrinsic factors is slightly higher compared to satisfaction with intrinsic factors [42]. The highest predictor of job satisfaction of Nepalese commercial banks' employees is salary, followed by cooperation among employees, working environment, and training and promotion [43]. The findings on South Koreas companies show that supervisor humility correlates positively to employee job satisfaction [44].

3. Materials and Methods

The research in this study was conducted by means of an anonymous survey that included a sample of bank employees at entry level positions (junior and senior bank officers performing supervisory and operative tasks) in both banks' in urban and rural areas in four banking institutions in the Republic of Serbia, which have branch offices in the capital city of Belgrade and in the rural areas. Confidentiality was ensured and the purpose of the research was explained. The questionnaires were distributed in May of 2019 and collected over a five-month period. Of the 500 questionnaires distributed, in the city of Belgrade, 478 were returned usable (a response rate of 95%), and of the 200 questionnaires sent to the branch offices in rural areas 189 were returned usable (a response rate of 94.5%). Therefore, the total number of usable surveys was 667.

The data was analyzed using the paired samples test, correlation, factor analysis, linear and multiple regression [45] from the statistical software package SPSS 19.0. The level of significance was set at $p = 0.05$.

The first part of the survey refers to socio-economic indicators (gender, age, years of service, and qualifications). In the second part of the survey, job satisfaction was measured, using an adjusted job satisfaction questionnaire [24]. Five facets of the Job Descriptive Index (work, pay, promotions, supervision, and co-workers) were selected. The considered facets of the Job Descriptive Index are salary (how satisfied the employee is with the level of employee salaries comparable to other salaries in the banking sector), cooperation with closest associates (how satisfied the employee is with his co-workers), the possibility of promotion (how satisfied the employee is with the opportunities for promotions), cooperation and good relationships with superiors (how satisfied the employee is with the supervisor's behavior), and nature of the job (e.g., job content, how satisfied the employee is with his/her job in general). The sixth facet (remuneration policy—how satisfied the employee is with the policy that encourages, motivates, and retains its employees) was included in the research as an important parameter of bank remuneration and benefits. Bank employees are usually expected to be motivated by salary, which is often higher than in other sectors. The employees were also asked to rate their loyalty to the bank by rating their turnover intentions (Likert scale 1–5) [25]. The measurement of job satisfaction was carried out using a model based on six variables: c_1—satisfaction with salaries, c_2—satisfaction with cooperation with closest associates, c_3—satisfaction with the possibility of promotion up the hierarchical structure, c_4—satisfaction with the remuneration policy, c_5—satisfaction with cooperation and relationships with superiors, and c_6—satisfaction with the nature of the job. A five-point Likert scale was used for the measurement of job satisfaction: 1—very dissatisfied, 2—dissatisfied, 3—not quite sure, 4—quite satisfied, 5—very satisfied. "If a measure is facet-based, overall job satisfaction is typically defined as a sum of the facets" [46] (p. 397). Therefore, the total job satisfaction (TJS) was calculated on the basis of the six different facets, therefore the satisfaction of the individual employee was obtained using the formula:

$$TJS_i = \frac{(c_{1i} + c_{2i} + c_{3i} + c_{4i} + c_{5i} + c_{6i})}{6} \tag{1}$$

where: $i = 1, \ldots, n$; while n represents the total number of employees.

This coefficient shows the average value of total job satisfaction on the basis of the scores of six variables. For the analysis of total job satisfaction for the banking sector, the total job satisfaction coefficient (TJS) is calculated as:

$$TJS = \frac{(c_1 + c_2 + c_3 + c_4 + c_5 + c_6)}{6}, \tag{2}$$

whereas:

$$c_j = \frac{\sum_{i=1}^{n} c_{ij}}{n}, \ (j = 1, \ldots, 6), \tag{3}$$

The first research objective was to investigate whether there were differences in job satisfaction of bank employees on the lower positions in banks' in urban and rural areas by means of Paired Samples Test.

The second objective of the research was to analyze the importance of every facet of job satisfaction and establish whether the job satisfaction parameters such as bank employees' satisfaction with salary, cooperation with closest associates, possibility of promotion, remuneration policy, cooperation and good relationships with superiors, and the nature of the job had a statistically significant effect on total job satisfaction. For the continuation of our second research objective the authors tested the relationship between the job satisfaction variables and total job satisfaction in order to find a linear regression equation that best predicts the dependent variable as a linear function of the independent variables, therefore linear regression was applied. To address the potential endogeneity problem in the linear regression model, the authors have applied two-stage least squares (2SLS) regression analysis.

The third research objective was to examine whether the teamwork factors influence total job satisfaction. The respondents were asked to rate various teamwork perception on a five-point interval scale (1 is the lowest grade and 5 is the highest). The teamwork questionnaire was adapted and conceptualized according to the methodology proposed by Bateman, Wilson, and Bingham [47]. The authors have then examined the perceived teamwork attitudes of the bank employees in urban and rural areas by factor analysis and the correlation of the obtained factors with total job satisfaction. The final purpose of the research on the teamwork and job satisfaction was to find a multiple regression equation that best predicts the dependent variable, therefore multiple regression was applied.

4. Results

The structure of the sample by gender, age, level of education, and the number of years spent in the organization is shown in Table 1. Of bank employees in the urban area, 56.3% are female and 43.7% male. The situation is similar in the rural area, with 54.5% female employees and 45.5% male employees. The average age of the employees in banks in the urban and rural area is in the age range of 35–50 years (respectively 65.9% and 66.7%). Regarding the level of education, the largest number of the employees in banks in the urban and rural area (respectively 69.9% and 68.8%) has a professional college diploma and has spent 6–15 years working in the bank (respectively 47.5% and 47.1%). The obtained descriptive socio-demographic characteristics in relation to both samples show that there are no significant deviations, and the obtained results can be considered relevant. In both samples, only 3.5% of the bank employees in the urban area and 3.7% of the bank employees in the rural area strongly agreed or agreed that they have the intention to leave the bank. The majority of bank employees are ready to remain with their current employer.

Table 1. Sample structure according to socio-economic indicators.

		Urban Bank Employees		Rural Bank Employees	
		Frequency	Percentage	Frequency	Percentage
Gender	Male	209	43.7	86	45.5
	Female	269	56.3	103	54.5
	Total	478	100.0	189	100.0
Age	Up to 35	126	26.4	52	27.5
	35–55	315	65.9	126	66.7
	Over 55	37	7.7	11	5.8
	Total	478	100.0	189	100.0
		Average value: 43.47 Std. Dev = 5.66 Coeff. Var = 13%		Average value: 43.16 Std. Dev = 5.50 Coeff. Var = 12.74%	
Education*	A	55	11.5	28	14.8
	B	334	69.9	130	68.8
	C	56	11.7	20	10.6
	D	33	6.9	11	5.8
	Total	478	100.0	189	100.0
The number of years spent in the organization	Less than 5	141	29.5	66	34.9
	6–15	227	47.5	89	47.1
	16–20	79	16.5	20	10.6
	Over 20	31	6.5	14	7.4
	Total	478	100.0	189	100.0
		Average value: 10.73 Std. Dev = 4.91 Coeff.Var = 45.76%		Average value: 10.56 Std. Dev = 4.94 Coeff.Var = 46.78%	

* A. High school diploma. B. Professional college diploma. C. Higher educational diploma (BA). D. Post–graduate qualification (MA; PhD). Source: Author's calculation.

The results presented in Table 2 indicate that the average level of satisfaction with salary (c_1) indicated by our respondents from urban areas is 3.03 (Std. Dev. 0.91), while those from rural areas ranked their satisfaction at 4.04 (Std. Dev. 0.79). Using the Paired Samples Test, the scores for the related units were tested to see whether the mean difference between the paired observations on a particular outcome was statistically significant. By comparing the values of average score of satisfaction with salary (Mean −1.101; Std. Dev. 1.278; Std. Error Mean 0.093 and $p = 0.000$), we can conclude that there is a statistically significant difference between the two groups in their satisfaction with the salary. Employees in banks in rural areas are more satisfied with their earnings, possibly because the cost of living in rural areas is lower than in the city, most of them grow some of their own food, and usually have much lower housing costs. They also have fewer opportunities of employment as the majority of companies are headquartered in the urban areas. The urban employees have higher living costs (housing, food, utilities, transportation costs, etc.) and are therefore more dissatisfied with their pay than their colleagues living and working in rural areas.

Table 2. Descriptive indicators for job satisfaction parameters.

Attitudes Related to Job Satisfaction	Scores	Urban Bank Employees		Rural Bank Employees		Total Bank Employees		p
		n	%	n	%	n	%	
Satisfaction with the salary	1	31	6.5	1	0.5	32	4.80	
	2	86	18.0	2	1.1	88	13.19	
	3	210	43.9	44	23.3	254	38.08	
	4	139	29.1	84	44.4	223	33.43	0.000
	5	12	2.5	58	30.7	70	10.49	
		Mean = 3.03 Std. Dev. = 0.91		Mean = 4.04 Std. Dev.= 0.79		Mean = 3.32 Std. Dev. 0.99		
Satisfaction with the cooperation with the closest associates	1	10	2.1	4	2.1	14	2.10	
	2	50	10.5	19	10.1	69	10.34	
	3	208	43.5	86	45.5	294	44.08	
	4	179	37.4	70	37.0	249	37.33	0.591
	5	31	6.5	10	5.3	41	6.15	
		Mean = 3.36 Std. Dev. 0.83		Mean = 3.33 Std. Dev. 0.81		Mean = 3.35 Std. Dev. 0.83		
Satisfaction with the possibility of promotion on the hierarchical level	1	16	3.3	22	11.6	38	5.70	
	2	32	6.7	48	25.4	80	11.99	
	3	166	34.7	91	48.1	257	38.53	
	4	197	41.2	24	12.7	221	33.13	0.000
	5	67	14.0	4	2.1	71	10.64	
		Mean = 3.56 Std. Dev. 0.93		Mean = 2.68 Std. Dev. = 0.91		Mean = 3.31 Std. Dev. = 1.01		
Satisfaction with the remuneration policy	1	38	7.9	0	0	38	5.70	
	2	130	27.2	7	3.7	137	20.54	
	3	189	39.5	55	29.1	244	36.58	
	4	100	20.9	91	48.1	191	28.64	0.000
	5	21	4.4	36	19.0	57	8.55	
		Mean = 2.87 Std. Dev. = 0.98		Mean = 3.83 Std. Dev. = 0.78		Mean = 3.14 Std. Dev. = 1.02		
Satisfaction with the cooperation and relationships with the superiors	1	14	2.9	4	2.1	18	2.70	
	2	62	13.0	34	18.0	96	14.39	
	3	205	42.9	73	38.6	278	41.68	
	4	168	35.1	67	35.4	235	35.23	1.000
	5	29	6.1	11	5.8	40	6.00	
		Mean = 3.28 Std. Dev. = 0.87		Mean = 3.25 Std. Dev. = 0.89		Mean = 3.27 Std. Dev. = 0.88		
Satisfaction with the nature of the job	1	21	4.4	3	1,6	24	3.60	
	2	104	21.8	3	1,6	107	16.04	
	3	226	47.3	36	19.0	262	39.28	
	4	110	23.0	91	48.1	201	30.13	0.000
	5	17	3.6	56	29.6	73	10.94	
		Mean = 3.00 Std. Dev. = 0.88		Mean = 4.03 Std. Dev. = 0.83		Mean = 3.29 Std. Dev. = 0.98		

Source: Author's calculation.

For urban bank employees, the average score for satisfaction with cooperation with their closest associates (c_2) was 3.36 (St. Dev. 0.83), while for staff in rural areas. this average score was 3.33 (Std. Dev. 0.81). The relative satisfaction of the respondents was again assessed using the Paired Samples Test. Comparing level of satisfaction with cooperation (Mean −0.048; Std. Dev. 1.217; Std. Error Mean 0.089 and $p = 0.591 > 0.05$), we can conclude that there was not a statistically significant difference between the groups on this issue. For the sample as a whole, the average score was almost identical, amounting to 3.32 (Std. Dev. = 0.989). In order to achieve the set organizational goals, it is necessary to nurture relationships of full trust between the members of a team and encourage cooperation.

The average score of the satisfaction with the possibility of promotion up the hierarchical ladder (c_0) among employees in urban banks was 3.56 (Std. Dev. 0.93), and in rural areas it was 2.68 (Std. Dev. 0.91). Using the Paired Samples Test once again (Mean 0.608; Std. Dev. 1.412; Std. Error Mean 0.103 and $p = 0.000$), we can conclude that there was a statistically significant difference. These results suggest that bank employees in the metropolitan area have greater opportunities for promotion as they are closer to bank headquarters and there are more branches spread throughout the whole city. Contact with top management is more frequent, their commitment can be noticed, and they can move more quickly to more senior management positions.

The average score of satisfaction with the remuneration policy (c_4) among bank employees in the urban area was 2.87 (Std. Dev. 0.98), and in rural areas, 3.83 (Std. Dev. 0.78). The results of the Paired Samples Test (Mean −1.079; Std. Dev. 1.284; Std. Error Mean 0.093 and $p = 0.000$), show that there is a statistically significant difference between the mean scores of satisfaction with the incentive reward system. The significantly higher value of the satisfaction with the remuneration policy among employees in banks in rural areas suggests that employees in urban banks believe their employers do not monitor, recognize, or incentivize good results well enough. This dissatisfaction with the incentive reward system is also related to their dissatisfaction with pay. In contrast, employees in rural areas expect less, are aware that they have less opportunity take on greater responsibility, and try to use every opportunity to offer the best service to clients because they live in small communities where everyone knows each other. They are therefore satisfied with more modest rewards.

Among the respondents employed in urban banks, the average score of satisfaction with cooperation and relationships with superiors (c_5) is 3.28 (Std. Dev.0.87), and in rural areas it is 3.25 (Std. Dev. 0.89). Using the Paired Samples Test, the results show (Mean 0.000; Std. Dev. 1.288; Std. Error Mean 0.094 and $p = 1.000 > 0.05$), that there is not a statistically significant difference between the average score of two related groups. In the entire sample of respondents, the average score is 3.27 (Std. Dev. = 0.088), which may be the result of a distinct formalization of working relationships through a clear hierarchical structure.

The satisfaction with the nature of the job (c_6) among respondents employed in banks in urban areas was rated with an average score of 3.00 (Std. Dev. 0.88), while in banks in rural areas this score is 4.03 (Std. Dev. 0.83). Comparison using the Paired Samples Test (Mean −1.190; Std. Dev. 1.311; Std. Error Mean 0.095 and $p = 0.000$), shows that there is a statistically significant difference between the job satisfaction scores. The content of the job is the determining factor when establishing an employment relationship. Although the nature of the job in the groups of respondents among whom the survey was conducted is not fundamentally different, we see that bank employees in rural areas are more satisfied with the nature of the job.

Linear regression was used to examine the influence of the job satisfaction facets on total job satisfaction (TJS) as shown in Table 3. In Table 3, column 2 shows the regression equation. The coefficient of determination (r^2) of all observed job satisfaction predictors to total job satisfaction indicates that the relationship is positive. By using the analysis of the variance (ANOVA), the variance of the observed data was tested to determine if a regression model can be applied. The empirical level of the F-distribution was higher than the critical value of F-significance in all parameters, therefore the regression equation can be accepted as a predictor of total job satisfaction. The T-test determined the significance of the regression coefficient in predicting the movement of TJS. Compared to the critical value t, the absolute value of t-statistics is higher, which means that all the job satisfaction parameters are statistically significant as predictors in determining total job satisfaction.

Table 3. Regression of the impact of job satisfaction variables on total job satisfaction (TJS).

N = 667	Regression Equation	R	r^2	Standard Error	F-Distribution
c_1	$1.329 + 0.5880 \times c_1$	0.8160	0.6658	0.413	1325.006
c_2	$1.154 + 0.6343 \times c_2$	0.7360	0.5417	0.483	786.027
c_3	$1.824 + 0.4398 \times c_3$	0.6198	0.3842	0.560	414.813
c_4	$1.550 + 0.5513 \times c_4$	0.7899	0.6239	0.438	1103.331
c_5	$1.339 + 0.5927 \times c_5$	0.7293	0.5319	0.488	755.556
c_6	$1.348 + 0.5876 \times c_6$	0.8082	0.6533	0.420	1252.899

c_1—satisfaction with the salary. c_2—satisfaction with the cooperation with the closest associates. c_3—satisfaction with the possibility of promotion on the hierarchical level. c_4—satisfaction with the remuneration policy. c_5—satisfaction with the cooperation and relationships with the superiors. c_6—satisfaction with the nature of the job. Source: Author's calculation.

The influence of respondents' satisfaction with earnings—c_1 and total satisfaction was examined using linear regression. Based on the coefficient of determination, it was concluded that 66.58% of the variance in TJS can be explained by the satisfaction of respondents with earnings (c_1). TJS increases on average by 0.5880 when c_1 increases by one degree. The same method suggests that 54.17% of the variance in TJS consumption can be accounted for by satisfaction with cooperation with closest associates (c_2). TJS increases on average by 0.6343 when c_2 increases by one degree. Variations in the predictor c_3 account for 38.42% of the TJS change, and TJS increases on average by 0.4398 when satisfaction with the possibility of promotion increases by one degree. It can be concluded that 62.39% of the change in TJS is explained by c_4 variations. TJS increases on average by 0.5513 when satisfaction with the remuneration policy increases by one degree. The variations in c_5 contribute to the 53.19% of the TJS change. TJS increases on average by 0.5927 when satisfaction with cooperation and good relationships with superiors increases by one degree. Based on the coefficient of determination r^2, it can be concluded that 65.33% of the TJS change is explained by the c_6 variations. TJS increases on average by 0.5876 when the satisfaction with nature of the job increases by one degree.

In address to potential endogeneity in the model, the authors have decided to apply Two-stage least squares (2SLS) regression analysis (Table 4). The authors have identified an instrumental variable which would be suitable for the X in the model. The employees' loyalty to the bank (employees' intent to turnover) has been employed as the instrumental variable used to determine an exogenous part of the variability from the endogenous predictor [48]. The employees without intent to turnover are more attached to the organization, and their job satisfaction is higher. The previous results have revealed significant associations between the job satisfaction and intent to turnover [25]. The study of relationship of satisfaction and loyalty has been shown that the dependence between employee satisfaction and loyalty is strong [49]. All the assumptions for the instrumental variable have been satisfied [50]. The relevance assumption (the instrument "employees' loyalty to the bank" has a causal effect on X), the exclusion restriction (the instrument "employees' loyalty to the bank" affects the outcome Y only through X), the exchangeability assumption (the instrument "employees' loyalty to the bank" does not share common causes with the outcome Y). Thus, we have inserted an instrumental variable (Z = "employees' loyalty to the bank") in our base specification.

All the assumptions of the two-stage least squares (2SLS) regression have been met. The two-stage least squares (2SLS) regression analysis has confirmed that all job satisfaction predictors are statistically significant in determining the total job satisfaction (Table 4). It is interesting that the values of the regression coefficient are higher than in the first regression model and values of r^2 differ in both models. The main conclusions stay the same, thus all the predictors are statistically significant in the model. The influence of respondents' satisfaction with earnings—c_1 and total satisfaction was examined by the statistical method of linear regression. Based on the coefficient of determination, it is concluded that 31.6% of the variance in TJS consumption is explained by the satisfaction of respondents with earnings

(c_1). TJS increases on average by 0.685 when c_1 increases by one degree. Based on the coefficient of determination, it is concluded that 16.20% of the variance in TJS consumption is explained by the satisfaction with cooperation with closest associates (c_2). TJS increases on average by 2.458 when c_2 increases by one degree. In regard to the predictor c_3, it can be concluded that 24.3% of the TJS change is explained by the c_3 variations. TJS increases on average by 1.059 when the satisfaction with the possibility of promotion increases by one degree. It can be concluded that 38.8% of the TJS change is explained by the c_4 variations. TJS increases on average by 0.749 when satisfaction with the remuneration policy increases by one degree. The variations in c_5 contribute to the 24% of the TJS change. TJS increases on average by 1.384 when the satisfaction with the cooperation and good relationships with superiors increases by one degree. Based on the coefficient of determination, it can be concluded that 36.7% of the TJS change is explained by the c_6 variations. TJS increases on average by 0.883 when the satisfaction with nature of the job increases by one degree.

Table 4. Two-stage least squares (2SLS) regression analysis of the impact of job satisfaction variables on total job satisfaction (TJS).

N = 667	Regression Equation	R	r^2	Standard Error	F-Distribution
c_1	$1.007 + 0.685 \times c_1$	0.562	0.316	0.424	100.097
c_2	$-4.955 + 2.458 \times c_2$	0.403	0.162	1.585	7.158
c_3	$-0.227 + 1.059 \times c_3$	0.493	0.243	0.838	25.630
c_4	$0.930 + 0.749 \times c_4$	0.623	0.388	0.482	77.447
c_5	$-1.253 + 1.384 \times c_5$	0.490	0.240	0.849	24.922
c_6	$0.375 + 0.883 \times c_6$	0.606	0.367	0.511	68.872

c_1—satisfaction with the salary. c_2—satisfaction with the cooperation with the closest associates. c_3—satisfaction with the possibility of promotion on the hierarchical level. c_4—satisfaction with the remuneration policy. c_5—satisfaction with the cooperation and relationships with the superiors. c_6—satisfaction with the nature of the job. Source: Author's calculation.

Teamwork and Job Satisfaction

The results of descriptive statistics in relation to the parameters of teamwork are shown in Table 5. The average scores for teamwork attitudes in the entire sample (N = 667) range from 3.33 to 3.89. "The team members' complaints are regularly monitored, and lessons are systematically applied in the work" (M = 3.89) received the highest rating from our respondents. With this in mind, all team members' complaints are taken into consideration to prevent further dissatisfaction and we can conclude that the team members' satisfaction is very important to the team. The second highest score, "Team goals are clearly defined" (M = 3.81), reflects the team members' opinion that team leaders play a significant role in teamwork and have the ability to acquaint all team members with the teams' goals. In this way, the employees can identify their individual goals within the organizational ones, which can contribute to greater synergy in the team. The team members follow and respect organizational rules in resolving customer complaints (M = 3.80). All team members are appropriately trained and competent to perform their job professionally (M = 3.79) indicates the importance bank staff give to the need for training and improving their competence in teamwork performance. They believe that it is important to have standards that define the teamwork (3.78). "Team members are flexible and willing to perform other tasks within the team" (3.33), "Every innovation in teamwork is valued and rewarded" (3.47), and "Problems related to business/clients are quickly discovered" (3.45) have the lowest average scores. We can conclude that team members are not ready to do other tasks that are not defined by the scope of the teamwork, which is closely related to the fact that employees believe their additional engagement in the work of the team is not sufficiently appreciated or that their tight schedules do not allow them to pursue additional tasks. They are more focused only on their own role in the team. Accordingly, they are not motivated to propose innovation as they are not adequately rewarded. The problems with the timely detection of problems in work with clients reflect the long and complex procedures that the bank employees have to follow, involving moreover other departments in the bank or credit committee, before the problem is detected.

Table 5. Descriptive indicators for teamwork parameters.

	Mean	Std. Dev.
A clearly defined team membership is present	3.6702	0.93917
Team goals are clearly defined	3.8081	0.79359
Team members' roles are clearly defined	3.6222	0.88599
Effective team communication is present	3.6297	0.92312
I have a sense of team members' value	3.7721	0.87109
Other organizational units of the company appreciate the team in which the individual works	3.6822	0.90595
I have a sense of pride of belonging to the team	3.7406	0.87175
Each member of the team maximally contributes to the work of the team	3.7646	0.85604
Team members are appropriately trained and competent to perform their duties professionally	3.7871	0.82725
Team members are appropriately trained in job-related procedures	3.6642	0.85140
There is a formal system for recognizing training needs and additional education	3.6162	0.91840
The needs for education and training are systematically identified	3.6207	0.98566
Based on the needs of the employees, additional training is provided	3.6432	0.99479
Team members are trained to perform a number of various tasks within the team	3.6762	0.86823
Team members are flexible and willing to execute additional work within the team	3.3268	0.94022
Team members highly appreciate additional education	3.6672	0.90884
Team members are encouraged to try new methods of work	3.6957	0.88125
The team is involved in new projects related to its products/services from the very beginning	3.6237	0.89085
Each innovation in teamwork is appreciated and rewarded	3.4708	0.93816
Problems related to business/clients are quickly discovered	3.4798	0.92875
Detected problems are quickly solved	3.6222	0.92089
Problem solving is perceived as learning and team development	3.6672	0.90221
Team members often propose innovations in work	3.6417	0.92298
Team members readily accept innovation in work	3.7001	0.90048
Team members are familiar with the needs of their clients	3.6897	0.83176
It has been clearly defined who the clients of each team are	3.6642	0.85316
The work standards within the team are clearly defined	3.7841	0.86898
Standards of teamwork are regularly monitored	3.6567	0.90409
Feedback information on teamwork monitoring is obtained on a regular basis	3.5787	0.92075
There are quantitative standards of teamwork efficiency that are followed	3.6132	0.87697
Team adheres to organizational standards for addressing customer complaints	3.8006	0.84406
Team members' complaints are regularly reviewed, and lessons are systematically applied in further work	3.8861	0.82867

Source: Author's calculation.

As a next step, by means of factor analysis, the authors wanted to analyze whether there is a set of factors that can explain the interrelationships among those variables for a collection of observed teamwork variables. The factor structure matrix presented in Table 6 contains factor loadings that represent the correlation coefficients between the extracted factors and the variables, and indicate the importance of each variable for a single factor [51,52].

The data were processed in the statistical package SPSS for Windows, version 22. In order to examine the latent structure of perceived teamwork attitudes, the factor analysis with the principal component method was applied. As the Kaiser-Meyer-Olkin measure of sampling adequacy was satisfied (KMO = 0.976), Bartlett's test of sphericity was significant (χ^2 = 17997.126, $p < 0.000$), therefore the factor analysis was conducted. Using the Cattell scree criterion, 3 factors were retained. The factors were rotated with Varimax rotation, the factor scores were determined, and we calculated the Cronbach's reliability coefficient for each factor. The Cronbach's reliability coefficients were: 0.949, 0.954, and 0.954. Taking into account the saturation shown in the circuit matrix (Table 7), the obtained factors are grouped into 3 units. The first factor accounts for 54.29% of the variance in the model (11 items), the second factor for 7.96% of variance (11 items) and the third factor for 4.39 % of variance (10 items). The first factor shall be labeled Workplace learning, and within that factor four statements with the highest factor loadings (0.802, 0.798, 0.796, and 0.795) most accurately describe it. The squares of the indicated correlation coefficients represent the variance proportions of certain variables that are attributed to the effect of a given factor. The second factor shall be labeled Team quality, and within it four factors with the highest factor loadings (0.838, 0.799, 0.762, and 0.669) accurately describe it. The third factor shall be labeled Team supervision, and within that factor four statements have the highest factor loadings (from 0.758 to 0.667).

Table 6. Exploratory factor analysis of perceived teamwork attitudes.

	Component		
	1	**2**	**3**
Team members are appropriately trained in job-related procedures	0.802		
The needs for education and training are systematically identified	0.798		
There is a formal system for recognizing training needs and additional education	0.796		
Team members highly appreciate additional education	0.795		
Based on the needs of the employees, additional training is provided	0.752		
Team members are encouraged to try new methods of work	0.694		
Team is involved in new projects related to its products/services from the very beginning	0.690		
Team members often propose innovations in work	0.674		
Problem solving is perceived as learning and team development	0.647		
Team members are trained to perform a number of various tasks within the team	0.603		
Team members are appropriately trained and competent to perform their duties professionally	0.554		
A clearly defined team membership is present		0.838	
Effective team communication is present		0.799	
I have sense of pride of belonging to the team		0.762	
Team members are flexible and willing to execute additional work within the team		0.669	
Problems related to business/clients are quickly discovered		0.640	
Team members' roles are clearly defined		0.625	
Each member of the team maximally contributes to the work of the team		0.589	
Team goals are clearly defined		0.583	
I have a sense of team members' value		0.581	
It has been clearly defined who the clients of each team are		0.580	
Other organizational units of the company appreciate the team in which the individual works		0.563	
There are quantitative standards of teamwork efficiency that are followed			0.758
Team members' complaints are regularly reviewed and lessons are systematically applied in further work			0.712
Feedback information on teamwork monitoring is obtained on a regular basis			0.706
The team adheres to organizational standards for addressing customer complaints			0.667
Standards of teamwork are regularly monitored			0.647
The work standards within the team are clearly defined			0.607
Team members are familiar with the needs of their clients			0.576
Each innovation in teamwork is appreciated and rewarded			0.524
Detected problems are quickly solved			0.523
Team members readily accept innovation in work			0.499
Initial Eigenvalues	17.375	2.549	1.406
Percentage of variation Cumulative	54.298	7.964	4.395
Cumulative Percentage	54.298	62.262	66.657
Cronbach's alphas	0.949	0.954	0.954

Extraction Method: Principal Component Analysis. Rotation Method: Varimax with Kaiser Normalization a. Rotation converged in 9 iterations. Source: Author's calculation.

Table 7. Correlation of obtained factors with total job satisfaction.

	Pearson Correlation	Total Job Satisfaction	Spearman's Correlation	Total Job Satisfaction
Total job satisfaction	Pearson Correlation Sig. (2-tailed) N	1.000 667	Correlation Coefficient Sig. (2-tailed) N	1.000 667
Workplace learning	Pearson Correlation Sig. (2-tailed) N	0.245 ** 0.000 667	Correlation Coefficient Sig. (2-tailed) N	0.255 ** 0.000 667
Team quality	Pearson Correlation Sig. (2-tailed) N	0.542 ** 0.000 667	Correlation Coefficient Sig. (2-tailed) N	0.512 ** 0.000 667
Team supervision	Pearson Correlation Sig. (2-tailed) N	0.258 ** 0.000 667	Correlation Coefficient Sig. (2-tailed) N	0.299 ** 0.000 667

** $p < 0.01$. * $p < 0.05$. Source: Author's calculation.

In order to establish whether there is a relationship between the factor scores and total job satisfaction and if so, how strong that relationship is, correlation analysis was applied. The correlation analysis further determined the direction of the relationship between the variables. The correlation between all factor scores and job satisfaction was statistically significant ($p < 0.05$).

With the aim of identifying whether the teamwork factors influence total job satisfaction, the data were analyzed using a multiple regression procedure, while the set of factors obtained in the factor analysis for each sample group were used as predictors. By performing the multiple regression, the authors wanted to investigate and explain the relationship of the independent variables to the dependent variables if those relationships prove to be linear [46,53,54]. Additionally, the authors wanted to investigate whether the regression method was appropriate for calculating higher-order factor scores in addition to primary [55]. A standard multiple linear regression was performed on the sample (Table 8). Before the regression was performed, the authors tested the assumption that the independent variables are not highly correlated with each other (r = 0.7 and above) [56]. Tolerance values were greater than 0.7, and VIF values not greater than 10, confirming that there is no multicollinearity. The model evaluation was then undertaken. Multiple regression was conducted to determine the best linear combination of all factors for predicting teamwork factors on total job satisfaction. The empirical level F of the distribution is 159.824 and indicates that the high value of F distribution is not accidental, and that the regression equation is applicable. This combination significantly predicted teamwork factors in total job satisfaction, with all three variables significantly contributing to the prediction. The beta values are as follows: The largest coefficient, indicating which independent variable has the greatest influence on the dependent variable, is found in Team quality, followed by Team supervision and Workplace learning. The adjusted R squared value was 0.417. This indicates that 42% of the variance in teamwork factors on total job satisfaction was explained by the model. TJS increases on average by 0.386 when the Team quality increases by one degree. TJS increases on average by 0.184 when Team supervision increases by one degree. TJS increases on average by 0.175 when Workplace learning increases by one degree. In our case, all three factors are statistically significant in predicting overall satisfaction. In the analyzed model, the multiple regression model is as follows:

Total satisfaction = 3.280 + 0.175 × *Workplace learning* + 0.386 × *Team quality* + 0.184 × *Team supervision*

Table 8. Multiple regression analysis model summary.

Model	Unstandardized Coefficients		Standardized Coefficients	t	Sig.	95.0% Confidence Interval for B	
	B	Std. Error	Beta			Lower Bound	Upper Bound
(Constant)	3.280	0.021		155.545	0.000	3.238	3.321
Workplace learning	0.175	0.021	0.245	8.285	0.000	0.133	0.216
Team quality	0.386	0.021	0.542	18.304	0.000	0.345	0.428
Team supervision	0.184	0.021	0.258	8.706	0.000	0.142	0.225

Source: Author's calculation.

The non-inclusion of the socio-demographic characteristics (gender, age, education, and the number of years spent in the organization) as confounding variables could cause an omitted variable bias. In order to prevent the omitted variable bias, which could have led to an overestimation or underestimation of the strengths of the effect, the confounding variables were included in the model. After including the confounding variables, the results indicated a statistical significance for all the confounding variables, with the exception of gender. The positive correlation was determined between the confounding variables of age and education with total satisfaction and the negative correlation with years spent in the organization with total satisfaction. The initial model including Workplace learning, Team quality, and Team supervision explained the 42% variance of Model 1. However, the sum of the squares of the residual is greater than the sum of the squares of the model. After the inclusion of three statistically significant confounding variables in Model 2, the sum of the squares of the residual is less than the sum of the squares of the model, and the subsequent model improvement explained 44.3% of the variance of Model 2.

The results of the Model 2 (Table 9) led to an increase in the impact of Workplace learning, Team quality, and Team supervision on overall job satisfaction. Team quality significantly affects job satisfaction (Beta = 1.048, $p < 0.001$), which indicates that this is one of the crucial factors that positively affect the improvement of overall job satisfaction. Team supervision and Workplace learning have a similar positive impact on overall job satisfaction (Beta = 0.495 and 0.437, $p < 0.001$).

Table 9. Multiple regression B-values for Model 1 and 2.

MODEL 1		MODEL 2	
	B		B
Constant	3.280 **	Constant	2.950 **
Workplace learning	0.175 **	Workplace learning	0.437 **
Team quality	0.386 **	Team quality	1.048 **
Team supervision	0.184 **	Team supervision	1.495 **
		Age	1.118 *
		Education	1.104 **
		The number of years spent in the organization	−0.084 **

** $p < 0.01$. * $p < 0.05$. Source: Author's calculation.

5. Discussion

The findings show that urban bank employees are only more satisfied with the possibility of promotion through the hierarchical structure, while the bank staff in rural areas is more satisfied with the nature of the job, with the salary, and with the remuneration policy. These findings show that the bank managers in urban areas need to enrich the content of their work and find more stimulating systems for rewarding employees, especially those employees who achieve good business results, both in terms of salaries and additional bonuses. A very similar level of satisfaction with cooperation with closest associates and with satisfaction with cooperation and relations with the superiors was found among urban and rural bank employees. The findings confirm that staff in rural areas is more satisfied with their jobs [57], with the personal qualities of their immediate supervisors and the immediate support at work they received from managers [58]. This conclusion indicates that the bank managers need to identify and eliminate the reasons for dissatisfaction with the work of bank employees in urban areas. In addition, they should formulate transparent support procedures that employees receive from their superiors. Although the salaries are generally lower in rural areas, rural workers are significantly more satisfied with the pay and the work itself, whereas urban workers are more satisfied with their co-workers [59]. These findings show that the bank managers in rural areas should redefine the policy related to interpersonal relationships, especially in the field of cooperation with top managers. Changes in pay level can be associated with increasing rates of pay satisfaction, which can contribute to a higher level of job satisfaction [60].

The regression results revealed that all the predictor variables, salaries, cooperation with closest associates, promotion up the hierarchical structure, remuneration policy, cooperation and relations with superiors, and the nature of the job are all statistically significant predictors of job satisfaction. This is confirmed by findings [61] that show that job, rewards and recognition, opportunities, teamwork, immediate supervisor, and communication have proven to be predictors of employee engagement. The satisfaction of bank employees with their salaries has the greatest influence on total job satisfaction. Therefore, bank managers must take into account the possibility of increasing the salaries of employees in accordance with their achieved results. This finding is in line with the findings of Sousa–Poza [62], where the importance of a high income as a determinant of job satisfaction is very significant. Job satisfaction is higher when pay satisfaction is higher, and the opposite holds as well [63]. The salary indicates how the worker is evaluated by the employer, and higher salary directly contributes to higher job satisfaction [39,64,65]. In that way, the bank managers would create greater loyalty and a sense of belonging to the bank among the employees, which would also increase the retention rate. On cooperation with closest associates, Oshagbemi [66] points out that individuals who had friendships with colleagues and supervisors reported higher levels of job satisfaction. This confirms the findings that relations at work, both with colleagues and with management, seem to be an important explanatory variable in job satisfaction equations [33,62,67,68]. The findings have shown that satisfaction with cooperation and relations with superiors had the greatest influence on job satisfaction [69]. This finding indicates that the bank managers should build good and harmonious interpersonal relationships, both between the employees themselves and between employees and their superiors. This can be achieved through various programs and seminars on the development of social skills (assertive communication, interpersonal relationships, constructive criticism, conflict management, and stress management). Peterson, Puia and Suess [70] have indicated that work satisfaction, supervision, coworker relationships, pay, and promotion potential were predictors of total job satisfaction. In the Two-stage least squares (2SLS) regression analysis model where the instrumental variable loyalty (employees' intent to turnover) has been introduced, it was shown that the satisfaction with the remuneration policy system has had the greatest impact on overall job satisfaction. The higher the remuneration given to employees, the higher the total job satisfaction of employees will be, and thus the lower turn-over, which is in line with the findings of Lindgren and Paulsson [71], Taylor [72], Vosloo, Fouche, and Bernard [73], and Naji [74]. These findings indicate that the bank managers should design new reward programs to enable employees at lower hierarchical levels to be additionally rewarded and thus more stimulated and satisfied.

The research results showed that in the total sample of bank employees, teamwork attitudes (average score from 3.33 to 3.89) were rated slightly better in relation to the job satisfaction parameters. These findings indicate that the importance of teamwork that can be enhanced by the bank managers that should additionally stimulate the most successful team leaders, because they have the greatest responsibility for the team work performance. The highest average scores of the perceived teamwork attitudes highlight the teams' ability to monitor team members' complaints so we can conclude that in this way the team members' satisfaction is nourished. Team leaders possess the ability to acquaint all team members with the team's goals. Team members have great respect for organizational rules in resolving customer complaints and consider themselves appropriately trained and competent to perform their jobs professionally with the teamwork standards clearly defined. By the means of factor analysis, we grouped the teamwork attitudes into several factors such as workplace learning, team quality, and team supervision. This analysis shows that the highest percentage of variance in the factor analysis model is found in workplace learning. In particular, this work adds to the findings of Rowden and Conine [40], as their findings show a statistically significant relationship between workplace learning and job satisfaction, predicting 41% of overall job satisfaction variance. Employees that have learning opportunities will experience greater satisfaction, and thus are prepared to value their job in the bank more. Kozlowski and Ilgen [75] point out that training can shape team processes to enhance team effectiveness, and learning was positively related to project performance [76]. Human resources managers should further devise models for predicting teamwork efficiency since "having tools to predict whether a set of individuals can become an effective team will provide a company with the opportunity to be proactive, rather than having to react to an ineffective team" [18] (p. 265). Wu and Chen [77] have pointed out that knowledge sharing had a positive effect on team performance. A professionally stimulating and challenging work environment proved to be directly correlated to the team performance [17]. Team-level training was also found to have a significant positive relationship to team member satisfaction [78]. Team supervision is also very important for the quality of teamwork as the "The supervisor encourages self-managing work team members to develop their performance standards, conduct self-evaluations and self-regulate their own behavior" [79] (p. 819). Reed's research [80] has shown that the level of employees' supervision had a statistically significant relationship with job satisfaction. From the above, it can be concluded that the profitability and full satisfaction of bank clients, as the main goal of the bank's managers, largely depends on team supervision that is able to increase the quality of banking services, innovate existing banking services, optimize resources, and continuously monitor costs and work efficiency. The multiple regression results indicate that team quality has the greatest influence on the bank employees' job satisfaction. This is confirmed by the findings of other authors. Effective quality teamwork significantly influenced job satisfaction [81]. Teamwork enhances job satisfaction levels in both public and private sector organizations as "teamwork is not only important for productivity in the organization, rather this is also important for generating a higher level of job satisfaction among employees" [79] (p. 828.). Quality teamwork directly positively influences the important issues of cost savings, higher workforce retention, and reduced turnover [82].

6. Conclusions

The quality of the bank-employee relationship is a key factor in creating competitiveness in the current business environment. The issue of trust and value in the employment relationship that can be built between a bank and its employees can contribute to higher levels of employees' performance. "The greater focus on outcomes that capture the quality of the employee-organization relationship itself in terms of fulfillment of needs, quality of interaction, adaptability and identification is needed" [83] (p. 28). This study, however, provides more substantive outcomes, as it reveals job satisfaction and teamwork attitudes of both the urban and rural bank employees' on various specificities. These include not just standard parameters of job satisfaction such as salaries, cooperation with closest associates, promotion, remuneration policy, cooperation and relationships with superiors, and the nature of the job, but some additional facets of teamwork attitudes and their relationship to job satisfaction. Bank employees' sense of job satisfaction can be strongly attributed to workplace learning opportunities. The employees should be provided with a chance for professional development, by creating an environment that encourages their initiative to enhance the business processes. The culture of successful teamwork that stimulates innovation should be further promoted. At the organizational level, job satisfaction should be permanently improved through various initiatives that can contribute to job satisfaction such as talent programs, performance appraisals, soft skills training, team coaching, bank community projects, internal and external education (funding of various types of licenses, certificates, professional exams, and memberships), rotation of jobs, comprehensive benefits, and more flexible organizational structure. Human resources managers should set a specific set of instruments aimed at determining and monitoring the employees' job satisfaction which is directly related to the quality teamwork.

Top managers create the business strategy of the organization and the middle management and lower levels of supervisory and operative managers are in charge of its' implementation. This research included a sample of lower level employees in the hierarchical structure of banks. The realization of the set organizational goals depends on their competencies, satisfaction, and motivation. The findings of this study reveal several practical implications that urge specific actions. The top management of the bank should recognize the preferences of the employees at the lower levels of the hierarchy and motivate, educate, and prepare them for the realization of business goals. Top managers are recommended to self-evaluate individual parameters through more frequent employee surveys, so that in accordance with the defined goals, they can redefine business strategies in the right way in relation to the needs of the business environment. The profitability and full satisfaction of clients can be achieved by increasing the level of service quality, by innovation of existing services, optimization of resources, and continuous monitoring of costs. The growth and development can be achieved through technical–technological development and improvement of business processes, development of marketing concept, increasing market share, and placement of new banking services. The strategies for improving corporate responsibility can be directed towards development of corporate responsibility, development of human potential, improvement of the business risk management, security, and resource protection levels.

Additional programs and trainings should be included in the bank human resources education plan, either by an external contractor (trainings, seminars, counseling, etc.) or by internal educational activities such as technology training (payment services, exchange operations, prevention of money laundering, etc.) and internal seminars on the development of skills (training coordinators for employees in different organizational units, etc.) in order to improve the employees' knowledge and competencies, improve the promotion and sale of banking services, employees' performance, and bank corporate image. The training program must be transparent, so that all employees have an insight into the offered types of education. Transparency of additional education can have a positive effect on employees, especially due to the avoidance of suspicion that certain organizational units are more privileged to apply for the additional training programs. The bank managers should promote the additional training of the employees', so that employees become more motivated. Employees' motivation and other aspects of organizational behavior that can contribute to the innovative ways of providing services in banks should be further investigated since the creation and development of innovations is connected to the special project teams or special functional groups.

These findings can help decision makers in the banks understand the issue of job satisfaction and bring about optimum alternatives for solving the human problems. Solid principles founded on recent research can answer management needs to improve employees' attitudes and deepen the quality of the employees–bank relationship. Further research needs to be conducted in different kinds of enterprises taking into the account different managerial levels. The findings on banks employees' job satisfaction could be then compared to employees' job satisfaction in other business sectors in order to crystallize other dimensions, which can contribute to organizational behavior practices. The findings show that future research should also go in the direction of determining the correlation between job satisfaction and additional education of employees. The future research can also focus on facilitation of teamwork by studying the impact of training on the team quality as well as other attitudes of teamwork that can predict a high performing team in the banking sector.

Differences in urban and rural bank employees perceived attitudes can be attributed to the regional differences since the work conditions may differ regionally. In some bank branches the work conditions might be better. The urban workers definitely have higher costs of living and more opportunities for finding new jobs in the banking sector, because the main banking institutions are concentrated in the urban areas. The rural bank employees have to deal with the scarcity of jobs offered in the rural areas suffering high levels of unemployment. The present findings can also be attributable to expectations, since banking sector employees are perceived as having a decent standard of living due to higher starting salaries than those offered in the other sectors.

This research is based on the sample of urban and rural bank employees in the Republic of Serbia, therefore its findings cannot be generalized to a wider range. A deeper understanding of job satisfaction of bank employees in countries that are adapting to the market economy postulates can constitute useful directions for human resource managers adapting to market economy management approaches. As there is no previous research on employees' job satisfaction in Serbia, the findings of this research may be of special interest to various stakeholders. The limitations that can distract respondents to complete such surveys are linked to the busy schedule, fear of information leakages, and perceived non-relevance of the survey to their professional career. This limitation of the study may point to the need for further studies in the broader research area, as it can lead researchers towards a deeper understanding of the employee–bank relationship. The empirical testing of the model can provide a modest contribution to fill in the literature toward generalizing similar findings.

Author Contributions: Conceptualization, methodology, formal analysis, supervision, and original draft preparation, S.L. and J.V.-T.; investigation and resources, S.M. and J.R.-M.; project administration, J.M. and N.L. All authors have read and agreed to the published version of the manuscript.

Funding: This research received no external funding.

Conflicts of Interest: The authors declare no conflict of interest.

References

1. Keltner, B.; Finegold, D. Adding Value in Banking: Human Resource Innovations for Service Firms. *Sloan Manag. Rev.* **1996**, *38*, 57–68.
2. Serbian Chamber of Commerce, Financial Sector. Available online: http://www.pks.rs/PrivredaSrbije.aspx?id=4 (accessed on 4 April 2020).
3. Gajić, T.; Vujko, A.; Cvijanović, D.; Penić, M.; Gagić, S. The state of agriculture and rural development in Serbia. *R Econ.* **2017**, *3*, 196–202. [CrossRef]
4. Prazan, J.; Koutna, K.; Skorpikova, A. Report on the Development of Organic Farming and the Policy Environment in Central and Eastern European Accession States, 1997–2002 (Deliverable 6), EU-CEEOFP Project Report. 2004. Available online: https://orgprints.org/4799/1/EUCEEOFP_D7_final_report.pdf (accessed on 5 March 2020).
5. Robbins, S.; Coulter, M. *Menadžment*, 8th ed.; Data Status: Beograd, Srbija, 2005.
6. Лекић, С.; Ерић, И. Организационо понашање. In Београдска Пословна Школа—Висока Школа, 2nd ed.; СтруковнихСтудија: Београд, Србија, 2018.
7. Alam, A.; Ramay, I. Antecedents of Organizational Commitment of Banking Sector Employees in Pakistan. *Serb. J. Manag.* **2012**, *7*, 89–102. [CrossRef]
8. Becker, H. Notes on the concept of commitment. *Am. J. Sociol.* **1960**, *66*, 32–40 [CrossRef]
9. Meyer, J.; Stanley, D.; Jackson, T.; McInnis, K.; Maltin, E.; Sheppard, L. Affective, normative, and continuance commitment levels across cultures: A meta-analysis. *J. Vocat. Behav.* **2012**, *80*, 225–245. [CrossRef]
10. De Menezes, L.M. Job Satisfaction and Quality Management: An Empirical Analysis. *Int. J. Oper. Prod. Manag.* **2011**, *32*, 308–328. [CrossRef]
11. Bae, K.B.; Yang, G. The Effects of Family-Friendly Policies Job Satisfaction and Organizational Commitment: A Panel Study Conducted on South Korea's Public Institutions. *Public Pers. Manag.* **2017**, *46*, 25–40. [CrossRef]
12. Demircioglu, M.A. Examining the Effects of Social Media Use on Job Satisfaction in the Australian Public Service: Testing Self-Determination Theory. *Public Perform. Manag. Rev.* **2018**, *41*, 300–327. [CrossRef]
13. Edmondson, A. Psychological safety and learning behavior in work teams. *Adm. Sci. Q.* **1999**, *44*, 350–383. [CrossRef]
14. Certo, S.C.; Certo, T.S. *Modern Management: Concepts and Skills*, 15th ed.; Pearson: New York, NY, USA, 2019.

15. Griffin, R.W.; Moorhead, G. *Organizational Behavior: Managing People and Organizations*, 11th ed.; South-Western College Pub: Boston, MA, USA, 2013.
16. Adams, S.G.; Simon, L.; Ruiz, B. A pilot study of the performance of student teams in engineering education. *Age* **2002**, *7*, 1.
17. Thamhain, H.J. Leading technology—Based project teams. *Eng. Manag. J.* **2004**, *16*, 35–42. [CrossRef]
18. Ross, M.T.; Jones, E.C.; Adams, S.G. Can team effectiveness be predicted? *Team Perform. Manag.* **2008**, *14*, 248–268. [CrossRef]
19. Locke, E.A. The Nature and Causes of Job Satisfaction. In *Handbook of Industrial and Organizational Psychology*; Dunnette, M.P., Ed.; Rand McNally: Chicago, IL, USA, 1976; pp. 1297–1350.
20. Baron, A.R.; Greenberg, J. *Organizational Behavior in Organization: Understanding and Managing the Human Side of Work*, 8th ed.; Prentice Hall: Upper Saddle River, NJ, USA, 2002.
21. Locke, E.A. What is Job Satisfaction? *Organ. Behav. Hum. Perform.* **1969**, *4*, 309–336. [CrossRef]
22. Smith, P.C.; Kendall, M.; Hum, C.L. *The measurement of satisfaction in work and retirement*; Rand McNally: Chicago, IL, USA, 1969.
23. Yeager, S.J. Dimensionality of the Job Descriptive Index. *Acad. Manag. J.* **1981**, *24*, 205–212. [CrossRef]
24. Gregson, T.H. Factor Analysis of Multiple–Choice Format for Job Satisfaction. *Psychol. Rep.* **1987**, *61*, 747–750. [CrossRef]
25. Gregson, T.H. Measuring job satisfaction with a multiple-choice format of the job descriptive index. *Psychol. Rep.* **1990**, *66*, 787–793. [CrossRef]
26. Saner, T.; Eyupoglu, S.Z. The job satisfaction of bank employees in North Cyprus. *Procedia Econ. Financ.* **2015**, *23*, 1457–1460. [CrossRef]
27. Ali, A.; Khan, I.H.; Ch, M.A.; Ch, A.S.A. Level of Job Satisfaction among Employees of Banking Industries at Lahore 2016. *Eur. Online J. Nat. Soc. Sci.* **2018**, *7*, 92–108.
28. Sakaya, A.J. The Supremacy of Financial and Non-Financial Rewards towards Employees' Job Satisfaction in Tanzania; Evidence from Commercial Banks—Mtwara Region. *Int. J. Innov. Sci. Res. Technol.* **2019**, *4*, 605–616.
29. Vallejo, R.D.; Vallejo, J.A.D.; Parra, O.P. Job satisfaction in banking workers. *Psicothema* **2001**, *13*, 629–635.
30. Crossman, A.; Abou-Zaki, B. Job satisfaction and employee performance of Lebanese banking staff. *J. Manag. Psychol.* **2003**, *18*, 368–376. [CrossRef]
31. Agustiningsih, H.N.; Thoyib, A.; Djumilah, H.; Noermijati, N. The Effect of Remuneration, Job Satisfaction and OCB on the Employee Performance. *Sci. J. Bus. Manag.* **2016**, *4*, 212–222. [CrossRef]
32. Khan, N.A.; Parveen, S.A. Comparative Study of Job Satisfaction of Employees in Public and Private Sector Banks in India with Reference to U.P. State. *Sci. Int.* **2014**, *26*, 813–820.
33. Chahal, A.; Chahal, S.; Chowdhary, B.; Chahal, J. Job satisfaction among bank employees: An analysis of the contributing variables towards job satisfaction. *Int. J. Sci. Technol. Res.* **2013**, *2*, 11–20.
34. Imam, A.; Raza, A.; Shah, T.F.; Raza, H. Impact of Job Satisfaction on Facet of Organizational Commitment (Affective, Continuance and Normative Commitment): A Study of Banking Sector Employees of Pakistan. *World Appl. Sci. J.* **2013**, *28*, 271–277. [CrossRef]
35. Islam, J.; Mohajan, H.; Datta, R. A study on job satisfaction and morale of commercial banks in Bangladesh. *Int. J. Econ. Res.* **2012**, *3*, 152–172.
36. Siswanto, S. Creating the Superior Islamic Banking through Improving Quality of Human Resources. *Pak. Journal Commer. Social Sci.* **2011**, *5*, 216–232.
37. Delgado-Piña, M.I.; Rodríguez-Ruiz, O.; Rodríguez-Duarte, A.; Sastre-Castillo, M.A. Gender Diversity in Spanish Banks: Trickle-Down and Productivity Effects. *Sustainability* **2020**, *12*, 2113. [CrossRef]
38. Metle, M.K. Education, job satisfaction and gender in Kuwait. *Int. J. Hum. Resour. Manag.* **2001**, *12*, 311–332. [CrossRef]
39. Gazioglu, S.; Tansel, A. Job satisfaction in Britain: Individual and job related factors. *Appl. Econ.* **2006**, *38*, 1163–1171. [CrossRef]
40. Rowden, R.W.; Conine, C.T., Jr. The impact of workplace learning on job satisfaction in small US commercial banks. *J. Workplace Learn.* **2005**, *17*, 215–230. [CrossRef]
41. Sowmya, K.R.; Panchanatham, N. Factors influencing job satisfaction of banking sectors employees in Chennai, India. *J. Law Confl. Resolut.* **2011**, *3*, 76–79.

42. Jie, C.T.; Hasan, N.A.; Bidin, R. Job satisfaction among employees: An investigation of government-linked bank institution in Malaysia. *J. Kemanus.* **2018**, *16*, 33–39.

43. Neupane, B. A Study on Factors Influencing the Job Satisfaction of Bank Employees in Nepal (With special reference to Kathmandu, Lalitpur, and Bhaktapur District). *NCC J.* **2019**, *4*, 9–15. [CrossRef]

44. Miao, S.; Fayzullaev, A.K.U.; Tohirovich Dedahanov, A. Management Characteristics as Determinants of Employee Creativity: The Mediating Role of Employee Job Satisfaction. *Sustainability* **2020**, *12*, 1948. [CrossRef]

45. Ryan, T. *Modern Regression Methods*, 2nd ed.; Wiley: New York, NY, USA, 2008; Available online: https://www.wiley.com/en-us/Modern+Regression+Methods%2C+2nd+Edition-p-9780470081860 (accessed on 5 March 2020).

46. Judge, T.A.; Klinger, R. Job satisfaction: Subjective well-being at work. In *The Science of Subjective Well-Being, Edited by Larsen*; Eid, M., Randy, J., Eds.; The Guilford Press: New York, NY, USA, 2008; pp. 393–413.

47. Bateman, B.; Wilson, C.F.; Bingham, D. Team effectiveness—Development of an audit questionnaire. *J. Manag. Dev.* **2002**, *21*, 215–226. [CrossRef]

48. Pokropek, A. Introduction to instrumental variables and their application to large-scale assessment data. *Large Scale Assess. Educ.* **2016**, *4*, 1–20. [CrossRef]

49. Strenitzerová, M.; Achimský, K. Employee Satisfaction and Loyalty as a Part of Sustainable Human Resource Management in Postal Sector. *Sustainability* **2019**, *11*, 4591. [CrossRef]

50. Lousdal, M.L. An introduction to instrumental variable assumptions, validation and estimation. *Emerg. Themes Epidemiol.* **2018**, *15*, 1–7. [CrossRef]

51. Kurnoga Živadinović, N. Defining the basic product attributes using the factor analysis. *Ekon. Pregl.* **2004**, *55*, 952–966.

52. Gašević, D.; Jovičić, D.; Tomašević, D.; Vranješ, M. Primena faktorske analize u istraživanju potrošačkog etnocentrizma. *Škola Bizn.* **2017**, *2*, 18–37.

53. Seber, G.A.F.; Lee, J.A. *Linear Regression Analysis*; Wiley: Hoboken, NJ, USA, 2003.

54. Diez, D.; Barr, C.; Rundel, M. *OpenIntro Statistics, Creative Commons License*; OpenIntro, Inc.: Boston, MA, USA, 2012.

55. Thompson, B. *Exploratory and confirmatory factor analysis: Understanding concepts and applications*; American Psychological Association: Washington, DC, USA, 2004.

56. Pallant, J. *SPSS Survival Manual (Version 15)*, 3rd ed.; McGraw Hill: New York, NY, USA, 2007.

57. Gellis, Z.D.; Kim, J.; Hwang, S.C. New York state case manager survey: Urban and rural differences in Job activities, job stress, and job satisfaction. *J. Behav. Health Serv. Res.* **2004**, *31*, 430–440. [CrossRef] [PubMed]

58. Grujičić, M.; Jovičić-Bata, J.; Novaković, B. Motivation and job satisfaction of healthcare professionals in urban and rural areas in the Autonomous Province of Vojvodina. *Med. Pregl.* **2018**, *71*, 33–41. [CrossRef]

59. Fossum, J.A. Urban-Rural Differences in Job Satisfaction. *Ind. Labor Relat. Rev.* **1974**, *27*, 405–409. [CrossRef]

60. Heneman, R.; Porter, G.; Greenberger, D.; Strasser, S. Modeling the Relationship between Pay Level and Pay Satisfaction. *J. Bus. Psychol.* **1997**, *12*, 147–158. [CrossRef]

61. Vijay Anand, V.; Vijaya Banu, C.; Badrinath, V.; Veena, R.; Sowmiyaa, K.; Muthulakshmi, S. A Study on Employee Engagement among the Bank Employees of Rural Area. *Indian J. Sci. Technol.* **2016**, *9*, 1–10. [CrossRef]

62. Sousa-Poza, A.; Sousa-Poza, A.A. Well-being at work: A cross national analysis of the levels and determinants of job satisfaction. *J. Socio Econ.* **2000**, *29*, 517–538. [CrossRef]

63. McCausland, W.D.; Pouliakas, K.; Theodossiou, I. Some are punished and some are rewarded: A study of the impact of performance pay on job satisfaction. *Int. J. Manpow.* **2005**, *26*, 636–659. [CrossRef]

64. Ghinetti, P. The public-private job satisfaction differential in Italy. *Labour* **2007**, *21*, 361–388. [CrossRef]

65. Jones, R.J.; Sloane, P. Regional differences in job satisfaction. *Appl. Econ.* **2009**, *41*, 1019–1041. [CrossRef]

66. Oshagbemi, T. How satisfied are academics with the behaviour/supervision of their line managers? *Int. J. Educ. Manag.* **2001**, *15*, 283–291. [CrossRef]

67. Clark, A.E. Job satisfaction in Britain. *Br. J. Ind. Relat.* **1996**, *34*, 189–217. [CrossRef]

68. Clark, A.E. Job satisfaction and gender: Why are women so happy at work? *Labour Econ.* **1997**, *4*, 341–372. [CrossRef]

69. Lekić, N.; Vapa-Tankosić, J.; Lekić, S.; Rajaković, J. An analysis of factors influencing employee job satisfaction in a public sector. *E M Ekon. Manag.* **2019**, *22*, 83–99. [CrossRef]

70. Peterson, D.K.; Puia, G.M.; Suess, F.R. Yo Tengo La Camiseta: An exploration of job satisfaction and commitment among workers in Mexico. *J. Leadersh. Organ. Stud.* **2003**, *10*, 73–88. [CrossRef]
71. Lindgren, L.; Paulsson, S. Retention: An Explanatory Study of Swedish Employees in the Financial Sector Regarding Leadership Style, Remuneration and Elements Towards Job Satisfaction. Bachelor's Thesis, Växjö Universitet, Växjö, Sweden, 26 February 2008. Available online: http://lnu.diva-portal.org/smash/get/diva2:205794/FULLTEXT01.pdf (accessed on 3 April 2020).
72. Taylor, J. Remuneration Policy in the Australian Public Service: Fairness and Trust. In Proceedings of the Australian Political Studies Association, Perth, Australia, 30 September–2 October 2013; Volume 1, p. 28.
73. Vosloo, W.; Fouche, J.; Barnard, J. The Relationship between Financial Efficacy, Satisfaction With Remuneration And Personal Financial Well-Being. *Int. Bus. Econ. Res. J.* **2014**, *13*, 1455–1470. [CrossRef]
74. Naji, A.H. Components of Remuneration and Employee Satisfaction: The Impact of Effort Rewards and career Advancement. *Int. J. Arts Sci.* **2014**, *7*, 425–436.
75. Kozlowski, S.W.J.; Ilgen, D.R. Enhancing the Effectiveness of Work Groups and Teams. *Psychol. Sci.* **2006**, *7*, 77–124. [CrossRef]
76. Kuthyola, K.F.; Liu, J.Y.C.; Klein, G. Influence of task interdependence on teamwork quality and, project performance. In Proceedings of the Big Data Analytics for Business and Public Administration, Poznań, Poland, 28–30 June 2017; Abramowicz, W., Ed.; Springer: Heidelberg, Germany, 2017; pp. 135–148. [CrossRef]
77. Wu, M.C.; Chen, Y.H. A factor analysis on teamwork performance—An empirical study of inter-instituted collaboration. *Eurasian J. Educ. Res.* **2014**, *55*, 37–54. [CrossRef]
78. Doolen, T.L.; Hacker, M.E.; Van Aken, E. Managing organizational context for engineering team effectiveness. *Team Perform. Manag.* **2006**, *12*, 138–154. [CrossRef]
79. Sharma, J.P.; Bajpai, N. Teamwork a Key Driver in Organizations and Its Impact on Job Satisfaction of Employees in Indian Public and Private Sector Organizations. *Glob. Bus. Rev.* **2014**, *15*, 815–831. [CrossRef]
80. Reed, T.L. The Relationship between Supervision, Job Satisfaction, and Burnout among Live-In and Live-On Housing and Residence Life Professionals. LSU Ph.D. Thesis, Louisiana State University and Agricultural and Mechanical College, Baton, LA, USA, 2015; p. 1036. Available online: https://digitalcommons.lsu.edu/gradschool_dissertations/1036 (accessed on 3 April 2020).
81. Musriha, H. Influence of teamwork, environment on job satisfaction and job performance of cigarette rollers at clove cigarette factories in East Java, Indonesia. *J. Copernic.* **2013**, *3*, 32–40.
82. Xyrichis, A.; Ream, E. Teamwork: A concept analysis. *J. Adv. Nurs.* **2008**, *61*, 232–241. [CrossRef]
83. Coyle-Shapiro, J.A.-M.; Shore, L. The employee-organization relationship: Where do we go from here? *Hum. Resour. Manag. Rev.* **2007**, *17*, 166–179. [CrossRef]

Article

Cognitive Component of the Image of a Rural Tourism Destination as a Sustainable Development Potential

Ksenija Leković [1], Slavica Tomić [1], Dražen Marić [1,*] and Nikola V. Ćurčić [2]

[1] Faculty of Economics in Subotica, University of Novi Sad, 24000 Subotica, Serbia; ksenija.lekovic@ef.uns.ac.rs (K.L.); slavica.tomic@ef.uns.ac.rs (S.T.)
[2] Institute of Agricultural Economics, 11000 Belgrade, Serbia; nikola_c@iep.bg.ac.rs
* Correspondence: drazen.maric@ef.uns.ac.rs; Tel.: +381-24-628-135

Received: 27 October 2020; Accepted: 5 November 2020; Published: 12 November 2020

Abstract: Sustainable tourism should maintain a high level of tourist satisfaction, so identifying components of tourism destination image plays an important role in destination management and marketing. This study aims to explore issues related to the image of a rural tourism destination, with the focus on the cognitive component. It also aims to analyze three dimensions of the cognitive component: functional, mixed, and psychological. Furthermore, this study gives the answer to the question of which dimension of the cognitive component makes the most significant impact on the general image of a rural tourism destination. The sample comprised 562 respondents. Data analysis included exploratory factor analysis (EFA), confirmatory factor analysis (CFA), and structural equation modeling (SEM). The results indicate the existence of three dimensions of the cognitive component, and it can be concluded that the psychological dimension of the cognitive component has the most significant impact on the general image of a rural tourism destination.

Keywords: sustainable development; sustainable tourism development; rural tourism; image of a rural tourism destination; cognitive component; dimensions of the cognitive component

---

1. Introduction

Sustainable development represents a concept that refers to the potential of productive activities to meet the contemporary needs without jeopardizing the opportunities of posterity [1]. Tourism, as one of the global engines of development, results in tourism impacts, which can be classified into three categories: economic, environmental, and sociocultural [2]. These impacts can either be positive (job creation, enhancement in the quality of life of local residents, an improved public image of the destination, protection of cultural heritage, etc.) or negative (overcrowding, disruption of local lifestyle, environmental damage, intensive use of resources, etc.) [1,3,4]. As a result of its dynamic nature, tourism offers new destinations and new forms of travel but on the other hand requires new approaches to management, new resources, and constant innovations [5]. Suitable balance between these impacts is reflected in sustainable tourism. Sustainable tourism respects the wholesomeness of natural, social, and economic setting by rationally exploiting natural and cultural heritage while offering posterity and equal opportunities to profit from the same resources [6]. Therefore, by reducing energy consumption and pollution rate, the protected environment and the preserved cultural heritage can lead to the sustainable tourism sector [7]. Principles of sustainable tourism development are applicable to all forms of tourism. Thus, the sustainable hospitality industry is a key indicator of sustainable development in rural regions [8].

According to Okech et al. [9], it is important to advance tourism in rural regions because it increases people's involvement in the advancement of the travel and hospitality industry and brings wider benefits to rural areas. Moreover, the travel and hospitality industry, especially rural tourism, can be beneficial to people who live in urban areas and look for relaxation. For those consumers, rural tourism is eco-friendly travel for entertainment and enjoyment in nature [10]. At the same time, rural tourism can be beneficial to people who live in rural areas as a means of earning a living. It also reduces the migration of population from these regions and creates new employment opportunities [11–13]. Rural areas can attract consumers from urban areas by the means of establishing a bond with their cultural, historical, ethnic, and geographical roots [14]. Furthermore, rural areas represent places where the traditions, heritage, and natural beauty meet [15]. Sustainability in rural regions should merge four factors: ecological, economic, social, and cultural [16]. Therefore, rural tourism can be considered a synonym for sustainability [12]. According to Irshad [17] and Lane [18], rural tourism can be regarded as an umbrella concept because it covers various forms of travel and hospitality industry, such as ecotourism, farm tourism, agritourism, adventure-seeking, equestrian tourism, and food and wine tourism. Thus, in line with the umbrella concept, rural tourism includes all forms of activities undertaken by consumers in rural areas that contain some elements related to traditions, culture, and hospitality [6]. Authenticity and tradition are the main characteristics that separate these forms of tourism from other selective forms of tourism [19].

Success of rural tourism depends on many factors, which must ensure tourist satisfaction [20]. First of all, this form of tourism, which includes small villages with attractive landscapes, thermal springs, rivers, and lakes, enables consumers to reunite with nature and local culture [21,22]. It also offers a peaceful experience, low costs, and healthy challenges [23]. Rural tourism destinations can be described as packages of many different facilities and services that determine their attractiveness. On the other hand, tourists perceive a rural tourism destination as a concept that comprises variety of products and services. During their stay, tourists "consume" the destination as an intrinsic experience composed of multiple individual experiences and external elements [24]. This process results in a composite image referring to the possible favorable or unfavorable evaluation of the destination. The image of a destination can be defined as a set of impressions, ideas, expectations, and emotional thoughts a tourist has of a specific destination [25,26]. Authors Lai and Li cite the most commonly used terms for defining the image of a destination: impression, perception, belief, idea, and representation [27]. According to Agapito et al. [28], the image of a destination plays a substantial part in tourist behavior in different phases of the buying process: (1) in the decision-making process with regard to opting for a particular destination, (2) in the process of evaluating the actual experience against previous expectations, and (3) in the process of revisiting the destination and spreading positive/negative word-of-mouth. Thus, the image of a rural tourist destination makes a notable impact on tourist's behavior, preferences, and the intention to revisit that particular destination.

The explanation of tourism destination image starts from the multi-attribute concept [25,29–33]. According to this concept, tourism destination image constitutes three interrelated components: (1) cognitive, (2) affective, and (3) conative or behavioral. The cognitive component is the aggregate of what a tourist knows or believes about a destination [34]. The second component of this concept, the affective component, denotes the emotional responses that reflect the tourist's feelings towards the destination [35]. The cognitive component guides the affective one. In other words, it is an antecedent of the affective component because tourists' evaluative responses stem from their acquaintance with the particular rural tourism destination [36]. Finally, the combination of these two components produces a general image related to the positive/negative evaluation of the destination—the conative component. This component represents how and why the tourist's knowledge and feelings contribute to the selection of a particular destination. It is formed when the destination selection is completed and a decision is made [37,38]. Thus, the cognitive and affective components represent tourists' subjective associations linked to a destination, whereas the conative component outlines the desired future situation—revisit, recommendation, and positive word-of-mouth. Besides own experience (as a primary source), opinions of other tourists communicated through word-of-mouth represent the main secondary source of information upon which decision can be made [39].

The objective of this particular study was to define and validate the dimensions that make up the cognitive component of the image of a rural tourist destination. According to Chen [40], dimensions of the cognitive component are easier to render and implement, so it is crucial for destination managers and marketers to know which dimensions have the most significant impact on tourists' future behavior. Findings of several studies show that both cognitive and affective components influence tourists' behavioral intentions. However, the immediate consequence of the cognitive component is greater than the impact of the affective one [20,29,41,42]. This study applied the composition of a cognitive component, which consists of three dimensions. These dimensions are: (1) functional—based on more tangible sensations, such as surroundings, lodging and price levels; (2) psychological—based on more abstract attributes, such as ambience and sociability; and (3) mixed—based on characteristics of a destination that are between functional and psychological, such as gastronomy [43].

2. Materials and Methods

The general objective of this study was to look into the structure of the cognitive image of a rural tourism destination, analyzing its dimensions (functional, psychological, and mixed) and the impact that they make on the general image. The general aim is indicated in two research objectives:

-   Research objective 1: To verify the existence of three dimensions of the cognitive image of a rural tourism destination: functional, mixed, and psychological dimensions.
-   Research objective 2: To analyze which dimension of the cognitive component of the image of a rural tourist destination makes the most significant impact on the general image.

A structured questionnaire designed for the purpose of this study was distributed to tourists who spent their holiday on a farm or in a winery in the Province of Vojvodina. The convenience sampling method was implemented having previously confirmed that each potential respondent had met the two requirements: (a) that he/she had visited a farm/winery in the Province of Vojvodina in the past two years and (b) that he/she was over 18 years of age. Farms and wineries in the Province of Vojvodina were selected as rural tourism destinations because Vojvodina is a region exemplary of rural tourism with its natural beauty in combination with culture, tradition, and specific gastronomy. Data were gathered in the months of April and May 2020. The questionnaire was distributed online. Initially, 572 questionnaires were collected, but some of them had to be disregarded because they were not filled in correctly, eventually producing a sample of 562 respondents.

The sample profile is presented in Table 1. Table 1 indicates that 62.6% of the respondents were male, and the remaining 34.4% were female. At the time of the research, most respondents were between 18 and 29 years of age (36.7%) and had a bachelor's degree (51.8%). The majority lived in four person households (33.6%) and had a household net monthly disposable income between 601 and 900 euros (24.6%).

The data showed that most of the respondents visit rural tourism destinations once or twice a year (34.2%), most often with their families (57.8%). The Internet is the most common source of information when planning a trip (49.6%).

The research was conducted by means of a structured questionnaire containing three parts. The first part dealt with socio-demographic characteristics, the second part contained statements about trip characteristics, and the third part referred to dimensions of the cognitive component of the image of a rural tourist destination and the general image of the destination. The third part was adapted from Polo Pena et al. and Alcaniz et al. [43,44] It included three dimensions (functional, mixed, and psychological) of the cognitive component of the image of a rural tourist destination. Each dimension comprised six items: (a) the functional dimension: It is a nice/beautiful place with attractive natural scenery, There are many historical monuments/sites/museums, Tourists can enjoy fairs/festivals/wine routes, Tourists can undertake different activities in the countryside, The availability of accommodation is good, The local transport is good; (b) the mixed dimension: This place is environmentally friendly, The cleanliness and hygiene are good, The offer of local products on the local markets is good, Partaking in the local gastronomic activities/tours is good, Tourists become familiar with the local culture, Tourists can engage in social interactions with members of the local community; and (c) the psychological dimension: It is a quiet/peaceful place, Staff and members of the local community provide tourists with warm welcome, It offers personalized service to tourists, It has quality services, It offers good value for money, There is high quality accommodation. This part contained 18 statements reflecting the three dimensions. The questionnaire used a five-point Likert-type scale, which ranged from 1 ("strongly disagree") to 5 ("strongly agree"). The general image was measured by means of a single five-point scale, from highly unfavorable (1) to highly favorable (5).

Table 1. Profile of survey respondents (n = 562).

Variables	Percent (%)
Gender	
Female	37.4
Male	62.6
Age	
18–29	36.7
30–39	29.2
40–49	19.8
50–59	10.1
≥60	4.3
Education level	
Elementary school	0.4
Secondary school	17.3
Bachelor's degree	51.8
Master's degree	18.5
Doctoral degree	12.1
Net monthly disposable income per household	
≤300 euro	5.9
301–600 euro	22.1
601–900 euro	24.6
901–1200 euro	19.2
1201–1500 euro	10.7
>1500 euro	17.6
Household composition (no. of members)	
1	9.8
2	19.8
3	22.6
4	33.6
≥5	14.2
Frequency of engaging in rural tourism	
Once every two or three years	26.9
Once or twice a year	34.2
Three or four times a year	15.1
More than four times a year	23.8
Travel companions	
Alone	1.6
Partner	18.3
Family	57.8
Friends	21.2
Other	1.1
Information sources used to plan the trip	
The Internet	49.6
Recommendations of friends and/or relatives	32.7
Own experience	13.9
Newspapers/magazines/catalogues	0.5
Television	0.4
Travel agencies	2.8

Data were analyzed in two steps. First, exploratory factor analysis (EFA) was undertaken to determine the structure of the cognitive component of rural tourist destination image. Second, confirmatory factor analysis (CFA) was conducted to investigate whether the established dimensionality and factor-loading pattern fit well. Structural equation modeling (SEM) was used to examine which dimension has the most significant impact on the general image of the rural tourist destination. Data were processed with SPSS 21.0 and AMOS 21.0 statistical packages.

3. Results

In order to address the first research objective, we started with EFA by considering all 18 items (see Table 2 for details) measuring the three dimensions of the cognitive component of the image of a rural tourist destination: functional, mixed, and psychological. Each dimension was measured using six items. The statistical criteria in this study were met. The KMO value for the cognitive component of the image of a rural tourist destination was 0.957 > 0.60 [45], which confirmed that there was an adequate number of items for each factor. Furthermore, the value of Bartlett's test of sphericity [$\chi2$ (113) = 9168.38; $p < 0.001$] was significant, rejecting the null hypothesis that the correlation matrix was an identity matrix. Based on the results of the initial exploratory factor analysis, there were five items (FUNCT 3, MIX 1, MIX 2, MIX 3, MIX 5) that loaded on two factors in the preliminary three-factor structure.

Table 2. Dimensions and items of the cognitive component of the image of a rural tourist destination.

Dimensions	Items
FUNCT: Functional	Natural environment (FUNCT1)
	Historical sights, monuments, museums (FUNCT2)
	Fairs, festivals, wine routes (FUNCT3)
	Offer of activities in nature (FUNCT4)
	Availability of accommodation (FUNCT5)
	Local transport (FUNCT6)
MIX: Mixed	Environment (MIX1)
	Cleanliness and hygiene (MIX2)
	Local products offer (MIX3)
	Gastronomy (MIX4)
	Local culture (MIX5)
	Connection with the local population (MIX6)
PSYCHO: Psychological	Tranquility (PSYCHO1)
	Friendliness/hospitality (PSYCHO2)
	Customized offer (PSYCHO3)
	Quality services (PSYCHO4)
	Value for money (PSYCHO5)
	Quality accommodation (PSYCHO6)

After deleting five items, which cross-loaded on two factors, the final three-factor structure in this study was composed of 13 items. As shown in Table 3, six items for factor 1 represented the psychological dimension, five items for factor 2 represented the functional dimension, and two items for factor 3 represented the mixed dimension. Two factors (the psychological and functional dimension) had eigenvalues greater than one. Although the third factor (the mixed dimension) had an eigenvalue below 1, more precisely 0.645, it contributed with 4.964% to the total explained variance. The percentages explained by the psychological dimension were 58.436%, whereas the functional dimension explained 12.156% of the variance of the cognitive component of the image of a rural tourist destination. The dimension matrix after promax rotation was applied to ascertain which items were more relevant to each factor. In this research, all proposed 13 items for measuring dimensions of the cognitive component of the image of a rural tourist destination were catalogued by high loading factors in a range stretching from 0.667 to 0.933 (>0.50).

Table 3. Percentage of variances, eigenvalues, and factor loadings of dimensions of the cognitive component of the image of a rural tourism destination.

Constructs	Items	Eigenvalue	% of Variance	Factor Loading
PSYCHO	PSYCHO4	7.957	58.436	0.933
	PSYCHO6			0.900
	PSYCHO3			0.897
	PSYCHO5			0.879
	PSYCHO2			0.857
	PSYCHO1			0.730
FUNCT	FUNCT6	1.580	12.156	0.894
	FUNCT5			0.872
	FUNCT4			0.828
	FUNCT2			0.724
	FUNCT1			0.667
MIX	MIX6	0.645	4.964	0.855
	MIX4			0.744

In this study, EFA suggested a three-factor structure for the dimensions of the cognitive component of the image of a rural tourism destination. CFA was conducted to verify the factorial validity of the dimensions of the cognitive component of the image of a rural tourism destination. CFA can produce additional evidence related to the appropriateness of the proposed model with regard to the configuration of the factors recognized via EFA. The authors compared the models by means of chi-square (χ^2), χ^2/df, root mean square error of approximation (RMSEA), goodness-of-fit index (GFI), normed fit index (NFI), Tucker–Lewis index (TLI) and comparative fit index (CFI). A value of χ^2/df ≤ 5 is considered acceptable on samples larger than 200 [46,47]. A model that has a CFI and TLI above 0.95, RMSEA of less than 0.08, and GFI and NFI above 0.9 is considered to fit the data well [48,49].

Table 4 lays out the model specifications for the post hoc CFA. Goodness-of-fit statistics for the initial first-order model reveal that incorporation of the error covariance between FUNCT5 and FUNCT6 (E5 and E6), as well as PSYCHO1 and PSYCHO2 (E13 and E14), made a substantially large improvement to model fit. The factor structure in three-factor model with invariance error achieved adequate model fit (χ^2 = 204.295, χ^2/df = 3.405, RMSEA = 0.065, GFI = 0.948, NFI = 0966, TLI = 0.968 and CFI = 0.975). Therefore, the model of CFA laid out in Figure 1 is the finalized measurement model that indicates the structure of the cognitive component of the image of a rural tourism destination. In Figure 1, each observed variable (FUNCT1, FUNCT2, FUNCT4, FUNCT5, FUNCT6, MIX4, MIX6, PSYCHO1, PSYCHO2, PSYCHO3, PSYCHO4, PSYCHO5, PSYCHO6) has an error term (E1–E18) or measurement error, which reflects their adequacy in measuring the related underlying factors (FUNCT, MIX, PSYCHO). All factor loadings of the three dimensions ranged from 0.63 to 0.94. The results showed that the factor loadings exceeded the desirable standard of 0.50 [50].

Table 4. Model fit statistics for each hypothesized factor model.

Model	χ^2	χ^2/df	RMSEA	GFI	NFI	TLI	CFI
First-order without invariance error	296.650	4.785	0.082	0.920	0.950	0.950	0.960
First-order with invariance error	204.295	3.405	0.065	0.948	0.966	0.968	0.975

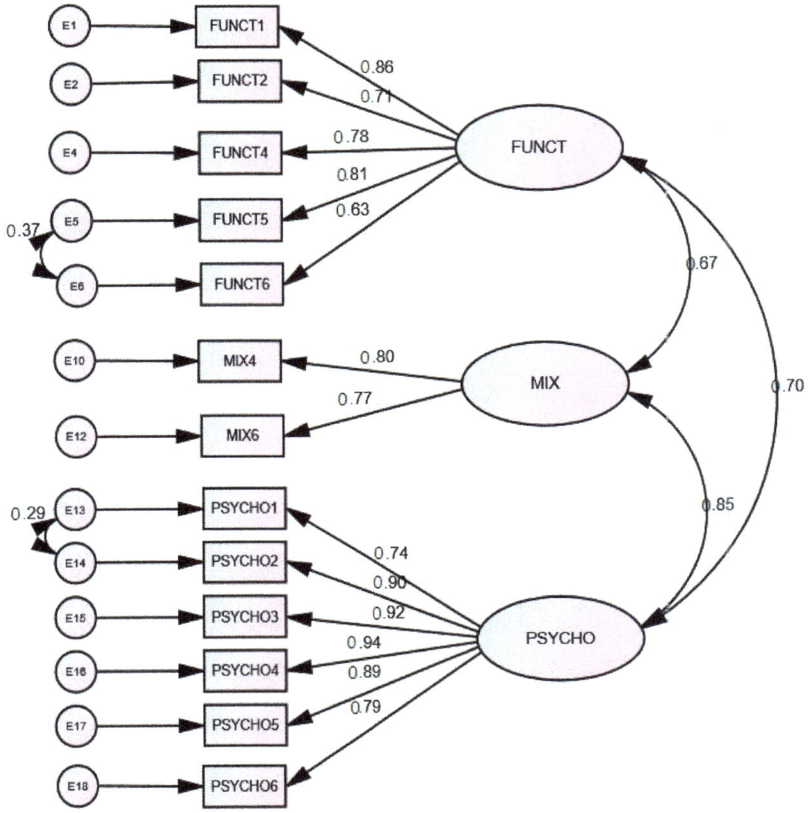

Figure 1. Confirmatory factor analysis—first-order model with invariance error.

The authors evaluated the measurement model through analysis of reliability, convergent, and discriminant validity (Table 5).

Table 5. Results of reliability, convergent, and discriminant validity testing.

	CA	CR	AVE	FUNCT	MIX	PSYCHO
FUNCT	0.880	0.899	0.643	*0.802*		
MIX	0.759	0.781	0.642	0.493	*0.801*	
PSYCHO	0.947	0.948	0.755	0.585	0.664	*0.869*

The authors assessed reliability by means of Cronbach's α (CA) and composite reliability (CR). Hair et al. [50] state that both Cronbach's α and composite reliability values for a construct must be higher than 0.7 for the construct to be deemed reliable. Table 5 clearly shows that all the constructs were reliable, as values for both Cronbach's α and composite reliability were well obviously above 0.7. The authors also assessed convergent validity by means of the average variance extracted (AVE). Hair et al. [50] claim that the AVE for every construct in the model must be above 0.5 for a measurement model to show sufficient convergent validity. It is evident from Table 5 that convergent validity was provided as AVEs for all constructs were higher than 0.5. Discriminant validity was analyzed using the Fornell–Larker criterion. The Fornell–Larker criterion prescribes that the AVE of all latent constructs should be higher than the highest squared correlations between any other construct, and the loadings of all indicators should be higher than all their cross-loadings [51]. Table 5 shows that each indicator loaded its highest values on its respective construct. The square root of the AVEs for all constructs was greater than the cross-correlation with other constructs. Therefore, the results laid out in the measurement model show that the psychometric properties for the latent constructs in the proposed model were good.

Based on the above findings, we can confirm the existence of three dimensions of the cognitive component of the image of a rural tourism destination, which addresses the first research objective.

A hierarchical factor structure was also examined in the research. Figure 2 illustrates the results of the second-order factorial structure for the general image of the rural tourism destination. In Figure 2, as well as Figure 1, each observed variable (FUNCT1, FUNCT2, FUNCT4, FUNCT5, FUNCT6, MIX4, MIX6, PSYCHO1, PSYCHO2, PSYCHO3, PSYCHO4, PSYCHO5, PSYCHO6) has an error term (E1–E18) or measurement error, which reflects their adequacy in measuring the related underlying factors (FUNCT, MIX, PSYCHO). Figure 2 also presents residual errors (E19–E21), which represent errors in prediction of endogenous factors (overall image) from exogenous factors (FUNCT, MIX, PSYCHO).

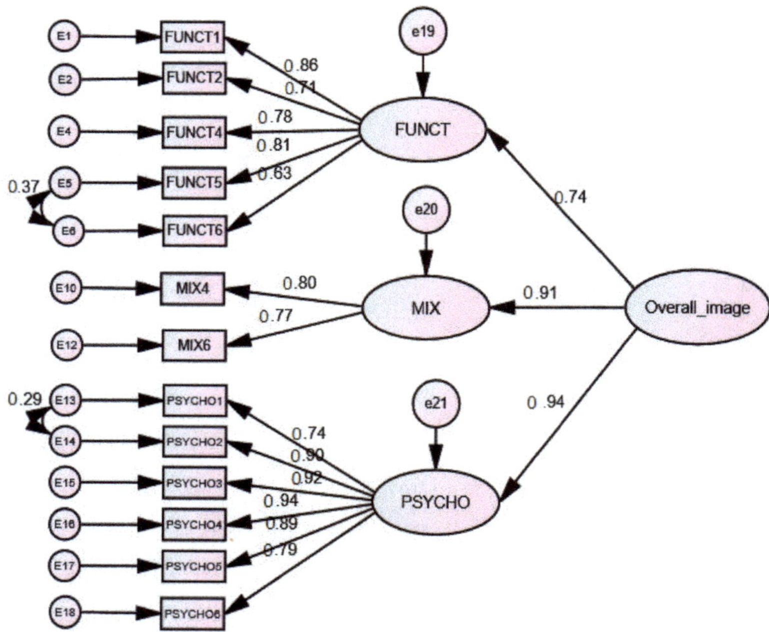

Figure 2. Confirmatory factor analysis—second-order measurement model for the general image of a rural tourism destination.

The path coefficients for the general image of the rural tourism destination varied among three dimensions: functional (0.77), mixed (0.91), and psychological (0.94). It can be concluded that the psychological dimension of the cognitive component of the image of a rural tourism destination makes the greatest, statistically significant ($p < 0.001$) impact on the general image of the rural tourism destination, which addresses the second research objective.

4. Discussion

The initial model in this study was generated based on earlier research, which revealed that the image of a rural tourism destination is essentially affected by the cognitive component and its dimensions. After deleting five items, the final three-factor structure was composed of 13 items:

six items representing the psychological dimension (It is a quiet/peaceful place, Staff and members of the local community provide tourists with warm welcome, It offers personalized service to tourists, It has quality services, It offers good value for money, There is high quality accommodation); five items representing the functional dimension (It is a nice/beautiful place with attractive natural scenery, There are many historical monuments/sites/museums, Tourists can undertake different activities in the countryside, The availability of accommodation is good, The local transport is good); and two items representing the mixed dimension (Partaking in local gastronomic activities/tours is good, Tourists can engage in social interactions with the members of the local community). From these results, we can confirm the existence of three dimensions of the cognitive component of the image of a rural tourism destination.

It should be noted that other authors too have considered many of the abovementioned dimensions. In their content analysis of four tourism destinations, Baloglu and Mangaloglu [52] measured the cognitive component using 14 items (good value for money, attractive surroundings and appealing landscapes, enjoyable climate, edifying cultural sites, comfortable lodging, appetizing local cuisine, exciting water sports, quality of infrastructure, personal safety, historical sites, unpolluted natural resources, good leisure activities, standard hygiene and cleanliness, friendly people). Quintal et al. [53] derived five factors (infrastructure, pastimes, perceived financial risk, local agricultural products, and climate) using 15 items, whereas Kim and Park [54] explored four factors (infrastructure; quality of experience; spots of interest; and value of surroundings, animation, and comfort) using 16 items for the cognitive component of the image of a tourism destination. In addition, while investigating the relationships of the cognitive, affective, and conative components of the tourism destination image, Basaran [41] defined 29 items within six dimensions of the cognitive component (scenery, cultural attraction, social setting and geographic location, infrastructure and facilities, accessibility, price and value). By examining the role of the image of a destination in tourist satisfaction Alcocer and Ruiz [20] composed the cognitive component using eight dimensions (natural resources; general infrastructure; tourism infrastructure; leisure and recreation tourism; culture, history, and art; natural environment; social environment; and political and economic factors).

This study also shows that the psychological dimension of the cognitive component of the image of a rural tourism destination has the most significant impact, followed by the functional dimension. These results are in line with the study of Alcaniz et al. [43], who proposed to explore in more detail the opinions on the functional and psychological continuum of the cognitive component. Similarly, Lin et al. [55] revealed in their study that not all dimensions representing the cognitive component have equal impact on tourists' destination choices. Thus, the final results of the EFA indicated that there were three cognitive dimensions explaining 59.9% of the variance (natural characteristics, amenities, and infrastructure).

5. Conclusions

Understanding the structure of the image of a rural tourism destination contributes significantly to developing a sustainable competitive advantage in the hospitality industry market. Due to the difficulty in measuring sustainability, it is essential to emphasize sustainability as the purpose of tourism and not necessarily the outcome [56]. Sustainable tourism should maintain a high level of tourist satisfaction and, at the same time, raise tourists' awareness about sustainability. Tourists tend to perceive various images that affect their behavior, so pinpointing the components of tourism destination image plays an important role in destination management and marketing. This study focused on the cognitive component of the image of a rural tourism destination, aiming to clarify its dimensions (functional, mixed, and psychological), because understanding these dimensions provides information that can help destination managers and marketers develop sustainable positioning strategies. The development of an appropriate image can contribute to further improvement in the development of the hospitality industry at a particular destination. Furthermore, creating the appropriate image is important because a positive image helps in positioning the destination in relation to its competitors. Creating a unique destination image has become imperative in order to survive in a competitive tourism market.

This study has several limitations. First of all, this research explored only the Province of Vojvodina as a typical rural tourism destination. Secondly, the sample included only domestic tourists as respondents. Finally, this study investigated only the cognitive approach to the image of a rural tourism destination. This fact limits the significance of tourism destination image as predictor of future tourist behavior. Future research should conduct this analysis repeatedly at a greater number of destinations and include foreign tourists and residents as well. It could also include more items such as "personal safety", especially having in mind the current situation with the COVID-19 pandemic. Future research could also incorporate the affective component of the image of a rural tourism destination so as to analyze its impact on the general image of a tourism destination.

Author Contributions: All of the authors formulated goals of the research and interpreted available literature; conducting and analyzing research was performed by K.L. and S.T.; while implications, limitations, and future research steps were developed by D.M. and N.V.Ć. All authors have read and agreed to the published version of the manuscript.

Funding: This research received no external funding.

Conflicts of Interest: The authors declare no conflict of interest.

References

1. Martinez, J.M.G.; Martin, J.M.M.; Fernandez, J.A.S.; Mogorron-Guerrero, H. An analysis of the stability of rural tourism as a desired condition for sustainable tourism. *J. Bus. Res.* **2019**, *100*, 165–174. [CrossRef]
2. Fennell, D.A. *Ecotourism*; Routledge: New York, NY, USA, 2007.
3. Almeida, F.; Pelaez, M.A.; Balbuena, A.; Cortes, R. Residents' perceptions of tourism development in Benalmadena (Spain). *Tour. Manag.* **2016**, *54*, 259–274. [CrossRef]
4. Andereck, K.L.; Nyaupane, G.P. Exploring the nature of tourism and quality of life perceptions among residents. *J. Travel Res.* **2011**, *50*, 248–260. [CrossRef]
5. Ivanović, V.; Gašić, M.; Simić, J. The Appliocation of Modern Information and Communication Technologies in Tourism. *Ann. Fac. Econ. Subot.* **2016**, *52*, 145–156.
6. Gavrila-Paven, I.; Barsan, M.C.; Lia-Dorica, D. Advantages and Limits for Tourism Development in Rural Area. *Procedia Econ. Financ.* **2015**, *32*, 1050–1059.
7. Gavrilović, Z.; Maksimović, M. Green Innovations in the Tourism Sector. *Strateg. Manag.* **2018**, *23*, 36–42. [CrossRef]
8. Simpson, M.C. Community benefit tourism initiative—A conceptual oxymoron? *Tour. Manag.* **2008**, *29*, 1–18. [CrossRef]
9. Okech, R.; Haghiri, M.; George, B.P. Rural tourism as a sustainable development alternative: An analysis with special reference to Luanda, Kenya. *Rev. Cult. Tur.* **2012**, *6*, 36–54.
10. Radović, G.; Košić, K.; Demirović, D. Finansiranje kao ključni element strategije održivog razvoja ruralnog turizma u Republici Srbiji. *Ekon. Poljopr.* **2018**, *65*, 413–426. [CrossRef]
11. Sivesan, S. Sustainable Tourism Development in Jaffna District. *J. Tour. Hosp.* **2020**, *9*, 1–6.
12. Ayazlar, G.; Ayazlar, R.A. Rural Tourism: A Conceptual Approach. In *Tourism, Environment and Sustainability*; St. Kliment Ohridski University Press: Sofia, Bulgaria, 2015.
13. Bulin, D. Sustainable rural tourism strategies. A tool for development and conservation. *Qual. Access Success* **2011**, *2*, 102–111.
14. Dimitrovski, D.D.; Todorović, A.; Valjarević, A. Rural tourism and regional development: Case study of development of rural tourism in the region of Gruža, Serbia. *Procedia Environ. Sci.* **2012**, *14*, 288–297. [CrossRef]
15. Yilmaz, G.O. The Role of Rural Tourism in Rural Development. In *Global Issues and Trends in Tourism*; St. Kliment Ohridski University Press: Sofia, Bulgaria, 2016.
16. Trukhachev, A. Metodology for Evaluating the Rural Tourism Potentials: A Tool to Ensure Sustainable Development of Rural Settlements. *Sustainability* **2015**, *7*, 3052–3070. [CrossRef]

17. Irshad, H. *Rural Tourism—An Overview*; Government of Alberta, Agricultural and Rural Development: Alberta, Canada, 2010.
18. Lane, B. Rural Tourism: An overview. In *Handbook of Tourism Studies*; SAGE Publications: Thousand Oaks, CA, USA, 2009.
19. Grubor, A.; Leković, K.; Tomić, S. Rural Tourism Marketing of the Danube Region. *Int. J. Econ. Theory Pract. Soc. Issues Ekonomika* **2019**, *65*, 18–19.
20. Alcocer, N.H.; Ruiz, V.R.L. The role of destination image in tourist satisfaction: The case of a heritage site. *Econ. Res.* **2020**, *33*, 2444–2461.
21. Lo, M.C.; Mohamad, A.A.; Songan, P.; Yeo, A.W. Positioning rural tourism: Perspectives from the local communities. *Int. J. Trade Econ. Financ.* **2012**, *3*, 59–65. [CrossRef]
22. Kulcsar, N. Rural tourism in Hungary: The key of competitiveness. In Proceedings of the FIKUSZ 09, Symposium for Young Researchers, Faculty of Economics, Budapest, Hungary, 13 November 2009.
23. Fons, M.V.S.; Fierro, J.A.M.; Patino, M.G. Rural tourism: A sustainable alternative. *Appl. Energy* **2011**, *88*, 551–557. [CrossRef]
24. Carballo, M.M.; Arana, J.E.; Leon, C.J.; Moreno-Gil, S. Economic valuation of tourism destination image. *Tour. Econ.* **2015**, *21*, 741–759. [CrossRef]
25. Stylos, N.; Vassiliadis, C.A.; Bellou, V.; Andronikidis, A. Destination images, holistic images and personal normative beliefs: Predictors of intention to revisit a destination. *Tour. Manag.* **2016**, *53*, 40–60. [CrossRef]
26. Assaker, G. Examining a hierarchical model of Australia's destination image. *J. Vacat. Mark.* **2014**, *20*, 195–210. [CrossRef]
27. Lai, K.; Li, X. Tourism destination image: Conceptual problems and definitional solutions. *J. Travel Res.* **2016**, *55*, 1–16. [CrossRef]
28. Agapito, D.; Valle, P.; Mendes, J. The cognitive—Affective—Conative model of destination image: A confirmatory analysis. *J. Travel Tour. Mark.* **2013**, *30*, 471–481. [CrossRef]
29. Michael, N.; James, R.; Michael, I. Australia's cognitive, affective and conative destination image: An Emirati tourist perspective. *J. Islamic Mark.* **2018**, *9*, 36–59. [CrossRef]
30. Zhang, H.; Fu, X.; Cai, L.A.; Lu, L. Destination image and touris loyalty: A meta-analysis. *Tour. Manag.* **2014**, *40*, 213–223. [CrossRef]
31. Huang, S.; Gross, M.J. Australia's destination image among mainland Chinese travelers: An exploratory study. *J. Travel Tour. Mark.* **2010**, *27*, 63–81. [CrossRef]
32. Kokkali, P.; Koutsouris, A.; Chrysochou, P. Cognitive components of rural tourism destination images: The case of lake Plastiras, Greece. *Tourismos* **2009**, *4*, 273–291.
33. Pike, S.; Ryan, C. Destination positioning analysis through a comparison of cognitive, affective and conative perceptions. *J. Travel Res.* **2004**, *42*, 333–342. [CrossRef]
34. Pike, S. *Destination Marketing*; Butterworth-Heinemann: Burlington, MA, USA, 2008.
35. Hallmann, K.; Zehrer, A.; Muller, S. Perceived destination image: An image model for a Winter sports destination and its effect on intention to revisit. *J. Travel Res.* **2014**, *54*, 94–106. [CrossRef]
36. San Martin, H.; Rodriguez del Bosque, I.A. Exploring the cognitive-affective nature of destination image and the roll of psychological factors in its formation. *Tour. Manag.* **2008**, *29*, 263–277. [CrossRef]
37. Asli, D.; Tasci, A. A semantic analysis of destination image terminology. *Tour. Rev. Int.* **2009**, *13*, 1–15.
38. Tasci, A.; Gartner, W. Destination image and its functional relationships. *J. Travel Res.* **2007**, *45*, 413–425. [CrossRef]
39. Marine-Roig, E. Destination Image Analytics Through Traveller-Generated Content. *Sustainability* **2019**, *11*, 3392. [CrossRef]
40. Chen, J.S. A case of Korean outbound travelers' destination images by using correspondence analysis. *Tour. Manag.* **2001**, *22*, 345–350. [CrossRef]
41. Basaran, U. Examining the Relationships of Cognitive, Affective, and Conative Destination Image: A Research on Safranbolu, Turkey. *Int. Bus. Res.* **2016**, *9*, 164–179. [CrossRef]
42. Qu, H.; Kim, L.H.; Im, H.H. A model of destination branding: Integrating the concepts of the branding and destination image. *Tour. Manag.* **2011**, *32*, 465–476. [CrossRef]
43. Alcaniz, E.B.; Garcia, I.S.; Blas, S.S. The functional-psychological continuum in the cognitive image of a destination: A confirmatory analysis. *Tour. Manag.* **2009**, *30*, 715–723. [CrossRef]

44. Polo Pena, A.I.; Jamilena, D.M.F.; Molina, M.A.R. Validation of cognitive image dimensions for rural tourist destinations: A contribution to the management of rural tourist destinations. *J. Vacat. Mark.* **2012**, *18*, 261–273. [CrossRef]

45. Kaiser, H. An index of factorial simplicity. *Psychometrika* **1974**, *39*, 31–36. [CrossRef]

46. Marsh, H.W.; Hocevar, D. Application of confirmatory factor analysis to the study of self-concept: First-and higher order factor models and their invariance across groups. *Psychol. Bull.* **1985**, *97*, 562–582. [CrossRef]

47. Wheaton, B.; Muthen, B.; Alwin, D.F.; Summers, G.F. Assesing reliability and stability in panel models. *Sociol. Methodol.* **1977**, *8*, 84–136. [CrossRef]

48. Hu, L.T.; Bentler, P.M. Cutoff criteria for fit indexes in covariance structure analysis: Conventional criteria versus new alternatives. *Struct. Equ. Model. Multidiscip. J.* **1999**, *6*, 1–55. [CrossRef]

49. Kline, R.B. Methodology in the Social Sciences. In *Principles and Practice of Structural Equation Modeling*; Guilford Press: New York, NY, USA, 2011.

50. Hair, J.F.; Black, W.C.; Babin, B.J.; Anderson, R.E. *Multivariate Data Analysis*; Cengage Learning: New York, NY, USA, 2019.

51. Fornell, C.; Larcker, D. Structural equation models with unobservable variables and measurement error. *J. Mark. Res.* **1981**, *18*, 39–50. [CrossRef]

52. Baloglu, S.; Mangaloglu, M. Tourism destination images of Turkey, Egypt, Greece, and Italy as perceived by US-based tour operators and travel agents. *Tour. Manag.* **2001**, *22*, 1–9. [CrossRef]

53. Quintal, V.; Phau, I.; Polczynski, A. Destination brand image of Western Australia's South-West Region: Perceptions of local versus international tourists. *J. Vacat. Mark.* **2014**, *20*, 41–54. [CrossRef]

54. Kim, S.; Park, E. First-time and repeat tourist destination image: The case of domestic tourists to Weh Island, Indonesia. *Int. J. Tour. Hosp. Res.* **2015**, *26*, 421–433. [CrossRef]

55. Lin, C.H.; Morais, D.B.; Kerstetter, D.L.; Hou, J.S. Examining the Role of Cognitive and Affective Image in Predicting Choice Across Natural, Developed, and Theme-Park Destinations. *J. Travel Res.* **2007**, *46*, 183–194. [CrossRef]

56. Arrage, J.A.; Hady, S.A. Ecotourism and sustainability: Practices of the Lebanese nature-based operators. *Hotel Tour. Manag.* **2019**, *7*, 11–23.

Publisher's Note: MDPI stays neutral with regard to jurisdictional claims in published maps and institutional affiliations.

 *sustainability*

Review

Green and Sustainable Public Procurement—An Instrument for Nudging Consumer Behavior. A Case Study on Romanian Green Public Agriculture across Different Sectors of Activity

Rocsana Bucea-Manea-Țoniș [1,2], Oliva Maria Dourado Martins [3], Dragan Ilic [4], Mădălina Belous [5,*], Radu Bucea-Manea-Țoniș [6], Cezar Braicu [6] and Violeta-Elena Simion [5]

1 Department of Economic Sciences, Spiru Haret University, 030167 Bucharest, Romania; rocsense39@yahoo.com
2 Doctoral School, University of Physical Education & Sports, C.P. 060057 Bucharest, Romania
3 Instituto Politécnico de Bragança, 5300-253 Bragança, Portugal; oliva.martins@ipb.pt
4 Faculty of Economics and Engineering Management, University Business Academy, 21000 Novi Sad, Serbia; prof.dragan.ilic@gmail.com
5 Faculty of Veterinary Medicine, Spiru Haret University, 030045 Bucharest, Romania; ushmv_simion.violeta@spiruharet.ro
6 Faculty of Economic Sciences, Hyperion University, 030615 Bucharest, Romania; radub_m@yahoo.com (R.B.-M.-Ț.); cezar_braicu@hotmail.com (C.B.)
* Correspondence: madalina.belous@spiruharet.ro; Tel.: +40-722-306-189

Abstract: Green Public Procurement (GPP) became an efficient instrument to achieve the objectives of environmental policy expressed by the European Commission in its Communications. At the same time, it must be addressed by the public authorities as a complex process, in which all purchased goods and services must integrate perfectly into an entire puzzle-like system of legislation, the construction field, innovation, healthcare, food, and education. Scientific references published in the Web of Science (WoS) mainly between 2017 and 2020 were investigated, and they analyze the implications of green public procurement in various fields, as presented by scientific communities. This article brings as a novelty in this context the identification of some barriers in the adoption of these processes, so that they can be overcome. Based on good practices and international standards and trends, the article shows how aspects related to the implementation of green procurement in society can be taken into account. In the second stage, we added a case study on Romanian green agriculture and discussions regarding inter-correlation between different fields and GPP.

Keywords: sustainable public procurement (SPP); green public procurement (GPP); consumer behavior; sustainability; GPP barriers; green Romanian agriculture

Citation: Bucea-Manea-Țoniș, R.; Dourado Martins, O.M.; Ilic, D.; Belous, M.; Bucea-Manea-Țoniș, R.; Braicu, C.; Simion, V. Green and Sustainable Public Procurement—An Instrument for Nudging Consumer Behavior. A Case Study on Romanian Green Public Agriculture across Different Sectors of Activity. *Sustainability* **2020**, *13*, 12. https://dx.doi.org/10.3390/su13010012

Received: 16 November 2020
Accepted: 18 December 2020
Published: 22 December 2020

Publisher's Note: MDPI stays neutral with regard to jurisdictional claims in published maps and institutional affiliations.

1. Introduction

The procurement of goods, services, and works, in particular green public procurement (GPP), must be done with as little impact on the environment as possible. Moreover, Roehrich shows in a study that green supply chain management (GSCM) should exist before moving to procurement [1]. There is a need for permanent information on suppliers and finding the best short-term solutions with long-term impacts [1]. Furthermore, Circular Supply Chain Management (CSCM) offers a new and compelling perspective on the field of supply chain sustainability. Farooque et al. identifies a number of important directions that are not sufficiently covered and require further study in the future [2]. Of these, the collaboration in the supply chain and factors and barriers of CSCM are the ones we considered important [2].

GPP, at the same time, is an indicator of the "CE monitoring framework" for a circular economy [3–5]. Other indicators were identified: self-sufficiency for raw materials; waste

257

generation; food waste; recycling rates; recycling/recovery for specific waste streams; contribution of recycled materials to raw materials demand; trade-in recyclable raw materials; private investments, jobs, and gross value added. In line with the European Commission's clarifications on green and sustainable public procurement, many EU public authorities are implementing GPP as part of a broader approach to sustainability in their procurement. This process also addresses economic and social issues [6]. However, it is not the policy objectives that are changing; tools and techniques are those that change, including in the procurement system where [6] points out that the awarding of public procurement contracts, having as an award criterion the social aspects, was an important objective for the integration in a bigger market. These possibilities need to be considered in the future; economic crises can bring major changes—they can be a boost or budgetary constraints can change direction [7].

The European Commission considers that Sustainable Public Procurement (SPP) is [8] a method used by public authorities to accomplish the best equilibrium between economic, social, and environmental pillars of sustainable development, during the different stages of procuring goods, services, or works [6]. Thus, SPP implementation consists of six different aspects: Green Public Procurement (GPP), Internal Social Criteria (ISC), Social Return on Investment (SROI), Bio-based Public Procurement (BPP), Circular Economy (CE), and Innovation-oriented Public Procurement (IPP). There were designs including specific toolboxes for supporting SPP, including practices, management, and inter-organizational dimensions [8].

The market experience—a survey on public procurers in Holland [8] on the importance of knowledge and competencies, skills in SPP, affective engagement to SPP, and organizational education and information capacity regarding SPP types—proved that they are generally good abilities for an organization to hold/own, but that they otherwise do not have a direct positive effect on GPP. The knowledge background regarding sustainability has a positive impact on implementing GPP. Organization operationalized as affective commitment did affect GPP, but none of the other types of SPP. Organizational learning capacity influenced most types of SPP, IPP, GPP, and CIE. In conclusion competencies, skills, motivation, and convenience influence GPP but not all aspects of SPP [9].

2. Theoretical Framework

GPP can potentially be a very effective tool to develop environmental policies for creating competitive advantage when costs are reduced and, most importantly, resulting in positive environmental impacts. Environmental considerations that are rarely applied in public procurement can be effective and must be included in the technical specifications, award criteria, and performance clauses of the GPP contract. Contract performance clauses may also be used by public contracting entities to introduce environmental considerations into the procurement process. The lack of adequate regulations at national level and the lack of trained staff in this field may be some of the causes of failure of GPP implementation [10]. Improving the technical knowledge, building capacity of the procurement workforce and the knowledge and competences of the procurement officials on the matter will facilitate the implementation of GPP with a positive impact on the environment [11].

Environmental policy instruments, such as GPP, should be focused on the impacts of implementation, but also evaluate the efficiency of the instrument. This can be assessed on the basis of cost and achievement of objectives, and its long-term value-added potential. The state has to nudge the eco-innovation through GPP to achieve SPP. [12] According to [13], after a systematic review (80 practices identified by the European Union and implemented by governments in 80 countries) GPP practices are based on several dimensions: geographical origin, government capacity to implement, criteria used for implementation as the period in which this is done, and the impact of these practices on the environment but also on the economy. According to the authors, in order to reduce costs and technical complexities, the tender can be divided into sections and thus integrate demand from small buyers extending the participation to small suppliers. Aiming SPP, GPP can develop a joint procurement of small Governments [13].

While seeking positive economic results, particularly in the short term, productive activities pay less attention to the common interest or the environment. Due to its short-term objectives, this productive methodology has not realized the importance of people and the environment for the future of activities and our planet. Good practices for sustainable development are crucial. Moreover, urban sustainability, food, healthcare, and education were considered some important areas in terms of the sustainability of the consumer's behavior. It is necessary to change the way we think and use an integrated approach to discover what it means to produce, process, deliver, use, and especially recover and regenerate products [14]. In our research, we chose the most recent research work (2017–2020), although important previous references were thus excluded so as to highlight the trend in research in this field.

At the same time, we aimed to fill the gap regarding the transition from GPP to SPP, focusing on urban sustainability in building design and construction, trend in food consumer behavior for a sustainable environment, healthcare services, good practices in education for sustainable development, GPP innovation and its influence on consumer behavior and especially green public procurement applied in organic agriculture in Romania. Table 1 presents a general description of the GPP applied in organic agriculture in Romania.

Table 1. Green public procurement (GPP) applied in organic agriculture with a focus on Romania.

Dimension	GPP Applied in Organic Agriculture
Supply chain management (SCM); Circular economy (CE)	SCM is a crucial element in the approach of nudging consumer behavior to SPP, as an information tool to find the best supplier and the best short-term solutions with long-term impact. Circular Supply Chain Management is even more important because provide information on green raw materials, food waste, trade-in recyclable raw materials, etc. [1–4]
GPP across different sectors	GPP is studied as a tool in the EC approach of meeting Green Deal objectives. GPP applied in different sectors of activity has a positive influence on organic agriculture. Regarding organic agriculture in Romania, we have identified two gaps: constant failure to support this topic through research and more scientific references in this field and the need to implement measures to ensure sustainability. As noted in the presentation of the case study, some government measures (e.g., subsidies) have had a positive impact on the development of this sector. Our research highlights the problems in the organic agriculture sector of the Romanian economy. It shows that profit losses for farmers can be minimized but also the need to implement green measures in the context. The study can contribute to taking measures on continuous review and completion of national legislation in this field; consumer information campaigns about organic food, maintaining the health of the population; promotion in schools through education programs in the field of organic agriculture; developing scientific research so as to provide green solutions to different sectors of activity [5–16]

Table 1. *Cont.*

Dimension	GPP Applied in Organic Agriculture
Trend in Food Consumer's Behavior	Agricultural production combines best environmental practices, maintains biodiversity, and contributes to the conservation of natural resources, supporting animal husbandry and welfare. The gap in this field has been identified in the measures by which agriculture can comply and respects consumers' preferences for healthy products. Ecolabels are seen as a legitimate tool to mitigate the consequences in the public procurement supply chain for the entire life of the cycle of product development (resource acquisition, manufacture, packaging, and transportation, use, end-of-life) [17–33]
The impact of GPP on Healthcare Services	In the healthcare sector, there are important concerns regarding green policies and procurement, the impact of products on the environment, inconsistent organizational strategies on GPP, a high value of cost/benefits balance of green products inefficient supplier value-chain, governmental law and nudges on GPP [34–40]
GPP Innovation and its Influence on Consumer Behavior	Government agencies use GPP as a tool to nudge eco purchasing to avoid waste and pollution. They offer detailed information on price, performance, and other criteria for products, services, and raw materials. They facilitate environmental innovation that reduces environmental pressure. Eco-labels are efficient tools for nudging green consumer behavior, based on environmental or non-environmental primary GPP standards for "off-the-shelf" raw material or services [41–58]
Urban Sustainability; Building Design and Construction	Ecological construction materials are usually associated with innovation in manufacturing processes having incenting GPP with positive effects on a sustainable environment. GPP facilitates sustainable infrastructure for the agri-food sector (e.g., in rooftop farming) [59–66]
Good Practices in Education for Sustainable Development	SPP within the education sector and the staff and academic behavior regarding GPP represents a good example for the next generations and important drivers for sustainable professional, economic, and ethical concerns [67–71]
Barriers	The principal barriers related to GPP and SPP refer to unused land, insufficient energy, and an inappropriate transportation system. They are manifested as a need to adapt the governance challenges and create an equidistant balance between socio-environmental and social cohesion aspects, for surpassing the dysfunctionality in the sustainability ecosystem [72–85]

After analyzing all the implications of GPP in different sector of activity, we decided to make a case study on Romanian agriculture, starting from data collected from the website of the Ministry of Agriculture and Rural Development regarding dynamics of operators and areas used in organic farming in the period 2010–2019. Thus, we analyzed the extent to which some instruments, such as legislation, supported the implementation of policies in this sector. Public procurement is influenced by regulations and directives, and government legislative reforms are mostly the starting point in guiding them. If this trend continues, they may at some point become better known and anchored in practice. Their analysis and impact can be done on more versatile backgrounds, with the perspective of evolution and maturation in the future [15].

Grandia and Voncken says that "ability, motivation, and opportunity affect GPP but not all types of SPP" [16]. SPP can be achieved mainly through GPP and IPP as a branch of GPP. The center of this research concerns on the relevance of implementing GPP. In this regard, we highlighted some barriers that can be found in implementation and suggested a five-step plan that can be done before implementing any change.

3. Methodology

Our study with references from the Web of Science (WoS) mainly from 2017 to the present opens interesting perspectives regarding the impact of green public procurement on the future market and research, building an interdisciplinary bridge between science and the economy to reach the target of sustainability. The main topics were public green procurement in construction, food, healthcare sector, education, and the circular economy (waste management).

For the study we used a percolated systematic review of highly quoted literature on GPP impact on sustainability, implementing Preferred Reporting Items for Systematic Reviews and Meta-analysis (PRISMA) guidelines. As criteria, we chose: (i) highly quoted papers, gathered especially from Scopus, and Web of Science databases, (ii) authentic experimental research and review publications, (iii) published between mainly 2017 and 2020, (iv) presented in English, and (v) having as main search term GPP. In our research, we excluded: (i) book chapters and (ii) encyclopedia. Paper percolating was implemented using one of the most performant tools designed to extract and manage articles—Systematic Review Data Repository (SRDR). No institutional ethics approval was launched because the papers were publicly accessible.

In our quantitative research the main topics searched on Scopus and WoS databases, were "GPP in construction", "GPP in healthcare and food", "GPP barriers", "GPP in education", "GPP and circular economy", and "GPP innovation". We selected 80 articles that meet the criteria, with consistent contribution and different content that cover the period 2017–2020 for GPP in different sectors of activity. Although the GPP has already been studied from different perspectives, there is a lot to understand and reflect on, especially if the objective is achieving a Sustainable Public Procurement. In this sense, this research considered unused land, lack of funding, lack of knowledge as the main GGP gap that remains to be explored, especially related to the organic agriculture sector in Romania. Regarding organic farming in Romania, research has not yet been able to provide the necessary support to ensure sustainability. The subsidies have had some positive impacts but there are many problems to be solved. There is a need to implement sustainable measures, and the contribution of this study can help to implement measures and continuously review legislation and promote sustainability.

Thus, the study presents the perspectives and impact of GPP in different fields, such as healthcare and food, innovation, education, etc., that will be detailed in the next sections. The implementation of GPP should be integrated at the same time, in all these sectors, by all the stakeholders, to have a concomitant positive impact on consumer behavior and sustainability.

In the particular situation of the ecological operators in Romania, which we pursued as an objective, we collected and processed data from the official authorities, in order to follow the dynamics and impact of the legislation. The collected data was arranged in tables and descriptive statistics, such as frequencies, mean percentages, and histograms were used for the analysis of variables. In the second stage, we designed a linear regression model that assumes the following: the total area of organic farming is represented especially by cereals and green harvested plants. Organic farming loses a lot of profit because of the uncultivated land. Other cultures (dried legumes and protein crops, tuberous and root plants, industrial crops, vegetables, and permanent crops) have very weak representation. The farmers should use all the land available and plant organic seeds bought through GPP. Our analysis emphasizes inappropriate management in the Romanian agri-food field, due to different factors, but mainly due to lack of funding and knowledge. Here we see the importance of education that represents an important pillar in forming performant managers, that will be able to cultivate the appropriate type of crops following market requests, implementing circular economy principles, and making GPP for the agri-food sector. Education has a tight connection with innovation: the interaction between HEIs and innovator economic agents results in a positive impact on the agri-food sector, through IPP tools. Construction fields provide an appropriate infrastructure in agri-food sectors. Thus, the interconnection between different fields creates all the premises for green sustainable agriculture, that will provide healthy ecological food. In the end, the commitment to gain sustainable agriculture will impact positively the population's health. Most of all our article shows that all the fields analyzed have something in common—from GPP to SPP—an instrument for nudging consumer behavior to gain sustainability and a healthy population.

4. Trend in Food Consumer's Behavior for a Sustainable Environment

The concept of GPP has been widely recognized in recent years as a useful tool for promoting green products and services and reducing the impact on the environment of the activities of public authorities. Guidelines are provided on procurement for sustainable development not only at the European level (e.g., European Union) but also internationally (e.g., United Nations) [17].

The implementation of GPP by governments as policy instruments is unpredictable in the effect, although it usually aims to increase the purchase of organic food by the public sector and is expected to have a positive effect on increasing the area of organic agricultural land in a country. To achieve that, it is necessary to take into account the high variability of some environmental factors such as climate, soil quality as well as the specificity of the market [18]. The development of these practices is difficult, often impossible, but requires the existence of a general framework and as many good practices as possible specific to the public sector. These practices need an integrated promotion and dissemination mechanism, in the context presented by Elkington's "triple bottom line", a term which states that "business success depends not only on profitability but also on environmental quality and social justice". After 25 years from this statement, in 2019 Elkington made a recall for an adjustment to this term starting from the fact that " . . . success or failure on sustainability goals cannot be measured only in terms of profit and loss. It must also be measured in terms of the wellbeing of billions of people and the health of our planet, and the sustainability sector's record in moving the needle on those goals has been decidedly mixed. While there have been successes, our climate, water resources, oceans, forests, soils, and biodiversity are all increasingly threatened. It is time to either step up—or to get out of the way." [19]. These three dimensions of sustainable development through economic and social perspectives and environmental impact are also found in the 17 interconnected Sustainable Development Goals (SDGs) of the United Nation's Agenda 2030 [20] and they need to be adopted, implemented, and evaluated continuously, depending on the international, national and regional context. At the same time, they can be included in the general terms of sustainable public procurement (SPP) [21]. Neto (2020) analyzes in a study the current practices regarding public procurement of food and food services in the EU. He analyzes how they cover sustainability aspects by using distinct schemes for food purchase and catering but also sales of equipment in this field. The schemes available to the public in 11 European countries show that approximately 30 different sustainability criteria can be identified and used in procurement schemes. The most important is the environmental criteria where the focus is on two main aspects: the type of food products where the criteria include organic products, seasonality, packaging including the recycling process and fishery products certified following current regulations; secondly, the service acquisition for which the criteria are mainly dedicated to staff training, waste management, menu planning, optimized transport and implementation of an environmental management system. Other aspects follow the source of food products for which the criteria refer mainly to the integrated production and typology of the equipment used for which it is required; for example, that they respect the energy and water efficiency labeling. In addition to environmental issues, other criteria are recommended in the context of GPP, namely ethical criteria (related to animal welfare, for example), social criteria (which meet the needs of less developed countries), and health criteria (food safety systems, product traceability, etc.) [5,17].

The literature analyzed also highlights the concept of Food Loss and Waste (FLW) regarding edible products from plants and animals that are lost or not consumed by people. This food still has enough quality to be consumed but is discarded and considered waste because it has no commercial value. It can result from negligence as well as a conscious decision to throw food away. On the one hand, cities offer a huge variety of food. On the other hand, food waste is a huge urban problem in terms of sustainability.

For the twenty-eight member countries of the European Union (EU), FLW has been estimated at eighty-eight million tons in 2012, including inedible waste, which is equivalent to one hundred and seventy-three kilograms of waste per person per year, only in the European Union. Considering that food production was 865 kg/year, the FLW was approximately twenty percent of the total food produced [22].

The European Commission (EC) is committed to fighting FLW. In terms of impact, the FLW has a huge impingement, namely regarding resource efficiency, consumption, and waste management. These impacts also have some repercussions on the final cost. Furthermore, due to the use of natural resources, such as water and energy, FLW has a large influence on climate change as well. Thus, it is important to rationalize resources, prevent waste, reuse, and recycle materials, in an effort to stimulate the transition from a linear to a circular economy. This approach would drive global competitiveness as well as promoting sustainable growth and development [23,24].

Many international research projects have proven the importance of prioritizing actions for an effective outcome. The FUSIONS Project was a European research endeavor sponsored by the EC that defined a hierarchy of efficient use of resources. This started with prevention at the source and the recovery of edible food, which prioritizes human food over animal feed or reprocessing into non-food products. These extended to recycling resources, energy recovery and conversion, and residue disposal [22]. This hierarchy can result in three dimensions of actions: the first one would be in the sphere of prevention; the last one in monitoring and control the FLW. In the middle are the actions which have the objective to reduce it.

To mitigate the problem, the objective to combat FLW should be based on prevention first. When that is not possible, the cities should look towards its reduction. Finally, the only thing that remains is to monitor and try to control waste. Several initiatives have been taken in recent years to prevent food waste (e.g., a multi-stakeholder platform dedicated to food loss and waste prevention, a food waste prevention calculator, based on life cycle thinking), but evaluation methods for prevention are still needed to identify the best practices [25]. Prevention is one of the important issues [26–28]. It is part of the new circular economy package, which aims to promote sustainable consumption patterns, mainly in restaurants, catering, and households. During the purchase, the consumer can reduce the amount expended to prevent FLW [29].

Reducing can be explored by better plans on the production, processing, and manufacturing, as well as on the distribution chain. Less impulsive buying and consumption behaviors can also reduce FLW. The motivating factors for preventing or reducing FLW refer to voluntary behavior. In this sense, the individual should take small steps to decrease waste. It is a predictive ability that can be described in terms of solving problems. Regarding voluntary behavior, motivation is the keyword for FLW reduction [30]. The conceptual framework combines explicit or conscious factors related to opportunities, with implicit or unconscious factors, mainly related to motivations [31].

The evidence showed that FLW occurs along the entire food supply chain. At this level of waste, since the data already collected by Eurostat has been recognized as manifestly insufficient, it is crucial to develop quantitative and qualitative data monitoring actions, in parallel with other actions to reduce or prevent waste [22]. Finally, at the monitoring level, FLW assumes high values within the EU as well as in other developed countries. To control FLW, the EU approved a common methodology for all members which intends to measure FLW. Unfortunately, it has not yet been initiated [32]. Romania agreed with the methodology but was not able to implement it.

The whole economy is tied to the intensification of consumption, which can increase waste. The main paradigm is between higher consumption and preserving the environment. However, rationalizing resources and implementing environmentally friendly practices are fundamental complementary attitudes for achieving long-term sustainability. The current generation must be concerned with leaving a better planet. At least, this generation would maintain the same level of well-being for future generations. Recovery, reduction, recycling, and reuse are needed for environmental sustainability and human survival [33].

5. The Impact of GPP on Healthcare Services

In the healthcare sector, there are some challenges frequently manifested in the process of implementing GPP (Figure 1).

Figure 1. GPP challenges in the healthcare field (Source: adaptation after Ahsan, 2017, [34]).

These challenges have specific manifestations in different countries.

For example, in Australia, the highest concerns are related to insufficient legislation on GPP, management, and backing for green projects, GPP government nudges, and inefficient financial assistance. They have a solid background in ecological issues knowledge and supplier issues. If the GPP strategic challenges are overpassed, the decision-making process and operational implementations are easy to be assured in the Australian health sector [34].

In the UK and Italy, GPP in the health sector has an important weight and a positive effect on return on investment (ROI) for the public purse. Supplier issues register improvement in the green performance area. In the UK GPP has become a cultural standard implemented voluntarily, while in Italy the regulations on green performances and safety have impacted societal (consumer and organizational) behaviors. The worth of green products (the cost/benefit balance of green products and services) seem to be an important challenge in these countries, especially the carbon footprint that local stakeholders try to mitigate as to gain a trustful brand, on one side, and the ethical concerns, on the other side [34]. Regarding packaging, UK organizations prefer to adopt their GPP national standards (ISO 26000, SA8000, and AA1000). Social requirements for suppliers (child labor, forced labor, discrimination, and freedom of association) are implemented on GPP international standards for both Italy and the UK. They focus on the good implementation of the ISO 50001 standard on energy administration and household gas mitigation [35]. In Romania, this field is sub-represented.

GPP innovation has an ethical and social purpose and a systemic impact on healthcare. Thus, procurement offices have been established with the aim of intermediating devices for brand evaluation to design criteria on purchasing conditions and measures to control markets over time. GPP innovation is not dependent on R&D intensive technological change, [36], but a result of market procurement processes that raise sectorial goals [37]. The procurement offices frequently use the instrument of "total costs" for evaluation and budgeting the green life cycle impacts and to expose the conflicts between objectives on special domains and social innovation [37]. Unfortunately, the procurement offices might not have authority regarding innovation opportunities for the "high tech" medical products, as they have in consumables and non-clinical products [38]. In some cases, the lowest cost green product, which will bring short-term cost savings, is not the best solution for the long term, neither financially nor from a healthcare perspective [39,40]. Thus, procurement offices should include specialists in different healthcare sectors to advise on long-term alignment in their procurement.

6. GPP Innovation and Its Influence on Consumer Behavior across Different Countries

GPP is an investment process based on public policy to answer consumer needs for services, goods, utilities, and works, boosting their advantages as well as impacting the whole world by protecting the environment at the same time. To avoid waste and pollution, governmental programs have been designed concerning environmental consequences when deciding on eco purchasing. These programs evaluate price, performance, and other criteria for products, services, raw materials [41]. Sustainable procurement, implicitly GPP, is one of the goals of The United Nations 2030 Agenda for Sustainable Development. In this regard, the European Commission's Green Public Procurement action established as objective the enhancement in the environmental, energy, and social achievements of products and services through innovative developments [42].

IPP is increasingly contributing to sustainability through its aim in nudging the uptake of EI (environmental innovation) having an impact on regulatory push-pull instruments for decarbonization. This role was confirmed by literature: EI reduces the environmental pressure [39], but the innovative procurement models are different for each procurement context and different types of goods, raw materials, and services. In each context, the experience of procurement managers is decisive [43]. Organizational performance is in direct positive correlation with internal environmental involvement, supplier cooperation, customer requests, competitive challenges, and management support. Thus, GPP is correlated with firm performance [44,45].

In GPP, innovation appears as a process, service, and policy [46]: in the GPP process, in public services employing GPP, and in the use of GPP as an instrument for demand-side sustainable policymaking. GPP is used as part of innovation policy and new strategies and models that are used in value creation through PP. At the same time, IT support is used for PP. GPP brings new management challenges in out-of-the-box acquisitions of innovation, involving technical knowledge, accurate information, insight overview, and funding resources. GPP is important in the innovation cycle, such as R&D management, commercialization, and project and risk management [46].

Green Supply Chain Management (GSCM) will overpass barriers in reducing waste, facing competitiveness through innovative strategies (e.g., in green logistics and green procurement), reducing carbon dioxide emissions (due to replacement of fossil fuels with alternative fuels) and reducing operating costs. GSCM is an instrument to nudge GPP and gain sustainability and fulfilling Green Deals objectives [47].

Eco-labels are designed for environmental or non-environmental primary standards of information when a company does GPP for "off-the-shelf" raw material or services. Other important standards refer to layers of environmental performance, conception, configuration, and security. The most common criteria for a contract to be awarded are environmental characteristics measured by "most economically advantageous tender" (MEAT), which include environmental life-cycle costing. Environmental criteria are required for all participating companies in order to implement an eligible project [48].

Ecolabels are seen as a legitimate tool to mitigate the consequences in the public procurement supply chain for the entire life cycle of product development (resource acquisition, manufacture, packaging, and transportation, use, end-of-life). They are included in the international standard (ISO 20400) regarding sustainable acquisitions. ISO 20400 is an important landmark for sustainable procurement for any organization. It promotes standardized guidance for all different stakeholders involved in internal and external PP, irrespective of whether they are contractors, consumers, suppliers, or local authorities. Sustainability regulations and ecolabels and certifications are methods to verify the sustainability of supply chain consequences regarding being appropriate for the objective, health social, environmental, or ethical issues [48–50]. In Romania, more and more organic crops have appeared in recent years and the green labeling system has started. Special stands have been created in stores for eco products, but the efforts are big, and the results are shy.

The expected benefits of applying this standard into PP are more adequate management practices, differentiation between the programs that sustain environmental, human rights, or ethical issues, and encouraging the launch of similar programs [51,52]. A very complex standard is Good Environmental Choice Australia (GECA). It contains 19 standards ranging from different fields. GECA standard requirements [52] are (Figure 2):

Figure 2. Good Environmental Choice Australia (GECA) standard requirements (Source. Adaptation Ecolabel Index, 2020 [52]).

When a company does pure basic research, idea exploration, or design a solution it substantiates its activity on a standards inventory. Then the company does basic research, a prototype, and/or applied research. In this phase, the company substantiates its activity on terminology standards. Before commercialization, the company applies measurement and testing standards for experimental development and scale-up. Government involvement in standardization and key market providers' standards adoption is very important in the initial stages of life crisis [53,54]. Public procurement supports the development of innovation markets through the facilitation of the expression of new demand, creating a demand, improving the innovation environment, increasing reciprocal collaboration between suppliers' and users' knowledge, facilitation of co-adaptation and bilateral learning, and providing information and incentives [55,56].

In Germany, Public Procurement with Contracted Innovation (PPCI) facilitates incremental innovations, but not market novelties, because there is an insufficient stimulus for radical innovations. Thus, the dissemination of new technologies is facilitated, to the disadvantage of radical inventions [57]. Pollution management, innovations in the ecosystem services supply, innovation in eco-procurement, transportation, and greening of guests are decisive factors in establishing a cleaner eco-environment and ecological industry [58].

7. Urban Sustainability in Building Design and Construction

Urban sustainability is a subject of interesting debates in articles within the construction field [59] such as designing a novel procurement scheme for converting empty houses through sustainable methods with governmental funds and thus avoiding multiple deprivations. The positive impact on the sustainable economy is the increase of economic attractiveness for financing, reducing unemployment, reducing the number of families with small salaries, mitigating their addiction to social benefit, and mitigating crime rates, better health, better educational fulfillment, the increased claim for workers, reducing community breakdown, good presence in civic activities and good public services [60]. Reducing the number of empty homes will ameliorate the symptoms above [59]. Building eco-houses and sustainable renovation of empty houses help the development of eco-cities through CO_2 emissions reduction, mitigating the pressure on the greenfield development, addressing sustainable policies, such as the zero-VAT policy for new construction (reduced rate of 5%), Renewable Heat Incentive (RHI) and Feed-in Tariffs (FITs) schemes. The beneficiaries of the financial incentives are the owners (the house occupants) for the implementation of sustainable technology [60,61]. Digital technologies seem to be an incentive for the property development field, stimulating sustainable urban consumer behavior [61]. The reduction in house maintenance costs could increase the disposable income available for purchasing green and sustainable food and adopting healthier practices.

Other sources for urban sustainability are raw materials (concrete, steel, wood, glass, etc.), products used in construction, temperature control equipment, etc. [63]. The lack of an integration framework for the promotion of e-procurement (eP) and sustainable procurement (SP) in the construction industry deprives opportunities to optimize the implementation of resources. Yu et al. (2019) propose the development of an integration framework for the promotion of eP and SP in the construction industry [64]. The circular economy principles are met in a sustainable city in all of the building's development and utilization stages: (a) construction materials development process (all the actions needed in production from supply to transport and manufacturing), (b) edifice construction stage (the same actions as in stage a), (c) usage stage (renovation, conservation, energy losses, and restoration) and, (d) end-of-life (clearance and recycling). The utilization stage seems to be more expensive, due to the heating and cooling energy requirement [60,61,65]. The operational energy needs may be reduced also by using fossil ammunition or using electricity for lighting, but an important percentage of the energy needs should be reduced by using sustainable building materials. Using green construction materials, we gain a sustainable environment, due to innovative extraction or manufacture processes [66]. These processes mitigate dust emissions and noise during construction and bring important benefits regarding greenhouse emissions. From the one-health perspective, we must highlight the positive effects on the human (occupants) health too [60]. To mitigate the construction impact on the environment, CE has as a priority the development of detailed GPP criteria for the sector.

The construction sector can provide a sustainable infrastructure for the agri-food sector, especially in rooftop farming and other similar solutions in big, crowded cities.

8. Good Practices in Education for Sustainable Development

A complex study made on HE (higher education) fields—on universities from the UK, Canada, the USA, and Australia—reveals the most frequent GPP categories that universities invest in [67,68]. The study reveals that HEIs invest especially in ecologic electricity, paper, office equipment, and food, as might be seen below:

- indoor lighting products and paper for educational purposes (48%),
- office IT equipment (e.g., computing and communications machines, used by academic, professional, and administrative staff—43%),
- food and catering services (e.g., packaged foods and drinks to be sold in the university's cafes and refectories/restaurants, and for preparation and cooking in the university's kitchens—33%),
- sanitation and washing products or services—34%,
- disinfectant: general and for removal of insect and rat substances—24%
- gardening products and services (e.g., electrical/battery and non-electrical items—24%)
- local and organic food acquisition scheme—24%
- acquisitions made from eco-friendly and socially responsible organizations—10%

HEIs (Higher Education Institutions) SPP (Sustainable Public Procurement) behavior is a good example for the next generations. HEIs most important drivers for implementation of a good SPP behavior are [61,62] professional, economic, and ethical concerns:

- HEIs tendency to lead best practice 43%
- Cost savings 39%
- Moral/ethical motivations 38%
- Government legislation lawmaking 29%
- HEI's stakeholder demands and/or expectations 29%
- Expected-anticipated reputational benefits 20%
- Third-party requirements 10%

SP in the educational sector is endorsed by good practices set by HEIs President's and/or Chancellor's Office 58%, internal regulation management or bottom-up initiatives of HEI employees or students 57%, the awareness of a job well done 52%, ethical and moral concerns of employees 43%, or stakeholder requirements 29%.

In Romania, HEI should have the main aim to sustain innovation and to form professional managers, able to come up with an adequate solution to agri-food management challenges. The urgent need for performant management in resource allocation on the different types of cultures/plantations, based on GPP, in Romania, can be surpassed with the support of HEI graduates.

In HE the adoption and implementation of SPP is well represented, reducing the negative impact the business has on the environment. However, most important is the high awareness and commitment among staff and students regarding SPP, behavior that will be extended on future generations as second nature. Staff and academic behavior regarding Green Public Procurement (GPP) represent good nudges for future generation sustainable consumer behavior [67,69].

HEIs face many barriers in implementing SPP, such as costs ("reducing stress in each division's budget, giving bonuses if the SP is used as an important factor for the purchase decision"), lack of understanding legislation, lack of skills and knowledge regarding GPP, sometimes lack internal social criteria (ISC), lack of understanding Bio-based Public Procurement (BPP) and Innovation-oriented Public Procurement (IPP). However, in HEIs the Social Return on Investment (SROI) is very evident, because of the academic and students' commitment and awareness regarding GPP [70] and due to the recommendations and research published by academia in journals with high impact. State or stakeholders' incentives in implementing GPP encourage HEIs to overcomes barriers [67,71].

9. Barriers to Sustainability

The transition of industries in general from linear to circular production is still far from happening. Achieving more sustainable production, distribution, commercialization, and residues management involves the paradigm of consuming versus optimizing resources and preserving the environment. Perhaps, this is one of the greatest barriers to sustainability. Thus, the objective is to organize these barriers in two dimensions: from the offer perspective (supply) and the market view (demand). Furthermore, the identified barriers can be structured according to the resources. The suggestion is to consider tangible and intangible resources. Nevertheless, human resources should be divided between the workforce, which is tangible, and knowledge, mainly tacit, which is intangible. Thus, the first barrier is knowing the specificities of each city to develop a capable model to develop a plan and to implement it, aiming to urban sustainability [72]. Failure to select the right suppliers is the most important risk factor for SCM, as supplier selection plays an important role in achieving social, environmental, and economic benefits [73].

There are five dimensions when tacking about indicators: economic, social, environmental, institutional, and other [71]. In developing countries, critical barriers were related to governmental, human, knowledge and information, market, and cost and risk barriers [74,75]. The core of sustainable urban transport indicators is fossil fuel consumption, length of motorway system, the number of vehicles, and HDI. Weak incentive policies, lack of legislation, and insufficient social awareness, the company's unawareness, unmatched vision and culture, insufficient top management commitment and competitors' inaction are the main barriers considered in the photovoltaic industry [75,76].

According to the United Nations (UN), one of the basic barriers is understanding long-term population trends; how they understand how to manage natural resources sustainably, how they will adapt to climate change, and how they will know how to overcome barriers and access markets, financial services, increasing access to information and education [77,78]. It is necessary to maximize the benefits of agglomeration and, at the same time, minimize environmental degradation. This barrier is considered the most relevant barrier, namely in low and middle-low income countries, where urbanization occurs more rapidly and without organization. A proper social balance needs to be found [72].

The business model used by companies, which aims for short-term results can be responsible for a lot of carbon dioxide emitted by industries, or diversification of products made from CO_2, or using a low technological level or have no concert about the energy source. All those aspects were considered as a factor of influence of urban sustainability, often associated negatively [79,80].

Maybe one of the most important barriers is the prevalence of a dysfunctional sustainability ecosystem [81]. According to the authors, interest groups installed exploit the system only for their benefit. Other barriers were related to the need to adapt to governance challenges. At the same time, the city has to balance socio-environmental and social cohesion aspects when implementing nature-based solutions (NbS) [82,83]. This dynamic is complex and difficult to follow. To identify the main barriers to urban sustainability, from the offer (supply) market (demand) views, (supply versus demand), this revision results in an organizational framework. It aims to structure the tangible versus intangible factors. Table 2 below structures only the main barriers related to the consumption paradigm.

Table 2. Principal barriers related to consumption.

Barriers	Offer	Demand
Tangible	Land, energy, and transportation system	The need for energy and transportation system requirement compared to lifestyle desired
Intangible	Need to adapt to governance challenges and have to balance socio-environmental and social cohesion aspects [82,83] The business model used by companies [81] Dysfunctional sustainability ecosystem [81]	The lifestyle desired Understanding long-term population trends [78] Knowing the specificities of each city

In most situations, it is easier to identify the need for tangible resources. Barriers on the supply side are also easier to overcome, especially when compared to barriers on the demand side. On the contrary, as the name says, intangible resources are much more difficult to achieve and overcome. To implement urban sustainability, the practical implication of this framework is the structure to develop a plan [80]. From Table 2, our research suggested a methodology to plan urban sustainability in five steps. Figure 3 shows the sequence of steps.

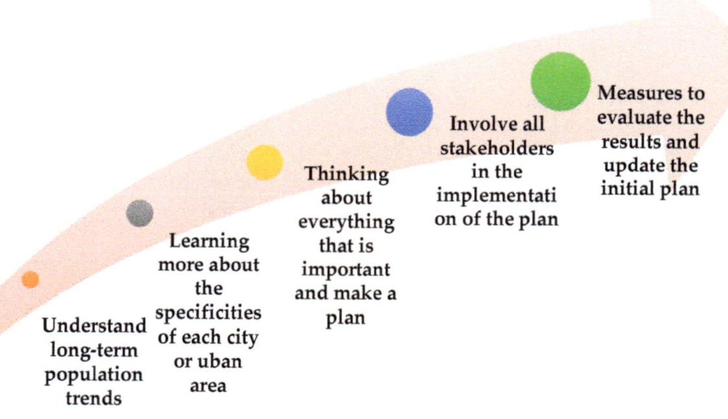

Figure 3. The five steps towards planning sustainability.

In the first step, it is suggested to understand long-term population trends. Following this step, it is suggested to learn more about the specificities of each city or urban area. How they work in terms of transportation and energy, demand and supply structure, and which lifestyle factors are the most significant to that community [66]. In the next step, it is suggested to think about everything important. It is suggested to make a plan before implementing any change. After the implementation, changing anything probably costs more. Finally, the objective should involve different stakeholders. So, it is important to evaluate the results. Perhaps the plan needs to be updated [72].

10. Green Public Procurement Applied in Organic Agriculture in Romania

Agricultural subsidy policies are important in the context of public food procurement. In Romania, the "Green public procurement guide" was approved in October 2018. This includes the minimum specification requirements regarding environmental protection for certain groups of products and services for GPP related to product groups and/or priority services, including food and catering services [84]. The main purpose of issuing the guide was to provide contracting authorities information on the mandatory minimum requirements to be provided when drafting documentation for the award of GPP contracts for products and services, according to the European Commission's Handbook on Green Public Procurement. These practices come in the context in which in Romania, the dynamics of operators and areas used in organic farming in the period 2010–2019 as provided on the website of the Ministry of Agriculture and Rural Development, show a significant increase in recent years (Figure 4), starting with 2017. These increases were highlighted for the total area allocated to organic farming, cereals, industrial crops, and green harvested plants.

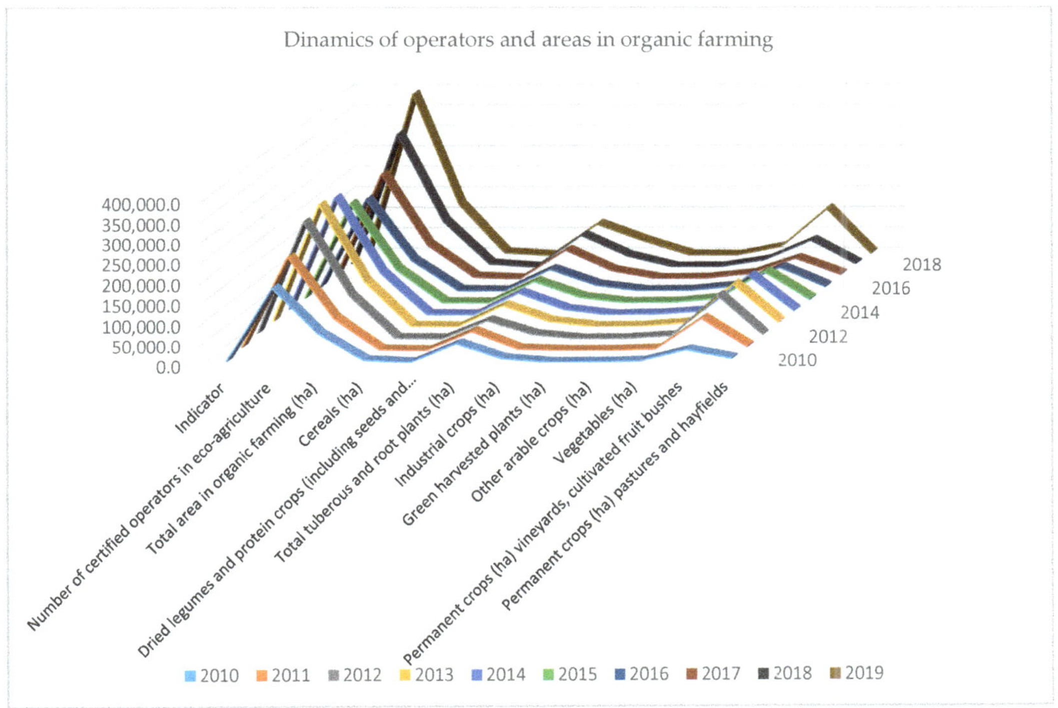

Figure 4. Dynamics of operators and areas in organic farming (Source: data processing following the website of the Ministry of Agriculture and Rural Development) [84].

It can be seen in Figure 4 that there was a peak in 2014, due to subsidies. The non-adaptation to these practices quickly led to a halving of the number of Romanian farmers who practice organic farming in just four years, even though subsidies for these crops have been higher than those for the land surface. The reason is the lack of a market and the strict obligations imposed on these beneficiaries of European subsidies. An important reason is that Romanian farmers do not easily understand or accept the importance of association. Organic vegetables and fruits last much less than those treated chemically, and if producers do not sell them fast enough, they end up as waste. Farmers cannot negotiate with supermarkets or wholesalers individually. They remain with unsold products because they cannot capitalize on their production in the markets. Organic products are more expensive while the purchasing power of the population is low. The best way forward would be to associate and set up agricultural cooperatives that constantly provide fresh goods. The status quo has begun to change in recent years. In 2019, 9821 operators were registered, and the total ecologically certified area was 395,227.97 ha, the largest in the last nine years. In 2018, 9008 operators were registered, the first increase after a significant decrease recorded for several years: from 15,544 in 2012 to 8434 in 2017, as specified by the Ministry of Agriculture and Rural Development.

In the second stage of the analysis, we designed a regression model that confirms our conclusion: organic farming is represented especially by cereals and green harvested plants. Organic farming loses a lot of profit because of the uncultivated land. This problem can be surpassed through GPP and IPP managed by Romanian specialists formed by innovative HEIs. The correlation matrix shows a strong positive correlation between the variable *Total area in organic farming* and *Cereals* (0.96), *Permanent crops (ha) vineyards, cultivated fruit bushes* (0.81), and *Green harvested plants* (0,80). The first conclusion is that most of the organic farming is represented by cereals, permanent crops and Green harvested plants. The second conclusion is that industrial crops are represented by Permanent crops (ha) vineyards, cultivated fruit bushes (0.89), and Dried legumes and protein crops (including seeds and mixtures of cereals and legumes) (0.80).

Our model is significant because the test assumptions are met, and the statistics are relevant. The test assumptions are: our variables are continuous, there is a linear relationship between the independent variables and dependent variables (graphic scatterplot), there are no extreme values, residuals are not correlated (Durbin–Watson coefficient 2214, included in the interval 1.5 < d < 2.5, Table 3), the dates demonstrate homoskedasticity and residuals have an approximately normal curve (Figure 5 P-P plot).

Table 3. Model Significance.

					Change Statistics					
Model Summary [b]										
Model	R	R Square	Adjusted R Square	Std. Error of the Estimate	RSquare Change	F Change	df1	df2	Sig. F Change	Durbin-Watson
1	0.985 [a]	0.971	0.957	12,444.97670	0.971	67.256	3	6	0.000	2.214

	ANOVA [b]					
	Model	Sum of Squares	Df	Mean Square	F	Sig.
1	Regression	31,249,080,149.865	3	10,416,360,049.955	67.256	0.000 [a]
	Residual	929,264,670.535	6	154,877,445.089		
	Total	32,178,344,820.400	9			

[a] Predictors: (Constant), Uncultivated, Cereals, Green_harvested; [b] Dependent Variable: Organic_farm.

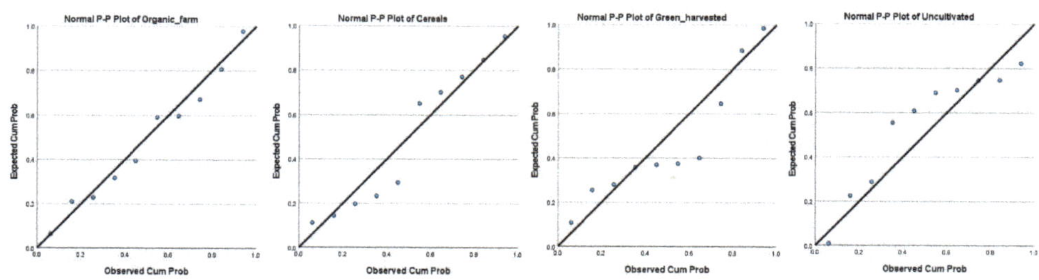

Figure 5. P-P plot for each analyzed variable.

Overall, the model is very representative (R^2 is 0.971). The variation of the independent variable (*cereals, green_harv, uncultivated*) explains 97% of the variation of the dependent variable (*total cultivated land*). The other percentages should be explained by other variables such as dried legumes and protein crops, tuberous and root plants, industrial crops, vegetables, and permanent crops.

The ANOVA test confirms the before mentioned findings because F > F crit and Sig F < 0.01 (Table 3).

The R coefficient is 0.99, bigger than the minimum limit of 0.95 as the model to be reliable. The T-test marginal significance level for estimated coefficients is less than only 0.01 for the cereals' coefficient, meaning that this coefficient is very well estimated, and the others could be better estimated.

Following the statistical test F, the resulting coefficient for *cereals* variable is 2.4, with a probability of guaranteeing results (Prob) of 0.01 (less than the sensitivity threshold of 0.05) so this variable's coefficient is well estimated. The coefficient for *green_harv* variable is 0.8 with 95% likelihood, it will be found in the interval [−1.1:2.7]. The coefficient for the *uncultivated* variable is −6.06, with 95% likelihood, it will be found in the interval [−16.16; 4.04]. The intercept value was estimated to be 86,081.9 with 95% likelihood, it will be found in the interval [−52.640; 224.804] (Table 4).

Table 4. Coefficients estimation and collinearity diagnostics.

Coefficients [a]

Model		Unstandard. Coeff.		Standard. Coeff.	T	Sig	95.0% Confidence Interval for B		Collinearity Statistics	
		B	Std. Err.	Beta			Lower	Upper	Tolerance	VIF
1	(Constant)	86,081.95	56,692.8		1.52	0.18	−52,640.4	224,804.3		
	Cereals	2.384	0.319	0.751	7.47	0.00	1.603	3.164	0.477	2.09
	Green_harv	0.807	0.790	0.130	1.02	0.34	−1.125	2.739	0.296	3.37
	Uncultiv.	−6.061	4.131	−0.180	−1.46	0.19	−16.168	4.046	0.319	3.14

Coefficient Correlations [a]

Model			Uncultiv.	Cereals	Green_harv
1	Correlations	Uncultivated	1.000	0.249	0.646
		Cereals	0.249	1.000	−0.357
		Green_harv	0.646	−0.357	1.000
	Covariances	Uncultivated	17.061	0.329	2.106
		Cereals	0.329	0.102	−0.090
		Green_harv	2.106	−0.090	0.623

Collinearity Diagnostics [a]

Model	Dimension	Eigenvalue	Condition Index	Variance Proportions			
				(Const.)	Cereals	Green_harv	Uncultiv.
1	1	3.748	1.000	0.00	0.00	0.00	0.00
	2	0.232	4.021	0.00	0.00	0.18	0.02
	3	0.017	15.020	0.00	0.65	0.61	0.13
	4	0.003	33.693	1.00	0.35	0.20	0.85

[a] Dependent Variable: Organic_farm.

The equation of the regression model is presented below:

$$\text{Total land organic farming} = 86{,}081.9 + 2.4 \times \text{Cereals} + 0.8 \times \text{Green\_harv} - 6.06 \times \text{Uncultivated land}$$

The "Collinearity Diagnostics" shows us the Eigenvalues of the scaled, un-centered cross-products matrix; the condition indexes; and the variance proportions. The lowest Eigenvalue (0.017), show that *Cereals* and *Green_harv* variables have some variance proportions (0.65 and 0.61) for Dimension 3, which means that 65% of the variance of the b-value for *Cereals* and 61% of the variance of the b-value for *Green_harv* were associated with the Eigenvalue 3 (the smallest Eigenvalue). This result suggests that there might be some dependency between these two variables. The largest VIF (3.37) is for *Green_harv*, but it is not greater than 10, so it is within tolerance. The corresponding Tolerance Statistic for *Green_harv* (0.29), which is not below 0.1, means that it is within tolerance. The average VIF is 2.87 ((2.098 + 3.373 + 3.138)/3 = 2.87), which is not substantially greater than 1. The average Tolerance Statistic is 0.35, which is not below 0.2. The VIF and Tolerance Statistics show that there is no multicollinearity in this data, but not very important collinearity exists between two of the predictors (*Cereals* and *Green_harv*). This collinearity is not sensible as to remove one of the variables from the proposed model.

Thus, the alternative hypothesis, H1, is accepted. H1 stands that organic farming is represented especially by cereals and green harvested plants. Organic farming loses a lot of profit because of the uncultivated land. The farmers should use all the land available and plant organically and adequate management in this field is required.

In the context of supply chains in Romanian agriculture, the phenomenon of integration is manifested more in the sphere of vegetable and animal production, vegetable residues, or even important parts of plant production being intended to feed meat animals, in particular, in the form of fodder. The propagating material is the subject of reproduction only in model greenhouses or farms that comply with high standards in the preservation and improvement of the quality of the genetic background. Storage and distribution are the responsibility of economic agents affiliated with the agri-food production sector, ex Danone, although a good trend is the appearance of high-capacity silos in the vicinity of crops, in order to reduce transport costs, or as a buffer warehouse in addition to the major shipping routes (e.g., Stanca, Braila) or rail transport (e.g., Boromir, Buzau). Fragmentation of the supply chain by the disappearance of research centers, seeds producers (agrosems; SCDL Buzau a rare and happy exception) as well as the lack of storage and raw processing capacities (drying, treatment for preservation, chopping/grinding, etc.) lead to the strong manifestation of the seasonality of production and price scissors at the expense of small agricultural entrepreneurs. The peculiarities of the Romanian agricultural system in the field of plant production and specific supply chains will be the subject of further research.

The private agricultural system dedicated to vegetable production is marked by two phenomena with economic impact: that of the procedure of concession of land in the public domain to private entrepreneurs and that of the amalgamation of individual properties of small size and dispersed in large agricultural holdings through the lease agreement. It stipulates lump sums and/or products in favor of the agricultural owner. A sharp trend, but of low economic impact, is the small biologically certified holding that produces extensively small quantities of products sought for export such as truffles, nuts, berries, herbs, and medicinal plants. However, the social role of this type of business is high, being in many cases the only source of income of a family.

11. Discussion

The United Nations 2030 Agenda for Sustainable Development includes as an important goal the sustainable acquisitions, based on specific programs able to measure the price-performance-impact on the environment balance regarding raw materials, services, products, technologies, etc. GPP and IPP have an important impact on sustainability, reducing the environmental footprint. GPP itself is also based on innovative policies, strategies, and acquisition model, following the type of product or service. Furthermore, eco-labels were designed to sustain the GPP process to gain social, health, environmental, and ethical sustainability. For a clear understanding and a good implementation of GPP international standards like GECA were set. These types of standards represent guidelines and restrictions regarding pollution management in different fields such as construction, transportation, services, tourism, health, education having a positive impact on the eco-environment (Figure 6).

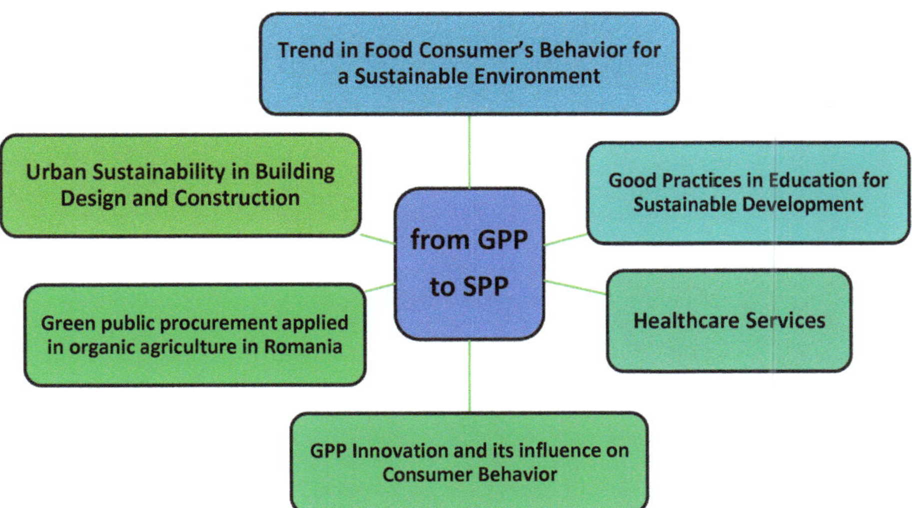

Figure 6. Integrated application of GPP and sustainable public procurement (SPP) in equilibrium between economic, social, and environmental pillars.

We chose organic farming in Romania as a case study because the EU Action Plan for organic farming comes with new regulations in the future which recommends initiatives to support consumer confidence, support for increased export opportunities for organic food, development of electronic import certification systems, encouraging the use of organic food in schools through GPP. Currently, over 40% of the land in Romania is worked by 0.56% of farmers, in farms larger than 300 hectares. Approximately the same area is worked by 97% of farmers (about 795,000 farmers), in small farms, under 50 hectares. This can be good from the perspective of ensuring sustainable local procurement and green agriculture. This approach can lead to foods that are better for the health and well-being of communities. However, these small, subsistence farms will not be able to exist if they do not associate. Therefore, they will have to support agriculture in terms of green procurement, supply chain management, circular economy; they will need legislation in this field and especially these measures will have to be implemented. Environmental protection is a major goal of organic farming. Agricultural production combines the best environmental practices, maintains biodiversity, and contributes to the conservation of natural resources, supports animal husbandry and welfare, and especially respects consumers' preferences for healthy products. The challenge of the sector is to ensure the constant increase of the supply and the demand of the consumers, keeping their confidence in quality. Legislation, the evolution of society played an important role. Research and innovation are tools for overcoming the challenges posed by ecological norms.

Building eco-houses, sustainable refurbishment of empty homes, and the use of eco-friendly construction materials and products have a positive impact on ecological footprints, such as CO_2 emissions, dust emissions, noise stress, and energy consumption, with a positive impact on health. The latest trend in sustainable landscape architecture is the establishment of horticultural crops on the roofs of public buildings in the center of cities to fix CO_2 and other substances with carcinogenic potential in the atmosphere, maintain a constant micro-climate inside the building and support several species of pollinators such as insects and birds. Supporting this type of activity requires minimal public allocations with seeds and fertilizers, since species resistant to drought, pollution and adapted to the local biosphere will be chosen. An incentive factor is the dissemination of ideas associated with maintaining ecological balance in large urban agglomerations through recreational and formative activities with small groups of students in such gardens in the vicinity of educational centers. In addition, employees in office buildings that have green spaces on the roof can benefit from a seating space and table without having to move to parks or restaurants far away from work.

It is recommended for HEIs to design and implement GPP planning and have quantitative and qualitative criteria for monitoring the GPP program: choosing the lowest price is not always a valid criterion; sometimes analyzing the life cycle cost of the green products/services procures will emphasize that the most expensive items can save on long-term costs. If most of the universities will implement SPP, a majority of organizations would follow on SPP implementation as the future organization's employees are today's students. Another important recommendation is to follow the Minimum Environmental Criteria. Improved communication regarding leadership engagement and sustainable routine practices are also important factors in gaining SPP in HEIs. Considering that, one might think that the basis of the problem lies in the costs. If one compared the effort implied with the short-term benefits associated, sustainability would not be a priority. Nevertheless, it is imperative to think long term. The reasons that lead people to think in the short term are very much associated with the financial conditions of families. Sometimes it is cheaper to buy than to reuse resources. Besides, there is no concern with the environment (included waste or residuals resources) when the associated costs are too high to bear. It means that people might be aware of the consequences but unwilling to change. This is a challenge that involves the underlying principles that drive humanity, as well as personal values.

Given the difficulty of evaluating principles, which involve the specificity of the cultural dimension of each community, the following question was raised: is it possible to overcome the existing barriers in the consumption paradigm? To answer this question, this work focused on identifying the main barriers involved in the paradigm of consumerism versus sustainability mainly through the lenses of cost in the implementation of sustainability.

Public procurement is not just an administrative procedure, but a significant actor in the market. Globally, 12% of GDP is spent on public procurement according to the World Bank [85], with as much as 20–25% in countries like Switzerland or the Netherlands. Thus, green practices in public procurement have the potential to make a truly large impact on environmental sustainability, provided that they are correctly used. Sustainability is not an individual task but a task for all of us. It is necessary to seek economic and social balance, as well as preserving the environment. It is also possible to bring together short and long terms objectives.

12. Conclusions

SPP can be achieved mainly through GPP and IPP, as a branch of GPP. The center of this research concerns the relevance of implementing GPP. In this regard, we highlighted some barriers that can be found in implementation and suggested a five-step plan that can be done before implementing any change.

Our research emphasizes, by analyzing 80 articles by the PRISMA method, the role of GPP in educating and changing consumer behavior, that has positive consequences on sustainability in different sectors of activity. Thus, educating towards "green" attitudes and competencies significantly impacts their willingness and ability to act according to GPP principles. Furthermore, specific legislation, policies, and procedures have to be implemented concerning each main area of public procurement interest: construction, healthcare, and food to have a concerted and concentrated impact. However, even more important is the use of GPP in driving sustainable innovation through IPP. The ability of governments to virtually create a market for specific innovative products, solutions, and ideas is decisive for the development of radical innovations.

The role of governments should be to stimulate the creation of environmentally friendly services and products by allocating their purchasing might towards those by relying more on Social ROI and less on traditional monetary measures. As it was observed from the case study of organic farming in Romania, organic farming operators have evolved only when society got involved for them: there were subsidies for land users, for the acquisition of raw materials, product promotion, etc. The weaknesses in the agrifood sector are the disorganization in management, as evidenced by large uncultivated territories and small investments for certain types of vegetables.

State intervention in agriculture is limited by the dominant form of ownership in Romanian agriculture, namely the private one. However, the State can provide subsidies directed to the type of investment (the foundation of culture, mechanization, fertilizer administration, weeding, according to the technological file), environmental taxes can be calibrated according to the certification of the culture (biological/ecological), and the ease of agricultural credit for certain investments and the proactive surveillance of the insurance sector in agriculture or the supplementation of the insurance fund can be implemented. Local Agricultural Consulting Directorates (ANCA) can provide the necessary documentation for the establishment and certification of organic production, financial advice from the granting to the completion of the investment plan. Local authorities can stimulate domestic producers of machinery for agriculture and food production by subsidizing production costs, facilitating fairs and exhibitions, there can be issued green certificates for consumption in green agriculture in the sector of energy production.

Managers of organically certified farms should be encouraged to participate in international agricultural fairs and exhibitions in order to benefit from the expertise of other countries in the promotion of green technologies. Graduates of agricultural technical education must benefit from training courses in research and production farms, where they can apply their new green-oriented knowledge acquired at school. Alternative forms of contractual work in agriculture due to seasonality and uneven distribution of the available and skilled labor force in agriculture by organizing shuttles, relocating the families of agricultural workers in decent conditions, and integrating them to the higher levels of the agricultural production chain should be encouraged.

For GPP to evolve into SPP it is necessary to overcome certain barriers, as we identified in our study, in five steps: understand long-term population trends; learn more about the specificities of each city or urban area; think about everything that is important and make a plan; involve all stakeholders in the implementation of the plan; use measures to evaluate the results and update the initial plan.

13. Limitations and Future Research Suggestions

Although we studied a large sample of articles, these were exclusively published in the Scopus and Web of Science databases. We could not cover other databases, due to the vastness of the field. In future research, we would analyze points of view from transdisciplinary areas to increase GPP and B2B capacity to operate at a transnational level, share ideas, and develop the best practices and methods. Future research should investigate whether GPP, through organizational sustainability practices, can shape operational environmental sustainability and sustainable organizational culture.

Author Contributions: Conceptualization, R.B.-M.-Ț. (Rocsana Bucea-Manea-Țoniș) and V.-E.S.; methodology, R.B.-M.-Ț. (Rocsana Bucea-Manea-Țoniș) and C.B.; software, R.B.-M.-Ț. (Rocsana Bucea-Manea-Țoniș); validation, R.B.-M.-Ț. (Radu Bucea-Manea-Țoniș), O.M.D.M.; formal analysis, M.B.; investigation, R.B.-M.-Ț. (Rocsana Bucea-Manea-Țoniș), V.-E.S.; resources, R.B.-M.-Ț. (Rocsana Bucea-Manea-Țoniș), O.M.D.M., M.B., D.I. and C.B.; data curation, R.B.-M.-Ț. (Radu Bucea-Manea-Țoniș); writing—original draft preparation, R.B.-M.-Ț. (Rocsana Bucea-Manea-Țoniș) and V.-E.S.; writing—review and editing, R.B.-M.-Ț. (Rocsana Bucea-Manea-Țoniș), O.M.D.M, and V.-E.S.; visualization, C.B.; supervision, R.B.-M.-Ț. (Rocsana Bucea-Manea-Țoniș), M.B.; project administration, R.B.-M.-Ț. (Rocsana Bucea-Manea-Țoniș). All authors have read and agreed to the published version of the manuscript.

Funding: This research received no external funding.

Conflicts of Interest: The authors declare no conflict of interest.

References

1. Roehrich, J.K.; Hoejmose, S.; Overland, V. Driving green supply chain management performance through supplier selection and value internalization: A self-determination theory perspective. *Int. J. Oper Prod. Manag.* **2017**, *37*, 489–509. [CrossRef]
2. Farooque, M.; Zhang, A.; Thürer, M.; Qu, T.; Huisingh, D. Circular supply chain management: A definition and structured literature review. *J. Clean. Prod.* **2019**, *228*, 882–900. [CrossRef]
3. Moraga, G.; Huysveld, S.; Mathieux, F.; Blengini, G.A.; Alaerts, L.; Van Acker, K.; de Meester, S.; Dewulf, J. Circular economy indicators: What do they measure? *Resour. Conserv. Recycl.* **2019**, *146*, 452–461. [CrossRef] [PubMed]
4. Marrucci, L.; Daddi, T.; Iraldo, F. The integration of circular economy with sustainable consumption and production tools: Systematic review and future research agenda. *J. Clean. Prod.* **2019**, *240*. [CrossRef]
5. Neto, B. Analysis of sustainability criteria from European public procurement schemes for food services. *Sci. Total Environ.* **2019**, *704*, 135300. [CrossRef] [PubMed]
6. Pircher, B. EU public procurement policy: The economic crisis as trigger for enhanced harmonization. *J. Eur. Integr.* **2019**, *42*, 509–525. [CrossRef]
7. European Commission. 2020-EU GPP Criteria. Available online: https://ec.europa.eu/environment/gpp/eu_gpp_criteria_en.htm (accessed on 1 September 2020).
8. European Commission. Green and Sustainable Public Procurement. Available online: https://ec.europa.eu/environment/gpp/versus_en.htm#:~{}:text=Sustainable%20Public%20Procurement%20(SPP)%20is,all%20stages%20of%20the%20project (accessed on 1 September 2020).
9. Trindade, P.C.; Antunes, P.; Partidário, P. SPP Toolbox: Supporting Sustainable Public Procurement in the Context of Socio-Technical Transitions. *Sustainability* **2018**, *10*, 67. [CrossRef]
10. Appolloni, A.; D'Amato, A.; Cheng, W. Is public procurement going green? Experiences and open Issues. In *The Applied Law and Economics of Public Procurement*; Piga, G., Treumer, S., Eds.; Routledge: Abingdon, UK, 2013; pp. 111–132.
11. Coppola, M.A.; Appolloni, A.; Piga, G. Implementation of Green Considerations in Public Procurement-A mean to promote sustainable development. In *Green Public Procurement Strategies for Environmental Sustainability*; Shakya, R.K., Ed.; IGI Global: Roma, Italy, 2019; pp. 23–44.
12. Cheng, W.; Appollonia, A.; D'Amato, A.; Zhu, Q. Green Public Procurement, Missing Concepts and Future Trends–A Critical Review. *J. Clean. Prod.* **2018**, *176*, 770–784. [CrossRef]
13. Litardi, I.; Fiorani, G.; Alimonti, D.; Appolloni, A. Green public procurement as a leverage for sustainable development documental: Analysis of 80 practices in European Union. In *Green Public Procurement Strategies for Environmental Sustainability*; Shakya, R.K., Ed.; IGI Global: Hershey, PA, USA, 2019; pp. 59–78. [CrossRef]
14. Hazen, B.T.; Mollenkopf, D.A.; Wang, Y. Remanufacturing for the circular economy: An examination of consumer switching behavior. *Bus. Strateg. Environ.* **2017**, *26*, 451–464. [CrossRef]
15. Patrucco, A.S.; Luzzini, D.; Ronchi, S. Research perspectives on public procurement: Content analysis of 14 years of publications in the journal of public procurement. *J. Public Procure.* **2017**, *17*, 229–269. [CrossRef]
16. Grandia, J.; Voncken, D. Sustainable Public Procurement: The Impact of Ability, Motivation, and Opportunity on the Implementation of Different Types of Sustainable Public Procurement. *Sustainability* **2019**, *11*, 5215. [CrossRef]
17. European Commission. GPP National Action Plans. Available online: https://ec.europa.eu/environment/gpp/action_plan_en.htm (accessed on 1 September 2020).
18. Lindström, H.; Lund, S.; Marklund, P.-O. How green public procurement can drive conversion of farmland: An empirical analysis of an organic food policy. *Ecol. Econ.* **2020**, *172*, 106622. [CrossRef]
19. Elkington, J. 25 Years ago I coined the Phrase "Triple Bottom Line". Here's Why It's Time to rethink It. *Harv. Bus. Rev.* **2018**, *25*, 2–5.
20. Sustainable Development Goals. Available online: https://www.un.org/sustainabledevelopment/sustainable-development-goals (accessed on 1 September 2020).
21. Stoffel, T.; Cravero, C.; La Chimia, A.; Quinot, G. Multidimensionality of sustainable public procurement (SPP)—Exploring Concepts and effects in Sub-Saharan Africa and Europe. *Sustainability* **2019**, *11*, 6352. [CrossRef]
22. Stenmarck, A.; Jensen, C.; Quested, T.; Moates, G. *Estimates of European Food Waste Levels*; Report from the FUSIONS Project; European Commission (FP7), Coordination and Support Action—CSA: Luxembourg, 2016; ISBN 978-91-88319-01-2.
23. *Estratégia Nacional e Plano de Ação de Combate ao Desperdício Alimentar*; Proposta Elaborada pela Comissão Nacional de Combate ao Desperdício Alimentar. (CNCDA) em Conformidade com o Despacho n.º 14202-B/2016, de 25 Novembro; Comissão Nacional de Combate ao Desperdício Alimentar; (CNCDA): Lisbon, Portugal, 2017.
24. *Combater o Desperdício Alimentar. Uma Responsabilidade do Produtor ao Consumidor*; Concertação, Sinergia, Envolvimento. Despacho n.º 14202-B/2016, publicado no DR, 2.ª série, N.º 227, de 25 de novembro de 2016; Comissão Nacional de Combate ao Desperdício Alimentar; (CNCDA): Lisbon, Portugal, 2016.
25. De Laurentis, V.; Caldeira, C.; Sala, S. No time to waste: Assessing the performance of food waste prevention actions. *Resour. Conserv. Recycl.* **2020**, *161*, 104946. [CrossRef]

26. Laschkolnig, A.; Habl, C.; Renner, A.T.; Bobek, J. EU Platform on Food Losses and Food Waste Terms of Reference (ToR); European Commission, Directorate-General for Health and Food Safety. Study on Big Data in Public Health, Telemedicine and Healthcare: Final Report. 2016. Available online: https://ec.europa.eu/health/sites/health/files/ehealth/docs/bigdata_report_en.pdf (accessed on 1 September 2020).
27. European Commission [EU]. *EU Actions against Food Waste*; European Commission: Luxembourg, 2018; Available online: https://ec.europa.eu/food/safety/food_waste/eu_actions_en (accessed on 1 September 2020).
28. European Commission [EU]. Food Waste Monitoring—Update. EU Platform on Food Losses and Food Waste. 2018. Available online: https://ec.europa.eu/food/sites/food/files/safety/docs/fw_eu-platform_20181206_flw_pres-03.pdf (accessed on 1 September 2020).
29. Goosens, Y.; Wegner, A.; Schmidt, T. Sustainability assessment of food waste prevention measures: Review of existing evaluation practices. *Front. Sustain. Food Syst.* **2019**, *3*, 90. [CrossRef]
30. Zeinstra, G.; Haar, S.V.D.; Bergen, G.V. Drivers, barriers and interventions for food waste behaviour change: A food system approach. *Food Biobased Res.* **2020**. [CrossRef]
31. van Geffen, L.; van Herpen, É.; van Trijp, H. Quantified Consumer Insights on Food Waste. Pan-European Research for Quantified Consumer Food Waste Understanding. Available online: http://eu-refresh.org/quantified-consumer-insights-food-waste (accessed on 1 September 2020).
32. *Commission Delegated Decision*; 2019/1597; Directive 2008/98/EC of the European Parliament: Brussels, Belgium, 2019; p. 77.
33. David, A.; Thangavel, Y.D.; Sankriti, R. Recover, Recycle and Reuse: An Efficient Way to Reduce the Waste. *Int. J. Mech. Prod. Eng. Res. Dev.* **2019**, *9*, 31–42.
34. Ahsan, K.; Rahman, S. Green public procurement implementation challenges in Australian public healthcare sector. *J. Clean. Prod.* **2017**, *152*, 181–197. [CrossRef]
35. Chiarini, A.; Opoku, A.; Vagnoni, E. Public healthcare, practices and criteria for a sustainable procurement: A comparative study between UK and Italy. *J. Clean. Prod.* **2017**, *162*, 391–399. [CrossRef]
36. Uyarra, E.; Zabala-Iturriagagoitia, J.; Flanagan, K.; Magro, E. Public procurement, innovation and industrial policy: Rationales, roles, capabilities and implementation. *Res. Policy* **2020**, *49*, 103844. [CrossRef]
37. Boon, W.; Edler, J. Demand, challenges, and innovation. Making sense of new trends in innovation policy. *Sci. Public Policy* **2018**, *45*, 435–447. [CrossRef]
38. Uyarra, E.; Flanagan, K.; Magro, E.; Zabala-Iturriagagoitia, J.M. Anchoring the innovation impacts of public procurement to place: The role of conversations. *Environ. Plan.* **2017**, *35*, 828–848. [CrossRef]
39. Meehan, J.; Menzies, L.; Michaelides, D.R. The long shadow of public policy; Barriers to a value-based approach in healthcare procurement. *J. Purch. Supply Manag.* **2017**, *23*, 229–241. [CrossRef]
40. Millera, F.A.; Lehouxb, P. The innovation impacts of public procurement offices: The case of healthcare procurement. *Res. Policy* **2020**, *49*, 104075. [CrossRef]
41. Vatalis, K.I.; Manoliadis, O.G.; Mavridis, D.G. Project Performance Indicators as an Innovative Tool for Identifying Sustainability Perspectives in Green Public Procurement. *Procedia Econ. Financ.* **2012**, *1*, 401–410. [CrossRef]
42. Ghisetti, C. Demand-pull and environmental innovations: Estimating the effects of innovative public procurement. *Technol. Forecast. Soc. Chang.* **2017**, *125*, 178–187. [CrossRef]
43. Ghisetti, C.; Quatraro, F. Green technologies and environmental productivity: A cross-sectoral analysis of direct and indirect effects in Italian regions. *Ecol. Econ.* **2017**, *132*, 1–13. [CrossRef]
44. Ghosh, M. Determinants of green procurement implementation and its impact on firm performance. *J. Manuf. Technol. Manag.* **2019**. [CrossRef]
45. Qiansong, Z.; Jieyi, P.; Yisa, J.; Taiwen, F. The impact of green supplier integration on firm performance: The mediating role of social capital accumulation. *J. Purch. Supply Manag.* **2020**, *26*, 100579.
46. Obwegeser, N.; Müller, S.D. Innovation and public procurement: Terminology, concepts, and applications. *Technovation* **2018**, *74*, 1–17. [CrossRef]
47. Teixeira, C.R.B.; Assumpção, A.L.; Correa, A.L.; Savi, A.F.; Prates, G.A. The contribution of green logistics and sustainable purchasing for green supply chain management. *Indep. J. Manag. Prod.* **2018**, *9*, 1002–1026. [CrossRef]
48. De Giacomo, M.R.; Testa, F.; Iraldo, F.; Formentini, M. Does Green Public Procurement lead to Life Cycle Costing (LCC) adoption? *J. Purch. Supply Manag.* **2019**, *25*, 100500. [CrossRef]
49. Harrisa, K.; Divakarlaa, S. Supply chain risk to reward: Responsible procurement and the role of ecolabels. *Procedia Eng.* **2017**, *180*, 1603–1611. [CrossRef]
50. Qiang, T.; Jianlei, M.; Regina, B.; Lianbiao, C.; Ying, F.; Yu, L. Achieving grid parity of solar PV power in China-The role of Tradable Green Certificate. *Energy Policy* **2020**, *144*, 111681. [CrossRef]
51. International Organisation for Standardisation (ISO). Sustainable Procurement One Step Closer to an International Standard. Available online: http://www.iso.org/iso/home/news_index/news_archive/news.htm?refid=Ref1873 (accessed on 2 September 2020).
52. Ecolabel Index—GECA. Available online: http://www.ecolabelindex.com/ecolabel/good-environmental-choice-australia (accessed on 2 September 2020).

53. Good Environmental Choice Australia. Our Standards. Available online: http://www.geca.org.au/standards/ (accessed on 2 September 2020).

54. Rainville, A. Standards in Green Public Procurement—A framework to enhance innovation. *J. Clean. Prod.* **2017**, *167*, 1029–1037. [CrossRef]

55. Bleda, M.; Chicot, J. The role of public procurement in the formation of markets for innovation. *J. Bus. Res.* **2018**, *107*, 186–196. [CrossRef]

56. Vijay, P.O.; Sanjib, P.; Joydeep, G. Recycling carbon tax for inclusive green growth: A CGE analysis of India. *Energy Policy* **2020**, *144*, 111708. [CrossRef]

57. Czarnitzki, D.; Hünermund, P.; Moshgbar, N. Public Procurement of Innovation: Evidence from a German Legislative Reform. *Int. J. Ind. Organ.* **2020**, *71*, 102620. [CrossRef]

58. Wang, F.; Wang, K.; Wang, L. An examination of a city greening mega-event. *Int. J. Hosp. Manag.* **2018**, *77*, 538–548. [CrossRef]

59. Ceranic, B.; Markwell, G.; Dean, A. Too many empty homes, too many homeless—A novel design and procurement framework for transforming empty homes through sustainable solutions. *Energy Procedia* **2017**, *111*, 558–567. [CrossRef]

60. Brusselaers, J.; Van Huylenbroeck, G.; Buysse, J. Green Public Procurement of Certified Wood: Spatial Leverage Effect and Welfare Implications. *Ecol. Econ.* **2017**, *135*, 91–102. [CrossRef]

61. Palander, T.; Haavikko, H.; Kärhä, K. Towards sustainable wood procurement in forest industry–The energy efficiency of larger and heavier vehicles in Finland. *Renew. Sustain. Energy Rev.* **2018**, *96*, 100–118. [CrossRef]

62. Low, S.; Ullah, F.; Shirowzhan, S.; Sepasgozar, S.M.E.; Lee, C.L. Smart Digital Marketing Capabilities for Sustainable Property Development: A Case of Malaysia. *Sustainability* **2020**, *12*, 5402. [CrossRef]

63. Bohari, A.A.M.; Skitmore, M.; Xia, B.; Teo, M.; Khalil, N. Key stakeholder values in encouraging green orientation of construction procurement. *J. Clean. Prod.* **2020**, *270*, 122246. [CrossRef]

64. Yu, A.T.W.; Yevu, S.K.; Nani, G. Towards an integration framework for promoting electronic procurement and sustainable procurement in the construction industry: A systematic literature review. *J. Clean. Prod.* **2020**, *250*, 119493. [CrossRef]

65. Schestak, I.; Spriet, J.; Styles, D.; Prysor Williams, A. Emissions down the drain: Balancing life cycle energy and greenhouse gas savings with resource use for heat recovery from kitchen drains. *J. Environ. Manag.* **2020**, *271*, 110988. [CrossRef]

66. Macke, J.; Sarate, J.A.R.; de Atayde Moschen, S. Smart sustainable cities evaluation and sense of community. *J. Clean. Prod.* **2019**, *239*, 118103. [CrossRef]

67. Filho, W.L.; Skouloudis, A.; Londero Brandli, L.; Lange Salvia, A.; Veiga Avila, L.; Rayman-Bacchus, L. Sustainability and procurement practices in higher education institutions: Barriers and drivers. *J. Clean. Prod.* **2019**, *231*, 1267–1280. [CrossRef]

68. O'Donoghue, R. Situated Learning in Relation to Human Conduct and Social-Ecological Change. In *Schooling for Sustainable Development in Africa*; LotzSisitka, H., Shumba, O., Lupele, J., Wilmot, D., Eds.; Springer: Berlin, Germany, 2016; pp. 25–38.

69. Pacheco-Blanco, B.; Bastante-Ceca, J.M. Green public procurement as an initiative for sustainable consumption. An exploratory study of Spanish public universities. *J. Clean. Prod.* **2016**, *133*, 648–656. [CrossRef]

70. Vapa-Tankosić, J.; Ignjatijević, S.; Kiurski, J.; Milenković, J.; Milojević, I. Analysis of Consumers' Willingness to Pay for Organic and Local Honey in Serbia. *Sustainability* **2020**, *12*, 4686. [CrossRef]

71. Fuentes-Bargues, L.J.; Ferrer-Gisbert, P.S.; Gonzalez-Cruz, M.C. Analysis of Green Public Procurement of Works by Spanish Public Universities. *Int. J. Environ. Res. Public Health* **2018**, *15*, 1888. [CrossRef] [PubMed]

72. Delmonico, D.; Jabbour, C.J.C.; Pereira, S.C.F.; Jabbour, A.B.L.D.S.; Renwick, D.S.; Thomé, A.M.T. Unveiling barriers to sustainable public procurement in emerging economies: Evidence from a leading sustainable supply chain initiative in Latin America. *Resour. Conserv. Recycl.* **2018**, *134*, 70–79. [CrossRef]

73. Song, W.; Ming, X.; Liu, H.C. Identifying critical risk factors of sustainable supply chain management: A rough strength-relation analysis method. *J. Clean. Prod.* **2017**, *143*, 100–115. [CrossRef]

74. Chan, A.P.C.; Darko, A.; Olanipekun, A.O.; Ameyaw, E.E. Critical Barriers to Green Building Technologies Adoption in Developing Countries: The Case of Ghana. *J. Clean. Prod.* **2018**, *172*, 1067–1079. [CrossRef]

75. Liu, J.; Xue, J.; Yang, L.; Shi, B. Enhancing green public procurement practices in local governments: Chinese evidence based on a new research framework. *J. Clean. Prod.* **2019**, *211*, 842–854. [CrossRef]

76. Fang, H.; Wang, B.; Song, W. Analyzing the Interrelationships among Barriers to Green Procurement in Photovoltaic Industry: An Integrated Method. *J. Clean. Prod.* **2019**, *249*, 119408. [CrossRef]

77. Wade, A.; Sinha, P.; Drozdiak, K.; Mulvaney, D.; Slomka, J. Ecodesign, Ecolabeling, and Green Procurement Policies-enabling more Sustainable Photovoltaics? In Proceedings of the IEEE 7th World Conference on Photovoltaic Journal Pre-Proof 34 Energy Conversion (WCPEC), Waikoloa Village, HI, USA, 10–15 June 2018; pp. 2600–2605.

78. Jensen, L. (Ed.) *The Sustainable Development Goals Report*; United Nations: New York, NY, USA, 2019; Available online: https://unstats.un.org/sdgs/report/2019/The-Sustainable-Development-Goals-Report-2019.pdf (accessed on 2 September 2020).

79. Philips, M.; Schuler, E.R.; Alejandra, M. Groningen Leads the World. Available online: https://industrielinqs.nl/wp-content/uploads/2019/10/Paris-is-not-enough.pdf (accessed on 2 September 2020).

80. Wang, J.; Zhang, Y.; Goh, M. Moderating the role of firm size in sustainable performance improvement through sustainable supply chain management. *Sustainability* **2018**, *10*, 1654. [CrossRef]

81. Martek, I.; Hosseini, M.R.; Shrestha, A.; Edwards, D.J.; Durdyev, S. Barriers inhibiting the transition to sustainability within the Australian construction industry: An investigation of technical and social interactions. *J. Clean. Prod.* **2019**, *211*, 281–292. [CrossRef]
82. Shen, L.; Zhang, Z.; Zhang, X. Key factors affecting green procurement in real estate development: A China study. *J. Clean. Prod.* **2017**, *153*, 372–383. [CrossRef]
83. Shen, L.; Zhang, Z.; Long, Z. Significant barriers to green procurement in real estate development. *Resour. Conserv. Recycl.* **2017**, *116*, 160–168. [CrossRef]
84. Pentru Aprobarea Ghidului de Achiziţii Publice Verzi Care Cuprinde Cerinţele Minime Privind Protecţia Mediului Pentru Anumite Grupe de Produse şi Servicii ce se Solicită la Nivelul Caietelor de Sarcini. Available online: http://anap.gov.ro/web/wp-content/uploads/2018/11/ORDIN-Nr-1068-Achizitii-verzi.pdf (accessed on 2 September 2020).
85. Bosio, E.; Djankov, S. How Large is Public Procurement? Available online: https://blogs.worldbank.org/developmenttalk/how-large-public-procurement (accessed on 2 September 2020).

Article

The Taxation of Agriculture in the Republic of Serbia as a Factor of Development of Organic Agriculture

Goran Milošević [1,*], Mirko Kulić [2], Zvezdan Đurić [3] and Olivera Đurić [4]

[1] Faculty of Law, University of Novi Sad, Trg Dositeja Obradovića 1, 21000 Novi Sad, Serbia
[2] Faculty of Law for Commerce and Judiciary in Novi Sad, University Business Academy in Novi Sad, Geri Karolja 1, 21000 Novi Sad, Serbia; mirkokulic@hotmail.com
[3] School, Belgrade Business and Art Academy for Applied Studies, Kraljice Marije 73, 11000 Belgrade, Serbia; zvezdan.djuric@bpa.edu.rs
[4] School, Academy for Applied Studies, South Serbia, 16000 Leskovac, Serbia; zvezdadj@mts.rs
* Correspondence: g.milosevic@pf.uns.ac.rs

Received: 18 March 2020; Accepted: 10 April 2020; Published: 17 April 2020

Abstract: Agricultural activity is very important for every country that strives to create a stimulating, stable, abundant, sustainable and equal business environment for all market participants. By striving for sustainable economic development and growth, as well as preservation of the ecosystem, organic agricultural production aims to produce high-quality food. Within the socio-economic space, the role of the state as a regulator of production and market relations is indispensable. The state often uses fiscal policy as an instrument for the regulation of relations in the sphere of production and trade. The level of the tax burden is of vital importance for achieving a sustainable level of agricultural development. From the aspect of the Serbian economy, the taxation of agriculture in the future period must be based on a system of tax incentives for organic production. In the long run, this can increase the volume of organic production and the use of healthy food in human nutrition.

Keywords: agriculture; organic production; land; taxation; tax incentives; income; revenue; property

---

1. Introduction

Agricultural production emerged with the onset of human society. It has existed in the past and will exist in the future; it is the precondition for the biological survival of the human community and the determinant of growth and development of a society. The process of agricultural activity has not fundamentally changed throughout the history of its development; the changes have been imperceptible. Yet, agricultural activity has matured in terms of tools and human knowledge and progress has been made in the "utilization of nature"; and this spontaneously through the human community formation processes [1].

Agricultural stakeholders use the land as a factor of production. Excessive exploitation of the land may have negative consequences accompanied by a decline in land fertility and crop yields [2]. This shows that the land is a production input with a fixed value and cannot be increased. Contrary to the supply, demand for land is a category derived from the demand for goods; that is, goods produced on that land as a factor of production [3,4].

Agricultural activity is performed by a vast number of stakeholders, of unequal economic power, as well as of unequal production conditions. Accordingly, agricultural activity can be organized in the form of individual agricultural production, individual farms and agricultural companies [5]. Regardless of the organization, agriculture consists of conventional and organic production. Organic production is "friendly" towards the natural resources and the environment, thus becoming a general interest of the economy and social community. Under such conditions, of utmost importance is the role of the state, which has a task to create a stimulating, stable, yielding, sustainable and equal business environment for all market participants [6].

From the aspect of agricultural activity, the role of the state is very important, whether it be regulating economic relations or creating institutional frameworks for the behavior of the stakeholders of this activity. As a regulator of various social processes, the state has a significant role in the sphere of taxation, thus establishing the market balance among agricultural producers. In an attempt to regulate its tax system most simply and comprehensively, without hindering the market space, the state must at the same time achieve fiscal, political, economic, social and other goals through the taxation of agriculture. The scope of the tax burden on agriculture as an economic activity is of vital importance for achieving a sustainable level of development of this industry. Therefore, the taxation of agriculture must be approached with full respect for the objectives and economic interests of the holders of this activity.

This requires an active role of the economic and political authorities that reflects in the application of the necessary regulatory instruments in the process of establishing the necessary balance in the agricultural market. Within the economic policy of taxation, the state plays the roles of a regulator in the distribution of national income, allocator of funds, a stabilizer of conjuncture oscillations and a catalyst of economic growth [7,8].

The article aims to analyze the current situation through the analysis of the existing norms regulating the taxation of agriculture in the Republic of Serbia *de lege lata*, and to propose measures for its improvement *de lege ferenda*, so that organic agricultural production would receive a special tax treatment.

The basic hypothesis of our research is based on the premise that the Serbian legislature, through the tax system, treats all forms of agricultural activity identically, thus not contributing to the lasting and long-term growth of organic agricultural production. Organic production growth can only be achieved by measures that improve the efficiency of resource allocation, which is one of the basic assumptions of economic growth. In this regard, the state is tasked with creating a better institutional environment for doing business, to accelerate the market allocation. To confirm the aforementioned hypothesis, the article will analyze the taxes imposed on farmers.

2. Organic Agricultural Production in the Republic of Serbia

2.1. The Share of Organic Production in Total Agricultural Production

Agricultural production refers to a process of production of plant and animal products on agricultural land. Within the Serbian classification of economic activities, agriculture comprises the cultivation of crops and plants and animal husbandry, as well as organic production and cultivation of genetically modified plants and animals [9]. Based on the above definition, agricultural production can be classified into plant and livestock production. Plant production includes crop growing, vegetable growing, fruit growing and viticulture, while livestock production includes cattle farming, pig farming, sheep farming, poultry farming, cuniculture, goat farming, fish farming, beekeeping, etc.

Concerning the technological process used and other specific features of plant and livestock production, the agricultural production process consists of conventional (classic) and organic production. Conventional agriculture is based on the maximization of yield per unit of agricultural land. Maximization of yield requires large quantities of non-renewable natural resources and energy, as well as the use of different chemicals. The results of such agricultural production are high yields, albeit accompanied by an increasing use of non-renewable natural resources and degradation of land, water and air.

Organic agriculture developed in reaction to the increasing environmental degradation, deterioration of food quality and the growing threat to the health of the human population [6]. Organic agriculture is aimed at producing high-quality food while developing sustainable agriculture through the conservation of the ecosystem and increasing soil fertility. This implies legal regulation with the fulfillment of precisely defined conditions, as follows: isolation of land parcels, livestock farms and processing facilities from possible pollution sources; the adequate quality of irrigation water; coordinated development of plant and livestock production; and qualification of experts and producers for organic agriculture with the obligation of permanent innovation of knowledge [6]. Organic agriculture implies the use of natural fertilizers in the production process, with the exclusion of the use of pesticides, herbicides and artificial fertilizers. Protection in organic plant production is based on preventive agro-technical and hygiene measures, and in the case when diseases and pests cannot be suppressed preventively, biological protection agents are used.

From the perspective of the Republic of Serbia, agricultural production is an important factor in economic development. The total number of agricultural holdings in the Republic of Serbia in 2012 amounted to 631,552 farms, and their average economic size was 5939 euros [10]. Its development and importance have been especially intensified during the last years, through the development of an institutional framework for organic production. Within the Serbian normative solutions, organic production is the production of agricultural and other products that is based on the use of organic production methods in all phases of production, through the application of crop rotation, green manure, compost and biological insect control, but which excludes the use of genetically modified organisms and products that consist of or are deriving from genetically modified organisms, as well as the use of ionizing radiation [11]. Organic production includes organic plant and livestock production methods.

The Serbian production flows can be tracked by applying the annual index of agricultural production, which reflects changes in the volume of agricultural production at the level of annual periodicity. The calculation of the agricultural production index is based on the data on agricultural production in 2013 (basic period) and in 2014, which are presented in quantitative form. The agricultural production index for 2014 shows the growth of the physical volume of agricultural production compared to 2013 (Table 1) by slightly less than 2%. This growth is also visible in plant crops. Plant production in 2014 increased by just less than six percent compared to the previous year. If we analyze the livestock production flows, it is noticeable that no positive trends were achieved as were in plant production. Positive growth trends in the physical volume of agricultural production in 2014 were achieved only in pig, cattle and sheep production, while the production of other livestock products remained at the same level or even decreased, as in beekeeping.

Table 1. Agricultural production indices, 2014/2013 [12,13].

Plant Production (1, 2, 3)	**105.59**
1. Crop growing and vegetable growing (total)	111.85
- Grains	119.85
- Industrial plants	117.81
- Vegetables	82.99
- Forage (fodder) plants	104.60
2. Fruit growing	79.54
3. Viticulture	72.78
Animal husbandry (1+2+3+4+5)	**100.39**
1. Cattle farming	100.8
2. Pig farming	102.9
3. Sheep farming	108.2
4. Poultry farming	96.1
5. Beekeeping	51.24
Agricultural production in total	101.98

Organic production accounts for around 0.3% of total agricultural production. The arable land under organic production (Table 2) is mostly used for the production of fruits, grains, industrial and fodder plants, and the least used for the production of medicinal and aromatic plants.

Table 2. Arable land under organic plant production (in hectares) [14].

Organic Plant Production	2017	2018
Grains	3661.7	3613.6
Industrial Plants	2290.5	1961.8
Vegetables	2300	1995
Fodder Plants	1210.9	1336.5
Fruits	4055.9	5883.4
Medicinal and Aromatic Plants	1146	1934
Other	3112	5356

According to the data for 2017 (Table 3), of the total organic livestock production, small and large livestock was the most represented with a total of 78%, while poultry accounted for about 12% and beekeeping for 10%.

Table 3. Organic livestock production [14].

Organic Livestock Production	Bees	Laying Hens	Cattle	Goats	Horses	Donkeys	Pigs	Sheep
2017	2307 beehives	4415	3094 livestock units	2048	177	47	87	4665

2.2. Organic Farming Market

In 2014, the European organic products market grew by approximately 7.6% compared to the previous year and amounted to over 26 billion euros. Consumers in the European Union spent the most money on organic food—around 24 billion euros—where Germany was leading with a share of 30% of retail sales, followed by France with a share of 18%, the United Kingdom with a share of 9% and Italy with a share of 8% [15]. In terms of value, the German market share was 7.9 billion euros, the French market share was 4.8 billion euros, the United Kingdom's market share was 2.3 billion euros and Italy's market share was 2.1 billion euros. On a global scale, the United States has the largest organic products market, which amounts to 27.1 billion euros [16]. In 2014, the country with the highest per-capita organic consumption was Switzerland with 221 euros, followed by Luxembourg, Denmark, Sweden, Liechtenstein, Austria and Germany. In 2011, the organic food share of the total food market was the highest in Denmark with 7.6%, followed by Switzerland with 7.1% and Austria with 6.5% [15].

In contrast to the European Union market, the Serbian organic products market is small and underdeveloped, with a pronounced upward trend when observed through the increase in the number of producers involved in organic production. In 2010, the number of organic producers was 137, and in 2014 it amounted to 1867 registered entities (Table 4). Certain differences are noticeable in domestic, local and foreign markets. The domestic local market is very unstable and uneven. In the local market, organic products are traded between local producers, small retail stores and consumers in green markets. At the national market level, the presence of large manufacturers, retail chains and large business entities that are import–export oriented, is noticeable. A certain number of domestic companies are successful in the placement of organic products in foreign markets; however, Serbia's potential in terms of production and export of organic products has not been sufficiently utilized. (Table 4).

Table 4. Producers of organic products [14,17].

Year	2010	2011	2012	2013	2014
Number of producers involved in organic production	137	323	1061	1281	1867

Organic production in Serbia is export-oriented and most of these products are placed in foreign markets. In 2012, the volume of exported organic products was 1,561,672.50 kg worth 3,740,801 euros, and in 2013 it was 7,101,301.24 kg worth 10,090,801.37 euros [17]. In 2016, the volume of organic product exports amounted to approximately 19 million euros. The most exported products are low-processed, namely, frozen raspberries—10.1 million euros; frozen blackberries—1.5 million euros; fresh organic apples—717,400 euros; apple concentrate—1.6 million euros; dried fruit—2.2 million euros; and purées made of cherries, blackberries, raspberries and plum—over 846,000 euros [18].

2.3. Trade Exchange of Organic Products between Serbia and the EU and CEFTA Countries

By signing the Stabilization and Association Agreement, Serbia has entered the process of accelerated adjustment to the EU rules. One of the elements of adjustment is the alignment of Serbia's agricultural policy with the common agricultural policy of the EU. For Serbia, this meant introducing the principle of economic efficiency in agriculture, especially through cost reductions and improvement of production conditions. In the case of Serbia, improving efficiency through the complete value chain of organic production requires large investments, i.e., significant public resources. The state must use taxes and other incentives in order to create a sustainable and competitive placement of organic products on the EU market.

The most important trade partner of Serbia is the European Union. In 2015, the volume of organic products in total Serbian export was 70.4%, it is 13,787.417 euros (Table 5). In 2018, the value of exported organic products was 27,419,347.71 euros [20]. The largest exports were in Germany—7,433,661.00 euros; the Netherlands—3,495,144.00 euros; Austria—3,077,104.00 euros; Italy—2,724,359.00 euros; France—1,897,039.00 euros; and Belgium—1,191,811.00 euros [20]. The structure of the exported organic products is dominated by raw products: fruits, cereals and vegetables; juices; herbs and spices; mushrooms; and industrial plants. Within this market, the most important destination for organic products from Serbia is Germany, which is a large organic food producer and large importer of these products. In addition to fruits and vegetables, organic protein crops, primarily soybeans used to feed organic livestock, are significant imported products in Germany; so, they can be significant export products for Serbian producers [17]. Import of meat into the European Union is restrictive because of a strictly regulated market, i.e., due to the fact that the production of organic meat and dairy products mainly depends on the production of cereals and oilseeds, which must have an organic origin.

Table 5. The share of export of organic products exports from Serbia to EU and CEFTA countries in 2015 [17,19].

Countries	%	Export Value
EU	70.4	13,787,417
CEFTA	1.0	195,519
Other	28.6	5,590,444
Total	**100.0**	**19,573,380**

The second most important Serbian foreign trade partner is the CEFTA countries. With CEFTA countries Serbia has a surplus of USD 2,051.6 million in trade exchange at the beginning of 2020 [21]. A single free-trade market of about 30 million inhabitants, without customs and other fiscal duties, was formed by the countries of Central and Eastern Europe in 1992 in Krakow, with the signing of the Central European Free Trade Agreement (CEFTA). Members of CEFTA are Serbia, Albania, Bosnia and Herzegovina, North Macedonia, Moldova, Montenegro as well as the United Nations Mission in Kosovo and Metohija UNMIK/Kosovo 1244. Under the CEFTA agreement, Serbia has a significant position in the foreign trade of agricultural products, which gives it the opportunity to increase competitiveness. However, due to the fact that market of the organic products is not developed within the CEFTA countries, the volume of trade of organic products is modest (around 1% in total export of Serbia). Inadequate foreign trade can be attributed to unresolved political issues.

As a consequence, CEFTA agreement does not lead to the desired level of exchange flows in recent years. So, in the coming period Serbia must implement measures that will contribute to the growth of productivity, quality of products, marketing activity, etc., which will overall contribute to growth in price competitiveness [22]. The competitive advantages, including tax incentives for domestic organic food production, can activate a significant proportion of unused resources in order to create the conditions for success in the CEFTA and EU markets.

3. Taxation of Agriculture in the Republic of Serbia

The Serbian tax system distinguishes agricultural companies and individual farms when taxing agriculture. In the sphere of taxation, agricultural legal entities have the same treatment as all other legal entities, and they fall under the corporate tax system, while individual farmers are natural persons, thus from the aspect of taxation, they fall under the individual income tax system [5]. The Serbian tax system makes no distinction between conventional and organic farming. The subjects within taxation of agricultural activities are individual income, corporate profits, added value and property.

3.1. Taxation of Agricultural Activity by Individual Income Tax

Individual income tax is paid by natural persons, including farmers, who generate income. This is a subjective tax that affects the overall income of a particular taxpayer [2,23]. Income represents the sum of taxable revenues generated in a calendar year. Taxable revenue means the difference between gross taxable revenue earned by a taxpayer and the expenses he/she had in generating and preserving it [24]. The payer of individual income tax is a resident of the Republic of Serbia, on the income earned in the territory of the Republic of Serbia or some other state [25]. Natural persons engaged in agricultural activity taxed by the individual income tax system are obliged, by the law, to determine and settle (pay) their tax liability for all forms of taxable revenues generated in a calendar year [2]. Following the law, the revenue from the agricultural activity is included in the taxable annual income.

3.1.1. Taxation of Agricultural Activity by the Tax on Revenue from Self-Employment

Revenue from self-employment means the revenue from a business, including agriculture and forestry activities, unless the tax on such revenue is paid on some other grounds following the law [26]. The taxable revenue from self-employment is the taxable profit. The taxable profit is determined in the tax balance sheet by adjusting the profit declared in the income statement drawn up in compliance with the regulations governing accounting for the taxpayer who is obligated to keep double-entry accounts, or in line with the regulation passed by the Minister responsible for finance if the taxpayer is keeping single-entry accounts [2,27]. The rate of tax on self-employment revenue is 10%.

The taxpayer of the revenue from self-employment based on the revenue from agriculture and forestry is a natural person—a holder of a family farm registered in the Register of Farms following the regulations governing this field [28]—who keeps account books following the law [29]. All this thus results in the two cumulative conditions that the natural person—the holder of the family farm—must fulfill in order to be registered in the Register of Farms and to have the status of an entrepreneur: (1) to be registered in the Register of Agricultural Households and (2) to keep account books [2]. If a natural person—a holder of a family farm—does not fulfill the abovementioned cumulative conditions, they will not have the status of an entrepreneur [2].

This means that a natural person who generates revenue from agricultural and forestry activities—that is, earns revenues from agricultural and forestry products [30], including a registered farm that does not have the status of an entrepreneur—is not obliged to keep the book of accounts and pay individual income tax. The mentioned category of natural persons is not considered a taxpayer concerning individual income tax but is obliged to pay contributions for pension insurance and healthcare insurance following the law regulating the system of compulsory pension and healthcare insurance, provided that it has applied for insurance based on performing an agricultural activity with the competent fund [2].

From the aforementioned legal solution, it is clear that the majority of farmers in the Republic of Serbia will not be obliged to pay individual income tax, which provides them with a privileged treatment from the aspect of income taxation. Privileged treatment of farmers' income taxation is very common, especially in less developed countries [31]. In this respect, the Republic of Croatia and Bosnia and Herzegovina have given a privileged position to farmers within their tax legal solutions [32,33]. From a tax point-of-view, the non-taxation of income generated by farmers is unfair. However, in conditions of underdevelopment of agriculture and significant migration of the agricultural workforce to industrial centers, incentive measures involving the non-payment of individual income tax by non-entrepreneurial farmers may contribute to the efficiency of the market allocation of economic resources.

3.1.2. Taxation of Agricultural Activity by the Salary Tax

An entrepreneur, thus also a natural person who performs agricultural activity as an entrepreneur, who pays tax on actual revenue from self-employment, may decide to pay personal earnings [34]. Personal earnings mean the monetary amount that an entrepreneur pays and records in its books of accounts as their monthly personal income increased by accompanying liabilities of the earnings. Salary means paid personal income of an entrepreneur determined following the law [35]. From the aspect of taxation, personal income of an entrepreneur also includes the premiums for all kinds of voluntary insurance, as well as contributions to the voluntary pension fund, which the employer pays for his/her employees—insured persons covered by voluntary insurance—in compliance with the law governing the voluntary insurance and voluntary pension funds and pension plans [36]. The payer of salary tax is a natural person who earns a salary. The base of the salary tax consists of the paid-out or actual earnings, reduced by the statutory non-taxable amount [37]. The salary is taxed at a rate of 10%. The salary taxation system provides for tax reliefs for entrepreneurs who employ a new employee in the form of the right to a refund of a part of the tax paid on the earnings of a newly employed person [2].

3.1.3. Taxation of Agricultural Activity by Tax on Other Revenues

Within the individual income tax system, from the aspect of one's work, other revenues are also revenues from the sale of agricultural and forest products and services, including revenues from the collection and sale of forest fruits and medicinal herbs, as well as revenues from the growing and sale of mushrooms, raising and sale of bee swarms (bees) and snails [38]. The taxable revenue realized by a natural person from the sale of agricultural and forest products and services, including also the revenue from collection and sale of forest fruits and medicinal herbs, as well as growing and sale of mushrooms, bee swarms (bees) and snails, consists of the gross revenue reduced by the standardized expenditures amounting to 90% [39]. A taxpayer is a natural person who generates these revenues.

Exceptionally, revenues from the sale of agricultural and forest products and services, also including revenues from the collection and sale of forest fruits and medicinal herbs, as well as from the growing and sale of mushrooms and raising and sale of bee swarms (bees) and snails, are not taxable if generated by the following natural persons: holders of farms; natural persons who pay contributions for compulsory social insurance according to the decision as the insured based on agricultural activity, following the law governing contributions for compulsory social insurance; and beneficiaries of agricultural pension [40]. The abovementioned exemption does not apply to the following natural persons: members of a family farm and farmers who are not insured based on agricultural activity, and who generate revenues from the sale of agricultural and forest products and services, [31] including also revenues from the collection and sale of forest fruits and medicinal herbs, as well as from growing and sale of mushrooms, bee swarms (bees) and snails; hence, in this sense they are the taxpayers. The tax rate on other revenues is 20%. Reduction of the tax base on other agricultural income is a good solution for several reasons [2]: First, due to its amount, it encourages the development of agriculture and leads to lower prices of agricultural products; secondly, it does not affect the behavior of the farmers as it is not paid on actual income; and thirdly, it provides certainty in terms of the amount of tax liability. A problem may occur in terms of its static nature, and it is, therefore, necessary to periodically conduct its revaluation.

3.2. Taxation of Agricultural Activity by Corporate Profit Tax

In the sphere of taxation, agricultural legal entities are treated the same as all other legal entities, and they fall under the corporate profit tax system. A taxpayer of the corporate profit tax is a company or enterprise, or another legal entity founded to perform activities to gain profit. A taxpayer is also a cooperative and non-profit organization (legal entity not founded for generating a profit) that earns revenues by selling products on the market or providing services for a fee. A taxpayer is a resident of the Republic of Serbia who is subject to taxation of profit it generates in the territory of the Republic of Serbia and outside it [41]. A non-resident of the Republic of Serbia is also subject to taxation for any profit it generates through a permanent operating unit in the territory of the Republic of Serbia unless an international agreement for the avoidance of double taxation stipulates otherwise [42].

The corporate profit tax base is taxable profit. Taxable profit is determined in the tax balance sheet by adjusting the taxpayer's profit declared in the income statement, which has been drawn up in compliance with the International Accounting Standards, International Financial Reporting Standards, International Financial Reporting Standard for Small and Medium-Sized Entities and regulations governing accountancy [43]. The taxable profit of a taxpayer, who according to the regulations governing accounting is not applying the International Accounting Standards, International Financial Reporting Standards, International Financial Reporting Standard for Small and Medium-Sized Entities, is determined in the tax balance sheet by adjusting the taxpayer's profit, declared following the method of recognizing, measuring and estimating the revenues and expenditures prescribed by the Minister of Finance, in the manner provided by the law [44].

The corporate income tax rate is proportional and uniform and is 15%. Unless an international agreement for the avoidance of double taxation stipulates otherwise, withholding tax is charged and paid at the rate of 20% on the revenue earned by a non-resident legal entity from a resident legal entity following the law [45]. For the withholding tax for each taxpayer and for each revenue earned or paid out, the payer needs to calculate, withhold and pay into the prescribed accounts within three days from the day when the revenue was generated or paid [46].

3.3. Taxation of Agriculture by Value-Added Tax

The objects of value-added tax taxation are as follows: [47]

- the delivery of goods and providing of services (trade of goods and services) that a taxpayer carries out for a fee, within the scope of performing an activity;
- the import of goods.

A taxpayer of a value-added tax is a person, including a person without a head office or permanent residence in the Republic of Serbia, who independently performs the trade of goods and services, within the framework of his/her registered activity. From the aspect of taxation of farmers, one distinguishes individual farmers, entrepreneurial farmers and agricultural companies. An individual farmer is not a taxpayer of value-added tax.

An entrepreneurial farmer is a value-added taxpayer for the turnover of goods and services, namely: [48];

- by force of law—a farmer whose total turnover of goods and services in the previous 12 months is higher than 8,000,000.00 dinars
- by their option—in the case of the holder of a family farm registered in the Register of Farms, who has opted to have the status of an entrepreneur. This option is done by filing a tax return to the competent tax authority.

Therefore, persons who are owners of registered farms are the taxpayers of the value-added tax if they keep books of accounts and if the annual revenue from agricultural holding is higher than 8,000,000.00 dinars; that is, if they file a tax return themselves, thereby gaining the status of an entrepreneur.

Depending on the realized turnover of goods and services, agricultural companies are taxpayers of a value-added tax:

1. By force of law, namely: [49]

- a person whose total turnover of goods and services, except the turnover of the equipment and facilities for the performance of business activities in the previous 12 months, is higher than 8,000,000.00 dinars;
- a person who, when starting a business, estimates that over the next 12 months he/she will achieve a total turnover higher than 8,000,000.00 dinars.

2. By their option—persons who may opt to be in the value-added tax system (to be taxpayers).

The tax base applicable to the trade of goods and services is the amount of compensation (in money, objects or services) that a taxpayer receives or should receive for the goods delivered or services provided from the recipient of goods or services or a third party, including subsidies and other income, into which the value-added tax is not included [50].

Natural persons who are owners, tenants and other users of agricultural and forest land and natural persons who are, as holders, or members of the farm, registered in the Register of Farms following a regulation that governs the registration of farms, are entitled to reimbursement based on value-added tax [51]. The reimbursement is granted to farmers who perform the trade of agricultural and forest products as well as agricultural services to taxpayers. If the farmers perform the trade of goods and services to the taxpayers, the taxpayer is obliged to calculate the value-added tax reimbursement at the amount of 8% of the value of the received goods and services, and issue an accounting document, as well as to pay the farmers the calculated reimbursement in money (by making payment to a current or savings account) [52]. The taxpayers of the value-added tax have the right to deduct the amount of the reimbursement as a preliminary tax, on condition that they have paid the value-added tax reimbursement and the value of the received goods and services to the farmer [53].

Providing the farmers with the opportunity to exercise the right to reimbursement based on value-added tax may contribute to a better allocation of agricultural resources. Specifically, the reimbursement indirectly compensates farmers for the costs they incur when paying taxes on the acquisition of various production inputs. By reducing production costs based on reimbursement of value-added tax, the competitive position of these farmers in the agricultural market increases.

In terms of agricultural taxation, the value-added tax is not paid on the trade of the following [54]:

- agricultural land, as well as the letting of such land;
- facilities for agricultural activities, except for the first transfer of the right of disposal over newly constructed building facilities or economically divisible units within such buildings and the first transfer of equity stake in the newly constructed building facilities or economically divisible units within these facilities, as well as the trade of buildings and economically divisible units within these buildings, including equity stakes in these goods, in cases when the contract under which the trade of those goods is conducted, which was concluded between the value-added taxpayers, stipulates that value-added tax will be calculated for this trade, provided that the acquirer may fully deduct the calculated value-added tax as preliminary tax.

3.4. Taxation of Agriculture by Property Taxes

The Serbian tax system distinguishes three forms of property taxes: property tax, inheritance and gift tax and absolute rights transfer tax.

3.4.1. Taxation of Agriculture by Property Tax

Property tax is levied on a property in its static form. The subject of taxation is certain rights relating to real estate. The basis of taxation is the ownership right or the right of ownership of land, but this may also include other rights. When in addition to ownership right some other right also applies to a real estate, property tax is paid on that right, not on ownership right [55]. Real estate is understood to mean the following: land, residential and business buildings, apartments, office premises, garages and other buildings or their parts [56].

The taxpayer of the property tax is a legal entity and a natural person who has some of the rights to a real estate in the territory of the Republic of Serbia. The property tax base is determined differently, depending on whether the taxpayer keeps or does not keep books of accounts. The property tax base for real estate of the taxpayer who does not keep books of accounts is the real estate value determined following the Property Tax Law [57]. The property tax for the taxpayer who does not keep books of accounts is determined by the decision of the local self-government unit's body.

The property tax base for real estate of the taxpayer who keeps books of accounts and whose value in the account books is stated using the fair value method following the International Accounting Standards (IAS) and International Financial Reporting Standards (IFRS) and adopted accounting policies is the fair value stated on the last day of the fiscal year of the taxpayer in the current year [31]. The property tax base for real estate of the taxpayer, whose value in the account books is not stated using the fair value method, for undeveloped land—consists of the land value; for other real estate—the value of the buildings increases due to the value of associated land. The taxpayer who keeps books of accounts determines the property tax by self-taxation.

The Serbian legislator intends that the property tax base for agricultural land should be its market value. Taking into account the situation on the real estate market in the Republic of Serbia, we believe that a large number of local self-government units will be faced with the lack of turnover data and that the income from taxation of agricultural land will not achieve the desired fiscal objective of the legislator [58]. In this regard, smaller and underdeveloped local self-government units will have an additional problem.

The tax rates for property taxes are fixed by the local self-government unit assembly; however, the legislator has limited their maximum amounts. For agricultural land, the central government has stipulated the maximum permissible rate (up to 0.3%), and within this limit, local self-government units independently determine its exact amount [59].

The legislator has stipulated many tax reliefs that in the coming period could be a strong incentive for taxpayers to reclaim agricultural and forest land [60].

From the aspect of the taxation of agriculture, property tax is not payable on the right to real estate, namely:

- reclaimed agricultural and forest land—five years from the commencement of reclamation;
- aquatic land and water facilities registered in the cadastre of water resources or water facilities register, except facilities for fish farming (fish farms);
- buildings of an individual income taxpayer for revenue generated from agriculture and forestry or a taxpayer whose registered predominant activity is agriculture, intended and used solely for primary agricultural production;
- on the total tax base of a taxpayer, which for all his/her real estate in that territory is not greater than 400,000.00 dinars.

The privileged treatment of agricultural land, in terms of property tax, is also applied in the tax systems of the Republic of Montenegro and the Republic of Srpska. For real estate owned or used by a person entered in the Register of Agricultural Producers, a legal person or an entrepreneur engaged in the production, finishing, packaging or processing of agricultural products produced in Montenegro, which are used for carrying out this activity, the tax rate may be reduced by up to 90% of the tax liability with the tax rate set by law [61]. In the Republic of Srpska, cultivated agricultural land and real estate serving for own agricultural production are exempted from paying property tax [62].

3.4.2. Taxation of Agriculture by Inheritance or Gift Tax

Property can be transferred without compensation between living persons (*inter vivos*), as well as in the case of death (*causa mortis*). Between living people, the property is transferred without compensation by way of gift, while in the case of death, the property is transferred by way of inheritance. In terms of the taxation of agriculture, inheritance and gift tax is paid on the total area of a real estate on which certain rights are constituted, which are inherited by heirs, or received by donees as a gift [63].

The taxpayer of inheritance and gift tax is a resident and non-resident of the Republic of Serbia who inherits or receives as a gift some right to real estate in the territory of the Republic of Serbia. The inheritance tax base is the market value of the inherited property, reduced by debts, costs and other encumbrances the taxpayer is obliged to pay or settle in some other way from the inherited property on the date of the onset of tax liability [64]. The gift tax base is the market value of the property received as a gift, on the date of the onset of tax liability, which is determined by the tax authority [65].

Inheritance and gift tax are paid by the taxpayers who, concerning a decedent or donor, are [66]:

- in the second order of succession according to the legal order of succession—at the rate of 1.5%;
- in the third and any subsequent order of succession, or the taxpayers that are unrelated to the decedent or donor—at the rate of 2.5%.

In terms of the taxation of agriculture, the legislator has foreseen that the inheritance and gift tax is not payable by the following [67]:

- an heir in the first order of succession, decedent's spouse and parent and a donee in the first order of succession and donor's spouse;
- a farmer heir or donee in the second order of succession who inherits or receives as a gift the property that serves him/her for agricultural purposes, if he/she has lived with the decedent or donor in the same household for at least one year before the decedent's death or before receiving the gift; if the mentioned farmer heir or donee in the second order of succession who inherits or receives as a gift the property that serves him/her for agricultural purposes changes his/her occupation before the expiry of five years from the date on which he/she inherited or received as a gift the property concerned, he/she is obliged to notify the competent tax authority of the change of occupation within 30 days from the date of the change; in this case, the heir or donee pays the inheritance and gift tax at the rate of 1.5%.
- a donee—on the property relinquished to him/her in the probate proceedings, which he/she would have received had the heir or donor waived inheritance;
- property jointly acquired by the spouses during their marriage, which is divided between former spouses, whereby they regulate their property relations regarding divorce.

3.4.3. Taxation of Agriculture by Absolute Rights Transfer Tax

In terms of the taxation of agriculture, the absolute rights transfer tax is paid on the transfer against compensation of the real estate ownership rights. The following are exempted from the absolute rights transfer taxation [68]:

- transfer or acquisition of the absolute right to which value-added tax is paid, following the law governing value-added tax;
- acquisition of the ownership right to separate parts of real estate by the division of co-ownership between co-owners in proportion to their co-owned parts on the date of division;
- absolute rights transfer based on expropriation.

The taxpayers of the absolute rights transfer tax are the following: the seller or transferor of the mentioned rights; the person to which the absolute right is transferred; the person who is the taxpayer of every right that is transferred—when an absolute right is transferred based on a contract of exchange [69]. The base of the absolute rights transfer tax is the contracted price or the market value of the transferred right. The absolute rights transfer tax is paid at the rate of 2.5%.

In terms of agricultural taxation, the absolute right transfer tax is not paid on the following [70]:

- land exchange by which at least one legal entity or a natural person whose predominant activity or occupation is agriculture, acquires agricultural or forest land for its grouping;
- when the property right to real estate is transferred to the provider of lifelong care, the spouse or a person who is in the first order of succession concerning the recipient of care, on the part of the real estate the provider of care would have inherited on the contract conclusion date in compliance with the law;
- the sale of a legal entity as adjudicated as bankrupt—in proportion to the participation of social or state capital in the total capital of the legal entity concerned;
- property acquisition or on the realization of compensation under the law governing the return of seized property and compensation for seized property, or the law governing return (restitution) of property to churches and religious communities;
- acquisition of the right to the land based on commission.

4. Tax System as a Growth Factor of Organic Production

A production function expresses the relationship between factors of production that result in final products. The ideal production volume implies an optimal combination of factors of production, economic efficiency and optimal balance in the market of goods and services [71]. In such conditions, any change in input price will affect the equilibrium market movements. Organic production in Serbia is characterized by rising costs of production factors. The high costs of organic production are resulting from acquisition of quality seed, quality control, certification, greater engagement of the workforce, additional costs of land preparation and other costs. As a result of the rising costs of domestic production, an increase in the price of organic primary and processed products is manifested in the organic products market, which is much higher than the price of conventional products. In addition to the aforementioned, due to the lack of competitiveness of domestic producers, prices of most imported processed organic products are lower than those of domestic producers, which additionally contributes to the inadequate allocation of economic resources.

The factor of growing production costs is also evident through the effect of organic production taxation. The tax burden represents an additional cost for organic food producers, which directly affects the increase in the price of these products. The increase in production costs due to distortions generated by taxes and contributions represents a decrease in the production efficiency [71] of organic production within the national economy, which further reduces their competitiveness concerning foreign producers with more efficient production. Faced with the inability to optimize their production, domestic producers give up on creating a production process based on organic production and direct the allocation of their resources to conventional production. The decrease in production efficiency and competitiveness of domestic organic production can be accompanied by a decrease in employment in the country, which means that a part of the human resources remains unused in the production process, [72] which in the long run may cause a slowdown in the gross domestic product movement.

In addition to striving for economic growth and fiscal welfare, the state must also take into account the non-fiscal interests and citizens' demands for maintaining a healthy environment, nutrition and well-being. Therefore, in addition to subsidies and grants, the state must also implement a system of tax incentives to stimulate organic agriculture.

The Serbian tax system does not recognize the difference between conventional and organic agricultural production; that is, it does not differentiate between a healthy environment, a healthy diet and fiscal goals. The taxation system must substantially include measures and instruments for reducing the cost of organic production. Reducing the cost of organic products through a system of tax incentives and exemptions necessarily leads to increased production efficiency and competitiveness. A lower tax burden on the labor costs of organic production employees, acquisition of quality seed, quality control and certification, cost of land preparation and other costs, produces an increase in the competitive position of organic producers. The positive effects of tax incentives can help reduce the gray area related to turnover and employees in organic production and its conversion into legal flows. In the long run, the system of tax incentives leads to the improvement of the position of persons working in the organic production system, but are not registered for compulsory social insurance, which will contribute to optimizing the balance in the labor market.

5. Conclusions

The Republic of Serbia has favorable land and climate conditions for organic production. The importance of this production is reflected in the fact that it enables better use of areas with less favorable land and climate conditions, including also lands of poor physical, chemical and other properties. Organic production can also be a significant source of income and employment opportunities for people, especially in small family farms. In this sense, the state must create a better institutional environment for doing business, to ensure a better allocation of economic resources. To achieve this goal, the state has to play an active role in the process of implementation of the taxation system, as a regulator of the production efficiency of organic farming within the national economy.

Through the taxation system, the Serbian legislator must strengthen the competitiveness and production capacity of organic producers, which will contribute to the lasting and long-term growth of this activity, and consequently to the economic growth and development of the country. To achieve a desirable level of development of organic production in the Republic of Serbia, it is necessary to implement additional measures through the system of tax incentives for organic production. In this way, through the taxation system, the positive effect in terms of raising the level of healthy food use could be achieved. Tax incentive measures can be implemented through the system of:

- a reduction of the fiscal burden in self-employment revenue taxes and corporate profit taxes;
- exemption from the taxation on the turnover of organic products in value-added tax;
- exemption from the taxation on the turnover of products intended for organic production in value-added tax.

Author Contributions: All authors have contributed equally to the idea, creation, research, and supervision of all activities regarding this manuscript. All authors have read and agreed to the published version of the manuscript.

Funding: This research was funded by the authors themselves.

Conflicts of Interest: The authors declare no conflict of interest.

References

1. Marković, L.; Sokić, S. *Ekonomija*; Zavod za udžbenike i nastavna sredstva: Belgrade, Serbia, 1996; p. 365.
2. Milošević, G.; Vuković, M.; Jovanović, D. Taxation of Farmers by the Income Tax in Serbia. *Econ. Agric.* **2018**, *65*, 683–696. [CrossRef]
3. Milošević, G. *Osnovi Ekonomije*; Faculty of Law: Novi Sad, Serbia, 2016; p. 300.
4. Mrdak, G.; Knežević, R. Land and Labor and Capital as a Condition for Development Economy. *Ekonomika* **2014**, *60*, 112–113. [CrossRef]
5. Milošević, G.; Kulić, M. Theoretical Basis And Practical Experiences of Tax Shifting with Reference to Agriculture. *Econ. Agric.* **2011**, *58*, 281–297. [CrossRef]
6. Đurić, Z.; Đurić, O. Management of Organic Production at Organic Family Farm in the Function of Economic Development of Serbia. In Proceedings of the 17th International Conference on Accounting and Management, Zagreb, Croatia, 9–10 June 2016; Belak, V., Jurić, D., Eds.; Hrvatski računovodja-Nezavisno udruženje računovođa poreskih savetnika i finansijskih radnika: Zagreb, Croatia, 2016; pp. 1–16.
7. Ristić, Ž. *Fiskalna Ekonomija*; Savremena administracija: Belgrade, Serbia, 1991; p. 480.

8.   Vukša, S.; Anđelić, D.; Kolarski, I. Uloga Ekonomske Politike u Ekonomskim Krizama. *Oditor* **2015**, *12*, 13–21. [CrossRef]

9.   Regulation on Classification of Activities, "Official Gazette of RS", No. 54/2010. Available online: https://www.paragraf.rs/propisi/uredba_o_klasifikaciji_delatnosti.html (accessed on 28 December 2019).

10. Cvijanović, D.; Subić, J.; Paraušić, V. Popis poljoprivrede 2012-Poljoprivreda u republici Srbiji poljoprivredna gazdinstva prema ekonomskoj veličini i tipu proizvodnje u republici Srbiji; Republički Zavod za Statistiku. Available online: https://publikacije.stat.gov.rs/G2014/Pdf/G201414007.pdf (accessed on 26 March 2020).

11. Article 3, Paragraph 1, Item 8 of the Law on Organic Production, "Official Gazette of RS", Nos. 30/2010 and 17/2019. Available online: https://www.paragraf.rs/propisi/zakon-o-organskoj-proizvodnji-republike-srbije.html (accessed on 20 December 2019).

12. Statistical Office of the Republic of Serbia. Saopštenje broj 035—god. Lxv, 20.02.2015.-Statistika Poljoprivrede. Available online: https://publikacije.stat.gov.rs/G2015/Pdf/G20151035.pdf (accessed on 20 December 2019).

13. Statistički Godišnjak Republike Srbije, 2017 (Statistical Yearbook of the Republic of Serbia). Available online: https://pod2.stat.gov.rs/ObjavljenePublikacije/G2017/pdf/G20172022.pdf (accessed on 12 December 2019).

14. Ministry of Agriculture, Forestry and Water Management of the Republic of Serbia. Organska Proizvodnja po Okruzima-2017 and 2018 and Organska Stočarska Proizvodnja. 2017. Available online: http://www.dnrl.minpolj.gov.rs/en/o_nama/organska/organska_proizvodnja_u_srbiji.html (accessed on 24 December 2019).

15. Organic Market Info, European Organic Market Grew to More than 26 Billion Euros in 2014. Available online: https://organic-market.info/news-in-brief-and-reports-article/european-organic-market-grew-to-more-than-26-billion-euros-in-2014.html (accessed on 26 March 2020).

16. Tabaković, M.; Simić, M.; Dragičević, V.; Brankov, M. Organska poljoprivreda u Srbiji. *Selekcija i Semenarstvo* **2017**, *23*, 45–53. [CrossRef]

17. Simić, I. Organska Proizvodnja—Neiskorišćen Potencijal Republike Srbije, Knowhow 2ACT, Norwegian Embassy, Support Programme "Strengthening Civil Society. 2015, p. 12. Available online: http://eukonvent.org/wp-content/uploads/2015/03/Neiskorisceni-potencijaliorganskeproizvodnje_Nakon-sednice_FIN.pdf (accessed on 26 March 2020).

18. Gulan, B. Svet i Srbija: *Organska Poljoprivreda i Hrana*, Makroekonomija, Ekonomske Analize, Srbija, Okruženje, i Međunarodna Ekonomija, Makroekonomija.org. 2018. Available online: https://www.makroekonomija.org/poljoprivreda/svet-i-srbija-organska-poljoprivreda-i-hrana/ (accessed on 26 March 2020).

19. Ministry of Agriculture; Forestry and Water Management of the Republic of Serbia. Izvoz Organskih Proizvoda u 2015. i 2018 Godini po Zemljama, Podaci Uprave Carina. Available online: http://www.dnrl.minpolj.gov.rs/download/organska/2018/7%20Organska%20proizvodnja%20u%20Srbiji/Izvoz%202018%20po%20zemljama.pdf (accessed on 6 April 2020).

20. Ministry of Agriculture; Forestry and Water Management of the Republic of Serbia. Izvoz organskih Proizvoda u 2018. Godini po Vrsti Proizvoda, Podaci Uprave Carina. Available online: http://www.dnrl.minpolj.gov.rs/download/organska/2018/7%20Organska%20proizvodnja%20u%20Srbiji/Izvoz%202018%20po%20vrsti%20proizvoda.pdf (accessed on 26 March 2020).

21. Spoljnotrgovinska Robna Razmena za Tekući Period, Februar 2020, Statistical Office of the Republic of Serbia. Available online: https://www.stat.gov.rs/sr-latn/vesti/20200331-spoljnotrgovinska-robna-razmena-za-teku%C4%87i-period-i-februar-2020/?s=1701 (accessed on 5 April 2020).

22. Tomić, G. Marketing Organskih Poljoprivrednih Proizvoda. Ph.D. Thesis, University of Kragujevac Faculty of Economics, Kragujevac, Serbia, 2015; pp. 55–57. Available online: http://www.ekfak.kg.ac.rs/images/Studije/das/DoktorskeDisertacije/Tomic%20Gordana%20%20Doktorska%20disertacija.pdf (accessed on 5 April 2020).

23. Gnjatović, D. *Finansije i Finansijsko Pravo*; Policijska Akademija: Belgrade, Serbia, 1999; p. 68.

24. Article 2 of the Individual Income Tax Law, "Official Gazette of RS", Nos. 24/01, 80/02, 80/02, 135/04, 62/06, 65/06, 31/09, 44/09, 18/10, 50/11, 91/11, 93/12, 114/12, 47/13, 48/13, 108/13, 57/14, 68/14, 112/15, 113/17; 7/2018, 95/2018, 4/2019 and 86/2019. Available online: https://www.paragraf.rs/propisi/zakon-o-porezu-na-dohodak-gradjana.html (accessed on 12 January 2020).

25. Article 7 of the Individual Income Tax Law, "Official Gazette of RS", Nos. 24/01, 80/02, 80/02, 135/04, 62/06, 65/06, 31/09, 44/09, 18/10, 50/11, 91/11, 93/12, 114/12, 47/13, 48/13, 108/13, 57/14, 68/14, 112/15, 113/17; 7/2018, 95/2018, 4/2019 and 86/2019. Available online: https://www.paragraf.rs/propisi/zakon-o-porezu-na-dohodak-gradjana.html (accessed on 12 January 2020).

26. Article 31 of the Individual Income Tax Law, "Official Gazette of RS", Nos. 24/01, 80/02, 80/02, 135/04, 62/06, 65/06, 31/09, 44/09, 18/10, 50/11, 91/11, 93/12, 114/12, 47/13, 48/13, 108/13, 57/14, 68/14, 112/15, 113/17; 7/2018, 95/2018, 4/2019 and 86/2019. Available online: https://www.paragraf.rs/propisi/zakon-o-porezu-na-dohodak-gradjana.html (accessed on 12 January 2020).

27. Milošević, G. *Teorija i Praksa Finansijskog Prava*; Kriminalističko-policijska akademija: Belgrade, Serbia, 2011; p. 225.

28. Article 32, paragraph 2 of the Individual Income Tax Law. Available online: https://www.paragraf.rs/propisi/zakon-o-porezu-na-dohodak-gradjana.html (accessed on 12 January 2020).

29. Article 43, Paragraph 2, in Connection with Article 32, Paragraph 2 of the Individual Income Tax Law. Available online: https://www.paragraf.rs/propisi/zakon-o-porezu-na-dohodak-gradjana.html (accessed on 12 January 2020).

30. Article 85, Paragraph 1, Item 14 of the Individual Income Tax Law. Available online: https://www.paragraf.rs/propisi/zakon-o-porezu-na-dohodak-gradjana.html (accessed on 12 January 2020).

31. Cvjetković, C.; Veselinović, J.; Nikolić, I. Tax Treatment of Farmers in the Republic of Serbia. *Econ. Agric.* **2015**, *62*, 737–749. [CrossRef]

32. The Ministry of Finance of the Republic of Croatia. Tax administration, Income Tax. Available online: https://www.porezna-uprava.hr/obrtnici/Stranice/Porez-na-dohodak_Poljoprivrednik.aspx (accessed on 27 March 2020).

33. Income Tax Law, "Official Gazette of BIH", Nos. 10/08, 9/10, 44/11, 7/13 i 65/13). Available online: https://advokat-prnjavorac.com/zakoni/Zakon_o_porezu_na_dohodak_FBiH.pdf (accessed on 27 March 2020).

34. Article 33a, Paragraph 1 of the Individual Income Tax Law. Available online: https://www.paragraf.rs/propisi/zakon-o-porezu-na-dohodak-gradjana.html (accessed on 12 January 2020).

35. Article 13, Paragraph 3 of the Individual Income Tax Law. Available online: https://www.paragraf.rs/propisi/zakon-o-porezu-na-dohodak-gradjana.html (accessed on 12 January 2020).

36. Article 14b, Paragraph 1 of the Individual Income Tax Law. Available online: https://www.paragraf.rs/propisi/zakon-o-porezu-na-dohodak-gradjana.html (accessed on 12 January 2020).

37. Article 15a, Paragraph 2 of the Individual Income Tax Law. Available online: https://www.paragraf.rs/propisi/zakon-o-porezu-na-dohodak-gradjana.html (accessed on 12 January 2020).

38. Article 85, Paragraph 1, Item 15 of the Individual Income Tax Law. Available online: https://www.paragraf.rs/propisi/zakon-o-porezu-na-dohodak-gradjana.html (accessed on 12 January 2020).

39. Article 85, Paragraph 4 of the Individual Income Tax Law. Available online: https://www.paragraf.rs/propisi/zakon-o-porezu-na-dohodak-gradjana.html (accessed on 12 January 2020).

40. Article 85, Paragraph 12 of the Individual Income Tax Law. Available online: https://www.paragraf.rs/propisi/zakon-o-porezu-na-dohodak-gradjana.html (accessed on 12 January 2020).

41. Article 2, Paragraph 1 of the Corporate Profit Tax Law, "Official Gazette of RS", Nos. 25/2001, 80/2002, 80/2002, 43/2003, 84/2004, 18/2010, 101/2011, 119/2012, 47/2013, 108/2013, 68/2014, 142/2014, 91/2015, 112/2015, 113/2017, 95/2018 and 86/2019. Available online: https://www.paragraf.rs/propisi/zakon_o_porezu_na_dobit_pravnih_lica.html (accessed on 12 January 2020).

42. Article 3, Paragraph 1 of the Corporate Profit Tax Law, "Official Gazette of RS", Nos. 25/2001, 80/2002, 80/2002, 43/2003, 84/2004, 18/2010, 101/2011, 119/2012, 47/2013, 108/2013, 68/2014, 142/2014, 91/2015, 112/2015, 113/2017, 95/2018 and 86/2019. Available online: https://www.paragraf.rs/propisi/zakon_o_porezu_na_dobit_pravnih_lica.html (accessed on 12 January 2020).

43. Article 6, Paragraph 2 of the Corporate Profit Tax Law. Available online: https://www.paragraf.rs/propisi/zakon_o_porezu_na_dobit_pravnih_lica.html (accessed on 12 January 2020).

44. Article 6, Paragraph 3 of the Corporate Profit Tax Law. Available online: https://www.paragraf.rs/propisi/zakon_o_porezu_na_dobit_pravnih_lica.html (accessed on 12 January 2020).

45. Article 40, Paragraph 1 of the Corporate Profit Tax Law. Available online: https://www.paragraf.rs/propisi/zakon_o_porezu_na_dobit_pravnih_lica.html (accessed on 12 January 2020).

46. Article 71, Paragraph 1 of the Corporate Profit Tax Law. Available online: https://www.paragraf.rs/propisi/zakon_o_porezu_na_dobit_pravnih_lica.html (accessed on 12 January 2020).

47. Article 3 of the Law on Value Added Tax, "Official Gazette of RS", Nos. 84/2004, 86/2004, 61/2005, 61/2007, 93/2012, 108/2013, 6/2014, 68/2014, 142/2014, 5/2015, 83/2015, 5/2016, 108/2016, 7/2017, 113/2017, 13/2018, 30/2018, 4/2019 and 72/2019. Available online: https://www.paragraf.rs/propisi/zakon-o-porezu-na-dodatu-vrednost.html (accessed on 5 January 2020).

48. Article 34, Paragraph 5 and 6 of the Law on Value Added Tax. Available online: https://www.paragraf.rs/propisi/zakon-o-porezu-na-dodatu-vrednost.html (accessed on 5 January 2020).

49. Article 33, Paragraph 1 of the Law on Value Added Tax. Available online: https://www.paragraf.rs/propisi/zakon-o-porezu-na-dodatu-vrednost.html (accessed on 5 January 2020).

50. Article 17, Paragraph 1 of the Law on Value Added Tax. Available online: https://www.paragraf.rs/propisi/zakon-o-porezu-na-dodatu-vrednost.html (accessed on 5 January 2020).

51. Article 34, Paragraph 1 of the Law on Value Added Tax. Available online: https://www.paragraf.rs/propisi/zakon-o-porezu-na-dodatu-vrednost.html (accessed on 5 January 2020).

52. Article 34, Paragraph 3 of the Law on Value Added Tax. Available online: https://www.paragraf.rs/propisi/zakon-o-porezu-na-dodatu-vrednost.html (accessed on 5 January 2020).

53. Article 34, Paragraph 4 of the Law on Value Added Tax. Available online: https://www.paragraf.rs/propisi/zakon-o-porezu-na-dodatu-vrednost.html (accessed on 5 January 2020).

54. Article 25, Paragraph 2 of the Law on Value Added Tax. Available online: https://www.paragraf.rs/propisi/zakon-o-porezu-na-dodatu-vrednost.html (accessed on 5 January 2020).

55. Article 2a of the Property Tax Law. "Official Gazette of RS", Nos. 26/01, 45/02, 80/02, 135/04, 62/06, 61/07, 31/09, 5/09, 18/10, 50/11, 101/10, 93/12, 24/11, 78/11, 48/13, 57/12, 47/13, 68/14, 112/15, 95/18; 7/2018, 99/18, 4/2019 and 86/19. Available online: https://www.paragraf.rs/propisi/zakon_o_porezima_na_imovinu.html (accessed on 10 January 2020).

56. Milošević, G.; Kulić, M. Tax Liability and Fulfillment of Tax Obligations. *Zbornik radova Pravnog fakulteta u Novom Sadu* **2015**, *49*, 597–616. [CrossRef]

57. Article 5, Paragraphs 1 of the Property Tax Law. Available online: https://www.paragraf.rs/propisi/zakon_o_porezima_na_imovinu.html (accessed on 10 January 2020).

58. Article 7, Paragraphs 1 of the Property Tax Law. Available online: https://www.paragraf.rs/propisi/zakon_o_porezima_na_imovinu.html (accessed on 10 January 2020).

59. Article 11 of the Property Tax Law. Available online: https://www.paragraf.rs/propisi/zakon_o_porezima_na_imovinu.html (accessed on 10 January 2020).

60. Cvjetković, C. Ekološki porezi i poljoprivreda. *Zbornik radova Pravnog fakulteta u Novom Sadu* **2015**, *49*, 769–783. [CrossRef]

61. Article 12, Paragraph 6 of the Law on Real Estate Tax, Official Gazette of the Republic of Montenegro, Nos. 25/2019. Available online: https://me.propisi.net/zakon-o-porezu-na-nepokretnosti/ (accessed on 27 March 2020).

62. Article 9, Paragraph 1, Item 12 of the Law on Real Estate Tax, Official Gazette of Republic of Srpska, Nos. 91/15. Available online: https://www.poreskaupravars.org/dokumenti/zakoni/Zakon%20o%20porezu%20na%20nepokretnosti%2001_01_2016%20Sl_91-15.pdf (accessed on 27 March 2020).

63. Article 14, Paragraphs 1 of the Property Tax Law. Available online: https://www.paragraf.rs/propisi/zakon_o_porezima_na_imovinu.html (accessed on 10 January 2020).

64. Article 16, Paragraphs 1 of the Property Tax Law. Available online: https://www.paragraf.rs/propisi/zakon_o_porezima_na_imovinu.html (accessed on 10 January 2020).

65. Article 16, Paragraphs 2 of the Property Tax Law. Available online: https://www.paragraf.rs/propisi/zakon_o_porezima_na_imovinu.html (accessed on 10 January 2020).

66. Article 19, Paragraphs 1 and 2 of the Property Tax Law. Available online: https://www.paragraf.rs/propisi/zakon_o_porezima_na_imovinu.html (accessed on 10 January 2020).

67. Article 21, of the Property Tax Law. Available online: https://www.paragraf.rs/propisi/zakon_o_porezima_na_imovinu.html (accessed on 10 January 2020).

68. Article 23a of the Property Tax Law. Available online: https://www.paragraf.rs/propisi/zakon_o_porezima_na_imovinu.html (accessed on 10 January 2020).

69. Article 25 of the Property Tax Law. Available online: https://www.paragraf.rs/propisi/zakon_o_porezima_na_imovinu.html (accessed on 10 January 2020).

70. Article 31 of the Property Tax Law. Available online: https://www.paragraf.rs/propisi/zakon_o_porezima_n a_imovinu.html (accessed on 10 January 2020).
71. Baturan, L. Economic Analysis of Legal Regimes Governing Real Estate Property Transfer in the Republic of Serbia. Ph.D. Thesis, University of Belgrade Law faculty, Belgrade, Serbia, 2016; pp. 152–158. Available online: https://uvidok.rcub.bg.ac.rs/bitstream/handle/123456789/1720/Doktorat.pdf?sequence=1&isAllow ed=y (accessed on 28 March 2020).
72. Kulić, Ž.; Milošević, G.; Baturan, L. Labour Market and Collective Bargaining in the Republic of Serbia. *Zbornik Radova Pravnog Fakulteta u Nišu* **2018**, *57*, 149–168. [CrossRef]

Article

Marketing Mix Instruments as Factors of Improvement of Students' Satisfaction in Higher Education Institutions in Republic of Serbia and Spain

Sandra Brkanlić [1,*], Javier Sánchez-García [2], Edgar Breso Esteve [3], Ivana Brkić [1], Maja Ćirić [1], Jovana Tatarski [4], Jovana Gardašević [1] and Marko Petrović [2]

[1] Faculty of Economics and Engineering Management in Novi Sad, University Business Academy in Novi Sad, 21000 Novi Sad, Serbia; ivana.j.milosevic@fimek.edu.rs (I.B.); majaciric79@yahoo.com (M.Ć.); gardasevic.jovana@gmail.com (J.G.)

[2] Faculty of Law and Economic Sciences, University Jaume I, 12071 Castellon de la Plana, Spain; jsanchez@uji.es (J.S.-G.); petrovic@uji.es (M.P.)

[3] Faculty of Health Sciences, University Jaume I, 12071 Castellon de la Plana, Spain; ebreso@gmail.com

[4] Department of Industrial Engineering and Management, Faculty of Technical Science, University of Novi Sad, 21000 Novi Sad, Serbia; j.tatarski@uns.ac.rs

* Correspondence: sbrkanlic@gmail.com; Tel.: +381-648252145

Received: 3 August 2020; Accepted: 16 September 2020; Published: 21 September 2020

Abstract: This paper explores the impact of marketing mix instruments on the students' satisfaction in faculties in the Republic of Serbia and Spain, with the aim of determining how significant the effects of each marketing mix tool and their combinations are in relation to satisfaction of students in Higher Education Institutions (HEIs). The detailed literature review is provided in the theoretical part, which contributes to a better understanding of terms like marketing in higher education, marketing mix instruments in higher education and students' satisfaction. Data were collected from 896 respondents, who are all students at the faculties in Serbia and Spain, and were obtained using the questionnaire purposefully composed for this research. The methods used to highlight any gaps in this marketing mix practice and the relative customer–student satisfaction in HEIs are statistical analyses (descriptive analysis, correlation analysis, multiple regression analysis and t-independent samples tests), leading to the general conclusions regarding the following: by improving marketing mix instruments (service, distribution, human factor, physical evidence, service process) we can, and by improving (price, promotion) we cannot, improve students' satisfaction in higher education institutions. The general conclusions clearly highlight what needs to be improved in practice in higher education institutions to improve students' satisfaction, especially students' loyalty, students' choices, students' satisfaction with the quality of the marketing mix instruments at the faculty, students' satisfaction with expectation which they had upon enrolment and student satisfaction with the public image of the faculty, which is the main goal of these institutions.

Keywords: marketing in higher education; marketing mix instruments in higher education; higher education institutions; students' satisfaction; Republic of Serbia; Spain

---

1. Introduction

The higher education sector is predominantly facing modernization, pluralism, and market orientation in education. Regarding the higher education sector in the countries where the research was carried out (Serbia and Spain), according to the National Report Regarding the Bologna Process implementation 2012–2015, there are 22 accredited universities in the Republic of Serbia (13 public and nine private), while the same report specifies that there are 82 registered and accredited universities in Spain (50 public and 32 private) [1]. Eleven private universities have been established in the Republic of Serbia in the last 20 years, five of which were accredited in 2011, and six more obtained accreditation by 2015. Nonetheless, in 2011, forty faculties and vocational colleges sought accreditation [2]. Additionally, the growing number of universities in Spain is the result of an increasing demand (from 170,000 students enrolling in 1960, and 700,000 in 1980, to more than double that figure in 2014), mirroring a high-paced modernization and democratization, simultaneously creating tensions which stem from a rise in inequalities generally, a decrease in autonomy and trends of instrumentalization [3].

The aforementioned current state of affairs in the education market in Serbia, Spain, and in the global market, which witnesses an increase in the number of faculties and universities, is sure to dictate the marketing orientation of the institutions. Education is defined as a transaction or a relationship according to the marketing orientation, and, therefore, engagement and marketing efficiency are emphasized as the most suitable criteria. Education turns into something of more than just human significance for marketing in the area when the two aforementioned criteria are met [4,5]. As well as playing a significant part in the process of achieving the goals of the higher education institutions and its students, the marketing approach in higher education is advantageous for higher, social goals. Being marketing-orientated suggests endeavoring for quality in every management segment of the faculty, because the education service market offers a wide selection of education institutions, forms and programs. In their abundance, prospective students opt for the education which offers features which agree best with them [5,6].

Further complications in the global trends in education are caused by the transition to the Bologna Process and pluralism in education, to which faculties adapt on the fly. It is significant to point out that higher education institutions are coming up against the requirement to magnify their business marketing orientation in view of a number of factors including the following: increasing the presence of international higher education institutions, the establishment of private universities, the introduction of the accreditation system, and control measures and quality guarantees of education. Marketing orientation includes endeavouring for quality in every management segment of the faculty because the educational service market offers a wide range of education institutions, forms and programmes. In their abundance, the prospective students opt for the education that offer features which agree with them best, and they are put into operation using the marketing mix instruments. Present-day marketing defines marketing mix as one of the most commonly used concepts, whose instruments are crucial for the satisfaction of consumers [7].

The marketing goal of each business organization, inclusive of higher education institutions, is to achieve satisfaction of service end-users–students. The attempt to improve the marketing mix instruments originates from the fact that all educational institutions in the contemporary market game strive to gain as high a level of student satisfaction as possible so as to obtain a competitive advantage in the education market. Accordingly, the research aspires to establish which marketing mix instruments are correlated with the student satisfaction the most, and which marketing mix instruments have the greatest effect on the student satisfaction. Therefore, the aim is to ascertain which marketing mix instruments need to be perfected so as to affect the enhancement of the student satisfaction, and hence, the enhancement of the business performance and positioning of the institutions in the market. Moreover, the research provides an outline of the arithmetic means for each marketing mix instrument and student satisfaction. These point to the current situation and the respondents' position with regard to marketing mix instruments and student satisfaction, and also the insight into which instruments/dimensions require improvement [7].

In this paper, the researchers applied descriptive analysis, correlation analysis, multiple regression analysis and *t*-independent samples tests to measure or quantify the results of marketing mix instruments and student satisfaction, in order to correlate marketing mix instruments with student satisfaction in higher education institutions, as well as the impact of marketing mix instruments on student satisfaction in higher education institutions, at faculties in Serbia and Spain. A detailed literature review is rendered in the theoretical section, which leads to a better understanding of terms such as marketing, marketing in higher education institutions, marketing mix instruments in higher education institutions and student satisfaction. Data collected from 896 respondents, who are all students at the faculties in Serbia and Spain, were obtained using the questionnaire purposefully composed for this research, and statistical analyses (descriptive analysis, correlation analysis, multiple regression analysis and t-independent samples tests) were carried out and general conclusions were reached for the territories of the two countries, which have not been compared in any research to date, and which deal with the same or similar relationships in this field. The paper has the following structure:

- Literature review (Section 2);
- Method and data (Section 3);
- Research results and discussion (Section 4);
- Concluding remarks and suggestions for further research (Section 5).

The theoretical part of the paper is in Section 2 and it contributes to a better understanding of terms such as marketing, marketing in higher education institutions, marketing mix instruments in higher education institutions and students' satisfaction. Section 3 comprises the sample structure, data collection procedure and structure of the questionnaire, as well as preliminary results in the form of descriptive analysis, followed by the correlation analysis, multiple regression analysis and t-independent samples test, which were all used in the research. Having addressed the methods used in the research, research results and discussion are shown (Section 4). Finally, conclusions and future research directions are given in Section 5.

2. Literature Review

In 1985, American Marketing Association (AMA)—the most influential marketing organization in the world—defined marketing as the process of planning and implementing concepts, prices, promotion and distribution of ideas, goods and services, so as to create an exchange that satisfies the needs of individuals and organizations [5,8,9]. The newly formulated definition of marketing, which is also defined by AMA, makes a strategic shift in terms of understanding the concept of marketing. In other words, marketing is now viewed as an organizational function and a set of processes used to create, communicate and deliver value to consumers and manage the customer relationship in a way that is beneficial to the organization and its stakeholders [10]. The present-day perception of the marketing concept brings the customer and his/her satisfaction to the center of attention as the ultimate goal of all marketing activities. Converted to the sphere of higher education and higher education institutions, the ultimate goal of all higher education institutions is to satisfy students.

Across their entire evolution, higher education institutions were thought of as the generator of social progress and the power source of new social changes and ideas. Nevertheless, higher education has been in the state of crisis as a result of an uneven society development, disproportions of economic, political and military powers on the global scale, the explosion of knowledge, globalization, etc., which question the role and function of universities in modern education [11]. As well as playing a significant part in the process of achieving the goals of the higher education institutions and its students, a marketing approach in higher education is advantageous for the higher, social goals as well. Being marketing-orientated suggests endeavoring for quality in every management segment of the faculty, since the education service market offers a wide selection of education institutions, forms and programmes. In their abundance, prospective students opt for the education offer features which agree best with them [5,6].

Authors address the in-depth analysis of marketing and its application in institutions of higher education; the following theoretical discussion will point out the significance of marketing mix instruments and their specific aspects for higher education institutions. Different combinations of marketing mix instruments and their application can give rise to a competitive advantage on a higher education market. Similarly, it creates greater student satisfaction within a higher education institution.

Hence, this study underlines each particular instrument, which evidently indicates the following: in accordance with the most frequently cited definitions, the service represents any activity or benefit that one party offers to another, and which is, in essence, intangible and does not result in having ownership over anything. Its production can, yet need not, be connected with tangible goods [9,12] and it should be regarded in terms of its general characteristics and components, specifically related to the nature of its particular characteristics in the higher education institutions. With the view of one of the definitions, the price represents the amount at which goods and services are exchanged, therefore, the price, in any case, represents expense for the consumer [9,13], and in the study, the price is viewed through the uniqueness of its formation in the higher education institutions. Distribution represents the channels of distributing goods or services to the consumer. In view of the fact that in present-day business, most producers are not directly in contact with the end user, they use the services of an intermediary so as to achieve a better product or service placement. Thus, the marketing intermediary creates the marketing or distribution channels which direct all aspect of the flow of goods or services, from production to consumption, and this forms the base of distribution [9,14]. Authors presented distribution as the sum of all of its characteristics in the field of higher education. In line with one of the definitions, promotion is the process of communication between the service provider (higher education institutions) and a client (student) with the goal of establishing a positive attitude towards the services of the organization in the buying process [9,15,16]. Authors also deals with the nature of promotion features within higher education institutions. The human factor represents one of the service marketing mix instruments and the notion includes each person included in the service process. On the one hand, these are the employees who ought to be adequately recruited, trained, motivated, awarded, trained in teamwork, etc. On the other hand, the human factor implies that service users can influence the entire perception of the service, therefore, it is important to note their behavior, the degree of their engagement, the level of contact between them, education, training, etc. [9,13,16]. In this study, the specific nature of the human factor within the field of higher education is also emphasized. Physical evidence consists of the exterior and interior of the facility with the associated furniture, equipment, etc. As well as this, physical evidence includes the atmosphere inside the facility and a number of other factors [9,13,16], which are explained within this study through the significance of physical evidence for the higher education institutions. The service process includes procedures, mechanisms, and the flow of activities which create the service. For this reason, the delivery of service to the users is greatly affected by decisions which include the processes [9,17]. The importance of this instrument is presented within this study through the review of its complexity and specific nature, as well as the process design and its characteristics within higher education institutions. By analyzing and understanding each of the aforementioned marketing mix instruments and their characteristics and potential, marketers can benefit greatly in the process of creating a service offer in the higher education institutions in the global higher education market, including the Republic of Serbia and Spain. The benefits will enable them to get a head start in the higher education market, and thus higher student satisfaction.

It is required to take certain measurements and do research, and also to quantify the results. Satisfaction was measured with the aim of showing the level of consumer satisfaction and achieving objectification and quantification of consumers' subjective perceptions [18,19]. Consequently, measurement and research of consumer satisfaction was conducted in the higher education institutions.

Many education institutions have a tendency of ameliorating their offer so as to attract more users. Focusing on customer satisfaction is what every institution should start from. Creating happier, satisfied users—whether they are students, their parents, donors, professors, or employers—should be the prime aim that will contribute to quality in education institutions. What is essential for the work of each higher education institutions is user satisfaction. Level of satisfaction is determined by the difference between the service features, how they were perceived by users and user expectations. There are three levels of satisfaction. If the service is below expectations, the user's dissatisfaction arises. In case a faculty does not provide what the students anticipate, they will change their view towards the faculty and may leave or switch to another department or spread negative information about the college. On the other hand, if the faculty responds to student's expectations, the student will be contented and will become the best promoter of the faculty. If the characteristics of the institution exceed student's expectations, the student will be very satisfied or delighted [20–22].

Due to these circumstances, it is essential for institutions to assess student satisfaction, because higher education institutions have to view students as users and key actors. Hill asserts that students are principal university users, and higher education implies a certain economic and social service. Researchers highlight the importance of these relationships and suggest that the overall orientation of the institution should be translated into the level of these relationships in order to be effective. As stated by Gronroos, building long-term relationships with students should be the marketing goal, since the students are the most valuable university resource [21,23]. Accordingly, this study attempts to evidence which marketing mix instruments, from the students' perspective, have the strongest influence on student satisfaction, in order for them to be used as a tool to improve it further.

The research question of this study is the following: Which marketing mix instruments affect the student satisfaction in higher education institutions to the largest extent? The corresponding hypothesis is the following: H—Marketing mix instruments affect student satisfaction in higher education institutions to a large degree.

3. Method and Data

3.1. Defining the Sample—Sociodemographic Characteristics of the Sample

The research was carried out at faculties in two countries—Serbia (703; 78.5%) and Spain (193; 21.5%). The entire sample included 896 respondents. As for the gender of the respondents, the sample included 386 (43.1%) males and 510 (56.9) females. The sample consisted of 20- to 60-year-olds, with an average age of 29.10 years (SD = 5.41). The sample involved 165 (19.3%) first-year students, 208 (23.2%) second-year students, 240 (26.8%) third-year students, 227 (25.3%) fourth-year students and 17 (1.9%) fifth-or-higher-year students. A total of 39 (4.4%) respondents did not provide an answer to the question referring to their year of study. The average grade of the respondents ranged from 5 to 10, with an average value of 7.85 (SD = 0.95).

In the Republic of Serbia, the research was carried out at four private faculties within the private University Business Academy in Novi Sad, which include the following: Faculty of Economics and Engineering Management in Novi Sad, Faculty of Law for Commerce and Judiciary in Novi Sad, Faculty of Stomatology in Pančevo and Faculty for Applied Management in Belgrade. Concerning the respondents' gender, the subsample from Serbia consisted of 322 (45.8%) male, and 381 (54.2) female respondents. The subsample included 20 to 59-year-olds, with an average age of 29.42 years (SD = 4.97). Regarding the faculty, 226 (32.1%) students attended the Faculty of Economics and Engineering Management in Novi Sad, 231 (32.9%) students attended the Faculty of Law for Commerce and Judiciary in Novi Sad, 164 (11.7%) students were from the Faculty of Stomatology in Pančevo, and 82 (11.7%) students were from the Faculty for Applied Management in Belgrade. The subsample was balanced where the year of study is concerned, with 156 (22.2%) first-year students, 161 (22.9%) second-year students, 156 (22.2%) third-year students, and 191 (27.2%) fourth-year students. The average grade of the respondents ranged from 5 to 10, with an average value of 8.07 (SD = 0.90). The subsample was sufficient, and a volunteer sampling method was used. The research was conducted during one semester (summer) in 2018/19 school year. The survey was anonymous and group.

In Spain, the survey was conducted at two public faculties within the public University Jaume I in Castellon de la Plana, which included the following: Faculty of Law and Economic Sciences and Faculty of Health Sciences. Concerning the gender of the respondents, the subsample from Spain included 64 (33.2%) male, and 129 (66.8) female respondents. The subsample consisted of 20 to 62-year-olds, with an average age of 26.28 years (SD = 4.97). As for the faculty, 94 (48.7%) students attended the Faculty of Law and Economic Sciences, whereas 99 (51.3%) students were from the Faculty of Health

Sciences. Most respondents were third-year students (84; 43.5%), second-year students (47; 24.4%), and fourth year students, respectively (36, 18.7%). First-year students were fewer (9; 4.7%), as well as fifth and higher-year students (17; 8.8%). The average grade of the respondents ranged from 5 to 10, with an average value of 7.24 (SD = 0.82). The subsample was sufficient, and the volunteer sampling method was used as in the Republic of Serbia. In Spain, the survey was carried out using the online questionnaire, while students in the Republic of Serbia responded to questions via the printed questionnaire in paper form. The subsample comprised students of all years of studies at the above faculties. The research was carried out during one semester (winter) in 2018/19 school year. The questionnaire was online and anonymous.

3.2. Procedure for Data Collection and Structure of the Questionnaire

The research was conducted using the method of theoretical analysis and the empirical method, the so-called research method. The empirical research method's primary goal was the gathering of facts about the ongoing situation, as well as their subsequent analysis. There were three phases in conducting the empirical research: data collection, preparation of data for analysis, and statistical analysis. To collect concrete data and facts from the respondents via printed and online questionnaires, the survey method was applied. The non-standardized survey questionnaire was used as an instrument for data collection. It was a closed-question one specifically designed for the research. The results obtained using this and other adequate statistical methods were compared with the information from the theoretical analysis method application, which ensured more comprehensive conclusions and recommendations.

The survey method used for the purpose of the research and the instrument used in the survey is a non-standardized questionnaire, created particularly for this research. Key sections, factors and features were formulated using a detailed analysis of theoretical data resources and consulting relevant authors' works [16,18,19,24–27]. On these grounds, the questionnaire was created to include questions whose answers would lead to the possibility of processing the data and obtaining the necessary results.

The first part of the questionnaire contains five general questions serving to obtain detailed information on respondents, i.e., the sample. The questions refer to the gender, birth year, the faculty the respondent is attending, year of study, and average grade of the respondents. This information was used only for the purpose of the descriptive analyses. The questionnaire consisted of 64 items with the respondents giving their response relative to the degree of agreement with the statement on the scale from 1 to 7 (1—I totally disagree, 7—I agree completely). As for the research variables, the questionnaire is divided into two topical sections. The first segment addresses the independent research variable—marketing mix instruments. Applying different combinations of marketing mix instruments can lead to improved competitive advantage in the education market, boosting the student satisfaction. For this reason, each instrument is highlighted separately within this segment. This segment includes 56 closed type questions, i.e., statements that can be graded on the scale from 1 to 7. The section is divided into seven subsections that address each marketing mix instrument individually. The second segment addresses the dependent research variable—student satisfaction. This segment includes eight closed-type questions, i.e., statements which can be graded on the scale from 1 to 7.

Internal consistency (reliability) of seven dimensions of the questionnaire marketing mix instruments and the questionnaire, students' satisfaction is estimated based on the Cronbach α coefficient. The α coefficient is utilised to estimate the internal consistency, i.e., reliability of the test/scale. It ranges from 0.00 to 1.00 and is an estimate of the extent to which the dimensions of a single dimension measure the same construct. The α coefficient values for the applied questionnaires and their dimensions are given in Table 1, ranging from 0.854 to 0.915. According to the criteria proposed by Cho and Kim [28], the reliability of all dimensions is very high.

Table 1. Reliability of the dimensions of the questionnaire's marketing mix instruments and student satisfaction.

Instrument	Dimension	α
	Service	0.854
	Price	0.893
	Distribution	0.875
Marketing mix instruments	Promotion	0.952
	Human factor	0.912
	Physical evidence	0.915
	Service process	0.903
	Student satisfaction	0.939

3.3. Statistical Analysis

The collected data are processed through the SPSS program for 291 Windows, v21 (SPSS Inc., Chicago, IL, USA, 2012), with the frequency sampling method, descriptive statistics (arithmetic mean, standard deviation, asymmetry and kurtosis), correlation analysis (the coefficient of Pearson's correlation), multiple regression analysis and independent sample tests *t*. The research results are given numerically and shown in the form of a graph or a table.

3.3.1. Descriptive Analysis: Analysis of the Assessment of Marketing Mix Instruments and Students' Satisfaction

Descriptive analysis was carried out so as to analyses the assessment of marketing mix instruments and students' satisfaction. Descriptive statistic parameters (arithmetic mean, minimum and maximum value and standard deviation), for seven marketing mix instruments and students' satisfaction, are given in Table 2.The value of parameters indicating the distribution form (skewness and kurtosis presented in Table 2), are within the range of recommended values (±1.5) [29]. Therefore. it can be inferred that the distribution of scores is within the normal distribution range, on all tested questionnaires.

Table 2. Descriptive statistics parameters.

		Min	Max	AM	SD	Sk	Ku
	Service	8	56	42.76	8.61	−0.52	0.26
	Price	8	56	36.34	12.11	−0.33	−0.54
Marketing	Distribution	8	56	46.23	8.61	−0.90	0.60
mix	Promotion	8	56	36.07	13.54	−0.42	−0.77
instruments	Human factor	8	56	42.61	9.73	−0.55	−0.12
	Physical evidence	8	56	42.81	11.08	−0.99	0.66
	Service process	8	56	41.98	9.84	−0.49	−0.14
	Students' satisfaction	8	56	42.96	10.76	−0.72	0.02

Legend. Min—minimum value. Max—maximum value. AM—arithmetic mean. SD—standard deviation. Sk—skewness. Ku—kurtosis.

3.3.2. Correlation Analysis: Analysis of the Correlation between Marketing Mix Instruments and Students' Satisfaction

Correlation analysis examines the correlation between variables and cannot be interpreted by causality link. Correlation does not imply that one variable is dependent and the other is independent; it suggests that the two variables are linearly correlated. It can be interpreted as symmetrical: the correlation between X and Y is the same as the one between Y and X. The correlation analysis was carried out so as to analyze the correlation between marketing mix instruments and student satisfaction. Pearson correlation coefficient was used to analyze the correlation (Table 3).

Table 3. The correlation between marketing mix instruments and student satisfaction.

	Students' Satisfaction
Service	0.715 *
Price	0.560 *
Distribution	0.620 *
Promotion	0.463 *
Human factor	0.735 *
Physical evidence	0.631 *
Service process	0.769 *

* Significant for the level of $p < 0.01$.

3.3.3. Multiple Regression Analysis: Analysis of the Effect of Marketing Mix Instruments on Students' Satisfaction

Multiple regression analysis is a statistical technique that allows examination of the relationship between several independent or predictor variables. It is based on the effect, and therefore it tells us about the effect, but not about the causality of the link between the predictor and the criterion variable. It gives insight into how, precisely, the set of variables (predictor) can predict the specific outcome (criterion variable), as well as which predictor variable offers the most accurate prediction (criterion variable) [30]. The aim of the analysis was to identify which marketing mix instruments most efficiently predict student satisfaction. In this case, marketing mix instruments present independent (predictor) variables, while student satisfaction is the criterion variable.

In Table 4, the adjusted R^2 (the coefficient of determination) indicates the percentage of the variance of criterion variable which is explained by the set of predictor variables. The regression model is statistically significant (F (895) = 261.0, $p < 0.001$), whereby the set of predictors can explain 67.0% of the variance of the criterion variable—hence, marketing mix instruments explain 67.0% (adjusted $R^2 = 0.670$) of the satisfaction variance. Based on the percentage of the explained variance, it is inferred that the set of predictors is strongly correlated to the criterion variable.

Table 4. Regression model parameters: criterion of student satisfaction.

	Sum of Squares	Degree of Freedom	Mean Square	F	F Level	R	R^2	Adjusted R^2
					Significance			
Regression	103,541.6	895	9953.763	261.003	0.000	0.820	0.673	0.670

The multiple regression analysis was carried out so as to analyze the effect of marketing mix instruments on student satisfaction (Table 5).

Table 5. Marketing mix instruments as predictors of students' satisfaction.

Predictors	B	SE	Beta	t-Test	p Value
Service	0.270	0.039	0.216	6.900	0.000
Price	0.045	0.026	0.051	1.776	0.076
Distribution	0.122	0.034	0.098	3.565	0.000
Promotion	0.006	0.021	0.008	0.296	0.767
Human factor	0.215	0.039	0.195	5.556	0.000
Physical evidence	0.125	0.029	0.129	4.300	0.000
Service process	0.287	0.043	0.262	6.600	0.000

Legend. B—non-standardised coefficient. SE—standard error. Beta—standardised coefficient.

3.3.4. Descriptive Analysis and *t*-Independent Samples Test: Analysis of the Differences between the Assessment of Marketing Mix Instruments and Students' Satisfaction by the Subsamples from Serbia and Spain

Descriptive statistical measures, and a series of *t*-independent samples tests were applied in order to examine the potential differences between subsamples from the two countries (Serbia and Spain), within the context of marketing mix instruments and student satisfaction. Descriptive analysis and *T*-tests were carried out with the aim of identifying potential differences between subsamples from Serbia and Spain. Differences were analysed based on the assessment of marketing mix instruments and students' satisfaction. The aim of the analysis was to determine the assessment of relevant marketing mix instruments in faculties in Serbia and in faculties in Spain, as well as to determine if there is a need for the improvement in the instruments. In the same way, the analysis provided insight into the difference between the assessment of the students' satisfaction of each subsample separately, which gave us a clear insight into the present-day state of affairs in faculties in two countries (Serbia and Spain).

As particular differences were noticed in the arithmetic means of almost all dimensions/instruments in the context of the country of origin of the faculty (Serbia and Spain), a set of *t*-independent samples tests was used—tests were applied to analyse the importance of differences in arithmetic means between the two groups. Group membership (Serbia or Spain) is the independent (grouping) variable in all analysis, whereas the seven dimensions of marketing mix and students' satisfaction represent the dependent variable in each group. The results are shown in Table 6.

Table 6. The differences between the assessment of marketing mix instruments and students' satisfaction.

Instrument		Country	N	AM	SD	*t* Test	DF	*p*
	Service	Spain	193	39.23	5.80			
		Serbia	703	43.73	9.00	−6.57	894	0.000
	Price	Spain	193	36.28	7.68			
		Serbia	703	36.35	13.07	−0.06	894	0.946
	Distribution	Spain	193	40.46	6.73			
		Serbia	703	47.82	8.39	−11.22	894	0.000
Marketing mix instruments	Promotion	Spain	193	39.21	8.68			
		Serbia	703	35.21	14.49	3.65	894	0.000
	Human factor	Spain	193	38.94	7.68			
		Serbia	703	43.62	9.99	−6.03	894	0.000
	Physical evidence	Spain	193	44.69	7.45			
		Serbia	703	42.29	11.84	2.68	894	0.008
	Service process	Spain	193	38.35	7.35			
		Serbia	703	42.98	10.20	−5.90	894	.000
	Students' satisfaction	Spain	193	38.55	8.60			
		Serbia	703	44.17	10.97	−6.58	894	0.000

Legend. AM—arithmetic mean. SD—standard deviation. DF—degree of freedom. *p*—*p* value.

4. Results and Discussion

The results of a descriptive analysis on the level of total sample, which are shown in Table 2., and present the arithmetic mean value of the observed sample, indicate that distribution received the highest assessment grade (AM = 46.23). Student satisfaction (AM = 42.96), physical evidence (AM = 42.81), service (AM = 42.76), human factor (AM = 42.61) and service process (AM = 41.98) received medium assessment grades. The lowest assessment grades presented in the table refer to price (AM = 36.34) and promotion (AM = 36.07).

From the Correlation analysis figures given in Table 3, it is evident tha student satisfaction achieved the largest correlation with service process (the Pearson correlation coefficient r = 0.77), human factor (the Pearson correlation coefficient r = 0.73) and service (the Pearson correlation coefficient r = 0.71). Fairly smaller, yet still moderately large, correlations are achieved using physical evidence (the Pearson correlation coefficient r = 0.63), distribution (the Pearson correlation coefficient r = 0.62), price (the

Pearson correlation coefficient r = 0.56) and promotion (the Pearson correlation coefficient r = 0.46). Even though it is evident that the lowest student satisfaction is achieved by price and promotion, it is also comprehensively observable that all the marketing mix instruments are more or less related to student satisfaction or correlate with student satisfaction.

Multiple regression analysis results, on the level of total sample (Table 5), show the effect of each separate marketing mix instrument on student satisfaction. The standardised coefficient beta indicates the level of contribution of each predictor of students' satisfaction. The best effects in the students' satisfaction prediction are attained by the service process ($\beta = 0.26$, $p < 0.001$) and service ($\beta = 0.22$, $p < 0.001$), A slightly smaller effect is attained by the human factor ($\beta = 0.19$, $p < 0.001$), and physical evidence ($\beta = 0.13$, $p < 0.001$), while the distribution ($\beta = 0.10$, $p < 0.01$) attains the smallest effect in student satisfaction. A positive correlation with the criterion variable is attained by all the significant predictors. Price and promotion are not considered significant predictors of student satisfaction. In line with the regression analysis results, the conclusion was reached that hypotheses H0 is partially confirmed.

The results of the correlation analysis obtained at the total sample level show the correlation (to a greater or lesser extent) of all marketing mix instruments with student satisfaction, whereas multiple regression analysis shows influence of all marketing mix instruments on student satisfaction but for price and promotion. When correlating the results of correlation analysis and multiple regression analysis with the results of descriptive analysis at the total sample level, it can clearly be observed that the highest assessment grade and received medium assessment grades are precisely the tools of the marketing mix (distribution, physical records, service, human factor, service process), which proved to be significant for achieving student satisfaction using the aforementioned analysis, while the worst rated ones, at the total sample level, were those marketing mix instruments (price, promotion) which did not prove to be significant for improving student satisfaction. Descriptive analysis also indicates a medium assessment of student satisfaction at the total sample level.

Comparing the obtained results with the viewpoints of the authors concerning these relationships, it can be seen that there are authors who also emphasise the significance of marketing mix instruments, which, in this study, proved to be vital for achieving student satisfaction. Thus, Gerson points to the importance of the study programme quality as one of the key service aspects by stating that it is necessary to conduct market research in order to design and develop study programmes. Otherwise, unanticipated difficulties may occur, there might be less interest for certain study programmes, students might become discontented with the teaching process and the curricula, or they might even decide to transfer to other courses [4]. Banwet and Datta [31] and Hill et al. [32] reached the same conclusions through their research. Their conclusions state that student satisfaction is largely affected by the service itself, as they specifically highlighted the importance of the service quality and the teaching quality. Gruber et al. came to the same conclusions [33]. When measuring the level of correlation between 15 different elements in relation with the business of higher education institution and students' satisfaction, they ascertained that the quality and the concept of study programmes are largely correlated with students' satisfaction in faculties, which is also in line with the research results when the service–student satisfaction relation is in question. Likewie, Banwet and Datta [31] and Hill et al. [32] point to the importance of the service distribution for achieving student satisfaction, which is in line with the obtained research results. Douglas et al. [34] noted that the current demand is for cutting-edge service distribution, using technical equipment and state-of-the-art computer technologies, which can attain student satisfaction. Gajić [35] initially views distribution in the higher education sector considering the location of the institutions as an important factor, and concurrently he emphasises the importance of distance learning studies with the application of advanced technology. He specifically highlights his opinion that institutions have to take particular care of the service delivery methods, the atmosphere in the faculty and the message they are sending to the prospective students in this way. This confirms the exceptional importance of this instrument for achieving students' satisfaction, which is, together with the views of the majority of authors, consistent with the results of the research when it comes to the impact of distribution on students' satisfaction. As far as the effect of human factor on students' satisfaction is concerned, authors have voiced their opinions only about some individual aspects of the human factor and their impact on students' satisfaction, therefore, Hill et al. [32] and Pozo-Munoz et al. [36] assert that teaching staff play a crucial part in the business of higher education institutions and that the responsibility for the level of student satisfaction rests solely upon them. Voss et al. [37] and Price et al. [38] consider that the enrolled students affect the satisfaction of other students to a large degree, while Deming [39], in the same manner, states that many students form their own opinion based on the opinions of their peers. When examining the role that student services play in the faculty, Galloway [40] determined that the staff at students' services, the non-teaching staff, have an immediate influence on the way a faculty as an institution is perceived by students. According to him, it is vital that the non-teaching staff be professional, appropriately dressed and always available to students, because they have a strong effect on students' satisfaction, which was confirmed by the results of this research. Although the authors dealt only with some aspects of the human factor and their effect on student satisfaction, the largest share of views are in accordance with the research results. The authors also emphasised the importance of physical evidence and the service process for achieving student satisfaction in their studies, in line with the results of the research, so Starck, Zadeh, Ekman and Olsson [41] carried out research on Bangkok University International College, where 36% of students pointed to the importance of physical evidence, while at Webster University 40% of students responded in the same way, highlighting the significance of physical evidence for making a decision concerning which institution to enroll in, whereas Dale [42] pointed out the importance of the service process for achieving student satisfaction by emphasizing that the most important factor affecting student satisfaction is the overall impression concerning service, with special emphasis on the organisation of service delivery, and Banwet and Datta [31] carried out research among students who attended four different lectures given by the same professor, and had different impressions regarding the lectures. The students formed their opinion, among other categories (the obtained knowledge, availability of literature and course materials), based on the interactive nature of lectures and the entire lecturing process, which indicated the importance of certain aspects of service process for students' satisfaction.

On the other hand, the results in this study indicate that price and promotion do not have an effect on student satisfaction. Even though the authors have dealt very little with these marketing mix instruments and their impact on student satisfaction, there are a few, such as McCollough and Gremler [43,44], who highlight the importance of the tuition fee and service quality ratio by asserting that students should get value for the tuition fee they pay, and that this is the method by which to achieve satisfaction. Consequently, provided value should be proportional to the tuition fee and, in this manner, the tuition fee is definitely considered as adequate. Likewise, Gajić [35] defines price in higher education institutions from the economic and psychological points of view by correlating them with students' satisfaction. It can clearly be seen that research results, which do not point to the price as a significant factor for student satisfaction, are completely compatible with the position of the aforementioned authors, who did not attribute too much significance to price and never analysed it separately, only as a marketing mix instrument which can only be observed in synergy with other instruments. Therefore, authors have not analysed marketing mix instruments a lot. Likewise, when it comes to promotion, certain authors still empasise the importance of promotion for students' satisfaction, although this study did not confirm promotion as being significant for student satisfaction. Hence, Kotler and Fox [20] assert that the marketing communicator has to identify the target audience, clarify the required response, create a message, choose medium or media, choose resource attributes and collect feedback in order to achieve student satisfaction. Ivy [45] points out that promotion is abundant in other tools which can be used to attract new service users and achieve student satisfaction. In contrast, some authors suggest online promotional activities are the most efficient with students [46]. Many authors [47] state that public relations are prevalent promotional activities in private higher education institutions, and that they include a number of activities such as media relations interviews, which have a role in raising awareness in public about the accomplishments of the institution [48].

As can be inferred, the majority of the aforesaid authors only analysed specific aspects of promotion and their effect on students' satisfaction, without taking into account the complete effect of promotion in these types of institutions.

The results (Table 6) of this study (*T*-independent samples test at the level of subsample) point to the current situation in higher education institutions in the Republic of Serbia and Spain. Faculties in Serbia have statistically higher scores ($p < 0.01$) within the following dimensions/instruments: service, distribution, human factor, service process, and student satisfaction. However, faculties in Spain show statistically higher scores ($p < 0.01$) within the following dimensions/instruments: promotion and physical evidence. There is no statistically significant difference between the two groups of respondents concerning the dimension of price. The above results clearly indicate the current situation and the possible need to improve the marketing mix instruments and student satisfaction at the mentioned faculties in two different countries, such as the Republic of Serbia and Spain.

5. Concluding Remarks and Suggestions for Future Research

The research goals (scientific and social) were entirely accomplished by coming to a series of scientific and practical findings.

The scientific goal of the research is achieved by partially confirming the validity of the hypothesis and, on the grounds of the scientifically based research process, reliable data were obtained in an area that had not been sufficiently explored by other authors. Namely, through a theoretical review of the literature and a discussion of the results, it has been proved that a small number of authors have dealt with this topic, and that there are both different and contradictory attitudes about the effect of particular marketing mix instruments on students' satisfaction, and that there is no research analyzing the effect of marketing mix instruments on students' satisfaction, and especially not at faculties in Serbia and Spain. Therefore, the scientific purpose and justification of this research are undoubtedly proven.

On the basis of the research results (correlation analysis) it was determined that the strongest correlation with students' satisfaction is attained by (in this order): service process, human factor, service, while a somewhat weaker, yet still moderately strong, correlation was attained by (in this order) physical evidence, distribution, price, promotion. A general inference that can be made, with regard to the correlation analysis results, is that all marketing mix instruments correlate with student satisfaction.

Whereas the results of the correlation analysis highlight the correlation between all marketing mix instruments and students' satisfaction, the results obtained within multiple regression analysis imply that student satisfaction is affected by service, distribution, human factor, physical evidence and service process, while price and promotion do not have an effect on student satisfaction. These results give rise to a partial confirmation of the hypothesis, which suggests that the student satisfaction in higher education institutions is not affected by all marketing mix instruments and that it cannot be enhanced by improving all marketing mix instruments, but only those that turned out to be significant.

Consequently, the multiple regression analysis results lead to the inference that the betterment of students' satisfaction can be achieved by refining the following:

- Service (the study programme quality, the professors' lectures quality, quality of teaching assistants' seminars, quality of the literature used in a study programme, opportunity to perform professional practice, opportunity for international student exchange and training abroad, contemporariness and practical applicability of competencies acquired in faculty);
- Distribution (conducting lectures, seminars, consultation, exams according to the defined schedule, information availability at the student service, over the phone line, on website or social media);
- Human factor (professors' lectures and teaching assistants' seminars, professors' and teaching assistants' effort in motivating students, quality work and kindness by non-teaching staff, quality of the enrolled students regarding their average grades, knowledge and education level, etc.);
- Physical evidence (faculty location, design of the building and entire exterior of the faculty, number of amphitheatres and classrooms, interior and well-equipped faculty rooms, computer equipment, functional and well-equipped library, dress code of the faculty staff);

- Service process (teaching process organization—lectures, seminars and consultations, organisation process of exams and colloquiums, professional practice organisation process, international exchange organisation process, organisation process of informing students in faculty and providing interactive lectures and seminars).

They also lead to the inference that the betterment of students' satisfaction cannot be achieved by improving the following:

- Price (tuition fee and the correction of tuition fee relative to the competition and what the faculty offers for the set price, terms of payments, additional study costs—application forms, certificates, course books, scholarships);
- Promotion (quality promotion on TV, radio, billboards, in print media, in other institutions, on faculty website and social media, organising interesting promotional events, seminars, congresses and conferences).

General inferences that these results gave rise to undoubtedly imply that marketing mix instruments need to be enhanced by higher education institutions so as to improve students' satisfaction.

By producing practically applicable guidelines for improving students' satisfaction in higher education institutions, the social goal of the research is achieved. Where marketing mix instruments, which affect student satisfaction the most, are concerned, on the level of the total sample, a high assessment grade was given for distribution, thus improvements are not necessary to enhance student satisfaction, since students are satisfied with this marketing mix instrument. Service, human factor, physical evidence and service process obtained medium grades on the level of total sample, which suggests that there is room for betterment within these instruments so as to improve students' satisfaction in higher education institutions. When it comes to marketing mix instruments that affect students' satisfaction, the general inference is that, on the level of total sample, service, human factor, physical evidence and service process leave space for further betterment to improve students' satisfaction, because they were medium-rated.

The results which we came to within descriptive analysis of the total sample were finalised and updated with the *t*-independent samples test of the research subsamples. The inferences are that faculties where research was conducted in Serbia have space for the enhancement of physical evidence and service process because they were medium-rated, to improve students satisfaction, whereas faculties in Spain where research was conducted have space for improving service, human factor, and service process, because they were medium-rated, to improve students' satisfaction.

The limitations of the research process which the author was faced with were related to the sample size (especially in Spain). The fact that data collection was conducted in two different countries imposed certain difficulties in both Serbia and Spain. In this respect, certain limitations were encountered concerning the adjustment of the questionnaire to the different territorial, cultural, language, economic and other specificities of the two countries. Moreover, the poor data availability and the insufficient degree of respondents' motivation to answer the entire set of questions were also present throughout the research. However, the comprehensive and studious approach led to successfully overcoming any dilemma or limitation.

Considering the present-day state of affairs in the education market in general, the increasing number of institutions of higher education creates the need for further research which would improve the business performance of certain institutions and lead to winning the market game. Owing to the constant competition between public and private institutions regarding the quality and overall business, it is necessary to carry out research which would determine the effect of marketing mix instruments on the student satisfaction in private and public faculties in other EU and non-EU countries. Considering the fact that the research had been conducted in faculties within the private University Business Academy in Novi Sad and the public University Jaume I in Castellon de la Plana, there is the possibility of conducting research which would be mostly based on the comparison which would provide insight into the similarities and differences in results obtained in public and private institutions, with an emphasis on this exact correlation.

To enhance the business performance of higher education institutions in general, research on the effects of marketing mix instruments on students' satisfaction could be conducted in universities in other countries or continents as well. Faculties in Europe and North America, for example, could be compared within the suggested research, which could use a comparison of the results to gain insight to be used for the scientific and practical purposes alike.

Author Contributions: Conceptualization, J.S.-G. and S.B.; methodology, E.B.E.; software, E.B.E.; validation, J.S.-G., E.B.E. and S.B.; formal analysis, M.Ć.; investigation, M.Ć.; resources, I.B.; data curation, J.T., M.P.; writing—original draft preparation, S.B.; writing—review and editing, J.T., M.P.; visualization, J.G.; supervision, J.S.G.; project administration, I.B.; funding acquisition, M.Ć. All authors have read and agreed to the published version of the manuscript.

Funding: This research received no external funding.

Conflicts of Interest: The authors declare no conflict of interest.

References

1. National Report Regarding the Bologna Process Implementation (Serbia, Spain) 2012–2015, EHEA. Available online: http://www.ehea.info/ (accessed on 3 January 2020).
2. Avramović, Z. Problemi modernizacije obrazovanja u Srbiji. *Nac. Interes:Časopis Za Državna I Prav. Pitanja* **2011**, *7*, 9–31. [CrossRef]
3. Montané, A.; Beltrán, J.; Gabaldón-Estevan, D. Higher Education in Spain. *Qual. Assur. High. Educ. A Glob. Perspect.* **2017**, *1*, 41–67.
4. Maringe, F.; Gibbs, P. *Marketing Higher Education: Theory and Practice*; McGraw Hill: Maidenhead, UK, 2009.
5. Dibb, S.; Simoes, C.; Wensley, R. Establishing the scope of marketing practice: Insights from practitioners. *Eur. J. Mark.* **2014**, *48*, 380–404. [CrossRef]
6. Miljković, J.; Kovačević, M.J. Elementi marketing miksa kao činioci izbora visokoobrazovne institucije. *Andragoške Studije* **2011**, *1*, 135–156.
7. Brkanlić, S. Marketing Mix as Factors of Improvement of Image of Higher Education Institutions and Student Satisfaction. Ph.D. Thesis, University Jaume I, Castellon de la Plana, Spain, 27 May 2019.
8. Milisavljević, M.; Maričić, B.; Gligorijević, M. *Osnovi Marketinga*; Ekonomski Fakultet: Beograd, Serbia, 2004.
9. Rodić-Lukić, V.; Lukić, N. Primena marketing miksa koncepta kod strategije regrutovanja studenata—Univerzitet u Novom Sadu, Srbija. *Megatrend Rev.* **2016**, *13*, 183–202.
10. American Marketing Asociation. Available online: https://www.ama.org/ (accessed on 1 February 2020).
11. Miljković, J. Professionalization of high educational institutions marketing incontext of crisis. In *Adult Education: The Response to Global Crisis–Strengths and Chalenges of the Profession*; Medić, S., Ebner, R., Popović, K., Eds.; IPA, EAEA, DVV: Beograd, Serbia, 2010; pp. 207–220.
12. Palmer, A. *Principles of Services Marketing*, 4th ed.; Pampaloni; McGraw-Hill: London, UK, 2005.
13. Veljković, S. *Marketing Usluga*; Ekonomski Fakultet: Beograd, Serbia, 2006.
14. Ljubojević, Č. *Marketing Usluga*, 3th ed.; Stylos: Novi Sad, Serbia, 2002.
15. Ćirić, M. *Upravljanje Odnosima sa Klijentima u Bankama: Monografija*; Fakultet za Ekonomiju i Inženjerski Menadžment u Novom Sadu: Novi Sad, Serbia, 2013.
16. Wilkins, S.; Huisman, J. Factors affecting university image formation among prospective higher education students: The case of international branch campuses. *Stud. High. Educ.* **2015**, *40*, 1256–1272. [CrossRef]
17. Jobber, D.; Fahy, J. *Osnovi Marketinga*, 2nd ed.; Data Status: Beograd, Serbia, 2006.
18. Veljković, S. *Marketing Usluga*; CID Ekonomskog Fakulteta: Beograd, Serbia, 2009.
19. Gallifa, J.; Batallé, P. Student perceptions of service quality in a multi-campus higher education system in Spain. *Qual. Assur. Educ.* **2010**, *18*, 156–170. [CrossRef]
20. Kotler, P.; Fox, K.F. *Strategic Marketing for Educational Institutions*; Prentice-Hall: Englewood Cliffs, NJ, USA, 1995.

21. Vázquez, J.L.; Aza, C.L.; Lanero, A. University social responsibility as antecedent of students' satisfaction. *Int. Rev. Public Nonprofit Mark.* **2016**, *13*, 137–149. [CrossRef]
22. Hrnjic, A. The transformation of higher education: Evaluation of CRM concept application and its impact on student satisfaction. *Eurasian Bus. Rev.* **2016**, *6*, 53–77. [CrossRef]
23. Jurkowitsch, S.; Vignali, C.; Kaufmann, H.R. A student satisfaction model for Australian higher education providers considering aspects of marketing communications. *Innov. Mark.* **2006**, *2*, 9–23.
24. Masserini, L.; Bini, M.; Pratesi, M. Do Quality of Services and Institutional Image Impact Students' Satisfaction and Loyalty in Higher Education? *Soc. Indic. Res.* **2018**, *146*, 91–115. [CrossRef]
25. Chen, Y.C. A study of the interrelationships among service recovery, relationship quality, and brand image in higher education industries. *Asia-Pac. Educ. Res.* **2015**, *24*, 81–89. [CrossRef]
26. Aghaz, A.; Hashemi, A.; Sharifi Atashgah, M.S. Factors contributing to university image: The postgraduate students' points of view. *J. Mark. High. Educ.* **2015**, *25*, 104–126. [CrossRef]
27. Osman, A.; Saputra, R.; Luis, M. Exploring mediating role of institutional image through a complete structural equation modeling (sem): A perspectve of higher education. *Int. J. Qual. Res.* **2018**, *12*, 517–536.
28. Cho, E.; Kim, S. Cronbach's coefficient alpha: Well known but poorly understood. *Organ. Res. Methods* **2015**, *18*, 207–230. [CrossRef]
29. Tabachnick, B.G.; Fidell, L.S. *Using Multivariate Statistics*, 6th ed.; Pearson: Boston, MA, USA, 2013.
30. *Electronic Statistics Textbook*; StatSoft: Tulsa, OK, USA, 2020; Available online: http://www.statsoft.com/textbook/ (accessed on 20 January 2020).
31. Banwet, D.K.; Datta, B. A study of the effect of perceived lecture quality on post-lecture intentions. *Work Study* **2003**, *52*, 234–243.
32. Hill, Y.; Lomas, L.; MacGregor, J. Students' perceptions of quality in higher education. *Qual. Assur. Educ.* **2003**, *11*, 15–20. [CrossRef]
33. Gruber, T.; Fub, S.; Voss, R.; Glaser-Zikuda, M. Examining Student Satisfaction with Higher Education Services—Using a New Measurement Tool. *Int. J. Public Sect. Manag.* **2010**, *23*, 105–123. [CrossRef]
34. Douglas, J.; Douglas, A.; Barnes, B. Measuring student satisfaction at UK University. *Qual. Assur. Educ.* **2006**, *14*, 251–267. [CrossRef]
35. Gajić, J. Značaj marketing miksa u visokoobrazovnim institucijama. *Singidunum J. Appl. Sci.* **2012**, *9*, 29–41.
36. Pozo-Munoz, C.; Rebolloso-Pacheco, E.; Fernandez-Ramirez, B. The 'ideal teacher': Implications for student evaluation of teacher effectiveness. *Assess. Eval. High. Educ.* **2000**, *25*, 253–263. [CrossRef]
37. Voss, R.; Gruber, T.; Szmigin, I. Service quality in higher education: The role of student expectations. *J. Bus. Res.* **2007**, *60*, 949–959. [CrossRef]
38. Price, I.; Matzdorf, F.; Smith, L.; Agahi, H. The impact of facilities on student choice of university. *Facilities* **2003**, *21*, 212–222. [CrossRef]
39. Deming, W.E. *Out of the Crisis*; Massachusetts Institute of Technology Deming's Quality Points: Cambridge, UK, 1982; Available online: http://www.ou.edu/class/busad4013/docs/pptch2/sld006.htm (accessed on 10 December 2006).
40. Galloway, L. Quality perceptions of internal and external customers: A case study in educational administration. *TQM Mag.* **1998**, *10*, 20–26. [CrossRef]
41. Starck, K.; Zadeh, S.H.; Ekman, P.; Olsson, E.M. Marketing within Higher Education Institutions: A Case Study of Two Private Thai Universities. Master's Thesis, Mälardalen University, Västerås, Sweden, 29 May 2013.
42. Dale, B.G. *Managing Quality*, 4th ed.; Blackwell Publishing: Oxford, UK, 2003.
43. Gremler, D.D.; McCollough, M.A. Student satisfaction guarantees: An empirical examination of attitudes, antecedents, and consequences. *J. Mark. Educ.* **2002**, *24*, 150–160. [CrossRef]
44. McCollough, M.A.; Gremler, D.D. Guaranteeing student satisfaction: An exercise in treating students as customers. *J. Mark. Educ.* **1999**, *21*, 118–130. [CrossRef]
45. Ivy, J. A new higher education marketing mix: The 7Ps for MBA marketing. *Int. J. Educ. Manag.* **2008**, *22*, 288–299. [CrossRef]

46. Cheung, A.C.; Yuen, T.W.; Yuen, C.Y.; Cheng, Y.C. Promoting Hong Kong's higher education to Asian markets: Market segmentations and strategies. *Int. J. Educ. Manag.* **2010**, *24*, 427–447. [CrossRef]
47. Friedman, S.M.; Villamil, K.; Suriano, R.A.; Egolf, B.P. Alar and apples: Newspapers, risk and media responsibility. *Public Underst. Sci.* **1996**, *5*, 1–20. [CrossRef]
48. Symes, C.; Meadmore, P.; Limerick, B. Secondary education in queens land: Never a matter of primary concern. *Discourse* **1994**, *14*, 92–102.

MDPI
St. Alban-Anlage 66
4052 Basel
Switzerland
Tel. +41 61 683 77 34
Fax +41 61 302 89 18
www.mdpi.com

Sustainability Editorial Office
E-mail: sustainability@mdpi.com
www.mdpi.com/journal/sustainability